Therapeutic Peptides and Proteins

Formulation, Processing, and Delivery Systems

Third Edition

Therapeutic Peptides and Proteins

Formulation, Processing, and Delivery Systems

Third Edition

Ajay K. Banga

CRC Press
Taylor & Francis Group
Boca Raton London New York

CRC Press is an imprint of the
Taylor & Francis Group, an **informa** business

First published in paperback 2024
First published in hardback 2019

First published 2015 by
CRC Press
2385 NW Executive Center Drive, Suite 320, Boca Raton FL 33431

and by CRC Press
4 Park Square, Milton Park, Abingdon, Oxon, OX14 4RN

CRC Press is an imprint of Taylor & Francis Group, LLC

Publisher's Note
The publisher has gone to great lengths to ensure the quality of this reprint but points out that some imperfections in the original copies may be apparent.

ISBN: 978-1-4665-6606-4 (hbk)
ISBN: 978-1-03-292056-6 (pbk)
ISBN: 978-0-429-09990-8 (ebk)

DOI: 10.1201/b18392

Visit the Taylor & Francis Web site at
http://www.taylorandfrancis.com

and the CRC Press Web site at
http://www.crcpress.com

This book is dedicated to my wife, Saveta, and my children, Anshul and Manisha, for their support and understanding during the preparation of this book.

Contents

SECTION II Drug Delivery Systems

Preface

During the publication of the first edition of this book, there were only 19 biotechnology medicines on the market, and that number grew to 100 when the second edition came out. Now, there are more than 500 biologics on the market, including more than 200 therapeutic proteins, making the predictions in the preface to the first two editions come true. These products include more than 40 monoclonal antibodies, for indications ranging from treatment or mitigation of various types of cancer to rheumatoid arthritis. More than 900 more biologics are in development, targeting more than 100 diseases, with monoclonal antibodies and vaccines being the predominant category. More than 8000 companies are working in the biotechnology field in the United States alone. The global market for these products is more than $160 billion.

However, the clinical application of these therapeutic peptides and proteins is limited by several problems, such as lack of physical and chemical stability or the lack of desirable attributes for adequate absorption or distribution. Thus, as these therapeutic peptides and proteins are made available, it will be essential to formulate these drugs into safe, stable, and efficacious delivery systems. The pharmaceutical scientist involved in this effort needs to call upon the knowledge of several disciplines, such as pharmaceutics, medicinal chemistry, biochemistry, microbiology, and chemical engineering and needs to keep abreast with the latest research in the published literature. This book provides a comprehensive overview of the field for scientists in industry and academia and for students, while also providing practical information on the challenges facing the formulation and delivery aspects of these unique macromolecules. The book should also be useful in a hospital setting to understand potential physicochemical stability problems that may result during the reconstitution or administration of the new recombinant proteins.

This third edition has thoroughly updated the information under different topics with references to recent literature, and several new figures have been added. In addition, new topics have been added or expanded, such as subvisible particles, advances in our understanding of protein aggregation and fibrillation, recent regulatory approval of inhaled insulin, the principles of quality by design approach, antibody–drug conjugates, biosimilars, and microneedle delivery systems. More than 20 of the top-selling biopharmaceuticals will come off patent by 2020, creating significant opportunity for biosimilars, with an estimated market size for biosimilars in excess of $50 billion. This is an exciting time for realizing the promise of recombinant proteins and other biotechnology medicines, with the biotechnology industry enjoying a return of investor enthusiasm.

The book has been divided into two main sections. Section I (Chapters 1 through 5) introduces the reader to therapeutic peptides and proteins and discusses several practical aspects of relevance to the pharmaceutical industry, which may be broadly categorized as formulation and pharmaceutical processing. Chapter 1 provides insights into the field of pharmaceutical biotechnology and explains how recombinant DNA techniques now allow us to produce therapeutic proteins in a commercially viable form. Chapter 2 introduces the issues unique to these drugs with respect to the structure and analysis aspects. The physical and chemical pathways of peptide and protein degradation are discussed, along with methods of stabilization (Chapters 3). Chapters 4 and 5 are devoted to the formulation development and pharmaceutical processing aspects for therapeutic peptides and proteins. Chapter 5 includes a detailed discussion of lyophilization. Section II (Chapters 6 through 10) of this book discusses the various drug delivery routes for these drugs. Chapter 6 provides an overview of the pharmacokinetic aspects of therapeutic peptides and proteins and discusses controlled delivery systems for parenteral administration, including microsphere formulations. Chapter 7 discusses research progress on oral delivery systems while Chapter 8 discusses transdermal and topical delivery. Chapter 9 focuses on mucosal delivery, including products marketed for pulmonary or nasal delivery. Finally, Chapter 10 discusses recombinant protein subunit vaccines and delivery methods.

Since this book is the work of a single author, I was able to minimize overlap between chapters and provide extensive cross-referencing to direct the reader to relevant parts of the book for related topics. I thank Pooja Bakshi and Saveta Banga for their extensive help with the preparation of this book. Finally, I would like to acknowledge the staff of CRC Press (a member of the Taylor & Francis Group) for their efforts in bringing this book to the readers.

Author

Dr. Ajay K. Banga, PhD, is professor, department chair, and endowed chair in the Department of Pharmaceutical Sciences at the College of Pharmacy, Mercer University, Atlanta, Georgia. He holds a PhD in pharmaceutical sciences from Rutgers—The State University of New Jersey. He has more than 250 peer-reviewed publications and research abstracts. Dr. Banga currently serves on the editorial board of 10 journals, serves as associate editor for 1 journal, and has served as the editor-in-chief for a drug delivery journal. He serves as codirector for the Center for Drug Delivery Research at the College of Pharmacy and is a fellow of the American Association of Pharmaceutical Scientists.

Section I

Formulation and Processing

1 Pharmaceutical Biotechnology
The Arrival of Recombinant Proteins

Advances in biotechnology have resulted in a significant increase in the number of therapeutic peptides and proteins that are reaching the market. This trend is expected to continue and escalate in the future. Applications of biotechnology are not limited to pharmaceuticals; they extend to agriculture, food processing, and chemicals, among others. However, the most significant activity to date has been in human health care, both therapeutics and diagnostics (Assenberg, Wan, Geisse, & Mayr, 2013; Montague, 1992; Sterling, 1990; Wordell, 1992). Although a detailed discussion of biotechnology is not within the scope of this book, this chapter provides a very brief overview of some of the principles of biotechnology that can be critical in the formulation and delivery of peptides and proteins. For example, the presence of host-cell protein (HCP) as impurity in the final product may be responsible for stability or immunogenicity problems of a formulation. Two things need to be pointed out here. First, biotechnology principles go beyond the production of proteins by recombinant DNA (rDNA) technology. For example, antisense agents are oligonucleotides, not proteins. Similarly, gene therapy is of direct benefit to the patient but does not involve the development of a drug as conventionally understood. Some of the biotechnology principles and products are briefly reviewed in this chapter, but the focus of this book in the remaining chapters is on the formulation and delivery of therapeutic peptides and proteins. Second, it must be realized that the basic concepts of formulation and delivery of these drugs are applicable to any peptide or protein, whether of biotechnology origin or not.

1.1 WHAT IS BIOTECHNOLOGY?

Biotechnology is the use of biological systems or living organisms to produce or modify materials or products. Biotechnology-derived drugs are produced by rDNA technology, known as genetic engineering or gene cloning, whereby human genes responsible for the production of a protein are transferred to an organism such as an *Escherichia coli* bacterium. The *E. coli* is then cultured by fermentation to achieve a large-scale production of therapeutic proteins in a commercially viable scale (Emtage, 1986; Harford, 1985).

Calcitonin can be considered as an example of the cost-effectiveness of the biotechnology process. Chemical synthesis of calcitonin would need linking one amino acid at a time. The 32-amino-acid sequence requires hundreds of individual steps to complete the synthesis. In contrast, recombinant technology can reduce the cost of the large-scale calcitonin production by about 20-fold (Levy, 1993). Recombinant technology has been very critical to meet the high demand for proteins, which is partly due to the low bioavailability of the new drug delivery systems, which may slowly replace parenteral delivery (Devogelaer et al., 1994; Levy, 1993). However, short peptides are generally prepared by chemical synthesis, and automated peptide synthesizers are also available. Impurity profile of peptides made by solid-phase synthesis would be different from that of proteins made by recombinant technologies (D'Hondt et al., 2014).

1.1.1 Recombinant DNA Technology

rDNA technology, or genetic engineering, involves combining genetic material from two sources in vitro and then introducing the recombined material into a host cell.

One aspect of rDNA technology (Figure 1.1) involves the transfer of a gene from a plant or animal cell into a bacterial plasmid. Enzymes called restriction endonucleases are used to cut DNA from the cell. For example, a DNA fragment equal to one gene that codes for a particular protein can be cut from a human cell for transfer to bacteria. The same enzyme is used to open a site in the DNA molecule for the bacterial cell, thus allowing easy integration of human genes into bacteria. A new hybrid, called rDNA molecule, results and is then incorporated into a new bacterium. The new strain is selectively grown and purified (cloned), and if the incorporated gene is expressed, then the bacteria would produce human protein. Human protein may then be manufactured through large-scale fermentation and purification. The native structure of the protein should be maintained during downstream processing. The advent of rDNA technology continues to drive the trend toward development of plasma-free protein therapeutics. This trend is also driven by the concern on the part of regulatory agencies about the inherent potential for transmission of infectious agents through plasma-derived proteins or additives, as well as their heterogeneity and lack of reliability of serum supply (Grillberger, Kreil, Nasr, & Reiter, 2009). Challenges experienced during the development of high-purity, plasma-derived factor VIII concentrate have been described in the literature (Schulte, 2013).

There are several disadvantages associated with the use of *E. coli*, and more sophisticated organisms such as yeasts and mammalian cells have also been utilized (Emtage, 1986; Walsh, 2010). These disadvantages primarily relate to glycosylation, the addition of sugar residue to a peptide backbone (Sola & Griebenow, 2010). Proteins produced in *E. coli* are not glycosylated. However, in eukaryotic cells, glycosylation is a common posttranslational modification made to proteins. Also, when *E. coli* is used, lysis of the bacteria is necessary to recover the protein, and since lysis releases many other proteins and cellular components, significant purification is required. Plant and insect cells have also been used as bioreactors for protein production (Cox, 2012; Crosby, 2003; Hellwig, Sack, Spiegel,

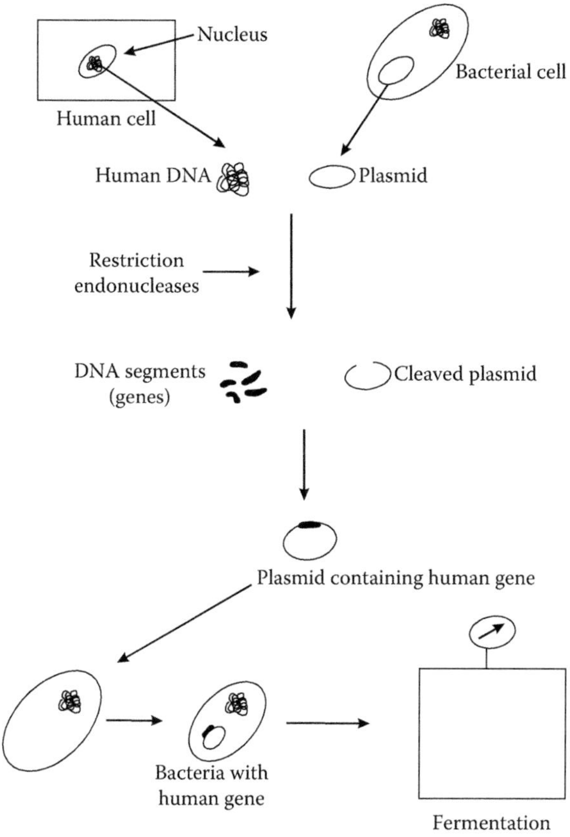

FIGURE 1.1 rDNA technology for the production of a human protein by a bacterial cell.

Drossard, & Fischer, 2003; Kwon, Verma, Singh, Herzog, & Daniell, 2013; Smith, Richter, Arntzen, Shuler, & Mason, 2003).

When yeast or cell cultures are used, protein may be released directly into the medium, and purification will be simpler, starting with removal of the cells themselves. However, mammalian cells do have drawbacks also related to bioprocessing and scale-up, which result in long processing times and elevated costs. Mammalian cell expression in Chinese hamster ovary (CHO) cells is used for the majority of products on the market today as well as for those in clinical development (Zhu, 2012). A strategy of in vitro evolution has been suggested by isolating protein variants with improved expression, solubility, and stability, including resistance to aggregation (Buchanan et al., 2012). The glycosylation status of a protein must be clearly established (Jenkins & Curling, 1994). Although most physiological proteins are glycoproteins, glycosylation is not necessarily required for bioactivity. Recombinant human interleukin 2 produced in *E. coli* is not glycosylated but retains full biological potency. However, immunogenicity issues should also be considered.

1.1.2 Monoclonal Antibodies

The normal antibody response to antigenic challenge is polyclonal since a large number of different immunoglobulin (Ig) molecules are produced by different lymphocytes. Polyclonals derived from the plasma obtained from nonimmunized donors are known as standard intravenous immunoglobulins (IVIGs). For persons specifically immunized with an antigen, high titers of antibodies specific for that antigen are produced and obtained as specific IVIGs or specific polyclonal antibodies. Immunoglobulin products are also available for intramuscular administration. While IVIGs are well established and continue to carry a prominent role in medicine, it is possible to produce immunoglobulin molecules that all possess the same structure (monoclonal). Monoclonal antibodies are produced by the fusion of an individual B lymphocyte with a cancer cell (usually from a myeloma). The fusion produces a hybrid cell that has both the antibody-producing character of a B lymphocyte and the immortality or continuous reproduction of a myeloma cell, allowing the commercial production of a unique antibody molecule.

Monoclonal antibodies are finding widespread use in diagnosis and therapeutics. The monoclonal antibody Orthoclone OKT3 was approved for use in kidney, heart, and liver transplant rejection in 1986. There was then a lag period of several years when not much was approved, but in recent years, several monoclonal antibodies have been approved for varying indications. Monoclonal antibodies constitute the fastest-growing segment of biopharmaceuticals, with many already on the market and many more in clinical trials. Table 1.1 lists the monoclonal antibodies currently on the market. Unlike most other therapeutic proteins, monoclonal antibodies have relatively high doses and long half-lives (Daugherty & Mrsny, 2006). Therefore, they are generally not suited to be delivered by some of the alternate delivery systems discussed in this book and are rather given by parenteral administration.

Monoclonal antibodies have many possible diagnostic and therapeutic applications in the management of human carcinomas, including colorectal, gastric, ovarian, endometrial, breast, lung, and pancreatic cancers. Cytotoxic drugs can be linked to monoclonal antibodies for targeting specific tumor-associated antigen (Schlom, 1988). Similarly, radioisotopes can also be directed to specific sites by the use of monoclonal antibodies. A monoclonal antibody has also been developed for imaging myocardial necrosis associated with acute myocardial infarction. The product (Myoscint®, Janssen Biotech, Inc.) was on the market and had a high degree of sensitivity for detecting infarction and specificity for excluding a recent ischemic event in patients admitted with chest pain syndrome but is now withdrawn from the market (Spada & Walsh, 2005). Monoclonal antibodies have also been linked to enzymes for the detection of other antibodies or antigens by enzyme-linked immunosorbent assay, which will be discussed in Chapter 2. Typically, monoclonal antibodies are prepared using mouse lymphocytes. and this had limited their effectiveness since there is often a human immunogenic response to the mouse antibody. Using genetic engineering, the mouse antibody can be humanized to produce a chimeric antibody, which is part human and part mouse in structure, and several of these have now been approved for marketing with several more in development. A stable human myeloma that may be useful to generate human monoclonal

TABLE 1.1
Examples of FDA-Approved Monoclonal Antibodies

Product Name	Company/Immunoglobulin Type	Drug Conc.; Inactives	Indication	Approval Date
Orthoclone OKT3 (Muromonab-CD3)	Janssen-Cilag/mouse IgG_{2a}	1 mg/mL soln contg. 5 mg[a]; sod. phosphate, NaCl, polysorbate 80 (pH 7.0)	Acute allograft rejection in transplant patients	1986
			Reversal of heart and liver transplant rejection	1993
ReoPro (abciximab)	Eli Lilly/chimeric human–murine F_{ab}	2 mg/mL soln contg. 10 mg; sod. phosphate, NaCl, polysorbate 80 (pH 7.2)	Complication of coronary angioplasty	1994
Rituxan (rituximab)	Biogen Idec/Genentech chimeric murine/human $IgG_{1\kappa}$	10 mg/mL concentrate contg. 100 or 500 mg; NaCl, sod. citrate, polysorbate 80 (pH 6.5)	Non-Hodgkin's lymphoma	1997
Zenapax (daclizumab)	Roche/Genentech/humanized IgG_1	5 mg/mL soln contg. 25 mg; sod. phosphate, NaCl, polysorbate 80 (pH 6.9 adjusted with HCl/NaOH)	Prophylaxis of acute kidney transplant rejection	1997
Simulect (basiliximab)	Novartis/chimeric $IgG_{1\kappa}$	Lyophilized (10 mg/2.5 mL or 20 mg/5 mL); monobasic pot. phosphate, disod. hydrogen phosphate (anhydrous), NaCl, sucrose, mannitol, glycine	Renal transplant rejection	1998
Remicade (infliximab)	Janssen Biotech/Merck/ chimeric $IgG_{1\kappa}$	Lyophilized powder (100 mg reconst. to 10 mg/mL); sucrose, polysorbate 80, sod. phosphate (pH 7.2)	Crohn's disease	1998
			Rheumatoid arthritis	1999–2002
Herceptin (trastuzumab)	Genentech/humanized $IgG_{1\kappa}$	Lyophilized powder (440 mg reconst. to 21 mg/mL with Bacteriostatic Water for Injection (BWFI)); L-histidine HCl, L-histidine, α, α-trehalose dehydrate and polysorbate 20, benzyl alcohol in BWFI (pH approx. 6)	Metastatic breast cancer	1998
Synagis (palivizumab)	MedImmune/humanized $IgG_{1\kappa}$	Lyophilized powder (100 mg/mL clear or slightly opalescent soln.); histidine, glycine, and mannitol	Respiratory syncytial virus	1998
Mylotarg (gemtuzumab ozogamicin)[b]	Wyeth/humanized $IgG_{4\kappa}$	Lyophilized powder (5 mg/20 mL); dextran 40, sucrose, NaCl, sod. phosphate	Acute myeloid leukemia	2000

(Continued)

TABLE 1.1 (*Continued*)
Examples of FDA-Approved Monoclonal Antibodies

Product Name	Company/Immunoglobulin Type	Drug Conc.; Inactives	Indication	Approval Date
Campath (alemtuzumab)	Genzyme/humanized $IgG_{1\kappa}$	30 mg/3 mL solution; NaCl, sod. phosphate, KCl, pot phosphate, polysorbate 80, disodium edetate (pH 6.8–7.4)	Lymphocytic leukemia	2001
Zevalin (ibritumomab tiuxetan)	Spectrum Pharmaceuticals/ murine $IgG_{1\kappa}$	3.2 mg/14 mL solution; NaCl, Na acetate, HSA, sod. phosphate, pot. phosphate, KCl, pentetic acid (pH 7.1 adjusted with HCl or NaOH)	Non-Hodgkin's lymphoma	2002
Humira (adalimumab)	AbbVie/rhIgG_1 mAB specific for human TNF	40 mg/0.8 mL soln; NaCl, sod. phosphate Na citrate, citric acid monohydrate, mannitol, polysorbate 80 and Water for Injection (WFI), USP. NaOH added to adjust pH to 5.2	Rheumatoid arthritis	2002
Xolair (omalizumab)	Genentech/Novartis $IgG_{1\kappa}$	Lyophilized powder (202.5 mg reconst. with SWFI to 150 mg/1.2 mL); sucrose, L-histidine, polysorbate 20	Allergy-related asthma	2003
Raptiva (efalizumab)	Genentech/Merck Serono/ recombinant humanized IgG_1	Lyophilized powder; reconst. to 100 mg/ml; sucrose, L-histidine, polysorbate 20 (pH 6.2)	Plaque psoriasis candidates for systemic therapy	2003
Bexxar (tositumomab and I^{131} tositumomab)	GlaxoSmithKline/mouse IgG_{2a}	14 mg/mL single-use vials with 10% maltose, 145 mM NaCl, 10 mM phosphate, and WFI	Non-Hodgkin's lymphoma	2003
Avastin (bevacizumab)	Genentech/Roche/humanized IgG1		Colorectal cancer	2004
Erbitux (cetuximab)	Bristol-Myers Squibb/Eli Lilly/ Merck KGaA/chimeric IgG1	2 mg/mL preservative-free injectable liquid	Colorectal cancer	2004
Vectibix (panitumumab)	Amgen/human IgG	20 mg/mL preservative-free solution	Colorectal cancer	2006
Lucentis (ranibizumab)	Genentech/Novartis/humanized $IgG_{1\kappa}$		Macular degeneration	2006
Tysabri (natalizumab)	Biogen Idec/Elan/humanized $IgG_{4\kappa}$		Multiple sclerosis and Crohn's disease	2006

(*Continued*)

TABLE 1.1 (*Continued*)
Examples of FDA-Approved Monoclonal Antibodies

Product Name	Company/Immunoglobulin Type	Drug Conc.; Inactives	Indication	Approval Date
Soliris (eculizumab)	Alexion Pharmaceuticals/ humanized IgG2		Paroxysmal nocturnal hemoglobinuria	2007
Cimzia (certolizumab pegot)	UCB/humanized Fab fragment	Lyophilized powder with sucrose, lactic acid, and polysorbate or prefilled syringe with sod. acetate, sod. chloride, and WFI	Crohn's disease	2008
Ilaris (canakinumab)	Novartis/human $IgG_{1\kappa}$	Injection with sucrose, L-histidine, and polysorbate 80	Cryopyrin-associated periodic syndrome	2009
Simponi (golimumab)	J&J/Merck/human $IgG_{1\kappa}$	Single-dose prefilled syringe or autoinjector providing 50 mg per 0.5 mL of solution	Rheumatoid arthritis, psoriatic arthritis, and ankylosing spondylitis	2009
Arzerra (ofatumumab)	GSK/human $IgG_{1\kappa}$	20 mg/mL in 5 or 50 mL single-use vials	Chronic lymphocytic leukemia	2009
Prolia (denosumab)	Amgen/human IgG2	60 mg/1 mL in single-use prefilled syringe or vial	Postmenopausal osteoporosis, solid tumor's bony metastasis	2010
Actemra and RoActemra (tocilizumab or atlizumab)	Humanized $IgG_{1\kappa}$		Rheumatoid arthritis	2010
Yervoy (ipilimumab or MDX-101)	Bristol-Myers Squibb/human $IgG_{1\kappa}$	5 mg/mL in single-use 50 mg or 200 mg vial	Melanoma	2011
Adcetris (brentuximab vedotin)	Chimeric IgG1		Anaplastic large cell and Hodgkin's lymphoma	2011
Entyvio (vedolizumab)	Takeda/humanized IgG1		Crohn's disease, ulcerative colitis	2014

[a] May contain a few fine translucent protein particles.
[b] Drug sensitive to light.
(compiled from information in reference, PDR Staff, [2014], Ho and Gibaldi [2003], and Spada and Walsh [2005]).

antibodies has also been described (Ho & Gibaldi, 2003). As just discussed, the failure of many antibodies in clinic in the past was due to human antimouse antibody response to mouse monoclonals. Chimeric antibodies solved this problem to a large extent. Also, companies like Abgenix (now Amgen) have used their Xenomouse™ transgenic mice to produce fully human antibody, panitumumab (Table 1.1). Human antibody genes have replaced the mice genes in these transgenic mice so that their genome can now produce fully human antibodies, and it is no longer required to humanize each individual antibody produced in mice. More recently, there has been increasing attention on fragments of monoclonal antibodies. These are not glycosylated but have the antigen-binding properties and be produced economically in easier to handle microbial organisms rather than in mammalian cells (Spadiut, Capone, Krainer, Glieder, & Herwig, 2014; Weisser & Hall, 2009). Similarly, chemically programmed antibodies are being explored, and they involve covalent conjugation of small molecules to monoclonal antibodies to enable targeting multiple target molecules, due to the chemical diversity of small molecules (Rader, 2014). Cocktails of monoclonals, each with unique binding and specificity, are also being investigated to improve clinical efficacy, a so-called recombinant polyclonal approach (Wang, Coljee, & Maynard, 2013).

1.1.3 Expanding Scope of Biotechnology

Although earlier developments were restricted to rDNA and monoclonal antibodies, the field of biotechnology is currently expanding into other areas. These new areas include proteomics, antibody–drug conjugates (ADCs), antisense technology, gene therapy, transgenic technology, and drug delivery technologies. The scope of biotechnology now runs the spectrum from designer organisms to tissue regeneration. The end products of some of the newer applications of biotechnology are oligonucleotides, and not therapeutic peptides and proteins. In this sense, they are outside the scope of this book, but a brief overview of these new applications will be provided.

Protein engineering techniques can also improve the clinical utility of therapeutic proteins (Carter, 2011). For example, site-directed mutagenesis can be used to achieve a precise alteration of the protein sequence by a point mutation of the DNA that encodes the amino acid sequence (McPherson & Livingston, 1989). Gene diagnostics is another promising area, and genetic counseling will become a routine part of the health-care system. The ability to amplify specific segments of DNA with a polymerase chain reaction has enhanced the analytical limits so that genetic defects can be determined with just a few cells (Sadee, 1988). An ideal situation for the delivery of a recombinant protein would be where a cell from the recipient can be isolated and grown. Recombinant genes can be introduced into this cell, and it can then be reimplanted into the host organism (Barr & Leiden, 1991; Dhawan et al., 1991).

Although many of the new therapeutic peptides and proteins are products of biotechnology, synthetic peptides or proteins continue to be important (Craik, Fairlie, Liras, & Price, 2013). For example, Salmon et al. (1993) have described a system for production and screening of large libraries of peptides to speed up drug discovery. All the peptides in this library were produced by solid-phase chemical synthesis. Techniques are now available to routinely generate synthetic peptide libraries with millions of sequence combinations, with minimal investment in equipment. Biologically active peptides or proteins can then be identified in this combinatorial library by using a monoclonal antibody, soluble receptors, or directed evolution approaches (Medynski, 1994; Urvoas, Valerio-Lepiniec, & Minard, 2012). Small peptides (–2 to 50 residues) and especially peptides containing unnatural amino acids will continue to be made by synthetic methods rather than by rDNA technology. The use of some of the more common new areas of biotechnology will now be discussed.

1.1.3.1 Antibody–Drug Conjugates

ADCs consist of a mAb chemically linked to a cytotoxic drug with a linker. ADCs take advantage of the specificity of mAbs to deliver potent cytotoxic drugs selectively to antigen-expressing tumor cells (Alley, Okeley, & Senter, 2010; Casi & Neri, 2012; Perez et al., 2014; Trail, 2013). They are designed to stay stable in circulation and reach target site where they undergo antigen-specific binding, internalization, and intracellular drug release. This reduces the systemic toxicity of the cytotoxic drugs. Though simple in concept, design of ADCs poses many challenge, including those related to mAb specificity, stability in circulation, complex pharmacokinetics, stoichiometry, and linker chemistry/technology. The reactive linker molecule needs to be carefully analyzed to ensure the purity of the conjugate (Li, Medley, Zhang, Wigman, & Chetwyn, 2014). The high cost of goods makes ADCs more relevant for severe diseases, and most successful applications are currently in the oncology area (Casi & Neri, 2012; Trail, 2013). Several advances made in the past few decades led to the first FDA approval in 2000 (Mylotarg®, gemtuzumab ozogamicin) for acute myeloid leukemia though the product was withdrawn a decade later for lack of improvement in overall survival. In recent years, two ADCs have been approved: Adcetris® (brentuximab vedotin) for the treatment of Hodgkin's and anaplastic large cell lymphomas and Kadcyla® (ado-trastuzumab emtansine) for the treatment of patients with HER2-positive breast cancer. This has generated considerable interest and investment, and currently, 30 ADCs are in clinical development (Perez et al., 2014). Other approaches to target monoclonal antibodies include immunotoxins and immunoliposomes (Sapra & Shor, 2013). In addition to monoclonals, other macromolecules such as polysaccharides have also been conjugated to cytotoxic drugs for cancer therapy (Goodarzi, Varshochian, Kamalinia, Atyabi, & Dinarvand, 2013).

1.1.3.2 Antisense Agents and RNAi

The principles involved in antisense technology exploit the genetic sequence of events in the cell. As is well known, the double-helix DNA molecule does not directly serve as an immediate template for protein synthesis. Instead, it directs the synthesis of a complementary messenger RNA in a process called transcription. The mRNA then directs protein synthesis by ribosomes, in a process called translation. The sequence of events by which a DNA genetic sequence produces a protein is called gene expression. If an RNA complementary to mRNA is introduced into a cell, it will pair up with the mRNA and prevent it from translating a gene into a protein. This agent, called an antisense RNA, will thus form an antisense duplex that will block protein synthesis and can have therapeutic applications in human diseases (Alper, 1993; Chrisey & Hawkins, 1991; Lares, Rossi, & Ouellet, 2010; Misra et al., 1993; Wu-Pong, 1994).

This technology can be used to turn off disease-related genes. Since the antisense agent targets the primary cause of the disease, it could potentially provide a cure. Antisense agents may suppress disease-causing damaged or mutated human genes or foreign genetic material such as viral or bacterial genes. Promising application would be to suppress (or downregulate) oncogenes, which are implicated in development of cancer (Chrisey & Hawkins, 1991) or treatment of tumors by blocking the overexpression of growth factor genes (Lenz & Mansson, 1991). However, antisense oligonucleotides may often be unsuitable for therapeutic application as they are susceptible to enzymatic degradation by various nucleases, and thus stabilization techniques such as by chemical modification, encapsulation into liposomes, or other approaches may be required (Chrisey & Hawkins, 1991; Sueishi et al., 1994).

Another problem relates to the low efficiency of penetration of oligonucleotides into intact cells. An oligonucleotide with 17 bases will have a molecular weight of about 5000, which is smaller than most therapeutic proteins but larger than most conventional drugs. The size places a limitation on oral absorption of oligonucleotides as well (Sherman, 1993). Interest later shifted to double-stranded short-interference RNA molecules (siRNAs or RNAi) that exist naturally in the cells of many species. Like antisense agents, RNAi's interfere with mRNA, but they provide a more natural way to silence genes unlike the totally synthetic antisense agents. However, delivery of RNAi is also very challenging, and thus the promise may be slow to realize. Conjugation of a SPACE peptide (see Section 8.2.4) to siRNA has been reported to cause enhanced absorption via skin and lead to the knockdown of the corresponding protein targets (Hsu & Mitragotri, 2011). Similarly, microneedles and carbon nanotubes have been shown to enhance delivery of siRNA into skin and lead to gene silencing in tumor tissue or silencing of reporter gene expression in skin (Gonzalez-Gonzalez et al., 2010; Siu et al., 2014).

1.1.3.3 Transgenic Therapeutics

The term *transgenic* applies to the transfer of genes between species. For example, human genes can be introduced into an animal model at an embryonic level, by microinjection of purified DNA into the pronucleus of a fertilized egg. The new genetic information gets integrated into the chromosomal DNA of the host organism, and the gene will be passed on to the offspring by the normal laws of heredity. By identifying the gene of interest in the offsprings and continued matings, a transgenic line of animals can be produced. The use of transgenic animals has the potential to limit the total number of animals used in research. Transgenic technology can be used for studying human genetic diseases and for the production of a human protein by the animal. A human gene expressed in the animal will produce a protein that will appear in all its body fluids. The protein can then be isolated from milk or blood, and the live animal becomes a living *factory* for the synthesis of a human protein (Chew, 1993; Garthoff, 1994; Houdebine, 2009). Human antithrombin III, isolated from milk of farm transgenic mammals, has received approval from the European Agency for the Evaluation of Medicinal Products (Houdebine, 2009). Since the cloning of a sheep (Dolly) in 1997, transgenic technology has received more attention and proteins have been produced in transgenic animals using mice, rabbits, pigs, goats, sheep, or other animals. Site-specific gene transfer may also potentially generate immune-tolerant humanized tissues.

1.1.3.4 Gene Therapy

Gene therapy involves the transfer of genetic material to the cells of an individual for therapeutic benefit, such as the cure for a human genetic disease. The introduced gene must then become a part of the system and respond to physiological changes. For example, an engineered insulin gene must respond to changes in blood glucose levels for gene therapy to be successful for diabetes (Morgan, 1993). Genetic material, usually cloned DNA, can be transferred to cells by physical, chemical, or biological means. Biological transfer by means of viruses is the most widely used method. Advances in vaccine technology have laid the groundwork for the development of viral vector gene therapy delivery systems (Summers & Cooney, 1994). However, currently there are certain barriers that need to be overcome. Nonviral vectors are being developed to use gene therapy to treat incurable diseases such as cancer and genetic disorders (Canine & Hatefi, 2010; Ibraheem, Elaissari, & Fessi, 2014).

In principle, an antisense oligonucleotide can be designed to target any single gene within the entire human genome. Thus, any disease whose causative gene is known may benefit from gene therapy. The barriers relate to the ability of the oligonucleotide to cross the cellular membrane in order to reach the cytoplasm or nucleus. Once inside the cell, the oligonucleotide must not degrade and must bind specifically and with high affinity to the RNA target. Several attempts that are being investigated to overcome these barriers are discussed in the literature (Apaolaza, Delgado, Pozo-Rodriguez, Gascon, & Solinis, 2014; Milligan, Matteucci, & Martin, 1993; Suk et al., 2014). Plasmids containing gene of interest or viral vectors are now being investigated. The field of gene therapy suffered a setback because of an unfortunate death in 1999, which changed the political and social climate with intense public attention. However, we have seen success stories in recent years (Wirth, Parker, & Yla-Herttuala, 2013) as well as a marketed product (alipogene tiparvovec, Glybera®).

The treatment of adenosine deaminase (ADA) deficiency is an ideal candidate for gene therapy. ADA deficiency is a rare genetic disease in which afflicted children lack the ADA enzyme necessary for the normal function of the immune system, resulting in severe combined immunodeficiency syndrome. The process of gene therapy for this disease involves the isolation of lymphocytes from a severely immunodeficient patient. These lymphocytes are then grown in the laboratory, when new genes are introduced into them. Finally, they are safely returned to the patient, and results have shown that patients have responded favorably to the therapy (Morgan, 1993; Moseley & Caskey, 1993). Gene therapy protocols have been investigated for diseases such as cystic fibrosis, HIV infection, type 1 diabetes, various hemoglobin disorders, and several cancers and cardiovascular diseases (Nwanguma, 2003).

1.1.3.5 Human Genome Project

The human genome project was launched in 1988 by the National Institute of Health (NIH, Bethesda, MD) and the Department of Energy (Washington, DC). The goal of the project was to sequence the entire six-billion-base-pair human genome by the year 2005. These base pairs were expected to code for approximately 100,000 genes in the human body. The project had brought together molecular biologists, geneticists, and medical researchers, who, with the aid of robotics and other tools, worked on this project (Pestka, 1994; Summers & Cooney, 1994; Uber, 1994). Based on this work, NIH had filed patents on many gene fragments but later dropped the claim. Such patenting of the human genome may allow some companies, universities, or governments to control the use of as yet undiscovered uses related to patented DNA sequences (Wuethrich, 1993). The project was completed ahead of schedule in 2003 with the unraveling of the entire human DNA sequence and the surprise observation that there are only about 30,000 genes in the human genome. Also, it was noted that interindividual variations are attributed to only 0.1% of the genome. Genetic variability among individuals is now being realized as a major determinant of variable drug effect, and the field of pharmacogenomics, which has some applications already, has been predicted to have far-reaching effects on medical practice in the coming years (Mesko, Zahuczky, & Nagy, 2012; Nwanguma, 2003). Though complete genome sequencing was done quickly, the functional characterization of the sequenced genome has proceeded more slowly (Galperin & Koonin, 2010). Antibodies for any protein encoded by these genes are being generated by in vitro methods to complete the human proteome (Dubel, Stoevesandt, Taussig, & Hust, 2010). Proteomics is expected to bridge the gap between genomics and drug discovery in the coming years.

1.2 BIOTECHNOLOGY INDUSTRY

The first biotechnology company was founded in 1971. Since then, the industry has made steady progress, with over a thousand biotechnology companies in the United States alone. In addition, the United Kingdom has several hundred companies, and other countries active in this area include Japan and the European Economic Community (Burrill, 1992b; Dibner, 1986; Dibner & White, 1989; Itoh, 1987; Small, 1992; Szkrybalo, 1987). Other countries are also active in the biotechnology area, and many are now getting active in the biosimilar area (see Section 3.4). Although it is often believed that biotechnology-derived products have been slow to arrive, their total development time, at least for the early products, was about 7 years. This is less than the average development time of 12 years for conventional drugs (Bienz-Tadmor & Brown, 1994).

Although biotechnology companies provide efficient discovery and development, they often need to collaborate with the pharmaceutical industry to bring a product to the market. Pharmaceutical companies provide superior distribution channels and clinical and regulatory assistance. These complementary strengths create a good atmosphere for strategic alliances (Burrill, 1992b; Lyons, Lang, & Guzzo, 1993). Similar collaborative arrangements between universities and the biotechnology industry also exist. Traditionally, such collaborative efforts existed between the universities and the pharmaceutical industry (Swann, 1986). The biotechnology industry has generally been able to maintain its independence and raise capital from the public. Some of the biggest players in the United States include Amgen, Gilead Sciences, Celgene, and Biogen Idec. Consolidation, mergers, and acquisitions have been commonly part of an industry-wide trend (Burrill, 1990; Das, 2003; Dibner, 1992; Spalding, 1993).

1.2.1 Biotechnology Products on the Market

Over 200 biotechnology-derived drugs are currently approved by the U.S. FDA. Follow-on products or biosimilars are discussed in Section 3.4. Many of these biotechnology-derived products are available in limited quantities, especially when first introduced. For example, when interferon-β-1b (Betaseron®, Bayer HealthCare) was first introduced, supplies were limited and it was allocated to patients by lottery. However, eventually rDNA technology allows the production of sufficient quantities in a commercially viable manner. As discussed, different indications of the same drug may receive regulatory approval at different times. Also, it should also be pointed out that the same drug produced in a different cell line will need separate regulatory approval and may actually be structurally different. Considering the example of β-interferon again, Betaseron discussed earlier is different from that developed by Biogen. Biogen's molecule is glycosylated, has one amino acid different from Betaseron®, and is produced in mammalian cell lines rather than *E. coli*. Many of the products introduced have been engineered in some way to improve their therapeutic or pharmacokinetic properties. Such second-generation biopharmaceuticals may have altered amino acid sequence, an altered glycocomponent, or a covalently attached chemical moiety such as polyethylene glycol (Walsh, 2004). Monoclonal antibodies was discussed in Section 1.1.2, and a few of the other recombinant products will be briefly discussed below. Please also see Tables 4.1 and 4.2 in Chapter 4 for a listing of some of the products on the market.

1.2.1.1 Human Insulins and Analogs

1.2.1.1.1 Human Insulin

Several million patients currently require insulin, with a worldwide market in billions. Human insulin (*Humulin*, Eli Lilly) was the first therapeutic recombinant polypeptide to be approved by the FDA in October 1982. Recombinant human insulin is synthesized in *E. coli* bacteria and is chemically and physically equivalent to pancreatic human insulin and biologically equivalent to human and pork insulin (Chance, Kroeff, Hoffmann, & Frank, 1981; Ladisch & Kohlmann, 1992). The advantage of human insulin lies in its lower antigenicity relative to animal-derived insulins (Brogden & Heel, 1987). Human insulin has now largely replaced the animal-derived insulins.

Before 1986, human insulin was prepared by the production of genetically engineered A and B chains in separate fermentations. These two chains were then isolated, purified, and joined together chemically. Recombinant insulin is currently produced by the production of proinsulin, followed by the enzymatic cleavage of the C-peptide to produce human insulin. Humulin® produced by these two manufacturing processes has been shown to have no pharmacokinetic or clinical differences.

Recombinant human insulin can be used by diabetic patients in much the same way as insulin derived from animal sources. Clinical advantages that may be attributed to the recombinant product are slight and may include less immunogenicity and quicker absorption, but these depend on patient factors as well. Several formulations of *Humulin* are available today: regular (R), NPH (N), lente (L), ultralente (U), *Humulin* 50/50, and *Humulin* 70/30. Using one of these formulations, rapid or intermediate or long action can be achieved. In 1996, human recombinant insulin made utilizing *Saccharomyces cerevisiae* (bakers' yeast) as the product organism was approved (Novolin, Novo Nordisk). Considerable effort has been directed to investigate noninvasive delivery systems for insulin and its analogs (Khafagy, Morishita, Onuki, & Takayama, 2007) as will be discussed in the subsequent chapters.

1.2.1.1.2 Human Insulin Analogs

In addition to basal insulin needs that control glucose levels between meals, most diabetics need fast-acting insulin after meals. Insulin analogs with reduced tendency for self-association, which have a rapid onset of action and can thus better mimic meal-stimulated pharmacokinetics of insulin observed in nondiabetics, are highly desired (Defelippis, Chance, & Frank, 2001). Modifications in the insulin sequence in the B26–30 region led to the approval of three rapid-acting recombinant insulins, insulin lispro (*Humalog*, Eli Lilly), insulin aspart (*NovoLog*, Novo Nordisk), and glulisine (Apidra, Sanofi-Aventis). Also, two long-acting analogs, insulin glargine (*Lantus*, Aventis) and detemir (Levemir, Novo Nordisk), have been marketed. Insulin glargine has a different isoelectric point, while insulin detemir has an added fatty acid residue (2005; Sheldon, Russell-Jones, & Wright, 2009). Insulin glargine had a U.S. market of $6.6 billion in 2012 (Bruno, Miller, & Lim, 2013).

The rapid-acting analogs mimic endogenous insulin profile closely and can be given at mealtime or even after a meal and thus provide an important advantage and are also less likely to cause hypoglycemia (Freeman, 2009). Both regular and glulisine insulins are equally effective during the treatment of diabetic ketoacidosis (Umpierrez et al., 2009). Insulin lispro is chemically a Lys (B28), Pro (B29) analog that was created when the amino acids at positions 28 and 29 on the insulin B chain are reversed. The gene for insulin lispro is then inserted into *E. coli* bacteria. Production is carried out under conditions designed to minimize the degree of microbial contamination. As per European Pharmacopoeia, tests for host-cell-derived proteins and single-chain precursor should be carried out prior to release. The praline group at position B28 in regular insulin is important for β-sheet conformation and formation of dimer units. Reversal of the proline–lysine sequence in insulin lispro eliminates important sites for hydrophobic interactions, thereby decreasing the tendency for dimerization. Zinc is added to commercial insulin preparations for stability, and they exist as hexameric units in their containers as marketed. However, unlike regular insulin, insulin lispro dissociates more rapidly to monomers due to decreased tendency for dimerization (Kucera & Graham, 1998). It thus has more rapid absorption, faster onset, and shorter duration of action compared to regular insulin, allowing more flexibility in dosing and mealtime scheduling (Hermans, Nobels, & De, I, 1999; Koivisto, 1998; Kucera & Graham, 1998; Nobels, Hermans, & De, I, 1999). The inversion of the proline and lysine apparently does not increase the immunogenicity of insulin lispro (Hermans et al., 1999). Insulin aspart was made by a single substitution of the amino acid proline by aspartic acid in position B28 and is produced utilizing baker yeast. Insulin aspart is also approved for use of subcutaneous insulin infusion via an external insulin pump.

Insulin glargine is made in *E. coli* and differs from human insulin in that the asparagine at position A21 is replaced by glycine and two arginines are added to the C-terminus of the B chain. Due to the addition of two amino acids, its molecular weight increases to 6.1 kDa. Insulin glargine is a long-acting analog and can provide nearly constant basal levels by a once-daily subcutaneous injection at bedtime (Ho & Gibaldi, 2003). Basal insulin needs of type 1 diabetic patients can be met either by using NPH human insulin or by using one of the newer long-acting insulin analogs (Monami, Marchionni, & Mannucci, 2009). In addition to clinical studies, safety evaluations now include investigating molecular characteristics of the insulin analogs by detailed analysis of receptor binding, activation, and in vitro mutagenicity tests (Hansen et al., 2012). An ultralong-acting basal insulin (insulin degludec, Tresiba) is being developed by Novo Nordisk (Heise et al., 2011; Wang, Surh, & Kaur, 2012).

1.2.1.2 Growth Hormone

Recombinant human growth hormone (GH) was first made available as somatrem, which includes an additional methionyl group at the N-terminus. This product is more immunogenic than the natural sequence that was produced later though the immunogenicity is believed to be due to *E. coli*–derived impurities and not the extra methionine (Patten & Schellekens, 2003). Later, human GH with the natural sequence (somatotropin) was made available (Humatrope®, Eli Lilly, and Nutropin AQ®, Genentech). Both products are approved for the treatment of GH deficiency in children. New formulations have also been investigated and compared for bioequivalence in human volunteers (Dao, Jacobs, Kuebler, Bakker, & Lippe, 2010). Prior to the availability of the recombinant products, GH was extracted from pituitary glands obtained at autopsy. The supply was limited, and furthermore, contamination with an infectious agent was responsible for the development of a viral-like condition in some patients (Jacob–Creutzfeldt syndrome). This extremely serious side effect, which led to a slow mental deterioration and death (Powell-Jackson et al., 1985), was eliminated when GH is obtained using recombinant techniques. The recombinant hormones can now be produced in sufficient quantities, and studies are now possible to explore a wider therapeutic use of GH. GH treatment is proving promising for short children who are not GH deficient, but such use is controversial since short stature by itself is not a disease. GH may also be promising for treatment of GH-deficient adults and to reverse catabolic conditions in patients. The availability of sufficient quantities of recombinant GH is also driving efforts to reassess the most effective and safe dosages and schedules for administration (Chalew, Phillip, & Kowarski, 1993). A recombinant version of bovine somatotropin has also been approved (Posilac®, Elanco) to boost milk production in dairy cows (Miller, 1992).

1.2.1.3 Interferons

Interferon was discovered in 1957 as a protein released by cells exposed to a virus that enables other cells to resist viral infection. Later, it was recognized that interferon is not a single protein but a large family of proteins. The great promise of interferons against viral diseases and cancers was recognized early on, but progress to fulfill the promise was slow for various reasons. The mode of action was not clearly understood, they were not available in enough quantity for effective clinical trials, and purification was difficult. The advent of rDNA technology allowed the isolation of the relevant gene and its cloning in *E. coli* to produce sufficient quantities for extensive studies (Pestka, 1983).

Interferons are now approved by the FDA for treatment of several diseases. These include genital warts, Kaposi's sarcoma, hairy cell leukemia, chronic granulomatous disease, and multiple sclerosis. Most interferons belong to one of three classes—alpha, beta, or gamma. The alpha family is largest with more than 20 members. Interferon-α and interferon-β are often classified as type I interferons, and these can be produced by essentially all cells in the body in response to a viral infection. Interferon-γ is a type II molecule, produced only by the T lymphocytes and natural killer cells of the immune system.

Together, interferons modulate the activity of almost every component of the immune system (Johnson, Bazer, Szente, & Jarpe, 1994). The interferon-α proteins contain 165 or 166 amino acids with molecular weights ranging from 17.5 to 23 kD. These proteins tend to be acidic, with isoelectric points of 5.7–7.0. Human interferon-β contains 166 amino acids, 21 of which are removed during secretion from cells. It is a natural glycoprotein with a molecular weight of 23 kD and isoelectric point between 6.8 and 7.8. Interferon-γ is composed of 143 amino acids and has no statistically significant sequence homology to interferon-α and interferon-β. It is acid labile and highly basic, with an isoelectric point of 8.6–8.7 (Wills, 1990). Interferon-α-2 (a and b) and interferon-v-1b concentrated solutions are official in the European Pharmacopoeia. The pharmacopoeia states that the limits for host-cell-derived proteins and host-cell or vector-derived DNA are approved by the competent authority. Identification tests based on peptide mapping, isoelectric focusing (for α-2), or N-terminal sequence analysis (for γ-1b) are included. Impurities of molecular mass differing from the native protein are examined by SDS polyacrylamide gel electrophoresis under both reducing and nonreducing conditions. Acceptance criteria based on the intensity of bands with molecular masses lower or higher than the principal band relative to principal bands obtained with reference solutions are specified. The monograph for interferon-γ-1b also states that covalent dimers and oligomers should not be greater than 2%, as determined by size-exclusion chromatography (2005).

1.2.1.4 Dornase Alfa (rhDNase)

Pulmozyme® (Genentech), an aerosolized recombinant version of human deoxyribonuclease I (rhDNase, dornase alfa) enzyme, was approved by FDA in March 1994, for cystic fibrosis. The purified glycoprotein is produced in genetically engineered CHO cells and contains 260 amino acids (MW, 37 kDa). In cystic fibrosis patients, extracellular DNA gets deposited in high concentrations (up to 15 mg/mL) in airway secretions. This causes a thickening of the mucus secretions in the respiratory and gastrointestinal epithelia. The thick and viscous secretions thus become difficult to expectorate and obstruct airways. DNase breaks down the accumulating DNA and thus causes a thinning of the infected mucus. Thus, aerosolized rhDNase becomes an effective method of improving lung function in patients with cystic fibrosis (Hubbard et al., 1992; Wallace & Lasker, 1993) and more indications are being explored (Zitter, Maldjian, Brimacombe, & Fennelly, 2013). Pulmozyme is administered through inhalation of an aerosol mist produced by a compressed air–driven nebulizer system. Each single-use ampul will deliver 2.5 mL of the solution to the nebulizer bowl (PDR Staff, 2014). Since mucus secretions also build up in pneumonia, bronchitis, and emphysema patients, DNase may have use in these conditions as well.

1.2.1.5 Antimicrobial Peptides

Some species of frogs are remarkable in their ability to resist wound infections, and it is now known to be due to the presence of bioactive magainin peptides that have broad activity against gram-positive and gram-negative bacteria (Berkowitz, Bevins, & Zasloff, 1990). Magainins have 23 residues, are devoid of cysteine, and form an amphipathic α-helix. Other antibacterial peptides such as defensins are more effective against gram-positive bacteria. Defensins contain 29–34 residues, have three intramolecular disulfide bonds, and are basic. Other antibacterial peptides responsible for insect immunity have also been identified (Boman, 1991). Antimicrobial peptides seem to be relatively harmless to eukaryotic cells and thus may have applications as topical microbicides for the prevention of HIV transmission though other drugs are also being investigated for such use (Rosenthal, Cohen, & Stanberry, 1998). Recently, it has been found that in addition to their role in infectious diseases, antimicrobial peptides may play a role in diseases as diverse as inflammatory disorders, autoimmunity, and cancer (Pinheiro da & Machado, 2012).

Subsets of these antimicrobial peptides are the arginine-rich peptide transporters. Oligomers of arginine composed of six or more amino acids seem to be able to cross biological membranes by an undefined mechanism. They may also transport any drugs covalently attached to them (Rothbard et al., 2002). It has been reported that a *oligoarginine*–cyclosporine conjugate penetrated the skin as well as the unmodified drug and then was able to enter targeted tissue T cells, presumably directly entering the cytosol by active transport across the lipid-rich cell membranes (Rothbard et al., 2000).

1.2.1.6 Tissue Plasminogen Activator

Activase® (tissue plasminogen activator, tPA or alteplase; Genentech) is a 527-amino-acid glycoprotein produced by rDNA technology in CHO cells and is identical to the naturally occurring molecule. It triggers clot-degrading system and is used in the management of acute myocardial infarction, acute ischemic stroke, and pulmonary embolism. While the existence of this endogenous enzyme has been known for decades, commercially viable quantities could be produced only when rDNA technology became available. Since its production in this manner, it has been administered to hundreds of thousands of patients (Absar, Choi, Yang, & Kwon, 2012; Grossbard, 1987; Sobel, Fields, Robison, Fox, & Sarnoff, 1985). Unmodified tPA is a glycoprotein with five functionally active structural domains. However, two engineered forms of tPA with extended half-lives are now available. Reteplase (*Retavase*, Roche Diagnostics) consists of only two native tPA domains and is unglycosylated as it is produced in *E. coli*. Tenecteplase (*TNKase*, Genentech) is another form with amino acid substitutions at three positions. It also has an increased specificity for fibrin binding (Walsh, 2004).

1.2.1.7 Fusion Proteins

A 91.4 kDa dimeric fusion protein (alefacept, *Amevive*, Astellas Pharma) produced in an engineered CHO cell line consisting of the extracellular CD2 binding portion of the human leukocyte function antigen 3 linked to the Fc region of human IgG is available for the treatment of chronic plaque psoriasis. Examples of other fusion proteins include denileukin diftitox (*Ontak*, Eisai, Inc.) and etanercept (*Enbrel*). Etanercept (*Enbrel*, Amgen), a 150 kDa dimeric fusion protein of TNF receptor linked to the Fc portion of a human IgG antibody, is expressed in an engineered CHO cell line and is indicated for treatment of rheumatoid arthritis. It had a market of US$ 7.9 billion in 2012 (Bruno et al., 2013). Denileukin diftitox, a 58 kDa fusion protein, consists of a modified form of diphtheria toxin fused to human IL-2 and expressed in an engineered *E. coli* cell line. It is provided as a frozen solution to be administered intravenously and is indicated for treatment of cutaneous T-cell lymphoma. The frozen solution is thawed before use but should not be vigorously shaken or refrozen (PDR Staff, 2014; Spada & Walsh, 2005; Walsh, 2004). An appreciation for such handling due to formulation, physical stability, and processing considerations will become more meaningful to the reader in the subsequent chapters of this book. Long half-life of albumin has also been used to develop an albumin fusion platform to develop fusion proteins. Albinterferon-α-2b, a genetic fusion of interferon with albumin, has been reported to have a half-life longer than pegylated interferon (Subramanian, Fiscella, Lamouse-Smith, Zeuzem, & McHutchison, 2007), while albumin-linker-erythropoietin fusion protein has been reported to be 7.8-fold more potent than erythropoietin for increasing the hematocrit of normal mice (Joung et al., 2009). Fusion proteins are being investigated for vaccine applications as well (Morefield, Touhey, Lu, Dunham, & HogenEsch, 2014).

1.2.1.8 Miscellaneous

Erythropoietin is a hematopoietic factor released by the kidney to stimulate erythrocyte production. Recombinant human erythropoietin can restore the hematocrit to normal in many patients with the anemia of end-stage renal disease (Eschbach, Egrie, Downing, Browne, & Adamson, 1987), though the product is also abused in sports. Currently, the product is approved for use in the terminal stages of renal failure and for anemia secondary to AZT and anemia associated with cancer chemotherapy. A monograph for erythropoietin concentrated solution is official in the European Pharmacopoeia. Identification tests based on peptide mapping, capillary zone electrophoresis, and polyacrylamide gel electrophoresis/immunoblotting are included. The solution should be stored below −20°C, and repeated freezing and thawing should be avoided (2005). An absorbable collagen sponge that contains recombinant human bone morphogenetic protein 2 (dibotermin alfa) and enhances bone healing is also available. Carriers are needed to maintain the concentration of osteogenic factors at the treatment site for enough time to allow bone-forming cells to migrate to the area of injury (Seeherman, Wozney, & Li, 2002). Bioactive mesoporous glass microspheres have been reported to be promising protein delivery systems as a filling material for bone regeneration and healing of defects (Wu et al., 2010).

Recombinant human collagen is available and provides an efficient scaffold for bone repair when combined with recombinant bone morphogenetic protein in a porous sponge (Yang et al., 2004). A hydrolase enzyme produced by rDNA technology (laronidase, *Aldurazyme*, Genzyme Corp.) is also available for patients with mucopolysaccharidosis, a rare, inherited genetic disease characterized by incomplete degradation of specific mucopolysaccharides in the body (Spada & Walsh, 2005).

1.3 REGULATORY ASPECTS OF BIOTECHNOLOGY-DERIVED DRUGS

Historically, biologics were extracted from natural sources and their chemical nature was not clearly defined. With the advent of highly purified biotechnology-derived agents, the distinction between drugs and biologics is less clear today. Nevertheless, the distinction is important from an administrative point of view since it determines which center of FDA will review the product. Until recently, drugs were regulated under the Food, Drug, and Cosmetic Act, while biologics are controlled by the Public Health Services Act and handled by the Center for Biologics Evaluation and Research (CBER). Under a reorganization within the FDA in 2003, well-characterized biologics are now handled by the Center for Drug Evaluation and Research (CDER). However, CBER continues to review blood products and vaccines and also evaluates gene therapy products as and when they become available (Marshall & Baylor, 2011). FDA publishes its current thinking on biotechnology products in the form of points to consider (PTC) bulletins, which provide some guidance to the manufacturers. This way, FDA can keep pace with the rapidly developing technology by periodically updating PTCs, which are draft documents (Beatrice, 2002; Conklin, 1993; Giddins, Dabbah, Grady, & Rhodes, 1987). Biotechnology manufacturers should follow current regulations and PTCs to submit an Investigational New Drug (IND) application to the FDA. In addition, the production of drugs or biologics by rDNA technology should follow the NIH guidelines for research involving rDNA molecules. For traditional drugs, a manufacturer finally submits a New Drug Application (NDA) to the FDA. However, for biologics, the manufacturer typically submits a Biologics License Application (BLA) to CDER or CBER. The equivalent application in Europe is called Marketing Authorization Application. A BLA may be considered as the equivalent of an NDA for a conventional drug molecule. For worldwide marketing of products, the pharmacopoeia with the most stringent guidelines and/or the guidelines of the International Conference on Harmonization (ICH) of technical requirements for registration of pharmaceuticals for human use (ICH) should be followed. Several ICH guidelines for quality of biotechnological products have been published, including those for stability testing. ICH guidelines are also available for viral safety evaluation of biotechnology products derived from cell lines of human or animal origin. A company should carefully evaluate products made in different sites or by different processes. For example, a change in manufacturing process for tPA was associated with a significant change in its clinically important characteristics (Beatrice, 2002). Ideally, the product should be fully characterized at early developmental stages. This may not happen due to time or cost constraints. Changes made later could then result in problems along the regulatory pathway. Also, poorly characterized products are not allowed in commercial manufacturing facilities. Therefore, material from bench-scale production may be utilized in preclinical or early clinical trials. Once well-characterized product made in commercial facilities is used in late-stage clinical trials, the results from the early clinical trials may be considered invalid if the new product is deemed different from the originally used product (Beatrice, 2002).

Many of the biopharmaceuticals have been developed under the 1983 Orphan Drug Act, which provides a 7-year market exclusivity for drugs, targeting diseases that affect fewer than 200,000 individuals in the United States. For these drugs, it becomes important to be the first to reach the market. For example, Amgen's Neupogen (G-CSF) was approved 2 weeks earlier than Leukine (Sanofi, GM-CSF) and was thus awarded a market exclusivity for several conditions characterized by neutropenia and became a billion dollar product. More recently, companies have been able to successfully argue that their product is different from the one previously approved under the Orphan Drug Act. For example, Biogen Idec successfully argued that their interferon product, Avonex, was not the same drug as Betaseron (Bayer HealthCare) since it has a better safety profile with regard to skin necrosis at injection sites. Currently, these two as well as another interferon product, Rebif (EMD Serono), are approved for treatment of multiple sclerosis (Ho & Gibaldi, 2003). For chronic use of biotechnology products, toxicity studies are also needed (Clarke et al., 2008). For a discussion of biosimilars, please see Section 3.4.

1.3.1 Preclinical and Clinical Studies

Preclinical studies conduct safety and efficacy evaluations of the drug, preferably in more than one test species. The test species used must have the same receptor as humans, and its metabolism and handling of the test drug should be similar to that of humans. Since human proteins are likely to be immunogenic to the animal species used, the immunogenicity induced in animals during preclinical studies may not be of a clinical concern. The immunogenicity potential in humans may only become clear once clinical trials start (see also Section 1.4). A common regulatory problem associated with INDs for biotechnology products relates to the manufacturing process. FDA requires a detailed characterization of the cell line and documentation to show the absence of fungi, bacteria, cellular DNA, and viruses in the final product. The manufacturing process should be designed to remove prion proteins and transmissible spongiform encephalopathy infectivity. A detailed description of the structure and a bioassay are also required. A description of some of the analytical techniques used to acquire this information is given in Chapter 2. For an IND application, some evidence of effectiveness, such as that based on animal studies, must be provided to allow trials in a patient population. Also, toxicity studies are required as the protein may be extremely toxic at pharmacological doses. Such toxicity studies will need less extensive testing if the protein is being used at physiological levels for replacement therapy. The conventional LD_{50} determination can be replaced by well-designed dose tolerance studies, but more toxicity studies are needed if there is evidence for development of neutralizing antibodies (Wong, 1998).

For traditional drugs, Phase I safety and tolerance studies are done in normal volunteers. However, for biotechnology products, the first human studies are done in patients. Phase II trials are conducted to establish the therapeutic effectiveness. The sponsor must define the primary indication and the methods to be used to assess the outcomes. The FDA prefers multicenter trials. While preclinical studies may have been done with GLP material, Good Manufacturing Practices (GMP) material is required for clinical trials. The process change required to produce GMP material can involve substantial investment and may result in a change in the characteristics of the drug or its purity profile. As a product moves through development cycle, it moves from R&D into pilot plant and then into manufacturing. Scale-up from laboratory to large-scale manufacturing needs process development. The product development cycle can be shortened by performing process development concurrent with production (Fahrner, 1993). Facility requirements for biotechnology plants also involve special considerations for facility and systems design, operating procedures, and validation (Hill & Beatrice, 1989a,1989b). FDA also asks manufacturers to submit an environmental assessment as a part of their application for licensure (Gibbs, 1989). Since biological products are often heat sensitive, certain manufacturing operations may need to be performed under controlled temperatures (below 10°C in some cases). Also, concern for microbial contamination must be considered at every step since these products are susceptible to such contamination. Good manufacturing practices are a must even during the initial steps to prevent any contamination by microorganisms (Hill & Beatrice, 1991). Preclinical and clinical studies with cytokines and growth factors focus on their utility not only as single agents but also in combination with other drugs such as chemotherapeutic agents (Talmadge, 1993).

1.3.2 Pharmacoeconomics and the Regulatory Process

The pharmacoeconomic considerations may also play a role in determining whether or not a biopharmaceutical reaches the market. This is especially true now with the focus on the cost of the health-care system. Even if a product is proven effective, FDA may consider the question whether it is effective from a cost-efficiency standpoint as well. Pharmacoeconomic considerations contributed significantly to the reluctance of the FDA to approve Janssen Biotech's expensive Centoxin for septic shock (Burrill, 1992a), though later safety issues were also identified. In contrast, another expensive product, erythropoietin, pays for itself since a number of dialysis patients can return to work (Fradd, 1993). Uncertainties of the regulatory process suggest that a company should not base its fortune on one blockbuster drug. Another dimension to the economics is that once a drug is approved, significant off-label uses are discovered (Burrill, 1992a; Fradd, 1993). The prices of biotechnology products are market driven, where the product efficacy and its competitiveness with currently available alternatives control the price. For example, Genentech's Activase (tPA) was criticized in the media when first approved since it was much more expensive than its competing product, Streptokinase. In addition, the insurance industry, ethical, and legal issues get involved in product pricing. Also, vaccines, therapeutics, and certain diagnostics face the risk of product liability suits. Biotechnology companies may seek liability insurance and recover the costs in pricing (Burrill, 1989). Pharmacoeconomics or more broadly outcome research considers the value for money for medications including considerations of efficacy, effectiveness, and efficiency (Peys, 1996). Biosimilars (see Section 3.4) have the potential to offer significant cost savings as compared to innovator products (Henry & Taylor, 2014).

1.3.3 Characterization and Purity of Biotechnology-Derived Proteins

Regulatory agencies have placed significant emphasis on fully characterizing proteins, especially aggregates/subvisible particles (see Section 3.2.2), including emphasis on analytical methods (den et al., 2011). Because of the techniques of rDNA technology, a potential of contamination of the final product by HCPs is very real. For example, a recombinant protein produced in the bacterium *E. coli* is released only when the bacteria are lysed. Cell lysis will release the protein of interest but also hundreds of other bacterial proteins, as well as other impurities such as product-related degradation products, fragments, aggregates, and DNA or adventitious agents. Purification of this mixture is a challenging task for the process engineer and other personnel. The purification process itself may add other impurities such as trace amounts of process chemicals used during the purification steps or may generate degradation products as the protein is exposed to stressful processing conditions. During growth, many microorganisms and animal cells secrete proteolytic enzymes, which may cause proteolysis of the therapeutic protein being produced, thus reducing yields. The use of a general serine protease inhibitor may be required during protein purification to reduce this proteolytic activity. For example, the irreversible serine protease inhibitor, 4-(2-aminoethyl)-benzenesulfonyl fluoride hydrochloride, was found to protect several proteins and enzymes, such as tPA, thrombin, plasma kallikrein, and subtilisin A, from unwanted proteolysis (Mintz, 1993). Even though a high-purity product is finally produced, minute components of bacteria, yeast, or other mammalian cell lines used in production may be carried over into the final formulation. The World Health Organization recommends that final products contain ≤100 pg of residual cellular DNA per single human dose. In addition, the product should be tested for mycoplasma and viral contamination. Viral inactivation and removal need to be demonstrated when recombinant proteins are expressed in mammalian cell lines, and this is very important for safety and regulatory approval. Serum required for the growth of mammalian cell cultures should be sterilized. Sterility and pyrogen tests can be performed, as required. Pyrogen testing can be replaced by Limulus amoebocyte lysate testing for endotoxin by the procedures described in U.S. or European pharmacopoeia. Also, safety testing is done in guinea pigs and mice by inoculating them with the final formulation and looking for weight loss or any unforeseen reactions. A final product lot will pass the safety test if all inoculated animals survive the test, if no unforeseen reactions occur, and if no weight loss takes place during the test (Schiff, Moore, Brown, & Wisher, 1992). An HCP impurity found in preparations of recombinant human acidic fibroblast growth factor was identified as the S3 ribosomal protein of *E. coli*.

A combination of N-terminal sequencing and Western blot analysis was used to identify this impurity. Such identification of impurities is important since a known description of physicochemical properties of the impurity will allow for suitable process modifications to eliminate the impurity from future lots. Also, it allows for specific in-process testing to produce a quality product consistently (O'Keefe, DePhillips, & Will, 1993). Written procedures should be in place to ensure the absence of adventitious agents.

1.3.4 Patents

Patents provide protection to the inventor by excluding others from practicing their invention for a period of time. This time period is 20 years from filing date in the United States. In order to be patentable, a discovery or invention must be novel, useful, and nonobvious to one of ordinary skill in the related art. A wide variety of biological materials such as DNA, RNA, plasmids, polypeptides, enzymes, hormones, and even transgenic animals can be patented (McDonnell, 1992). Several patent disputes have been reported in the biotechnology industry, such as Amgen versus former Genetics Institute (McDonnell, 1992), Amgen versus Chugai (Farber & Brotman, 1990; Warburg, 1993), and Scripps Clinic and Research Foundation versus Genentech (Griffen, Irving, & Sung, 1992). Litigation often results because there can be dual routes to get a therapeutic peptide—either from natural sources by traditional methods or by genetic engineering. It seems that claims made in terms of the function of the peptide or protein may be stronger than claims that specify a particular amino acid sequence. For example, amino acid substitutions can be made, and if these have no effect on structure or function, the new amino acid sequence may not have patent coverage. Since damages for patent infringement can run as high as $4 billion, claims to the function of a peptide or protein must be made, in addition to a coverage of specific amino acid sequences and possible changes (Farber & Brotman, 1990). Biological products are somewhat more difficult to protect with patents due to their complexity and several *similar* effective variants. However, the 2009 Biologics Price Competition and Innovation Act can now provide a 12-year exclusivity to biologics innovator (Stroud, 2013).

1.4 IMMUNOGENICITY OF PROTEINS

The body generates an immune response if it considers the administered protein to be foreign. Thus, proteins derived from animal sources such as bovine or porcine insulin are expected to be immunogenic. However, in reality, even recombinant human proteins are generally immunogenic, though to a lesser extent. This could be due to various reasons. The protein may have product-related impurities such as oxidized forms or aggregates. Another major source of immunogenicity is the presence of impurities such as those from the media or HCPs. The dose and route of administration also play a role, with subcutaneous routes generally being more immunogenic than intramuscular or intravenous routes (Narasimhan, Mach, & Shameem, 2012). The genetic makeup of the patient population also determines if the protein would be immunogenic (Stein, 2002), and thus adverse affects may be race dependent.

Thus, the immunogenicity of a protein cannot easily be predicted and is controlled by a complex multitude of factors and not a simple consideration of *self* versus *nonself.*

Some proteins are intrinsically more immunogenic, while some do not seem to induce detectable immune responses. Administration of self proteins to the body can break tolerance due to the presence of impurities or adjuvants or due to exposure to high levels of even *natural* sequences and may induce immunogenicity, while sometimes, the administration of even animal-derived proteins may not induce immunogenicity in most patients (Editorial, 2003; Patten & Schellekens, 2003). More structurally complex proteins are generally better antigens. However, sequence divergence from native proteins seems to play only a minor role (Patten & Schellekens, 2003). The degree and type of posttranslational modifications of the protein such as glycosylation can directly affect the immunogenicity of the protein (Jenkins & Curling, 1994). Thus, a protein produced in *E. coli*, which will not be glycosylated, is likely to induce a different immune response than the same protein produced in a mammalian cell line as it will be glycosylated in the latter case. Antibodies produced against an immunogenic protein may neutralize the efficacy of the protein, though this is not always the case. Generally, these antibodies do not cause any serious allergic or anaphylactic reactions, but if they neutralize the native protein, they can lead to devastating consequences (Patten & Schellekens, 2003; Rosenberg, 2003).

Human proteins being tested in animals are likely to be immunogenic to the animals, and this can be a major obstacle to preclinical studies and especially to long-term toxicology studies. Human proteins are expected to be immunogenic to animals, but a high immune response in all tested species may be predictive of immunogenicity in humans, though by no means certain. The availability of immune-tolerant transgenic mice has been very beneficial in this respect (Patten & Schellekens, 2003). Studies in nonhuman primates can more accurately predict immunogenicity prior to initiating clinical trials. Other predictive models such as MHC binding of peptides, computer algorithms, or in vitro naive T-cell assays are also being developed, but the accuracy of prediction would determine regulatory approval (Mire-Sluis, 2003). The development of analytical methods to detect and characterize antibodies is an active area of research, and improvements in these capabilities will have a significant impact on our ability to monitor clinically relevant antibodies, measure low binding affinity antibodies, and distinguish between neutralizing or binding antibodies. Similarly, animal-derived proteins are likely to be immunogenic to humans. Examples of the latter include the use of mouse-derived monoclonal antibodies or the use of bovine or porcine insulin in humans. Even different animal strains may differ in their immunogenic response, due to differing expression of major histocompatibility complexes. For a discussion of effect of antibody induction on pharmacokinetics of protein drugs, see Chapter 6. All new potential protein products need to be carefully evaluated for their immunogenic potential, and once marketed, they should be reevaluated each time there is a significant manufacturing change or if the patient population changes (Stein, 2002). The need to monitor immunogenicity even after a product is marketed was highlighted by the development of pure red cell aplasia in patients treated with Eprex® (epoetin alfa) in several countries where the product was marketed. The incidence was associated with subcutaneous administration, but the reasons for the sudden increase in immunogenicity after being on the market for several years were not clear. Manufacturing and formulation changes such as removal of human serum albumin from the formulation were initially implicated. Other possible reasons that may be implicated could be a shift in the route of administration to SC or a change in product container to syringes (Editorial, 2003; Rosenberg, 2003). Other factors that may be involved include changes in patient population such as those caused by marketing in different countries or changes in route of administration, formulation, storage, and handing (Bader, 2003). It has also been shown that a minute amount of epoetin may be associated with micelles of surfactant used in the formulation. It has been hypothesized that multiple epoetin molecules present in one micelle could lead to increased immunogenicity due to the presence of multiple epitopes exposed on the micellar surface (Hermeling, Schellekens, Crommelin, & Jiskoot, 2003). However, investigations by the manufacturer later revealed that the cause involved the generation of leachates from the interaction of polysorbate used in the formulation with the uncoated rubber syringe stoppers (see also Section 2.4.3.1), and conversion to a FluroTec®-coated stopper eliminated the problem (Sharma, 2007).

The preparation of a pure protein that has identical structure to a body protein and does not have any aggregates or other product impurities and is free of process impurities will generally produce a protein that is likely to be nonimmunogenic. Aggregation is a major factor that may induce immunogenicity (Hermeling, Crommelin, Schellekens, & Jiskoot, 2004; Rosenberg, 2006; Wang et al., 2012), and many existing analytical methods do not critically assess the presence of noncovalent aggregates. More novel approaches such as identifying relevant epitopes and creating modified proteins that can be tested in relevant assay systems are also underway (Tangri, LiCalsi, Sidney, & Sette, 2002). Impurities can come from the media, HCPs, or the columns used in purification or from the product itself such as aggregates or fragments or chemically modified forms such as oxidized forms. In order to detect these product or process-related impurities, specific and sensitive assays need to be developed. Immunogenicity considerations are important to the clinical application of protein delivery systems (Jiskoot et al., 2012). Protein aggregation during manufacture and bioprocessing can be minimized by carefully controlling the environment (Cromwell, Hilario, & Jacobson, 2006).

1.5 STRUCTURE-BASED DRUG DESIGN

The first structure-based search in 1975 discovered the antihypertensive agent, captopril. Over the next 15 years, millions were spent in this area with little results. This has begun to change now due to advances in NMR, x-ray crystallography, and computer modeling (Hendry, 1993; Kaminski, 1994; Peters & Mckinstry, 1994; Sun & Cohen, 1993), as well as computational protein design (Pantazes, Grisewood, & Maranas, 2011). Traditionally, pharmaceutical companies have discovered new drugs by screening extensive collection of chemical compounds. This screening stage of drug discovery is followed by toxicity studies and clinical trials. The screening stage can be speeded up by structure-based drug design. Once it was realized that receptors are proteins that bind ligands primarily by shape interactions, it became possible to search compounds that can be accommodated in the receptor's 3D shape (Gillmor & Cohen, 1993; Whittle & Blundell, 1994). Genentech has used rational drug design to discover anticoagulant agents from snake venom. Researchers found that three amino acids forming an RGD (arginine, glycine, and aspartic acid) motif were essential for biological activity. Several small RGD-containing cyclic peptides were studied to establish the conformational criteria. However, instead of using peptides, orally absorbable traditional small molecules were then synthesized, which had similar shape and activity (Peters & Mckinstry, 1994). Computer modeling has also been used in conjunction with experimental chemistry to develop a Lys-Pro analog of human insulin, which does not undergo self-association, unlike neutral regular insulin (see Section 1.2.1.1.1). The development of this analog was initiated by the observation that insulin-like growth factor-I (IGF-I) has many structural similarities to insulin but does not self-associate. The important difference between insulin and IGF-I was on the B chain, where the B28-B29 sequence was Pro-Lys for insulin but Lys-Pro for IGF-I. Molecular modeling was then used to ascertain that reversal of the Pro-Lys sequence will increase the free energy barrier to self-association (Galloway, 1993). Algorithms and programs are also being developed to find a relationship between the activities of proteins and peptide homologs and the characteristic of some regions in the primary structure of these molecules. Such programs may also be useful to find correlations between peptide physicochemical properties and immunogenicity and as a tool to improve some properties of the protein (Eroshkin, Zhilkin, & Fomin, 1993; Parker, Choi, Griswold, & Bailey-Kellogg, 2013). The final product should be tested for physical appearance, native protein structure, and bioassay. Several membrane translocalizational signals have been identified, and these peptide signals, typically 9–30 amino acid residues in length, have the capability of crossing plasma membrane and could potentially be used as delivery vectors for other drug molecules (Snyder & Dowdy, 2004; Tung & Weissleder, 2003).

1.6 CONCLUSIONS

The advent of rDNA technology is bringing physiological peptides and proteins to the market as therapeutic peptides and proteins. In addition, other biotechnology techniques such as monoclonal antibodies, antisense agents, transgenic technology, and gene therapy are bringing novel pharmaceutical agents and novel approaches to the treatment of diseases. With the completion of the human genome project, advances in functional genomics, proteomics, and pharmacogenomics will continue to advance therapeutics based on technology. More than 200 biotechnology-derived products have been marketed, and several hundreds are in various stages of clinical development. Insulin analogs with improved pharmacokinetics have been developed and marketed. Currently, most biotechnology-derived therapeutic proteins are available for administration by the parenteral route. As research progresses, alternative routes may become more common. Since these routes will provide a formidable barrier to protein delivery, the bioavailability is likely to be very low. This will increase the demand on quantities of the drug required even though there has been no increase in the number of users. Efficient production of proteins by the techniques of rDNA will then become even more important. Computer modeling will help the structure-based drug design for drugs of the future. Several unique regulatory and patent considerations exist for these biotechnology drugs. The immunogenicity potential of these proteins should also be carefully considered.

REFERENCES

Absar, S., Choi, S., Yang, V. C., & Kwon, Y. M. (2012). Heparin-triggered release of camouflaged tissue plasminogen activator for targeted thrombolysis. *J. Control. Release, 157,* 46–54.

Alley, S. C., Okeley, N. M., & Senter, P. D. (2010). Antibody-drug conjugates: Targeted drug delivery for cancer. *Curr. Opin. Chem. Biol., 14,* 529–537.

Alper, J. (1993). Oligonucleotides surge into clinical trials. *Bio/Technology, 11(11),* 1225.

Apaolaza, P. S., Delgado, D., Pozo-Rodriguez, A. D., Gascon, A. R., & Solinis, M. A. (2014). A novel gene therapy vector based on hyaluronic acid and solid lipid nanoparticles for ocular diseases. *Int. J. Pharm., 465,* 413–426.

Assenberg, R., Wan, P. T., Geisse, S., & Mayr, L. M. (2013). Advances in recombinant protein expression for use in pharmaceutical research. *Curr. Opin. Struct. Biol., 23,* 393–402.

Bader, F. G. (2003). *Case study: Immunogenicity of r-Hu-erythropoietin.* Bethesda, MD: IBC Life Sciences.

Barr, E. & Leiden, J. M. (1991). Systemic delivery of recombinant proteins by genetically modified myoblasts. *Science, 254,* 1507–1509.

Beatrice, M. G. (2002). Regulatory considerations in the development of protein pharmaceuticals. In S. L. Nail & M. J. Akers (Eds.), *Development and manufacture of protein pharmaceuticals* (14th ed., pp. 405–456). New York: Kluwer Academic/Plenum Publishers.

Berkowitz, B. A., Bevins, C. L., & Zasloff, M. A. (1990). Magainins: A new family of membrane-active host defense peptides. *Biochem. Pharmacol., 39,* 625–629.

Bienz-Tadmor, B. & Brown, J. S. (1994). Biopharmaceuticals and conventional drugs: Comparing development times. *BioPHARM, 7(2),* 44–49.

Boman, H. G. (1991). Antibacterial peptides: Key components needed in immunity. *Cell, 65,* 205–207.

Brogden, R. N. & Heel, R. C. (1987). Human insulin: A review of its biological activity, pharmacokinetics and therapeutic use. *Drugs, 34,* 350–371.

Bruno, B. J., Miller, G. D., & Lim, C. S. (2013). Basics and recent advances in peptide and protein drug delivery. *Ther. Deliv., 4,* 1443–1467.

Buchanan, A., Ferraro, F., Rust, S., Sridharan, S., Franks, R., Dean, G. et al. (2012). Improved drug-like properties of therapeutic proteins by directed evolution. *Protein Eng. Des. Sel., 25,* 631–638.

Burrill, G. S. (1989). Biotechnology product pricing: Issues, concerns, and strategies. *BioPHARM, 2(6),* 12–18.

Burrill, G. S. (1990). The Genentech-Roche partnership: Responses and repercussions. *BioPHARM, 3(5),* 62–66.

Burrill, G. S. (1992a). Biotech midyear update: New economic realities. *BioPHARM, 5(7),* 46–49.

Burrill, G. S. (1992b). Biotech's impact on the U.S. drug industry. *BioPHARM, 5(5),* 62–67.

Canine, B. F. & Hatefi, A. (2010). Development of recombinant cationic polymers for gene therapy research. *Adv. Drug Deliv. Rev., 62,* 1524–1529.

Carter, P. J. (2011). Introduction to current and future protein therapeutics: A protein engineering perspective. *Exp. Cell Res., 317,* 1261–1269.

Casi, G. & Neri, D. (2012). Antibody-drug conjugates: Basic concepts, examples and future perspectives. *J. Control. Release, 161,* 422–428.

Chalew, S. A., Phillip, M., & Kowarski, A. (1993). Plasma integrated concentration of growth hormone after recombinant human growth hormone injection—Implications for determining an optimal dose. *Am. J. Dis. Child, 147,* 274–278.

Chance, R. E., Kroeff, E. P., Hoffmann, J. A., & Frank, B. H. (1981). Chemical, physical and biologic properties of biosynthetic human insulin. *Diabetes Care, 4,* 147–154.

Chew, N. J. (1993). Emerging technologies: Transgenic therapeutics. *BioPHARM, 6(3),* 24–26.

Chrisey, L. A. & Hawkins, J. W. (1991). Antisense technology: Principles and prospects for therapeutic development. *BioPHARM, 4,* 36–42.

Clarke, J., Hurst, C., Martin, P., Vahle, J., Ponce, R., Mounho, B. et al. (2008). Duration of chronic toxicity studies for biotechnology-derived pharmaceuticals: Is 6 months still appropriate? *Regul. Toxicol. Pharmacol., 50,* 2–22.

Conklin, J. J. (1993). Biotechnology: ELAs and PLAs. In R. A. Guarino (Ed.), *New drug approval process* (2nd ed., pp. 133–144). New York: Marcel Dekker, Inc.

Cox, M. M. (2012). Recombinant protein vaccines produced in insect cells. *Vaccine, 30,* 1759–1766.

Craik, D. J., Fairlie, D. P., Liras, S., & Price, D. (2013). The future of peptide-based drugs. *Chem. Biol. Drug Des., 81,* 136–147.

Cromwell, M. E., Hilario, E., & Jacobson, F. (2006). Protein aggregation and bioprocessing. *AAPS J., 8,* E572–E579.

Crosby, L. (2003, April). Commercial production of transgenic crops genetically engineered to produce pharmaceuticals. *BioPharm Int., 16,* 60–68.

D'Hondt, M., Bracke, N., Taevernier, L., Gevaert, B., Verbeke, F., Wynendaele, E. et al. (2014). Related impurities in peptide medicines. *J. Pharm. Biomed. Anal., 101,* 2–30.

Dao, L., Jacobs, J., Kuebler, P., Bakker, B., & Lippe, B. (2010). Bioequivalence studies for three formulations of a recombinant human growth hormone: Challenges and lessons learned. *Growth Horm. IGF Res., 20,* 367–371.

Das, R. C. (2003, October). Progress and prospects of protein therapeutics. *Am. Biotechnol. Lab., 21,* 8–12.

Daugherty, A. L. & Mrsny, R. J. (2006). Formulation and delivery issues for monoclonal antibody therapeutics. *Adv. Drug Deliv. Rev., 58,* 686–706.

Defelippis, M. R., Chance, R. E., & Frank, B. H. (2001). Insulin self-association and the relationship to pharmacokinetics and pharmacodynamics. *Crit Rev. Ther. Drug Carrier Syst., 18,* 201–264.

den, E. J., Garidel, P., Smulders, R., Koll, H., Smith, B., Bassarab, S. et al. (2011). Strategies for the assessment of protein aggregates in pharmaceutical biotech product development. *Pharm. Res., 28,* 920–933.

Devogelaer, J. P., Azria, M., Attinger, M., Abbiati, G., Castiglioni, C., & Dedeuxchaisnes, C. N. (1994). Comparison of the acute biological action of injectable salmon calcitonin and an injectable and oral calcitonin analogue. *Calcif. Tissue Int., 55,* 71–73.

Dhawan, J., Pan, L. C., Pavlath, G. K., Travis, M. A., Lanctot, A. M., & Blau, H. M. (1991). Systemic delivery of human growth hormone by injection of genetically engineered myoblasts. *Science, 254,* 1509–1512.

Dibner, M. D. (1986). Biotechnology in Europe. *Science, 232,* 1367–1372.

Dibner, M. D. (1992). U.S. biotechnology and pharmaceuticals. *BioPHARM, 5(8),* 24–28.

Dibner, M. D. & White, R. S. (1989). Biotechnology in the United States and Japan: Who's on first? *BioPHARM, 2(2),* 22–29.

Dubel, S., Stoevesandt, O., Taussig, M. J., & Hust, M. (2010). Generating recombinant antibodies to the complete human proteome. *Trends Biotechnol., 28,* 333–339.

Editorial (2003). Unwanted immunogenicity of therapeutic biological products: Problems and their consequences. *BioProcess Int., September,* 60–62.

Emtage, S. (1986). Biotechnology and protein production. In S. S. Davis, L. Illum, & E. Tomlinson (Eds.), *Delivery systems for peptide drugs* (pp. 23–33). New York: Plenum Press.

Eroshkin, A. M., Zhilkin, P. A., & Fomin, V. I. (1993). Algorithm and computer program pro-anal for analysis of relationship between structure and activity in a family of proteins or peptides. *Comput. Appl. Biosci., 9,* 491–497.

Eschbach, J. W., Egrie, J. C., Downing, M. R., Browne, J. K., & Adamson, J. W. (1987). Correction of the anemia of end-stage renal disease with recombinant human erythropoietin. *N. Engl. J. Med., 316,* 73–78.

European Directorate for the Qualifying Medicines. *European Pharmacopoeia* (2014). (8th ed.) (vols. 8.0) France: Council of Europe.

Fahrner, R. (1993). Making concurrent process development work. *BioPHARM, 6(9),* 18–22.

Farber, M. B. & Brotman, H. F. (1990). The scope of patent protection for physiologically important proteins and peptides. *BioPHARM, 3(4),* 22–27.

Fradd, R. B. (1993). Searching for miracle biopharmaceuticals. *Bio/Technology, 11,* 870–871.

Freeman, J. S. (2009). Insulin analog therapy: Improving the match with physiologic insulin secretion. *J. Am. Osteopath. Assoc., 109,* 26–36.

Galloway, J. A. (1993). New directions in drug development—Mixtures, analogues, and modeling. *Diabetes Care, 16,* 16–23.

Galperin, M. Y. & Koonin, E. V. (2010). From complete genome sequence to 'complete' understanding? *Trends Biotechnol., 28,* 398–406.

Garthoff, B. (1994). The use of transgenic animals in drug research. *Eur. J. Pharm. Sci., 2,* 22–23.

Gibbs, J. N. (1989). Biopharmaceuticals, FDA, and the environment. *BioPHARM, 2(7),* 52–56.

Giddins, M. D., Dabbah, R., Grady, L. T., & Rhodes, C. T. (1987). Scientific and regulatory aspects of macromolecular drugs and devices. *Drug Dev. Ind. Pharm., 13,* 873–968.

Gillmor, S. A. & Cohen, F. E. (1993). New strategies for pharmaceutical design. *Receptor, 3,* 155–163.

Gonzalez-Gonzalez, E., Speaker, T. J., Hickerson, R. P., Spitler, R., Flores, M. A., Leake, D. et al. (2010). Silencing of reporter gene expression in skin using siRNAs and expression of plasmid DNA delivered by a soluble protrusion array device (PAD). *Mol. Ther. J. Am. Soc. Gene Ther., 18,* 1667–1674.

Goodarzi, N., Varshochian, R., Kamalinia, G., Atyabi, F., & Dinarvand, R. (2013). A review of polysaccharide cytotoxic drug conjugates for cancer therapy. *Carbohydr. Polym., 92,* 1280–1293.

Griffen, S. H., Irving, T. L., & Sung, L. M. (1992). Will you be liable for patent infringement? *BioPHARM, 5(6),* 26–30.

Grillberger, L., Kreil, T. R., Nasr, S., & Reiter, M. (2009). Emerging trends in plasma-free manufacturing of recombinant protein therapeutics expressed in mammalian cells. *Biotechnol. J., 4,* 186–201.

Grossbard, E. B. (1987). Recombinant tissue plasminogen activator: A brief review. *Pharm. Res., 4,* 375–378.

Hansen, B. F., Glendorf, T., Hegelund, A. C., Lundby, A., Lutzen, A., Slaaby, R. et al. (2012). Molecular characterisation of long-acting insulin analogues in comparison with human insulin, IGF-1 and insulin X10. *PLoS One, 7,* e34274.

Harford, S. (1985). Genetic engineering and the pharmaceutical industry. *Can. Pharm. J., 118,* 467–469.

Heise, T., Tack, C. J., Cuddihy, R., Davidson, J., Gouet, D., Liebl, A. et al. (2011). A new-generation ultra-long-acting basal insulin with a bolus boost compared with insulin glargine in insulin-naive people with type 2 diabetes: A randomized, controlled trial. *Diabetes Care, 34,* 669–674.

Hellwig, S., Sack, M., Spiegel, H., Drossard, J., & Fischer, R. (2003, October). Using plants as a production system for N-glycosylated proteins of therapeutic relevance. *Am. Biotechnol. Lab.,* 50–53.

Hendry, L. B. (1993). Drug design with a new type of molecular modeling based on stereochemical complementarity to gene structure. *J. Clin. Pharmacol., 33,* 1173–1187.
Henry, D. & Taylor, C. (2014). Pharmacoeconomics of cancer therapies: Considerations with the introduction of biosimilars. *Semin. Oncol., 41(Suppl. 3),* S13–S20.
Hermans, M. P., Nobels, F. R., & De, L., I. (1999). Insulin lispro (Humalog), a novel fast-acting insulin analogue for the treatment of diabetes mellitus: Overview of pharmacological and clinical data. *Acta Clin. Belg., 54,* 233–240.
Hermeling, S., Crommelin, D. J., Schellekens, H., & Jiskoot, W. (2004). Structure-immunogenicity relationships of therapeutic proteins. *Pharm. Res., 21,* 897–903.
Hermeling, S., Schellekens, H., Crommelin, D. J., & Jiskoot, W. (2003). Micelle-associated protein in epoetin formulations: A risk factor for immunogenicity? *Pharm. Res., 20,* 1903–1907.
Hill, D. & Beatrice, M. (1989a). Biotechnology facility requirements, part I: Facility and systems design. *BioPHARM, 2(9),* 20–26.
Hill, D. & Beatrice, M. (1989b). Biotechnology facility requirements, part II: Operating procedures and validation. *BioPHARM, 2(10),* 28–32.
Hill, D. & Beatrice, M. (1991). Facility requirements for biotech plants. *J. Parenter. Sci. Technol., 45(3),* 132–137.
Ho, R. J. Y. & Gibaldi, M. (2003). *Biotechnology and biopharmaceuticals: Transforming proteins and genes into drugs.* Hoboken, NJ: Wiley-Liss.
Houdebine, L. M. (2009). Production of pharmaceutical proteins by transgenic animals. *Comp. Immunol. Microbiol. Infect. Dis., 32,* 107–121.
Hsu, T. & Mitragotri, S. (2011). Delivery of siRNA and other macromolecules into skin and cells using a peptide enhancer. *Proc. Natl. Acad. Sci. U.S.A., 108,* 15816–15821.
Hubbard, R. C., McElvaney, N. G., Birrer, P., Shak, S., Robinson, W. W., Margaret, C. J. et al. (1992). A preliminary study of aerosolized recombinant human deoxyribonuclease in the treatment of cystic fibrosis. *N. Engl. J. Med., 326,* 812–815.
Ibraheem, D., Elaissari, A., & Fessi, H. (2014). Gene therapy and DNA delivery systems. *Int. J. Pharm., 459,* 70–83.
Itoh, T. (1987). Biotech trends in the Japanese pharmaceutical industry. *Biotechnology, 5,* 794–799.
Jenkins, N. & Curling, E. M. A. (1994). Glycosylation of recombinant proteins—Problems and prospects. *Enzyme Microb. Technol., 16,* 354–364.
Jiskoot, W., Randolph, T. W., Volkin, D. B., Middaugh, C. R., Schoneich, C., Winter, G. et al. (2012). Protein instability and immunogenicity: Roadblocks to clinical application of injectable protein delivery systems for sustained release. *J. Pharm. Sci., 101,* 946–954.
Johnson, H. M., Bazer, F. W., Szente, B. E., & Jarpe, M. A. (1994). How interferons fight disease. *Sci. Am., 270,* 68–75.
Joung, C. H., Shin, J. Y., Koo, J. K., Lim, J. J., Wang, J. S., Lee, S. J. et al. (2009). Production and characterization of long-acting recombinant human albumin-EPO fusion protein expressed in CHO cell. *Protein Expr. Purif., 68,* 137–145.
Kaminski, J. J. (1994). Computer-assisted drug design and selection. *Adv. Drug Del. Rev., 14,* 331–337.
Khafagy, E., Morishita, M., Onuki, Y., & Takayama, K. (2007). Current challenges in non-invasive insulin delivery systems: A comparative review. *Adv. Drug Deliv. Rev., 59,* 1521–1546.
Koivisto, V. A. (1998). The human insulin analogue insulin lispro. *Ann. Med., 30,* 260–266.
Kucera, M. L. & Graham, J. P. (1998). Insulin lispro, a new insulin analog. *Pharmacotherapy, 18,* 526–538.
Kwon, K. C., Verma, D., Singh, N. D., Herzog, R., & Daniell, H. (2013). Oral delivery of human biopharmaceuticals, autoantigens and vaccine antigens bioencapsulated in plant cells. *Adv. Drug Deliv. Rev., 65,* 782–799.

Ladisch, M. R. & Kohlmann, K. L. (1992). Recombinant human insulin. *Biotechnol. Prog., 8,* 469–478.

Lares, M. R., Rossi, J. J., & Ouellet, D. L. (2010). RNAi and small interfering RNAs in human disease therapeutic applications. *Trends Biotechnol., 28,* 570–579.

Lenz, G. R. & Mansson, P. E. (1991). Growth factors as pharmaceuticals. *Pharm. Technol., 15(1),* 34–40.

Levy, R. S. (1993). Efficient calcitonin production. *BioPHARM, 6(4),* 36–39.

Li, Y., Medley, C. D., Zhang, K., Wigman, L., & Chetwyn, N. (2014). Limiting degradation of reactive antibody drug conjugate intermediates in HPLC method development. *J. Pharm. Biomed. Anal., 92,* 114–118.

Lyons, P. D., Lang, J., & Guzzo, M. A. (1993). Consolidated alliances: How biotechnology and pharmaceutical companies are finding ways to prosper together. *Bio/Technology, 11,* 1529–1532.

Marshall, V. & Baylor, N. W. (2011). Food and Drug Administration regulation and evaluation of vaccines. *Pediatrics, 127(Suppl. 1),* S23–S30.

McDonnell, J. J. (1992). Patenting biotechnology inventions. In M. E. Klegerman & M. J. Groves (Eds.), *Pharmaceutical biotechnology: Fundamentals and essentials* (pp. 249–253). Buffalo Grove, IL: Interpharm Press, Inc.

McPherson, J. M. & Livingston, D. J. (1989). Protein engineering: New approaches to improved therapeutic proteins, part I. *Pharm. Technol., 13(5),* 22–32.

Medynski, D. (1994). Synthetic peptide combinatorial libraries. *Biotechnology, 12,* 709–710.

Mesko, B., Zahuczky, G., & Nagy, L. (2012). The triad of success in personalised medicine: Pharmacogenomics, biotechnology and regulatory issues from a Central European perspective. *N. Biotechnol., 29,* 741–750.

Miller, H. I. (1992). Putting the bST human-health controversy to rest. *Bio/Technology, 10,* 147.

Milligan, J. F., Matteucci, M. D., & Martin, J. C. (1993). Current concepts in antisense drug design. *J. Med. Chem., 36,* 1923–1937.

Mintz, G. R. (1993). An irreversible serine protease inhibitor. *BioPHARM, 6(2),* 34–38.

Mire-Sluis, A. R. (2003). Regulatory interests and the development of predictive models of immunogenicity, a symposium on *Immunogenicity: Resolving regulatory and scientific issues in unwanted immunogenicity*. Bethesda, MD: IBC Life Sciences.

Misra, P. K., Haq, W., Katti, S. B., Mathur, K. B., Raghubir, R., Patnaik, G. K. et al. (1993). Enkephalin antisense peptides—Design, synthesis, and biological activity. *Pharm. Res., 10,* 660–661.

Monami, M., Marchionni, N., & Mannucci, E. (2009). Long-acting insulin analogues vs. NPH human insulin in type 1 diabetes. A meta-analysis. *Diabetes Obes. Metab, 11,* 372–378.

Montague, M. J. (1992, January). Biotechnology and the future of medicine. *Pharm. Pract. News,* 8.

Morefield, G., Touhey, G., Lu, F., Dunham, A., & HogenEsch, H. (2014). Development of a recombinant fusion protein vaccine formulation to protect against *Streptococcus* pyogenes. *Vaccine, 32,* 3810–3815.

Morgan, R. A. (1993). Human gene therapy. *BioPHARM, 6(1),* 32–35.

Moseley, A. B. & Caskey, C. T. (1993). Human genetic disease and the medical need for somatic gene therapy. *Adv. Drug Del. Rev., 12,* 131–142.

Narasimhan, C., Mach, H., & Shameem, M. (2012). High-dose monoclonal antibodies via the subcutaneous route: Challenges and technical solutions, an industry perspective. *Ther. Deliv., 3,* 889–900.

Nobels, F. R., Hermans, M. P., & De, L., I. (1999). Insulin lispro (Humalog), a novel fast-acting insulin analogue: Guidelines for its practical use. *Acta Clin. Belg., 54,* 246–254.

Nwanguma, B. C. (2003). The human genome project and the future of medical practice. *Afr. J. Biotechnol., 2,* 649–656.

O'Keefe, D. O., DePhillips, P., & Will, M. L. (1993). Identification of an *Escherichia coli* protein impurity in preparations of a recombinant pharmaceutical. *Pharm. Res., 10,* 975–979.

Pantazes, R. J., Grisewood, M. J., & Maranas, C. D. (2011). Recent advances in computational protein design. *Curr. Opin. Struct. Biol., 21,* 467–472.

Parker, A. S., Choi, Y., Griswold, K. E., & Bailey-Kellogg, C. (2013). Structure-guided deimmunization of therapeutic proteins. *J. Comput. Biol., 20,* 152–165.

Patten, P. A. & Schellekens, H. (2003). The immunogenicity of biopharmaceuticals. Lessons learned and consequences for protein drug development. *Dev. Biol. (Basel), 112,* 81–97.

PDR Staff. (2014). *Physicians' Desk Reference* (2014). (68th ed.) Montvale, NJ: PDR Network, LLC.

Perez, H. L., Cardarelli, P. M., Deshpande, S., Gangwar, S., Schroeder, G. M., Vite, G. D. et al. (2014). Antibody-drug conjugates: Current status and future directions. *Drug Discov. Today, 19,* 869–881.

Pestka, S. (1983). The purification and manufacture of human interferons. *Sci. Am., 249,* 36–43.

Pestka, S. (1994). YACs, YACs, YACs, and the human genome project. *Pharm. Technol., 18,* 24.

Peters, R. & Mckinstry, R. C. (1994). Three-dimensional modeling and drug development. *Biotechnology, 12,* 147–150.

Peys, F. (1996). Pharmacoeconomics: Where is the link with pharmacokinetics and biopharmaceutics? *Eur. J. Drug Metab. Pharmacokinet., 21,* 189–200.

Pinheiro da, S. F. & Machado, M. C. (2012). Antimicrobial peptides: Clinical relevance and therapeutic implications. *Peptides, 36,* 308–314.

Powell-Jackson, J., Kennedy, P., Whitcombe, E. M., Weller, R. O., Preece, M. A., & Newsom-Davis, J. (1985). Creutzfeldt-Jakob disease after administration of human growth hormone. *Lancet, 2,* 244–246.

Rader, C. (2014). Chemically programmed antibodies. *Trends Biotechnol., 32,* 186–197.

Rosenberg, A. (2003). FDA perspective on immunogenicity testing—A risk based analysis, a symposium on *Immunogenicity: Resolving regulatory and scientific issues in unwanted immunogenicity*. IBC Life Sciences: IBC Life Sciences.

Rosenberg, A. S. (2006). Effects of protein aggregates: An immunologic perspective. *AAPS J., 8,* E501–E507.

Rosenthal, S. L., Cohen, S. S., & Stanberry, L. R. (1998). Topical microbicides: Current status and research considerations for adolescent girls. *Sex. Transm. Dis., 25,* 368–377.

Rothbard, J. B., Garlington, S., Lin, Q., Kirschberg, T., Kreider, E., McGrane, P. L. et al. (2000). Conjugation of arginine oligomers to cyclosporin A facilitates topical delivery and inhibition of inflammation. *Nat. Med., 6,* 1253–1257.

Rothbard, J. B., Kreider, E., VanDeusen, C. L., Wright, L., Wylie, B. L., & Wender, P. A. (2002). Arginine-rich molecular transporters for drug delivery: Role of backbone spacing in cellular uptake. *J. Med. Chem., 45,* 3612–3618.

Sadee, W. (1988). Genes and the pharmaceutical sciences. *Pharm. Res., 5,* 199–200.

Salmon, S. E., Lam, K. S., Lebl, M., Kandola, A., Khattri, P. S., Wade, S. et al. (1993). Discovery of biologically active peptides in random libraries—Solution-phase testing after staged orthogonal release from resin beads. *Proc. Natl. Acad. Sci., 90,* 11708–11712.

Sapra, P. & Shor, B. (2013). Monoclonal antibody-based therapies in cancer: Advances and challenges. *Pharmacol. Ther., 138,* 452–469.

Schiff, L. J., Moore, W. A., Brown, J., & Wisher, M. H. (1992). Lot release—Final product safety testing for biologics. *BioPHARM, 5(5),* 36–39.

Schlom, J. (1988). Monoclonal antibodies in cancer therapy: The present and the future. *BioPHARM, 1(8),* 44–48.

Schulte, S. (2013). Challenges for new haemophilia products from a manufacturer's perspective. *Thromb. Res., 134,* S72–S76.

Seeherman, H., Wozney, J., & Li, R. (2002). Bone morphogenetic protein delivery systems. *Spine, 27,* S16–S23.

Sharma, B. (2007). Immunogenicity of therapeutic proteins. Part 2: Impact of container closures. *Biotechnol. Adv., 25,* 318–324.

Sheldon, B., Russell-Jones, D., & Wright, J. (2009). Insulin analogues: An example of applied medical science. *Diabetes Obes. Metab, 11,* 5–19.

Sherman, M. I. (1993). Oligonucleotide-based therapeutics. *BioPHARM, 6(7),* 38–41.

Siu, K. S., Chen, D., Zheng, X., Zhang, X., Johnston, N., Liu, Y. et al. (2014). Non-covalently functionalized single-walled carbon nanotube for topical siRNA delivery into melanoma. *Biomaterials, 35,* 3435–3442.

Small, W. E. (1992). Biotechnology in the '90s. *BioPHARM, 5(4),* 17–20.

Smith, M. L., Richter, L., Arntzen, C. J., Shuler, M. L., & Mason, H. S. (2003). Structural characterization of plant-derived hepatitis B surface antigen employed in oral immunization studies. *Vaccine, 21,* 4011–4021.

Snyder, E. L. & Dowdy, S. F. (2004). Cell penetrating peptides in drug delivery. *Pharm. Res., 21,* 389–393.

Sobel, B. E., Fields, L. E., Robison, A. K., Fox, K. A. A., & Sarnoff, S. J. (1985). Coronary thrombolysis with facilitated absorption of intramuscularly injected tissue-type plasminogen activator. *Proc. Natl. Acad. Sci., 82,* 4258–4262.

Sola, R. J. & Griebenow, K. (2010). Glycosylation of therapeutic proteins: An effective strategy to optimize efficacy. *BioDrugs., 24,* 9–21.

Spada, S. & Walsh, G. (2005). *Directory of approved biopharmaceutical products.* Boca Raton, FL: CRC Press.

Spadiut, O., Capone, S., Krainer, F., Glieder, A., & Herwig, C. (2014). Microbials for the production of monoclonal antibodies and antibody fragments. *Trends Biotechnol., 32,* 54–60.

Spalding, B. J. (1993). Biopharmaceutical firms up R&D spending 71%. *Bio/Technology, 11,* 768–770.

Stein, K. E. (2002). Immunogenicity: Concepts/issues/concerns. *Dev. Biol., 109,* 15–23.

Sterling, J. (1990). The next decade of biotechnology—Where are we going? *J. Parenter. Sci. Technol., 44,* 66–73.

Stroud, J. (2013). Power without a patent: 12-year biologics data exclusivity period and a totality-of-the-evidence standard for biosimilarity. *Second International Conference and Exhibition on Biowaivers & Biosimilars,* September 23–25, Raleigh, NC.

Subramanian, G. M., Fiscella, M., Lamouse-Smith, A., Zeuzem, S., & McHutchison, J. G. (2007). Albinterferon alpha-2b: A genetic fusion protein for the treatment of chronic hepatitis C. *Nat. Biotechnol., 25,* 1411–1419.

Sueishi, T., Seki, T., Juni, K., Hasegawa, T., Saneyoshi, M., & Kawaguchi, T. (1994). Novel stabilizing method for antisense oligodeoxynucleotides. *Pharm. Res., 11,* 455–457.

Suk, J. S., Kim, A. J., Trehan, K., Schneider, C. S., Cebotaru, L., Woodward, O. M. et al. (2014). Lung gene therapy with highly compacted DNA nanoparticles that overcome the mucus barrier. *J. Control. Release, 178,* 8–17.

Summers, N. M. & Cooney, C. L. (1994). Gene therapy: Biotech's n + 1 technology. *Bio/Technology, 12,* 42–45.

Sun, E. & Cohen, F. E. (1993). Computer-assisted drug discovery—A review. *Gene, 137,* 127–132.

Swann, J. P. (1986). Insulin: A case study in the emergence of collaborative pharmacomedical research. *Pharm. Hist., 28,* 3–13.

Szkrybalo, W. (1987). Emerging trends in biotechnology: A perspective from the pharmaceutical industry. *Pharm. Res., 4,* 361–363.

Talmadge, J. E. (1993). The pharmaceutics and delivery of therapeutic polypeptides and proteins. *Adv. Drug Del. Rev., 10,* 247–299.

Tangri, S., LiCalsi, C., Sidney, J., & Sette, A. (2002). Rationally engineered proteins or antibodies with absent or reduced immunogenicity. *Curr. Med. Chem., 9,* 2191–2199.

Trail, P. A. (2013). Antibody drug conjugates as cancer therapeutics. *Antibodies, 2,* 113–129.

Tung, C. H. & Weissleder, R. (2003). Arginine containing peptides as delivery vectors. *Adv. Drug Deliv. Rev., 55,* 281–294.

Uber, D. C. (1994). Robotics and the human genome project. *Bio/Technology, 12,* 80–81.

Umpierrez, G. E., Jones, S., Smiley, D., Mulligan, P., Keyler, T., Temponi, A. et al. (2009). Insulin analogs versus human insulin in the treatment of patients with diabetic ketoacidosis: A randomized controlled trial. *Diabetes Care, 32,* 1164–1169.

Urvoas, A., Valerio-Lepiniec, M., & Minard, P. (2012). Artificial proteins from combinatorial approaches. *Trends Biotechnol., 30,* 512–520.

Wallace, B. M. & Lasker, J. S. (1993). Stand and deliver—Getting peptide drugs into the body. *Science, 260,* 912–913.

Walsh, G. (2004). Second-generation biopharmaceuticals. *Eur. J. Pharm. Biopharm. 58,* 185–196.

Walsh, G. (2010). Post-translational modifications of protein biopharmaceuticals. *Drug Discov. Today, 15,* 773–780.

Wang, F., Surh, J., & Kaur, M. (2012). Insulin degludec as an ultralong-acting basal insulin once a day: A systematic review. *Diabetes Metab. Syndr. Obes., 5,* 191–204.

Wang, W., Singh, S. K., Li, N., Toler, M. R., King, K. R., & Nema, S. (2012). Immunogenicity of protein aggregates—Concerns and realities. *Int. J. Pharm., 431,* 1–11.

Wang, X. Z., Coljee, V. W., & Maynard, J. A. (2013). Back to the future: Recombinant polyclonal antibody therapeutics. *Curr. Opin. Chem. Eng, 2,* 405–415.

Warburg, R. J. (1993). Recombinant conception: Does it require the DNA? *Bio/Technology, 11,* 1306–1307.

Weisser, N. E. & Hall, J. C. (2009). Applications of single-chain variable fragment antibodies in therapeutics and diagnostics. *Biotechnol. Adv., 27,* 502–520.

Whittle, P. J. & Blundell, T. L. (1994). Protein structure-based drug design. *Annu. Rev. Biophys. Biomol. Struct., 23,* 349–375.

Wills, R. J. (1990). Clinical pharmacokinetics of interferons. *Clin. Pharmacokinet., 19,* 390–399.

Wirth, T., Parker, N., & Yla-Herttuala, S. (2013). History of gene therapy. *Gene, 525,* 162–169.

Wong, L. C. K. (1998). Safety evaluation of peptides and proteins. *Acta Pharmacol. Sin., 19,* 175–181.

Wordell, C. J. (1992). Strategic planning for biotechnology products. *Hosp. Pharm., 27,* 521–528.

Wu, C., Zhang, Y., Ke, X., Xie, Y., Zhu, H., Crawford, R. et al. (2010). Bioactive mesopore-glass microspheres with controllable protein-delivery properties by biomimetic surface modification. *J. Biomed. Mater. Res. A, 95,* 476–485.

Wu-Pong, S. (1994). Oligonucleotides: Opportunities for drug therapy and research. *Pharm. Technol., 18,* 102–114.

Wuethrich, B. (1993). All rights reserved: How the gene-patenting race is affecting science. *Sci. News, 144,* 154–157.

Yang, C., Hillas, P. J., Baez, J. A., Nokelainen, M., Balan, J., Tang, J. et al. (2004). The application of recombinant human collagen in tissue engineering. *BioDrugs., 18,* 103–119.

Zhu, J. (2012). Mammalian cell protein expression for biopharmaceutical production. *Biotechnol. Adv., 30,* 1158–1170.

Zitter, J. N., Maldjian, P., Brimacombe, M., & Fennelly, K. P. (2013). Inhaled Dornase alfa (Pulmozyme) as a noninvasive treatment of atelectasis in mechanically ventilated patients. *J. Crit. Care, 28,* 218. e1–7.

2 Structure and Analysis of Therapeutic Peptides and Proteins

This chapter discusses the unique structural and analytical aspects of peptides and proteins. A detailed discussion of these topics is beyond the scope of this book. Instead, the discussion will emphasize the basic concepts that will have application to the rest of the book as formulation, pharmaceutical processing, and delivery of therapeutic peptides and proteins are discussed.

2.1 AMINO ACIDS: BUILDING BLOCKS OF PROTEINS

The 20 different naturally occurring amino acids are the building blocks of peptides and proteins. Amino acids are commonly represented by three-letter abbreviations, but sometimes, one-letter abbreviations are also used. Although the human body needs all 20 amino acids, it cannot synthesize some amino acids known as essential amino acids. Essential amino acids must be obtained from the diet. As the name implies, an amino acid contains an amino group and an acid group in the same molecule. Amino acids have a central carbon atom (alpha-carbon) to which are attached a carboxyl group, an amino group, a hydrogen atom, and a side chain. Different amino acids differ with respect to the side chain (R) only:

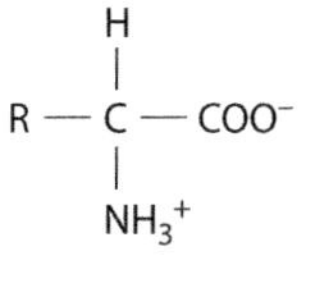

Gly is the simplest amino acid with no side chain. The side chain of Pro is unique in that it is bonded covalently to the nitrogen atom of the peptide group. The amide forms of Asp and Glu (Asn and Gln, respectively) occur naturally and are incorporated into proteins. Ionizable side chains of amino acids vary from acidic to basic. Aspartic and glutamic acids have a negative charge, while lysine and arginine have a positive charge at the physiological pH of 7.4. The imidazole group in histidine carries a partial positive charge at pH 7.4. Serine and threonine have side chains that carry no charge at any pH, but are polar in nature. In contrast, tryptophan, phenylalanine, and isoleucine have side chains that are more hydrocarbon like in character. The lipophilicity of amino acids is a property relevant for protein folding (van de Waterbeemd, Karajiannis, & El Tayar, 1994) as will be discussed in this section.

All amino acids (except glycine, where R=H) have a chiral carbon, since R, H, COOH, and NH_2 are four different groups. Thus, each amino acid (except glycine) exists as two enantiomers: L- and D-amino acids. The proteins in our body are made of only L-amino acids.

Amino acids are zwitterions, both in solution and in solid state. Because amino acids are zwitterions, they are solids with high melting points and have a good aqueous solubility. Each functional group has a characteristic dissociation constant, K, whose negative logarithm is called the pK_a. Since pK_a values were measured at fixed ionic strength, they are sometimes called *apparent dissociation constants*, pK_a'. The constant pK_1' usually refers to the most acidic group (Damm, Besch, & Goldwyn, 1966). For example, the dissociation constant for glycine pK_1'(COOH) and $pK_2'(NH_3^+)$ are 2.34 and 9.60, respectively. The pH at which all the molecules are in the zwitterionic form so that there is no charge on the molecule is called the isoelectric point (pI) of the amino acid. Of the 20 amino acids, 15 have pI near 6.0. The three basic amino acids have higher pI, while the two acidic ones have lower pI.

The amino acid cysteine exhibits some unique properties that play a crucial role in the stability of proteins. The thiol group of Cys is the most reactive of any amino acid side chain and will be readily oxidized to form a dimer, cystine:

$$2HS-CH_{2^-} \leftrightarrow -CH_2-S-S-CH_{2^-}$$

The resulting S–S bond is called a disulfide bridge and serves to hold the protein in its unique conformation, as will be discussed in Section 2.3.

2.2 STRUCTURE OF PEPTIDES AND PROTEINS

Amino acids join with each other by peptide bonds to form polymers referred to as peptides or proteins. Thus, the carboxyl and amino groups of each amino acid are participating in the peptide bond so that it will have a charge in the peptide only if it has an ionizable side chain. The distinction between peptides and proteins is somewhat arbitrary. Typically, peptides contain fewer than 20 amino acids, while proteins contain 50 or more amino acids. Between these two categories are polypeptides that contain about 20–50 amino acids. Again, it must be emphasized that the nomenclature is not well defined and different authors use these terms differently. As used in this book, a peptide may be considered to have a molecular weight less than 5000 while a protein lies above this value. By this definition, insulin (Molecular Weight (MW) 5808) is one of the smallest proteins available, while monoclonal antibodies are larger structurally more complex proteins (Figure 2.1) (Mellstedt, 2013). Therapeutic proteins are generally referred to as globular proteins as they are of nearly spherical shape in solution. A protein has several levels of structure. These are generally referred to as the primary, secondary, tertiary, and quaternary structures.

2.2.1 Primary Structure

The sequence of covalently bonded amino acids in the polypeptide chain is known as the primary structure. The primary structure of a protein is determined genetically

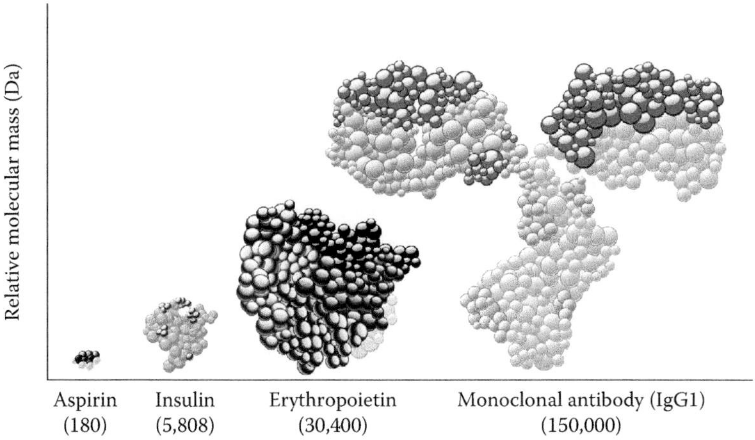

FIGURE 2.1 **(See color insert.)** Structural complexity of monoclonal antibodies relative to small-molecule agents and lower-molecular-weight biologics. (Reproduced from Mellstedt, H., *EJC Suppl.*, 11, 1, 2013. With permission.)

by the sequence of nucleotides in DNA. The amino acids in a protein chain are often called residues. Once a polypeptide chain is formed, there is only one residue with a free amino group (N-terminal) and only one with a free carboxyl group (C-terminal). Peptide and protein chains are always written with the N-terminal residue on the left.

2.2.2 Secondary Structure

The turns and loops of the polypeptide chain constitute the secondary structure. Alpha-helices and beta-sheets are the common conformations of the secondary structure, with the rest being β-bends, small loops, and random coils. In the α-helix, a single protein chain twists like a coiled spring. Each turn contains 3.6 amino acid residues, with the side chains pointing out from the helix. The backbone carbonyl carbon oxygen of each residue hydrogen bonds to the backbone NH of the fourth residue along the chain. Thus, all NH and CO groups are hydrogen bonded except for the first NH and last CO groups, which occur at the end of helices. Thus, the ends of helices are polar and are likely to be at the surface of protein molecules (Ji & Li, 2010). It may be noted that Pro residues are incompatible with this conformation, as their side chain is bonded to the backbone N atom, preventing its participation in hydrogen bonding.

In the β-sheet conformation, the protein backbone is nearly fully extended. Beta-sheets can be either parallel or antiparallel. In the parallel β-sheet, two or more peptide chains align their backbones in the same general direction. The antiparallel β-sheet has adjacent chains having parallel but opposite orientations. Hydrogen bonding in β-sheets also involves carbonyl oxygen and amide hydrogens, but the bonding is between chains, not along the same chain as in the case of α-helix. Again, proline cannot participate in β-sheet structure as it has no NH groups to participate in hydrogen bonding (King, 1989).

A protein may contain about 30% of the residues in α-helices and 30% in β-sheets (though this will vary widely from one protein to another), and the rest in other conformations, such as the random coils or β-bends. Some proteins such as growth hormone are largely α-helical in structure. On the other hand, the primary and secondary structural elements in immunoglobulins are the antiparallel β-sheets and random coil conformations, though some short stretches of α-helices and β-turns are present (Vermeer, Bremer, & Norde, 1998). A fully extended polypeptide chain of about 60 residues may be about 200 Å long, but the folded globular protein may actually be just about 30 Å in diameter. An α-helix generally contains 10–15 residues, while a β-sheet may contain only about 3–10 residues. At the end of these regular structures, the polypeptide chain generally makes sharp bends, thus reversing the direction of the polypeptide chain. These reverse turns, also called hairpin bends, β-bends, or β-turns, often connect antiparallel β-strands and are important to maintain the globular shape of the protein. Proline, because of its special geometry and *cis*-configuration, is often found in β-turns (Creighton, 1983b). These elements of secondary structure help the protein to neutralize its polar atoms through hydrogen bonds.

2.2.3 Tertiary Structure and Protein Folding

Tertiary structure of a protein is the overall packing in space of the various elements of secondary structure. Folding of proteins is governed by certain rules, and these rules along with several examples have been discussed in the literature (Denton, Konishi, & Scheraga, 1982; King, 1989; Rossmann & Argos, 1981; Sadana, 1987; Sali, Shakhnovich, & Karplus, 1994). Protein folds in a highly specific fashion, but it is believed that specific folding is determined by the amino acid sequence or the primary structure of the protein. If a protein was to fold by searching through all possible conformations in a random manner to find the lowest energy in the conformation of the native state, it would take a protein of 150 residues about 10^{48} years to fold correctly! In reality, the folding time for protein is more like 0.1–1000 s. Thus, the folding process is directed in some way.

It is known that the folding is guided by some general rules: the overall shape is spherical; polar groups are on the surface, while hydrophobic groups are buried in the interior; and the interior is closely packed. These steric constraints vastly reduce the number of conformations that need to be searched. Guided by these rules, the protein folds into a unique structure. However, it is currently not possible to precisely predict the 3D structure of a protein from its amino acid sequence.

As a protein folds, it will typically go through an intermediate state called a molten globule. This molten globule, which forms within milliseconds, has most of the secondary structure of the native state but has a less compact loose tertiary structure. In the second step, which can take 1 s, the native-like elements of tertiary structure begin to develop. Within the cell, the folding process is guided by molecular chaperones, and disulfide bridge-forming enzymes assist the proper formation of disulfide bridges. The chaperones are a class of proteins shaped like short cylinders. They have hydrophobic residues in the interior and can bind to any unfolded protein to shield it and protect it from aggregation with other protein molecules.

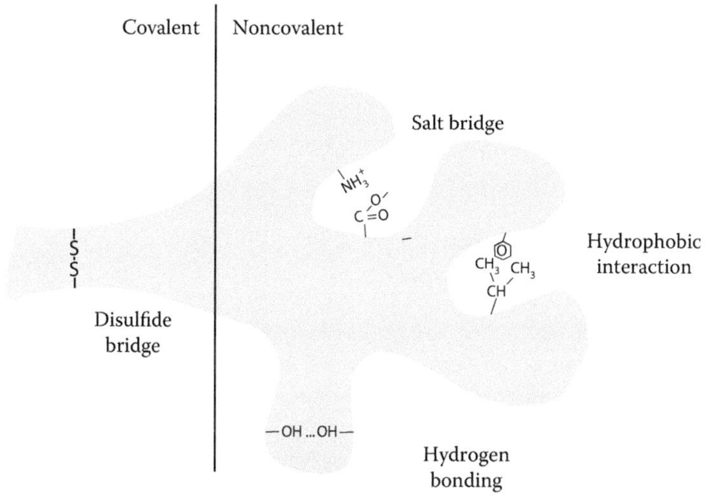

FIGURE 2.2 Schematic of the forces that stabilize the folded structure of a protein. The covalent forces include a disulfide bridge, while noncovalent forces include a salt bridge, hydrophobic interactions, and hydrogen bonding.

In essence, a protein cannot have two substantially different conformations. Although one comes across the term *conformational change* in the literature, these changes are localized alterations of the conformation. The forces that may stabilize the folded structure of a protein may be covalent or noncovalent, as schematically represented in Figure 2.2. Disulfide bridges between cysteine residues provide a covalent linkage that can hold together two chains or two parts of the same chain. The disulfide bonds are very hydrophobic relative to the thiol groups. The noncovalent forces that can stabilize the protein include hydrogen bonding, salt bridges, and hydrophobic interactions between the side chains of amino acid residues. Typically, a protein would have maximum conformational stability at its pI, but this may shift due to burial of nontitratable groups and the presence of counterions in the formulation (Branchu, Forbes, York, & Nyqvist, 1999). Unlike proteins, peptides may display several conformations in solution. This is because they lack the hydrogen bonds and disulfide bridges that stabilize the 3D structure of proteins. After exposure to denaturing conditions, peptides may rapidly reequilibrate to native conformations. However, peptides may assume a secondary structure in solution and this structure may be important for its biological activity (Johnson et al., 1993).

2.2.4 Quaternary Structure

Quaternary structure of a protein refers to the specific association of separate protein chains (subunits or monomers) to form a well-defined structure. The subunits are held together by noncovalent forces such as hydrophobic interactions, hydrogen bonds, or van der Waals' interactions. Protein–protein associations usually involve the more hydrophobic portions on the surface of a molecule (Young, Jernigan, & Covell, 1994).

Normally, the hydrophobic residues are buried in the interior, but they can also be closer to the exterior. This is especially true when the protein has a quaternary structure since the subunits form part of a larger aggregate whose geometry influences the location of the hydrophobic residues such as tryptophan in α-crystallin (Augusteyn, Ghiggino, & Putilina, 1993). The quaternary structure of insulin is discussed in Section 4.5.1.

2.3 PROTEIN CONFORMATION AND OTHER STRUCTURAL FEATURES

To understand the physicochemical and biological properties of a protein, knowledge of its 3D structure is essential. The 3D structure of a protein can be determined by x-ray diffraction of crystalline proteins (Blundell, Dodson, Dodson, Hodgkin, & Vijayan, 1971). Normally, the atoms that make up the protein do not have sufficient mass to interact well with radiation. However, in a protein crystal, all molecules are held in regular positions so that their diffraction pattern may add up. In essence, the crystal lattice amplifies the diffraction pattern. Since protein crystals may contain as much as 40%–60% solvent, proteins in solution generally retain the conformation determined within a crystal lattice. However, small proteins (less than about 50 amino acid residues) may not be able to maintain a single conformation in solution. Once the primary sequence of a protein is known, the full 3D structure can be determined by x-ray crystallography and nuclear magnetic resonance (NMR) spectroscopy, though characterizing the 3D structure of a protein is not a regulatory requirement. A repository of protein structures is available in the protein data bank (http://www.wwpdb.org). In addition to NMR spectroscopy, another technique gaining increasing recognition for determining the higher-order protein structure is hydrogen–deuterium exchange mass spectrometry (HDX-MS) (Challener, 2014).

Protein crystallization requires a substantial amount of work and considerable judgment and skills to interpret the electron density maps (Chou & Fasman, 1978; Creighton, 1983b). Protein crystallization can be made easier by new innovations, and creation of data banks for successful crystallization conditions, and crystal analysis software and computer modeling (Jen & Merkle, 2001). Large-scale crystallization for improved formulation and/or delivery may also be possible and is discussed later in this book (see Section 6.5.2).

It should be realized that conformational changes can take place in crystals, similar to those in solutions, in response to environmental changes such as pH change. Gursky, Badger, Li, and Caspar (1992) used cubic insulin crystals to show that while the positions of most protein and well-ordered solvent atoms are conserved, about 30% of the residues undergo changes in their positions as the pH is changed. As the crystals were titrated over the pH range of 7–11, the net negative charge on the protein was increased from about −2 to about −6. Insulin zinc crystals are used for the preparation of long-acting suspensions. In order to assure consistent kinetic profile and bioequivalency from one batch to another, it is important to maintain consistent crystal size and habits in insulin suspensions (Elarini & Tawashi, 1994). NMR is another technique that has been used to determine the solution structure of some peptides and small proteins. It can provide data that in many ways are complementary to that obtained from x-ray crystallography (Kwan, Mobli, Gooley, King, & Mackay, 2011; Wuthrich, 1989). However, the technique is complex and requires high concentrations of protein and long data acquisition times. Information from a variety of NMR experiments has been used to analyze the solution structure of human interleukin (hIL)-4 (Redfield et al., 1994), somatostatin (van den Berg, Jans, & Binst, 1986), lysozyme (Buck, Radford, & Dobson, 1993), and human granulocyte colony-stimulating factor (G-CSF) (Zink, Ross, Ambrosius, Rudolph, & Holak, 1992). The relationship between protein conformation and its tendency toward aggregation has been described (Chaudhuri, Cheng, Middaugh, & Volkin, 2014).

When a protein consists of more than 200 residues, the structure may be organized into loosely defined regions called domains. The domain is generally recognized as a stable folding unit. The polypeptide segments linking independent structural domains may be susceptible to proteolytic cleavage. This provides us with means to test the presence of domains as well as to separate and isolate them. Human plasma fibronectin is an example of a high-molecular-weight (530,000), multidomain glycoprotein consisting of two nearly identical subunits disulfide-bridged to each other (Lai et al., 1993). The structure of immunoglobulin G (IgG) also has rather well-defined domains. Antibody domains have been utilized as model systems to analyze the folding of all β-proteins (Thies, Kammermeier, Richter, & Buchner, 2001). Immunoglobulins are a heterogeneous family of proteins that circulate in body fluids and have antibody activity. Immunoglobulins are glycoproteins, but the polypeptide component possesses almost all the biological properties of the molecule. There are five classes of immunoglobulins, designated as IgG, IgA, IgM, IgD, and IgE. Furthermore, there are subclasses within each class. Immunoglobulins constitute about 20% of the total plasma proteins, and IgG constitutes about 75% of the total serum immunoglobulins. IgG is the only class that can cross human placenta, thus providing protection to the newborn during the first few months of life. The subclasses may exist as monomers or in higher association states, for example, IgM is a pentamer and IgA may exist in dimeric form. Each basic unit or monomer contains four polypeptide chains. The basic structure of IgG (MW approx. 150,000) is a Y-shaped molecule with two identical heavy (H) chains and two identical light (L) chains. For example, rituximab is an IgG1 chimeric antibody composed of two heavy chains of 451 amino acids and two light chains of 213 amino acids, with an approximate molecular weight of 145 kDa. It is produced by mammalian cell (Chinese hamster ovary [CHO]) suspension culture, from where the CD_2O antibody is then purified (PDR Staff, 2014).

The polypeptide chains in IgG are folded by disulfide bonds into globular regions or domains. Each domain contains about 100–110 amino acid residues. The heavy chain is composed of four domains, while the light chain is composed of two domains. Each polypeptide chain has an amino-terminal portion, the variable (V) region, and a carboxy-terminal portion, the constant (C) region. A small number of amino acids in the V regions of H and L chains form the antigen-binding site. Digestion of the molecule at the hinge region by the enzyme papain produces two F_{ab} (antigen-binding) fragments and one F_c (crystallizable) fragment (Creighton, 1983b; Goodman, 1991). These fragments behave as independent subunits within the IgG molecule, with the F_{ab} part responsible for antigen binding and the F_c part responsible for adsorption to receptors/tissues. Also, the F_{ab} part has been reported to be comparably stable to the complete molecule and may thus have potential therapeutic uses by itself (Vermeer, Norde, & van Amerongen, 2000; Welfle, Misselwitz, Hausdorf, Hohne, & Welfle, 1999). If the IgG molecule is adsorbed to a surface, the antigen-binding sites must be accessible for the molecule to stay immunologically active (Vermeer et al., 1998).

A few examples will illustrate some other structural concepts for proteins. Bovine serum albumin (BSA) is a protein that is very stable. This may be partly explained by the presence of 17 disulfide (Cys–Cys) bridges in the native structure. These disulfide bonds contribute to the formation and stability of the helical structure. If the disulfide bonds are reduced, the helical portion decreases dramatically. Disulfide bonds can be reduced by a chemical such as dithiothreitol, and the reduced structure can then be unfolded by the addition of urea or guanidine. The unreduced protein, however, resists unfolding by urea or guanidine. Chaotropic agents such as urea or guanidine can interact with the protein, while a surfactant such as sodium dodecyl sulfate (SDS) can disrupt this interaction. Thus, on addition of SDS, the helicity of BSA lost by urea denaturation may be mostly recovered. A rise in temperature may also cause a loss of helicity, which may recover to some degree upon cooling (Moriyama, Sato, & Takeda, 1993; Takeda, Hamada, & Wada, 1993; Takeda, Sasa, Kawamoto, Wada, & Aoki, 1988; Takeda & Wada, 1991; Takeda, Wada, Yamamoto, Moriyama, & Aoki, 1989). The addition of SDS has also been reported to accelerate the formation of helical structure in chymotrypsinogen and α-chymotrypsin in solutions of urea and guanidine hydrochloride (Takeda, Wada, & Moriyama, 1990). Urea or SDS has also been reported to completely dissolve insoluble aggregates of lyophilized bovine IgG (Sarciaux, Mansour, Hageman, & Nail, 1999). Human plasma fibronectin has been reported to undergo a transition from the native compact conformation to a more expanded form upon increasing ionic strength or glycerol content (Lai et al., 1993). During protein synthesis in humans or other eukaryotes, carbohydrate molecules may be covalently attached to the protein. The resulting protein is called a glycoprotein. Some proteins may also be modified after synthesis, and they are then called posttranslationally modified. The carbohydrate part of the glycoprotein may or may not be essential to bioactivity but will often affect the pharmacokinetics of the protein. Protein–water interactions play an important role in the folding, stability, and biological function of proteins.

Analogous to amino acids, the pI of a protein occurs at the pH where the positive and negative charges are balanced. At their pI, proteins behave like zwitterions. At any pH below its pI, the protein has a positive charge, and at pH above its pI, it has a negative charge. If a protein has an equal number of acidic and basic groups, its pI is close to 7.0. If it has more acidic than basic groups, the pI is low, and vice versa. Proteins vary in their pI value, and this information is important for formulation development as well as delivery technologies. The importance of pI to the formulation of the protein will be discussed in Chapter 4. Malamud and Drysdale (1978) have compiled a table on the pI of several proteins, but now several tools are available on the Internet to calculate the approximate pI of a protein. The basic theoretical problem in calculating the pK_as of ionizable groups in protein is to predict how the environment of the protein shifts the pK_a of an amino acid in the protein from that of the isolated amino acid in solution. Several equations and models are available for such calculations (Yang, Gunner, Sampogna, Sharp, & Honig, 1993). The presence of acid contaminants in peptides should be considered if using the potentiometric method for pK_a determinations (Fang, Fernando, Ugwu, & Blanchard, 1995).

2.4 ANALYTICAL METHODS

Because of the complexity of proteins, no single analytical method can detect all possible chemical, physical, and immunological changes in the protein structure (Jaksch & Tay, 2014). Thus, several analytical techniques such as electrophoresis, spectroscopy, chromatography, thermal analysis, immunoassays, and bioassays may be required to completely characterize a protein and examine its degradation profile. A second analytical technique may often be needed to confirm the conclusions made by a first technique, for example, the combined use of differential scanning calorimetry (DSC) and circular dichroism (CD) spectroscopy provides useful information about the heat-induced denaturation of proteins. Various analytical methods also have limitations when used to characterize protein aggregation, for example, size-exclusion chromatography (SEC) may result in solubilization of reversible aggregates due to dilution effects and also cannot characterize larger aggregates (Hamrang, Rattray, & Pluen, 2013).

While recombinant DNA technology and other biotechnology principles have revolutionized the production of proteins, their analysis is still largely dependent on traditional assay methodology, though many instrumental advances have been made and innovative methods are being developed (Mach & Arvinte, 2011). Many of the analytical procedures described in the following sections require the use of reference standards, from sources such as USP, World Health Organization (WHO), National Institutes of Health (NIH), and Food and Drug Administration (FDA). The USP includes a section on biotechnology-derived articles. This section discusses the scope of biotechnology in the development of pharmacopeial articles and various analytical methodologies for the quality control of biotechnology-derived proteins (United States Pharmacopoeia Convention, 2013). Analytical procedures should be validated for specificity, linearity, accuracy, recovery rate, sensitivity, precision, and limit of detection.

2.4.1 Electrophoresis

Commonly used electrophoretic analytical techniques for the analysis of peptides and proteins include SDS–polyacrylamide gel electrophoresis (PAGE) (including blotting methodologies), isoelectric focusing (IEF), and capillary electrophoresis (CE). These three techniques will be discussed in this section for their analytical use though it should be noted that electrophoresis is widely used also for separation of mixtures of proteins. Proteins, because of their charge, migrate under the influence of an applied electric field. The use of electrophoresis as an analytical and investigational tool for peptides and proteins is not new (Goldenberg & Creighton, 1984; Hartree, Lester, & Shownkeen, 1985; West, Wu, & Bonner, 1984). The movement of protein under electrophoresis depends on their mass-to-charge ratio.

A drawback of electrophoresis of proteins in its native state is that a single band on the gel may not necessarily mean that the sample is pure. This is because the band could have several components that have the same mass-to-charge ratio. A more common form of electrophoresis denatures the protein by SDS or other detergents (Hjelmeland & Chrambach, 1981) before carrying out PAGE. Proteins bind to a relatively constant amount of SDS (1.4 g/g of protein), and the mixture thus has a constant negative charge (that of SDS). Since charge becomes constant for all proteins, SDS-PAGE separates proteins based on their size (molecular weight) only. By varying the concentration of the acrylamide gel and the cross-linker, bisacrylamide, separation of proteins ranging in MW from 10 to 300 kDa can be achieved. SDS-PAGE is thus commonly used to determine the aggregation status of a protein. For proteins that contain disulfide bonds, the protein will migrate differently if the disulfide bonds are reduced (Jones, 1993). Following electrophoresis, the protein can be visualized and quantitated by Coomassie Brilliant Blue or silver staining. For quantitation of degradation products, Coomassie Blue is used. Silver staining is about 10 times more sensitive to detect impurities, but it cannot be used for quantitation. The SDS-PAGE separation of a protein can be combined with immunoblotting in a Western blot. Following electrophoresis, the resolved proteins are electrophoretically transferred onto a nitrocellulose or polyvinylidene difluoride (PVDF) membrane and reacted with an antibody after blocking all other binding sites. The complex can then be visualized using an enzymatically labeled (e.g., with horseradish peroxidase or alkaline phosphatase) or radiolabeled antibody (United States Pharmacopoeia Convention, 2013).

If a protein migrates in a pH gradient under an electric field, it will move toward the electrode of opposite sign until it reaches a region where the pH equals its pI. At this point, the net charge on the protein will become zero and thus migration will stop. This technique of IEF can be used to find the pI of a protein or to characterize its purity or to accomplish protein separations (Kohnert, Schmid, Zuhlke, & Fiedler, 1973; Kosecki, 1988; Pritchett, 1996; Welinder, 1971). It can also be used to test the stability of a protein since protein deamidation leads to the production of a new carboxylic acid group, resulting in a shift of pI toward the acidic side. Similarly, it can be used to separate different glycoforms of proteins that differ in the degree of sialylation of the complex carbohydrates. IEF technique can separate proteins that differ in pI by as little as 0.02 pH units. Electrophoretic separation is voltage dependent, but voltage can be increased only to a certain extent since it generates heat that can damage the matrix. In CE, the narrow diameter of capillaries allows dissipation of heat. Also, the capillaries eliminate the need for an anticonvective medium. Another advantage of CE is that the amount of sample required is very low due to the small dimensions of the capillary. Unlike reversed-phase high-performance liquid chromatography (RP-HPLC), CE presents mild nondenaturing conditions for the analysis of peptides and proteins (Dupin, Galinou, & Bayol, 1995; Strege & Lagu, 1993a). A combination of HPLC and electrophoresis thus provides us with high-performance capillary electrophoresis (HPCE), a rapid, precise, and efficient analytical technique, though it is not widely used anymore. Since CE is an instrumental procedure, online detection methods such as ultraviolet (UV) or fluorescence detection can be used (Jones, 1993; Rabel & Stobaugh, 1993). The most widely used CE method is free-zone capillary electrophoresis (CZE). A high voltage is applied to a fused silica capillary, and separation occurs due to electrophoretic migration and electroosmotic flow. This technique has been used for the separation and quantitation of the human growth hormone (hGH), human insulin, and proinsulin (Arcelloni, Fermo, Banfi, Pontiroli, & Paroni, 1993). When using CE with protein drugs, the protein may adsorb onto the capillary and this may be prevented by using surface-modified capillaries (Rabel & Stobaugh, 1993). Strege and Lagu (1993b) have reported that CE can be used for protein separations of high efficiency and resolution by using polyacrylamide-coated silica capillaries and buffers containing ionic surfactants. However, CE can be a difficult, variable, and time-consuming method for proteins, though it may work well for some other compounds.

2.4.2 Spectroscopy

2.4.2.1 UV Spectroscopy

The absorption spectra of a protein in the UV range (175–350 nm) are the net result of absorption of light by the carbonyl group of the peptide bond (190–210 nm), the aromatic amino acids (250–320 nm), and the disulfide bonds (250–300 nm). In 1962, Wetlaufer (1962) published a comprehensive discussion of the UV spectra of proteins and amino acids. Using experimentally determined specific absorptivity values, the routine measurement of protein concentration can be done by UV spectroscopy. Thus, UV spectroscopy can be used to measure protein content without the need for any calibration with standards, providing a convenient and accurate measure of protein concentration. Any protein with at least one tryptophan residue can be detected by UV spectroscopy at concentrations of 0.1 mg/mL or more. Phenylalanine and tyrosine contribute about one-sixth the molar absorptivity of that of tryptophan. Protein content assay is very important since other types of assay, such as potency assay, are also dependant on them. Protein content can also be measured by Kjeldahl nitrogen analysis or by colorimetric methods, which require reference standards. Absorption of UV light is not very sensitive to the conformation of the protein, except when the aromatic rings of Phe, Tyr, and Trp may be exposed to the surface. UV spectroscopy could be useful to monitor changes in the tertiary structure, but other techniques are generally more sensitive. Derivative spectroscopy can help with resolution of overlapping peaks. When derivatized, the maxima and minima of the original function take zero values and derivative curves can identify tiny differences between the original spectra, which enhances signal and resolves overlapped peaks (Ojeda & Rojas, 2013). Second derivative UV spectroscopy has been used to monitor protein conformation (Dasnoy, Le Bras, Preat, & Lemoine, 2012; Mach & Middaugh, 2011). High-throughput spectroscopy combined with measurement of intrinsic and extrinsic fluorescence techniques to detect changes in protein conformation has been used with calcitonin to select and characterize physically stable formulations (Capelle, Gurny, & Arvinte, 2009). Some proteins also contain prosthetic groups (e.g., heme group in hemoglobin) that absorb strongly in the UV–visible light. When using UV spectroscopy to measure proteins that may have some aggregates, errors may result due to light scattering, and these should be corrected (Winder & Gent, 1971). Since insoluble protein aggregates can scatter light in the 300–350 nm range, UV–visible spectroscopy may also be used to detect protein aggregation (insoluble aggregates).

2.4.2.2 Fluorescence Spectroscopy

The aromatic amino acids Phe, Trp, and Tyr exhibit fluorescence. The fluorescence is in the order of Trp > Tyr > Phe, with the result that most fluorescent spectra are typical of Trp. Fluorescence of Tyr is generally observed only in the absence of Trp, while that of Phe is observed only in the absence of both Tyr and Trp. Most proteins exhibit fluorescence in the 300–400 nm range when excited at 250–300 nm. Fluorescence is much more sensitive to protein conformation as compared to UV spectroscopy. The magnitude of fluorescence by itself may not be very informative, but it could serve as a probe to monitor changes in the protein conformation. The indole side chain of Trp is the largest and most fluorescent of all amino acids. Also, most proteins have only one or just a few residues of Trp, and these hydrophobic residues are generally buried in the interior of the folded protein. Therefore, unfolding of a protein molecule (e.g., prior to aggregation) is often accompanied by an increase in the fluorescence intensity. Also, a buried tryptophan will fluoresce with a wavelength maximum in the 325–330 nm range, while an exposed residue will fluoresce at about 350–355 nm. Tryptophan fluorescence can be used to monitor changes in the tertiary structure of a protein (Bhambhani et al., 2012) and can also be used as a probe of protein–peptide interactions. The quenching of intrinsic fluorescence of serum albumin by binding of a ligand can be used to study such interactions. For instance, Leu-enkephalin and its metabolite were found to quench the tryptophyl fluorescence of BSA (Jain, Kumar, & Kalonia, 1992; Parsons, 1982). Time-resolved fluorescence spectroscopy can be used to study proteins in solution, especially those that contain a single tryptophan, such as recombinant human interferon-γ or human albumin (Brochon, Tauc, Merola, & Schoot, 1993).

2.4.2.3 Circular Dichroism Spectroscopy

Amino acids (except glycine) are asymmetric due to the presence of the chiral carbon. The optical properties of polypeptides are due to the asymmetric centers of its constituent amino acids. Polypeptides thus interact differently with right and left circularly polarized light. CD is a technique that measures the unequal distribution of left and right circularly polarized light. The CD spectra of a protein are very sensitive to its conformation. In the far-UV region (below 250 nm), it can detect changes in the secondary structure, especially in α-helical content. Alpha-helical proteins typically show two minima, at 208 and 222 nm. The β-sheet gives weak CD signals, with minima around 215–217 nm. The CD spectra can be deconvoluted to calculate the percent of α-helix, β-sheet, bends, turns, and random coil components of the secondary structure. However, the information may be influenced by the solvent system and other factors so that it may be more useful to determine structural changes rather than an accurate prediction of absolute structural content. In the near-UV region (250–300 nm), CD spectroscopy can detect changes in the tertiary structure of proteins as this region reflects the environment of aromatic amino acids (Johnson, 1988). Therefore, the near-UV CD spectra can be used to confirm that the correct tertiary structure is present or that refolding was successful. CD spectroscopy has

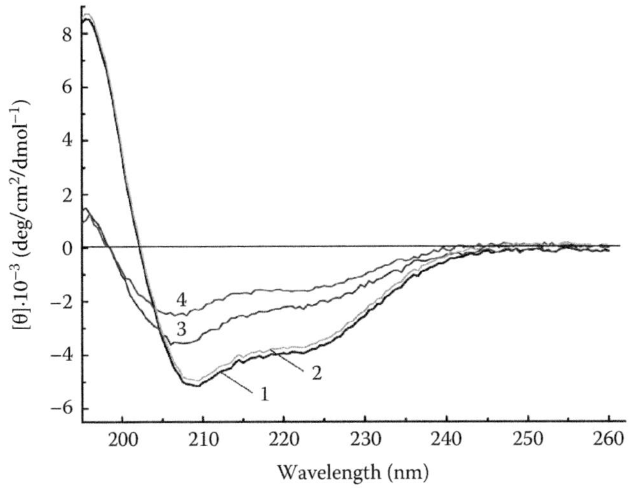

FIGURE 2.3 Effect of arginine on the secondary structure of insulin (0.5 mg/mL) monitored by far-UV CD (190–260 nm). CD spectra were taken alone (1) or incubated with 15 mM Arg (2) or with 5 mM DTT in the absence (3) or presence (4) of 15 mM Arg. (Reproduced from Smirnova, E. et al., *Int. J. Pharm.*, 471, 65, 2014. With permission.)

been used to study the optical activity of insulin (Ettinger & Timasheff, 1971a, 1971b) and leuprolide acetate, a luteinizing hormone–releasing hormone (LHRH) agonist (Powers, Adjei, Lu, & Manning, 1994). Far-UV CD spectroscopy has been used to show the potency of Arg to destabilize the conformational stability of insulin (Figure 2.3) (Smirnova, Safenkova, Stein-Margolina, Shubin, & Gurvits, 2014). Near-UV CD spectra have been used to show that the tertiary structure of hGH can be disturbed by cyclohexanol as the spectra for control hGH and hGH-containing cyclohexanol did not overlap between 268 and 285 nm (Maa & Hsu, 1996). CD spectroscopy can be used in regulatory filings to show that a change in formulation or a change in manufacturing procedure has not produced a change in the conformation of the protein (Bhambhani et al., 2012; Li et al., 2011).

Ellipticity is the typically used unit for CD spectra and is often expressed as mean residue ellipticity (degree cm^2 $dmol^{-1}$). The differential absorption of left and right circularly polarized light is very small, but the instrumentation used is very sensitive and can measure it accurately. Ellipticity is generally recorded as mdeg by the instrument. The instruments typically use a nitrogen purge as oxygen absorbs strongly below about 200 nm. Protein should be pure, and solutions should be filtered through a fine filter and degassed to improve the signal-to-noise ratio. Buffers that do not absorb in the wavelength range being investigated should be used and in low concentration. The optimum cell path length (typically 0.1–0.5 mm) and protein concentration (e.g., 0.5 mg/mL) will have to be worked out for getting good output. Several scans, as high as 64, are typically accumulated to reduce noise (Vermeer et al., 1998) and make the spectra smoother. For near-UV studies, a higher concentration (e.g., 2 mg/mL) and a cell with a longer path length (typically, 1–10 cm) are used.

2.4.2.4 Infrared Spectroscopy

Infrared (IR) spectroscopy relies on the vibrational bond energy and has been used for studies of protein conformation for a long time. However, it has low sensitivity, and it is difficult to obtain spectra and extract information. Water, which is the medium in which proteins are dissolved, absorbs strongly in the spectra. Interest in IR spectroscopy thus decreased but then revived with the availability of Fourier transform infrared (FTIR) spectrometers, which have overcome many of the disadvantages discussed. FTIR uses special deconvolution methods to separate and integrate overlapping amide I absorption bands associated with secondary and random structures to produce high-quality spectra from very small amounts of protein (about 100 μg) in the solid state. IR spectra of polypeptides and proteins exhibit a number of amide bands, which represent different vibrations of the peptide moiety, and are responsive to changes in the secondary structure and conformation of the polypeptide backbone (Arrondo, Muga, Castresana, & Goni, 1993; Jiang et al., 2011; Surewicz, Mantsch, & Chapman, 1993). FTIR spectroscopy has been used to monitor the changes in secondary structure of proteins, resulting from interfacial adsorption and aggregation (Bauer, Muller, Goette, Merkle, & Fringeli, 1994; Miller, Bourassa, & Smith, 2013). Changes in the secondary structure of lysozymes were analyzed by FTIR, and a maxima seen at 1650 cm^{-1} was assigned to the α-helix. An absorption band about 1620 cm^{-1} was assigned to aggregation as it was believed to be due to intermolecular β-sheets and was used to monitor aggregation-related changes (Malzert et al., 2002). The strong band near 1620 cm^{-1} and an associated weak band near 1685 cm^{-1} in D_2O have been reported to be indicative of aggregation for other proteins as well (Dong, Prestrelski, Allison, & Carpenter, 1995). In a study with recombinant hGH, a largely α-helical protein, a strong absorption band was observed at 1654 cm^{-1} and was relatively stable to changes in pH. However, this band decreased with temperature, while a band at 1620 cm^{-1} increased with temperature, suggesting the formation of aggregates (Cauchy, D'Aoust, Dawson, Rode, & Hefford, 2002). The IR amide I spectrum of recombinant human factor XIII in aqueous solution showed an absorbance maximum near 1642 cm^{-1}, indicating a predominantly β-sheet structure (Dong et al., 1997). FTIR analysis also indicated a loss of α-helical structure following adsorption of three model proteins (Brandes, Welzel, Werner, & Kroh, 2006). Combined use of DSC and FTIR has also been reported to monitor protein stability (Lin & Wang, 2012).

Because water absorbance can confound the FTIR spectra in the 1700–1600 cm^{-1} region, background absorption by water must be subtracted or the protein can be reconstituted in D_2O. Alternatively, spectra can be recorded on solid form of the drug. IR spectroscopy is the most common method to monitor protein denaturation during lyophilization. Lyophilization-induced structural changes can be monitored in the amide I, II, or III region. For example, protein unfolding during lyophilization may broaden and shift amide I component peaks to higher wave numbers. Lyophilization often leads to a conversion of α-helices to β-sheets for many proteins, which in turn is an indication of protein aggregation and/or increased intermolecular interaction (Wang, 2000). The amide I, II, and III bands differ in their sensitivity to changes in protein conformation, and it may thus be beneficial to look at amide II and III bands also in addition to the commonly used amide I band (van de, Haris, Hennink, & Crommelin, 2001).

2.4.2.5 Light Scattering

Rayleigh or static light scattering detectors have been used for a while, but relatively recent innovations have now allowed for dynamic laser light scattering detectors. Dynamic light scattering (DLS), also known as photon correlation spectroscopy (PCS), or quasi-elastic light scattering (QELS) measures fluctuations of the intensity of scattered light at very short time scales and is used to measure diffusion coefficients that are then used to calculate the Stokes radius or hydrodynamic radius (R_H) of macromolecules in the size range of 1–1000 nm. The molecular weight can be estimated from the measured R_H by using MW versus R_H calibration curves developed from standards of known MW and size. The advantages of this technique are that the sample can be analyzed without dilution or addition of buffers or other maneuvers usually required with other techniques. Due to Brownian motion, light scattered from particles fluctuates with time. DLS measures these fluctuations across very short time intervals to produce a single correlation curve that is then deconvoluted using algorithms to obtain size distribution information. A laser light source is typically used to provide a polarized monochromatic laser beam. The dynamic signal that is representative of random fluctuations about a mean value appears to be smooth when measured over a very short time scale (Mattison, Morfesis, & Kaszuba, 2003). Static and dynamic light scattering have been used to investigate the aggregation behavior of lysozyme solutions (Georgalis, Umbach, Raptis, & Saenger, 1997). The R_H is also influenced by particle shape and solvation. The R_H will thus vary depending on several factors, but examples of approximate values are 1.9 nm for lysozyme, 3.6 nm for BSA, and about 6 nm for IgG. For insulin, R_H may be in the 1.3–2.6 range as the molecule changes from monomeric to hexameric as a function of a pH change from 2.0 to 7.0. Monoclonal antibodies are generally around 12–14 nm, and the majority of proteins (in monomeric form) do not exceed 20 nm (Das, 2012). For most proteins, a single collection angle of 90° is sufficient, but for very large molecules (>500 kDa) or for aggregates larger than 10 nm, collecting the scattered laser light at two angles (90° and a low angle of 15°) will be needed. More angles may give more accuracy, and multiangle light scattering (MALS) instruments having three angles of measurement are commercially available. The additional angle extends the MW determination to about 40 million Daltons.

Light scattering in the visible range has also been used as a technique to study insoluble aggregation. This noninvasive technique thus allows the examination of particle size in essentially native form and has been used to study the aggregation of insulin formulations (Bohidar & Geissler, 1984; Kadima et al., 1993; Martindale, Marsh, Hallett, & Albisser, 1982). While special instrumentation is available, it may also be feasible to simply measure the turbidity of precipitating solutions by a UV–visible spectrophotometer. Light scattering may also be measured fluorimetrically (Brewster, Hora, Simpkins, & Bodor, 1991). Aggregation behavior of zinc-free insulin has also been studied by small-angle neutron scattering. It was found that detailed information on the aggregation behavior of insulin can be obtained from small-angle scattering experiments and that the results are derived independently of the light scattering results and models (Pedersen, Hansen, & Bauer, 1994). Turbidimetric measurements were used as a rapid screening procedure to identify excipients that had the potential to stabilize acidic fibroblast growth factor (aFGF) against heat-induced aggregation (Tsai et al., 1993). A sharp increase in the optical density in the range of 325–500 nm has been reported for reconstituted freeze-dried bovine IgG due to scattering from insoluble aggregates (Sarciaux et al., 1999). However, protein aggregates are often translucent and may not provide sufficient difference in refractive index, thus limiting the application of methods based on light scattering analysis (Zolls et al., 2012). Detection by the Coulter method is not dependent on optical properties, and this method has thus been shown to consistently detect more particles than microflow imaging and light obscuration (Barnard, Rhyner, & Carpenter, 2012).

2.4.2.6 Colorimetric Assay

Absorbance of light in the visible range is due to conjugated double-bond systems and forms the basis for colorimetric assays. Although peptides and proteins do not absorb visible light, they can react with reagents to form colored compounds. Ninhydrin is a widely used reagent that reacts with amino groups of amino acids and peptides to produce an intensely colored product, with a maximum absorbance at 570 nm. A relatively more specific but less sensitive method of quantitatively determining proteins is the biuret reaction. Biuret reaction of proteins with copper in a basic solution constitutes the commonly used Lowry assay for determination of protein content. The resulting blue product is quantitated in the visible region at 540–560 nm, and the assay is linear at microgram protein levels (United States Pharmacopoeia Convention, 2013). The Bradford colorimetric assay depends on the binding of the dye Coomassie Brilliant Blue to the product in an acidic environment (Creighton, 1983a). The reaction can take place in just 2.0 min with good color stability for 1.0 h to allow measurement at 595 nm. Fluorescent methods can also be used and usually use either fluorescamine or *o*-phthaldialdehyde (Creighton, 1983b). Colorimetric assays run by robotics and linked to plate readers can assay thousands of compounds in a day (Fisher & Higgins, 1994).

2.4.3 Chromatography

2.4.3.1 Reversed-Phase Chromatography

Several chromatographic techniques are based on HPLC, previously known as high-pressure liquid chromatography. In RP-HPLC, the hydrophobic interactions between the column packing and the hydrophobic regions of the protein are exploited. A majority of separations are performed on large-pore-sized silica-based stationary phases containing chemically bonded alkyl chains of various lengths, typically C_4, C_8, or C_{18}. In a study by Welinder, Linde, and Hansen (1985), it was reported that 80–100 Å silica-based C_{18} columns are the best for separation of insulin from insulin-related substances. However, for proteins with molecular weights greater than about 10,000, supports with pore size of at least 300 Å are preferred. In general, wide-pore packings in a short column are likely to give the best separations for proteins (Koyoma, Nomura, Shiojima, Ohtsu, & Horii, 1992). In order to avoid long empirical optimization procedures, a system for predicting peptide retention time is desirable. This may be possible by summing group retention coefficients of the amino acids present in the peptide (Schoneich et al., 1993). RP-HPLC can also be a useful tool for the study of peptide conformations resulting from peptide–peptide interactions (Blondelle, Buttner, & Houghten, 1992; Lebl, 1993).

Gradient elution RP-HPLC is capable of differentiating between insulins of bovine, porcine, or human origin and is reproducible and stability indicating (Adams & Haines Nutt, 1986; Damgaard & Markussen, 1979; Knip, 1984; Linde, Welinder, & Hansen, 1986; Lloyd & Corran, 1982; Ohta, Tokunaga, Kimura, Satoh, & Kawamura, 1984; Szepesi & Gazdag, 1981). The assay gave results that were more precise and more reproducible than those obtained by the mouse blood glucose method (Fisher & Smith, 1986). Also, the time required for analysis by HPLC can be as little as 1/60 of that required for a bioassay (Smith, Venable, & Collins, 1985). The use of RP-HPLC to characterize the degradation products of insulin has also been reported in the literature (Hamel, Peavy, Ryan, & Duckworth, 1986; Yonezawa et al., 1986). Most of these RP-HPLC determinations of insulin controlled the degree of ionization on insulin by using a mobile phase with low pH and high salinity. An alternative approach to get good peak shape and reproducible retention is to use an ion-pairing reagent. Examples of anionic ion-pair reagents are phosphate anion and alkanesulfonic acids, while those of cationic ion-pair reagents are tetramethylammonium bromide and tetrabutylammonium bromide (Peter, Szepesi, Balaspiri, & Burger, 1987; Vigh, Varga-Puchony, Hlavay, & Papp-Hites, 1982). In a typical RP-HPLC analysis, proteins are eluted with aqueous acetonitrile gradients containing 0.1% trifluoroacetic acid.

An RP-HPLC assay for hGH was reported to resolve hGH from its sulfoxide and desamido derivatives and was better than several alternative chromatographic and electrophoretic techniques (Riggin, Dorulla, & Miner, 1987). The sensitivity of an RP-HPLC assay can be increased with fluorescence detection of suitably derivatized peptides. Under optimal conditions, a sensitivity similar to that of immunoassay or bioassay can be achieved. Such sensitivity is critical for the measurement of peptides in biological fluids in support of preclinical and clinical studies (Ohno, Kai, & Ohkura, 1989). A drawback of reversed-phase columns for proteins is possible denaturation of protein due to the hydrophobic stationary phase. RP-HPLC can also be used to detect leachables coming from the container, for example, Eprex from prefilled syringes with uncoated rubber stoppers resulted in appearance of a series of extra peaks (Figure 2.4), and these contaminants caused immunogenic problems for patients (see Section 1.4) (Sharma, 2007). Another analytical technique, ultrahigh performance liquid chromatography (UPLC), is a variant of HPLC using columns packed with particle size less than 2 μm. This provides for better separation and faster analysis. Higher pressure is required in the instrument to achieve this analysis. The term *UPLC* is used loosely for this technique, but actually it is a trademark of Waters Corporation.

2.4.3.2 Size-Exclusion Chromatography

Gel permeation chromatography (GPC) or SEC is a chromatographic technique that, unlike other types of chromatography, depends on the relative size or hydrodynamic

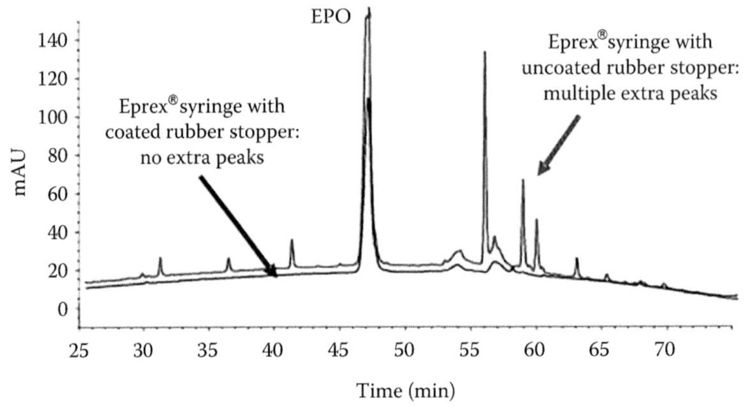

FIGURE 2.4 RP-HPLC of Eprex® from prefilled syringes with uncoated or FluroTec®-coated rubber stoppers. Uncoated stoppers had a series of extra peaks in addition to the peaks for the drug and polysorbate 80 used as excipient. (Reproduced from Sharma, B., *Biotechnol. Adv.*, 25, 318, 2007. With permission.)

volume of a macromolecule with respect to the size and shape of the pores of the packing. Once the SEC column is calibrated with molecular weight standards, the molecular weight of the compound being investigated can be found and it can also be used as a quantitative technique (Barth & Boyes, 1992). The technique of GPC uses a gel filtration column packed with particles of defined size. Smaller molecules penetrate more into the gel and thus elute later from the column. Molecular weight standards can be used to calibrate the column. GPC was initially done on columns packed with soft polymeric supports such as cross-linked dextrans, polyacrylamide, or agarose. These are suitable for low-pressure applications.

Supports with increased mechanical strength are now commercially available to be used in high-performance size-exclusion chromatography (HPSEC or SEC-HPLC). SEC-HPLC is an excellent technique for study of protein aggregation (Defelice, Hayakawa, Watanabe, & Tanaka, 1993; Vlaanderen, Van Grondelle, & Bloemendal, 1993; Watson & Kenney, 1988; Welinder, 1980) or for detection of low MW degradation products. The mobile phase used is usually aqueous. SEC-HPLC is typically used for study of covalently linked or stable noncovalently linked protein aggregates. However, it is generally not used for noncovalent protein aggregates that rapidly dissociate upon dilution, due to shifting retention times and altered apparent molecular weights. SEC-HPLC was found to be an ideal technique for analysis of hGH since it would differentiate the hGH dimer from its monomer. Other techniques such as RP-HPLC or electrophoresis led to the dissociation of the dimer. Even though the hGH dimer may have immunochemical activity, its biological activity is very low and thus SEC-HPLC assay became important to determine the potency of hGH (Riggin, Shaar, Dorulla, Lefeber, & Miner, 1988). Figure 2.5 shows the time evolution of aggregation of a monoclonal antibody as a function of time at pH 3.4 and 6.1 at 55°C. The SEC-HPLC peak shifts to the left as the aggregate grows from monomer to a dimer and eventually to a pentamer. A haze was observed visually, but

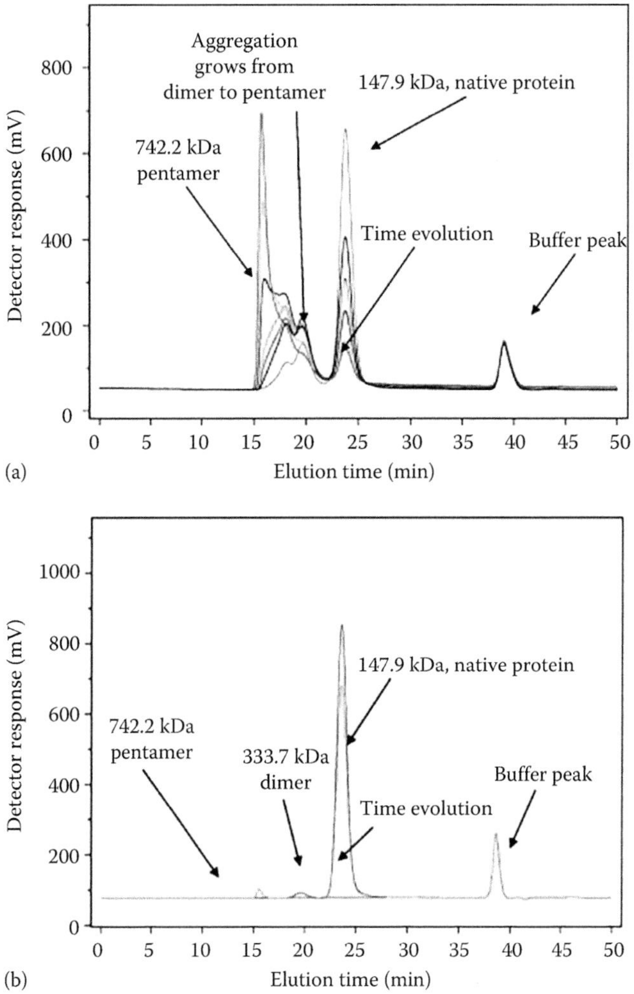

FIGURE 2.5 SEC-HPLC showing the time evolution of aggregation from monomer to dimer and further to pentamer for the protein Abbott-X buffered at pH 3.4 (a) and pH 6.1 (b) at 55°C. (Reproduced from Zhu, D. et al., *Thermochim. Acta*, 499, 1, 2010. With permission.)

higher-order aggregates cannot be detected by SEC-HPLC due to their low solubility in solution (Zhu et al., 2010).

Hydrophobic proteins may bond tightly to the silica solid-phase column matrix often used for SEC. This may be overcome by the addition of a surfactant to the mobile phase. The anionic surfactant SDS is sometimes used. Since SDS can dissociate noncovalent aggregates, caution must be used to interpret the results. A method has been described to use a rapid SEC-HPLC procedure that can determine both thermodynamics and kinetic parameters that describe rapid and reversible protein aggregation. A small volume-sized exclusion column allowed the study of monomer–dimer equilibrium of hGH. Due to a short retention time and use of computer simulation, it was possible to study aggregates with short half-lives ($t_{1/2}$ = 5 s) of dissociation (Patapoff, Mrsny, & Lee, 1993).

2.4.3.3 Liquid Chromatography–Mass Spectrometry

Mass spectrometry (MS)-based approaches are widely used for analysis of peptides, proteins, and their fragments, such as for metabolism studies (Guo & Carta, 2014; Katsila, Siskos, & Tamvakopoulos, 2012; Mesmin, Fenaille, Ezan, & Becher, 2011). Liquid chromatography–mass spectrometry (LC-MS) combines the physical separation capabilities of LC with mass analysis by MS. LC-MS has very high sensitivity in the pictogram range and is a powerful technique for the quantification of peptides and proteins in complex biological samples (Bronsema, Bischoff, & van de Merbel, 2013). Liquid chromatography–tandem mass spectrometry (LC-MS/MS) is a powerful technique for peptide mapping and overall protein structural characterization, including providing a glycan profile on each individual glycosylation site (Shah, Jiang, Chen, & Zhang, 2014).

2.4.3.4 Other Chromatographic Techniques

These include ion-exchange chromatography (IEXC), hydrophobic interaction chromatography, and affinity chromatography. In IEXC, the protein is resolved based on charge rather than hydrophobicity. The protein is allowed to bond to a matrix of opposite charges. The passage of a salt gradient over this column will then elute the protein as the salt takes the place of the protein to bind to the matrix. IEXC can also be performed as HPLC by using an ion-exchange column. A method useful for peptides in general has been reported and uses a strong cation-exchange column that is eluted at a pH of around 3 (Alpert & Andrews, 1988). IEXC is useful for measuring charge heterogeneity in protein preparations.

In hydrophobic interaction chromatography, the protein is loaded on a hydrophobic column at a very high salt concentration since hydrophobic interactions are strengthened by increasing salt concentration. The protein can then be eluted by running a reverse salt gradient. This technique may sometimes be preferred to RP-HPLC as the mobile phase used is aqueous and thus less likely to denature the protein. Hydrophobic interaction chromatography has been used for protein purification (Vorauer, Skias, Trkola, Schulz, & Jungbauer, 1992). Affinity chromatography is based on the principle of protein–ligand binding. The principle is similar to other types of chromatography except that the interaction between protein and the column resin is very specific (Pina, Lowe, & Roque, 2014).

2.4.4 New Techniques to Characterize Subvisible Particles

The use of SEC to characterize protein aggregates is limited to soluble aggregates, and aggregation to the extent of 0.1%–0.5% of the total protein may be required for reliable quantification. The limitations of SEC were found to be overcome by microflow imaging to characterize subvisible particles (Barnard, Singh, Randolph, & Carpenter, 2011). This technique relies on digital imaging of a flowing sample stream for analyzing particles in suspension. Images of individual particles are then analyzed by the system software, allowing detection in the range of 2–300 μm at low magnification and 0.75–75 μm at high magnification (Sharma, Oma, & Krishnan, 2009). Electron microscopic techniques allow low nm resolution and include scanning electron microscopy and transmission electron microscopy. Atomic force microscopy has also been used to visualize proteinaceous subvisible particles (Mach & Arvinte, 2011) and even generates a 3D picture. A variety of analytical methods may be needed to fully characterize protein aggregates of all sizes and types (Philo, 2006).

2.4.5 Thermal Analysis

DSC monitors the absorption or release of energy when a material undergoes changes due to a change in temperature, for example, during melting. DSC has been used as a tool to study protein folding and stability (Johnson, 2013; Wen et al., 2012). Two basic types of equipment are available. In power compensation DSC, the power difference required to maintain the sample and reference at the same temperature in isolated furnaces is measured. In heat flux DSC, the sample and reference are heated in the same furnace and the temperature difference between them is directly measured. In both cases, heat flow is plotted as a function of temperature. DSC is gaining importance as a useful tool to investigate changes in protein conformation. As a protein is heated, the transition from the native state to the unfolded state is accompanied by the appearance of an endothermic peak on DSC. The transition temperature, T_m, is equivalent to the melting of a crystal and is affected by the environmental conditions (such as pH) and the presence of pharmaceutical excipients. Also, thermograms of monoclonal antibodies may exhibit multiple transition temperatures for the heavy and light chains or different domains. These may be followed by an exothermic peak indicative of aggregation (Bliznukov et al., 2001; Gonzalez, Murature, & Fidelio, 1995; Vermeer et al., 1998, 2000). However, in some cases, as was observed for pertussis filamentous hemagglutinin, if the protein aggregates while being heated, the data may be confounded by a broad and noisy exothermic event and may not be interpretable (Krell et al., 2003). Models for protein degradation have been used to investigate the effect of pH or excipients on DSC scans for lysozyme (Cueto, Dorta, Munguia, & Llabres, 2003).

Thermal scans can be performed twice to investigate the reversibility of thermal unfolding. Conventional DSC does not have enough sensitivity to detect thermal transitions in proteins at concentrations typically used in marketed products. For these low concentrations, microcalorimetry would be preferred (Privalov & Dragan, 2007). Microcalorimetry has been used as a high-throughput screening tool for protein formulations, and it has been shown that measurement of unfolding temperatures is a good tool to predict long-term storage stability of protein formulations (Youssef & Winter, 2013). Also, differential scanning fluorimetry (DSF) has been used to screen protein formulations for stability and is based on measuring the fluorescence of a dye when bound to the hydrophobic regions of a denatured protein (Goldberg, Bishop, Shah, & Sathish, 2011; Malik, Matejtschuk, Thelwell, & Burns, 2013). In general, several extrinsic fluorescent dyes have been used as tools for protein characterization (Bhambhani et al., 2012; Capelle et al., 2009; Hawe,

Sutter, & Jiskoot, 2008). Similarly, fast-scan DSC can be used, and it utilizes faster heating and cooling rates than conventional DSC (Ford & Mann, 2012). Modulated DSC (MDSC) may also be used and would be more sensitive than conventional DSC (Vermeer & Norde, 2000). MDSC uses a software to apply a sine wave modulation to a standard, linear temperature program that allows separation of overlapping phenomena and deconvolution of complex transitions, resulting in greater resolutions. The heating rates are typically 1°C–5°C/min (rather than 10°C–20°C/min typically used in conventional DSC). Temperature modulation is typically ±0.1°C to ±1.0°C over a 30–80 s period of time so that about six modulations are included over the duration of the thermal event. MDSC separates the total heat flow observed in conventional DSC into a heat capacity component and a kinetic component. The heat capacity component is heating rate dependent and appears as a reversing signal, for example, glass transition temperature and melting endotherm. The kinetic component is a function of time at constant temperature and appears as a nonreversing heat flow signal, for example, for events such as crystallization and some melting phenomena (Coleman & Craig, 1996; Micard & Guilbert, 2000; Nail, Jiang, Chongprasert, & Knopp, 2002; Verdonck, Schaap, & Thomas, 1999). MDSC provides high resolution due to a slow heating rate while also providing a high sensitivity due to a high instantaneous heating rate from the modulation.

DSC is also commonly used in lyophilization studies to find the eutectic and glass transition temperatures for protein formulations. A eutectic melt is detected as a small endotherm, while a glass transition is observed as an abrupt change in the baseline, which represents an increase in heat capacity as the system passes from a glass to a fluid. However, the results should be interpreted carefully since they could be confounded by baseline drift or other noise or may be hard to observe due to the small change in heat capacity associated with glass transition. The results of thermal analysis are affected by the experimental conditions/formulation, and thus effects of factors such as drug concentration, pH of formulation, and heating and cooling rates should be investigated and reported (Peters, Molnar, & Ketolainen, 2014).

2.4.6 Immunoassays and Bioassays

Radioimmunoassay (RIA) was first developed in the mid-1950s for the assay of insulin but gained widespread acceptance only in the 1970s (Ceska, Grossmuller, & Lundkvist, 1970; Colt, Becker, & Quatrale, 1971; Dwenger, 1984; Genuth, Frohman, & Lebovitz, 1965; Morgan & Lazarow, 1963; Szabo & Mahler, 1970; Yalow, 1973). The assay exploits the specific and tight association of antibody with a peptide or protein drug as the antigen to determine very low concentrations of the drug. In carrying out the assay, the unlabeled protein is exposed to its antibody in the presence of a known amount of radiolabeled protein. The concentration of protein is determined by measuring the unbound hot ligand after separating from the bound ligand. RIA has been used as an analytical probe to assay and to detect structural changes in proteins such as cholecystokinin (Rehfeld, 1978), insulin (Ceska et al., 1970; Colt et al., 1971; Musial, Duran, & Smith, 1986), somatostatin (Mackes, Itoh, Greene, & Gerich, 1981), and vasopressin (Hoglund, 1988).

Another popular immunoassay is the enzyme-linked immunosorbent assay (ELISA). This is a solid-phase assay that utilizes the fact that small amounts of protein can adsorb to plastic surfaces. A commonly used form of ELISA is the double-antibody sandwich assay that utilizes two antibodies. The first antibody is allowed to adsorb on microtitration plates that typically have 96 wells to allow assay of multiple samples. After incubation, unadsorbed antibody is washed off, and any residual protein binding sites are blocked with a blocking agent such as another inert protein. The protein of interest (antigen) is then added to the wells and binds to the coated antibodies. Now, unbound antigen is washed off, and then antigen bound to the antibody is treated with the second antibody. The second antibody has an enzyme such as horseradish peroxidase or alkaline phosphatase that, with an appropriate substrate, can develop color and can be analyzed spectrophotometrically (Oellerich, 1984; Vashist, Marion, Lam, Hrapovic, & Luong, 2014). Thus, ELISA does not need a radiolabeled protein for the assay. The absorbance in all 96 wells of the microtitration plate can be read simultaneously and rapidly using instruments known as ELISA plate readers. Other approaches are also feasible where the plates may be coated with antigen and then treated with an antibody to be quantitated and finally with the secondary antibody. This way, an antibody can be quantitated, for example, to determine the immune response to administration of the antigen in the body.

Since the interaction between protein and antibody involves specific sites, immunoassays are sensitive to conformational changes in the protein. These assays indicate immunoactivity but not necessarily bioactivity. A conformational change at the immunoactive site may not affect the bioactivity of the protein, but it will lead to a drop in its assay value by immunoassay. Similarly, a change in structure or conformation that reduces the bioactivity of the protein may not affect its assay value by RIA, as long as the antigenic site is intact and capable of interacting with the antibody. As an example, the sites on the insulin molecule that act as immunogenic determinants and sites that bond to biological receptors are different. This led to the development of a radioreceptor assay for insulin that uses the biological receptor for insulin as a basis for assay (Trethewey, 1989). ELISA may be more sensitive and reproducible, but bioassays are more predictive of pharmacodynamic effects of the drug. Thus, a correlation between these two must be shown if ELISA is to be used. Also, lack of interference by endogenous proteins and antibodies must be demonstrated. Historically, animal model assays were the most commonly used bioassays, for example, the potency of hGH has long been determined by a weight gain in vitro bioassay using hypophysectomized rats. However, these assays have the drawbacks of animal use and poor reproducibility. In vitro bioassays have gained increasing acceptance in recent years (Foster et al., 1993), though these may have some variability as well. Physicochemical assays that are biomimetic may also be used. For example, the potency of alteplase (t-PA) has been measured by an in vitro clot lysis assay. The assay is based on measuring the time taken for a fixed amount of alteplase to dissolve a fibrin clot. Several in vitro bioassays have also been developed for routine analysis of hIL-1 activity. Similarly, the biological activity of interferons is measured by their dose-dependent inhibition of the cytopathic effect (CPE) due to virus inhibition of cell cultures in microtiter plates.

2.4.7 Other Quality Control Procedures

In addition to the specific analytical methods discussed previously, quality control of biotechnology-derived protein drugs may also need amino acid analysis, protein sequencing, and peptide mapping by HPLC. Amino acid analysis involves complete hydrolysis of the protein or peptide to its component amino acids (Thannhauser, Sherwood, & Scheraga, 1998). These amino acids are then separated and analyzed by chromatography with either precolumn or postcolumn derivatization. Protein sequencing provides information on the primary structure and confirms the complementary DNA (cDNA)-predicted amino acid sequence. Tests for residual host cell DNA should also be performed. DNA hybridization assays can be very expensive and time-consuming and may not be needed for routine batch release if the process is adequately validated. Similarly, the absence of bacteria, fungi, mycoplasma, and viruses should be demonstrated. Several analytical techniques may be required for this purpose (see Table 2.1). The validation of these techniques is important to assure quality assurance during biopharmaceutical manufacturing and subsequent formulation and processing of therapeutic peptides and proteins (McEntire, 1994). Several advanced analytical techniques that are currently assisting the characterization of recombinant proteins include (see also Section 2.4.3.3) matrix-assisted laser desorption ionization mass spectrometry (MALDI-MS), where the resulting ions are analyzed in a time-of-flight mass spectrometry (TOF-MS) that can detect the protein in picomole amounts very quickly, and electrospray ionization mass spectrometry (ESI-MS) (Krell et al., 2003; van Veelen et al., 1994; Watson, Shah, DePrince, Hendren, & Nelson, 1994), HPLC in line with MS (Mock, 1993), and sequential arrangement of RP-HPLC and high-performance liquid affinity chromatography (Gaillard, Martial, Marc, Engasser, & Fick, 1994).

2.4.7.1 Peptide Mapping

Peptide mapping involves the enzymatic or chemical cleavage of the protein into discrete peptides, which are then resolved by chromatographic techniques, such as RP-HPLC. The HPLC pattern results in a *fingerprint* that is characteristic for a given protein, making peptide mapping a very powerful and widely applicable tool for protein characterization. Peptide mapping is used to compare the structure of a specific lot to a reference standard to confirm correctness and lot-to-lot consistency of the primary structure. The use of peptide mapping techniques to analyze the primary structure of recombinant human interferon α-2, IL-2, and granulocyte macrophage colony-stimulating factor has been discussed (Allen et al., 1996; Stein, 1989). Peptide mapping is increasingly being used to characterize monoclonal antibodies for product identity and even as a stability indicating assay (Dick, Mahon, Qiu, & Cheng, 2009). Peptide maps can also be used to determine the identification/integrity of the N- and C-terminal ends of proteins by tracking the corresponding peptides in the maps. By stress testing of the protein, peptide mapping can also be validated to monitor other degradation pathways such as oxidation or deamidation. A combination of SEC, RP-HPLC, and CE has been used for peptide mapping (Stromqvist, 1994). After proteolytic digestion, the size differences of obtained fragments were first characterized by SEC. The fractions were then separated on the basis of hydrophobicity by

TABLE 2.1
Potential Impurities and Contaminants in Biotechnology-Derived Products

Impurities or Contaminants	Detection Method
Impurities	
Endotoxin	Bacterial endotoxins test, pyrogen test
Host cell proteins	SDS-PAGE[a], immunoassays
Other protein impurities (media)	SDS-PAGE, HPLC[b], immunoassays
DNA	DNA hybridization, UV spectrophotometry, protein binding
Protein mutants	Peptide mapping, HPLC, IEF[c], MS[d]
Formyl methionine	Peptide mapping, HPLC, MS
Oxidized methionines	Peptide mapping, amino acid analysis, HPLC, Edman degradation analysis, MS
Proteolytic cleavage	IEF, SDS-PAGE (reduced), HPLC, Edman degradation analysis
Aggregated proteins	SDS-PAGE, HPSEC[e]
Deamidation	IEF, HPLC, MS, Edman degradation analysis
Monoclonal antibodies	SDS-PAGE, immunoassays
Amino acid substitutions	Amino acid analysis, peptide mapping, MS, Edman degradation analysis
Contaminants	
Microbial (bacteria, yeast, fungi)	Microbial enumeration tests, test for specified microorganisms, sterility tests, microbiological testing
Mycoplasma	Modified 21 CFR method[f], DNAF[g]
Viruses (endogenous and adventitious)	CPE[h] and HAd[i] (exogenous virus only), reverse transcriptase activity, MAP[j]

Source: © 2013 The United States Pharmacopeial Convention. Used with permission.

[a] Sodium dodecyl sulfate–polyacrylamide gel electrophoresis.
[b] High-performance liquid chromatography.
[c] Isoelectric focusing.
[d] Mass spectrometry.
[e] High-performance size-exclusion chromatography.
[f] Draft guidelines relating to the Code of Federal Regulations, Title 21.
[g] DNA-binding fluorochrome.
[h] Cytopathic effect.
[i] Hemadsorption.
[j] Murine antibody production.

RP-HPLC, and finally, CE was used for separation based on charge differences. The utility of trypsin digestion and chromatography procedures to assess the reaction kinetics of recombinant porcine growth hormone at elevated temperatures has been demonstrated (Charman, McCrossin, & Charman, 1993).

2.4.7.2 Analytical Ultracentrifugation

In this technique, a high-speed centrifuge is used in conjunction with an optical system that scans the sample from top to bottom at predetermined intervals, thereby allowing quantification of protein aggregates present at each particular height. For measurement of sedimentation velocity, the rotor is spun at very high speed and rate at which the protein pellets are measured as it is a function of both mass and shape. Any aggregates will be separated based on the relative sedimentation coefficient of each species (Gabrielson et al., 2010; Krayukhina, Uchiyama, & Fukui, 2012). When measuring sedimentation equilibrium, a much lower rotor speed is used, and the sample is spun for a long time (typically 12–36 h) to reach an equilibrium between sedimentation and diffusion. At sedimentation equilibrium, aggregates can be readily detected as the concentration distribution depends only on molecular mass and is independent of molecular shape. It can also be used to confirm the expected molecular mass or degree of glycosylation or conjugation to polyethylene glycol (PEG). Electron paramagnetic resonance (EPR) spectroscopy has also been used in conjunction with analytical ultracentrifugation (AUC) to investigate the binding of surfactants to proteins (Jones, Cipolla, Liu, Shire, & Randolph, 1999). Several analytical techniques used together are required to characterize the physical or chemical degradation of a protein. For example, the denaturation and aggregation of basic fibroblast growth factor in solution formulations were analyzed by RP-HPLC, SEC, heparin affinity chromatography, and fluorescence spectroscopy (Sluzky, Shahrokh, Stratton, Eberlein, & Wang, 1994).

2.4.7.3 Analysis of Glycoproteins

Recombinant proteins produced in eukaryotic cells such as the CHO cells are capable of undergoing posttranslational modification to produce glycoproteins. For glycoproteins, the carbohydrate composition must also be characterized. Glycoproteins with identical polypeptide chains made in different cell lines may differ significantly in their carbohydrate structure (United States Pharmacopoeia Convention, 2013). In these proteins, while carbohydrate analysis is important to characterize the product, it is complicated due to the microheterogeneity of these proteins. These proteins may exhibit several bands on an IEF gel due to the heterogeneity of charge distribution, resulting from a varying number of sialic acid residues. However, the glycan structures on a protein are typically stable and do not change during formulation and storage. Therefore, glycosylation analysis is typically not performed as a part of stability studies and the multiple bands on gel electrophoresis may not be a cause for concern. Though most circulating human proteins are heterogeneously glycosylated, production of proteins by recombinant methods produces homogeneous human glycoproteins (Rich & Withers, 2009).

2.5 CONCLUSIONS

The unique structural and analytical aspects of therapeutic peptides and proteins have been discussed in this chapter. There are more than 20 different naturally occurring amino acids, some of which can be linked by peptide bonds to form peptides and proteins. A protein has several levels of structure, referred to as primary, secondary, tertiary, and quaternary structures. A change in one of these structures may lead to loss of biological activity of the protein. However, these changes may not be always detected by any particular analytical technique. For this reason, several analytical techniques are often used to completely characterize a peptide or protein drug.

These assay methods include electrophoretic, spectroscopic, chromatographic, and thermal analytical techniques. In addition, immunoassays and bioassays provide important information.

REFERENCES

Adams, P. S. & Haines Nutt, R. F. (1986). Analysis of bovine, porcine and human insulins in pharmaceutical dosage forms and drug delivery systems. *J. Chromatogr., 351,* 574–579.

Allen, D., Baffi, R., Bausch, J., Bongers, J., Costello, M., Dougherty, J. et al. (1996). Validation of peptide mapping for protein identity and genetic stability. *Biologicals, 24,* 255–275.

Alpert, A. J. & Andrews, P. C. (1988). Cation-exchange chromatography of peptides on poly(2-sulfoethyl aspartamide)-silica. *J. Chromatogr., 443,* 85–96.

Arcelloni, C., Fermo, I., Banfi, G., Pontiroli, A. E., & Paroni, R. (1993). Capillary electrophoresis for protein analysis—Separation of human growth hormone and human insulin molecular forms. *Anal. Biochem., 212,* 160–167.

Arrondo, J. L., Muga, A., Castresana, J., & Goni, F. M. (1993). Quantitative studies of the structure of proteins in solution by Fourier-transform infrared spectroscopy. *Prog. Biophys. Mol. Biol., 59,* 23–56.

Augusteyn, R. C., Ghiggino, K. P., & Putilina, T. (1993). Studies on the location of aromatic amino acids in alpha-crystallin. *Biochim. Biophys. Acta, 1162,* 61–71.

Barnard, J. G., Rhyner, M. N., & Carpenter, J. F. (2012). Critical evaluation and guidance for using the Coulter method for counting subvisible particles in protein solutions. *J. Pharm. Sci., 101,* 140–153.

Barnard, J. G., Singh, S., Randolph, T. W., & Carpenter, J. F. (2011). Subvisible particle counting provides a sensitive method of detecting and quantifying aggregation of monoclonal antibody caused by freeze-thawing: Insights into the roles of particles in the protein aggregation pathway. *J. Pharm. Sci., 100,* 492–503.

Barth, H. G. & Boyes, B. E. (1992). Size exclusion chromatography. *Anal. Chem., 64,* 428R–442R.

Bauer, H. H., Muller, M., Goette, J., Merkle, H. P., & Fringeli, U. P. (1994). Interfacial adsorption and aggregation associated changes in secondary—Structure of human calcitonin monitored by ATR-FTIR spectroscopy. *Biochemistry, 33,* 12276–12282.

Bhambhani, A., Kissmann, J. M., Joshi, S. B., Volkin, D. B., Kashi, R. S., & Middaugh, C. R. (2012). Formulation design and high-throughput excipient selection based on structural integrity and conformational stability of dilute and highly concentrated IgG1 monoclonal antibody solutions. *J. Pharm. Sci., 101,* 1120–1135.

Bliznukov, O. P., Kozmin, L. D., Klimovich, V. B., Tsybovsky, Y. I., Kravchuk, Z. I., & Martsev, S. P. (2001). Thermodynamic stability and functional activity of tumor-associated antibodies. *Biochemistry (Mosc.), 66,* 27–33.

Blondelle, S. E., Buttner, K., & Houghten, R. A. (1992). Evaluation of peptide-peptide interactions using reversed-phase high-performance liquid chromatography. *J. Chromatogr., 625,* 199–206.

Blundell, T. L., Dodson, G. G., Dodson, E., Hodgkin, D. C., & Vijayan, M. (1971). X-ray analysis and the structure of insulin. *Recent Prog. Horm. Res., 27,* 1–40.

Bohidar, H. B. & Geissler, E. (1984). Static and dynamic light scattering from dilute insulin solutions. *Biopolymers, 23,* 2407–2417.

Branchu, S., Forbes, R. T., York, P., & Nyqvist, H. (1999). A central composite design to investigate the thermal stabilization of lysozyme. *Pharm. Res., 16,* 702–708.

Brandes, N., Welzel, P. B., Werner, C., & Kroh, L. W. (2006). Adsorption-induced conformational changes of proteins onto ceramic particles: Differential scanning calorimetry and FTIR analysis. *J. Colloid Interface Sci., 299,* 56–69.

Brewster, M. E., Hora, M. S., Simpkins, J. W., & Bodor, N. (1991). Use of 2-hydroxypropyl-β-cyclodextrin as a solubilizing and stabilizing excipient for protein drugs. *Pharm. Res., 8,* 792–795.

Brochon, J. C., Tauc, P., Merola, F., & Schoot, B. M. (1993). Analysis of a recombinant protein preparation on physical homogeneity and state of aggregation. *Anal. Chem., 65,* 1028–1034.

Bronsema, K. J., Bischoff, R., & van de Merbel, N. C. (2013). High-sensitivity LC-MS/MS quantification of peptides and proteins in complex biological samples: The impact of enzymatic digestion and internal standard selection on method performance. *Anal. Chem., 85,* 9528–9535.

Buck, M., Radford, S. E., & Dobson, C. M. (1993). A partially folded state of hen egg white lysozyme in trifluoroethanol: Structural characterization and implications for protein folding. *Biochemistry, 32,* 669–678.

Capelle, M. A., Gurny, R., & Arvinte, T. (2009). A high throughput protein formulation platform: Case study of salmon calcitonin. *Pharm. Res., 26,* 118–128.

Cauchy, M., D'Aoust, S., Dawson, B., Rode, H., & Hefford, M. A. (2002). Thermal stability: A means to assure tertiary structure in therapeutic proteins. *Biologicals, 30,* 175–185.

Ceska, M., Grossmuller, F., & Lundkvist, U. (1970). Solid-phase radioimmunoassay of insulin. *Acta Endocrinol., 64,* 111–125.

Challener, C. (2014, February). Tackling the challenge of higher-order structure determination. *Biopharm Int.,* 20–24.

Charman, S. A., McCrossin, L. E., & Charman, W. N. (1993). Validation of a peptide map for recombinant porcine growth hormone and application to stability assessment. *Pharm. Res., 10,* 1471–1477.

Chaudhuri, R., Cheng, Y., Middaugh, C. R., & Volkin, D. B. (2014). High-throughput biophysical analysis of protein therapeutics to examine interrelationships between aggregate formation and conformational stability. *AAPS J., 16,* 48–64.

Chou, P. Y. & Fasman, G. D. (1978). Empirical predictions of protein conformation. *Annu. Rev. Biochem., 47,* 251–276.

Coleman, N. J. & Craig, D. Q. M. (1996). Modulated temperature differential scanning calorimetry: A novel approach to pharmaceutical thermal analysis. *Int. J. Pharm., 135,* 13–29.

Colt, E. W. D., Becker, K. L., & Quatrale, A. C. W. (1971). A rapid radioimmunoassay for insulin. *Am. J. Clin. Pathol., 55,* 40–42.

Creighton, T. E. (1983a). Chemical nature of polypeptides. In *Proteins: Structures and molecular principles* (pp. 1–60). New York: W.H. Freeman & Co.

Creighton, T. E. (1983b). *Proteins: Structures and molecular properties.* New York: W.H. Freeman & Co.

Cueto, M., Dorta, M. J., Munguia, O., & Llabres, M. (2003). New approach to stability assessment of protein solution formulations by differential scanning calorimetry. *Int. J. Pharm., 252,* 159–166.

Damgaard, U. & Markussen, J. (1979). Analysis of insulins and related compounds by HPLC. *Horm. Metab. Res., 11,* 580–581.

Damm, H. C., Besch, P. K., & Goldwyn, A. J. (1966). Chemistry of amino acids. In *The handbook of biochemistry and biophysics* (pp. 39–51). Cleveland, OH: The World Publishing Company.

Das, T. K. (2012). Protein particulate detection issues in biotherapeutics development—Current status. *AAPS Pharm. Sci. Technol., 13,* 732–746.

Dasnoy, S., Le Bras, V., Preat, V., & Lemoine, D. (2012). High-throughput assessment of antigen conformational stability by ultraviolet absorption spectroscopy and its application to excipient screening. *Biotechnol. Bioeng., 109,* 502–516.

Defelice, C., Hayakawa, K., Watanabe, T., & Tanaka, T. (1993). High-performance gel-permeation chromatographic analysis of protein aggregation—Application to bovine carbonic anhydrase. *J. Chromatogr., 645,* 101–105.

Denton, J. B., Konishi, Y., & Scheraga, H. A. (1982). Folding of ribonuclease a from a partially disordered conformation. Kinetic study under folding conditions. *Biochemistry, 21,* 5155–5163.

Dick, L. W., Jr., Mahon, D., Qiu, D., & Cheng, K. C. (2009). Peptide mapping of therapeutic monoclonal antibodies: Improvements for increased speed and fewer artifacts. *J. Chromatogr. B Anal. Technol. Biomed. Life Sci., 877,* 230–236.

Dong, A. C., Kendrick, B., Kreilgard, L., Matsuura, J., Manning, M. C., & Carpenter, J. F. (1997). Spectroscopic study of secondary structure and thermal denaturation of recombinant human factor XIII in aqueous solution. *Arch. Biochem. Biophys., 347,* 213–220.

Dong, A. C., Prestrelski, S. J., Allison, S. D., & Carpenter, J. F. (1995). Infrared spectroscopic studies of lyophilization- and temperature induced protein aggregation. *J. Pharm. Sci., 84,* 415–424.

Dupin, P., Galinou, F., & Bayol, A. (1995). Analysis of recombinant human growth hormone and its related impurities by capillary electrophoresis. *J. Chromatogr. A, 707,* 396–400.

Dwenger, A. (1984). Radioimmunoassay: An overview. *J. Clin. Chem. Clin. Biochem., 22,* 883–894.

Elarini, S. K. & Tawashi, R. (1994). Morphological and fractal-based methods describing insulin zinc crystal habits. *Drug Dev. Ind. Pharm., 20,* 1–10.

Ettinger, M. J. & Timasheff, S. N. (1971a). Optical activity of insulin. I. On the nature of the circular dichroism bands. *Biochemistry, 10,* 824–830.

Ettinger, M. J. & Timasheff, S. N. (1971b). Optical activity of insulin. II. Effect of nonaqueous solvents. *Biochemistry, 10,* 831–840.

Fang, X. J., Fernando, Q., Ugwu, S. O., & Blanchard, J. (1995). An improved method for determination of acid dissociation constants of peptides. *Pharm. Res., 12,* 1423–1429.

Fisher, B. V. & Smith, D. (1986). HPLC as a replacement for the animal response assays for insulin. *J. Pharm. Biomed. Anal., 4,* 377–387.

Fisher, D. K. & Higgins, T. J. (1994). A sensitive, high-volume, colorimetric assay for protein phosphatases. *Pharm. Res., 11,* 759–763.

Ford, J. L. & Mann, T. E. (2012). Fast-scan DSC and its role in pharmaceutical physical form characterisation and selection. *Adv. Drug Deliv. Rev., 64,* 422–430.

Foster, C. M., Borondy, M., Padmanabhan, V., Schwartz, J., Kletter, G. B., Hopwood, N. J. et al. (1993). Bioactivity of human growth hormone in serum—Validation of an in vitro bioassay. *Endocrinology, 132,* 2073–2082.

Gabrielson, J. P., Arthur, K. K., Stoner, M. R., Winn, B. C., Kendrick, B. S., Razinkov, V. et al. (2010). Precision of protein aggregation measurements by sedimentation velocity analytical ultracentrifugation in biopharmaceutical applications. *Anal. Biochem., 396,* 231–241.

Gaillard, I., Martial, A., Marc, A., Engasser, J. M., & Fick, M. (1994). Analytical method optimization for protein determination by fast high-performance liquid chromatography. *J. Chromatogr. A, 679,* 261–268.

Genuth, S., Frohman, L. A., & Lebovitz, H. E. (1965). A radioimmunological assay method for insulin using insulin-^{125}I and gel filtration. *J. Clin. Endocrinol. Metab., 25,* 1043–1049.

Georgalis, Y., Umbach, P., Raptis, J., & Saenger, W. (1997). Lysozyme aggregation studied by light scattering. I. Influence of concentration and nature of electrolytes. *Acta Cryst., D53,* 691–702.

Goldberg, D. S., Bishop, S. M., Shah, A. U., & Sathish, H. A. (2011). Formulation development of therapeutic monoclonal antibodies using high-throughput fluorescence and static light scattering techniques: Role of conformational and colloidal stability. *J. Pharm. Sci., 100,* 1306–1315.

Goldenberg, D. P. & Creighton, T. E. (1984). Gel electrophoresis in studies of protein conformation and folding. *Anal. Biochem., 138,* 1–18.
Gonzalez, M., Murature, D. A., & Fidelio, G. D. (1995). Thermal stability of human immunoglobulins with sorbitol. A critical evaluation. *Vox Sang., 68,* 1–4.
Goodman, J. W. (1991). Immunoglobulin structure and function. In D. P. Stites & A. I. Terr (Eds.), *Basic and clinical immunology* (pp. 109–121). Appleton & Lange. New York: McGraw Hill.
Guo, J. & Carta, G. (2014). Unfolding and aggregation of a glycosylated monoclonal antibody on a cation exchange column. Part II. Protein structure effects by hydrogen deuterium exchange mass spectrometry. *J. Chromatogr. A, 1356,* 129–137.
Gursky, O., Badger, J., Li, Y., & Caspar, L. D. (1992). Conformational changes in cubic insulin crystals in the pH range 7–11. *Biophys. J., 63,* 1210–1220.
Hamel, F. G., Peavy, D. E., Ryan, M. P., & Duckworth, W. C. (1986). High performance liquid chromatographic analysis of insulin degradation by rat skeletal muscle insulin protease. *Endocrinology, 118,* 328–333.
Hamrang, Z., Rattray, N. J., & Pluen, A. (2013). Proteins behaving badly: Emerging technologies in profiling biopharmaceutical aggregation. *Trends Biotechnol., 31,* 448–458.
Hartree, A. S., Lester, J. B., & Shownkeen, R. C. (1985). Studies of the heterogeneity of human pituitary LH by fast protein liquid chromatography. *J. Endocrinol., 105,* 405–413.
Hawe, A., Sutter, M., & Jiskoot, W. (2008). Extrinsic fluorescent dyes as tools for protein characterization. *Pharm. Res., 25,* 1487–1499.
Hjelmeland, L. M. & Chrambach, A. (1981). Electrophoresis and electrofocusing in detergent containing media: A discussion of basic concepts. *Electrophoresis, 2,* 1–11.
Hoglund, U. (1988). Use of perfluorated carboxylic acids in the separation of metabolites of vasopressin prior to radioimmunoassay. *J. Chromatogr., 430,* 128–134.
Jain, S., Kumar, C. V., & Kalonia, D. S. (1992). Protein-peptide interactions as probed by tryptophan fluorescence: Serum albumins and enkephalin metabolites. *Pharm. Res., 9,* 990–992.
Jaksch, F. L. & Tay, S. (2014). Editorial overview: Analytical biotechnology: New technologies for quantitative analysis of biological specimens and natural products. *Curr. Opin. Biotechnol., 25,* 4–6.
Jen, A. & Merkle, H. P. (2001). Diamonds in the rough: Protein crystals from a formulation perspective. *Pharm. Res., 18,* 1483–1488.
Ji, Y. Y. & Li, Y. Q. (2010). The role of secondary structure in protein structure selection. *Eur. Phys. J. E Soft Matter, 32,* 103–107.
Jiang, Y., Li, C., Nguyen, X., Muzammil, S., Towers, E., Gabrielson, J. et al. (2011). Qualification of FTIR spectroscopic method for protein secondary structural analysis. *J. Pharm. Sci., 100,* 4631–4641.
Johnson, C. M. (2013). Differential scanning calorimetry as a tool for protein folding and stability. *Arch. Biochem. Biophys., 531,* 100–109.
Johnson, W. C. (1988). Secondary structure of proteins through circular dichroism spectroscopy. *Ann. Rev. Biophys. Biophys. Chem., 17,* 145–166.
Johnson, W. C., Jr., Pagano, T. G., Basson, C. T., Madri, J. A., Gooley, P., & Armitage, I. M. (1993). Biologically active Arg-Gly-Asp oligopeptides assume a type II β-turn in solution. *Biochemistry, 32,* 268–273.
Jones, A. J. S. (1993). Analysis of polypeptides and proteins. *Adv. Drug Deliv. Rev., 10,* 29–90.
Jones, L. S., Cipolla, D., Liu, J., Shire, S. J., & Randolph, T. W. (1999). Investigation of protein-surfactant interactions by analytical ultracentrifugation and electron paramagnetic resonance: The use of recombinant human tissue factor as an example. *Pharm. Res., 16,* 808–812.

Kadima, W., Ogendal, L., Bauer, R., Kaarsholm, N., Brodersen, K., Hansen, J. F. et al. (1993). The influence of ionic strength and pH on the aggregation properties of zinc-free insulin studied by static and dynamic laser light scattering. *Biopolymers, 33,* 1643–1657.

Katsila, T., Siskos, A. P., & Tamvakopoulos, C. (2012). Peptide and protein drugs: The study of their metabolism and catabolism by mass spectrometry. *Mass Spectrom. Rev., 31,* 110–133.

King, J. (1989, April). Deciphering the rules of protein folding. *Chem. Eng. News, 10,* 32–54.

Knip, M. (1984). Analysis of pancreatic peptide hormones by reversed-phase high-performance liquid chromatography. *Horm. Metab. Res., 16,* 487–491.

Kohnert, K. D., Schmid, E., Zuhlke, H., & Fiedler, H. (1973). Isoelectric focusing of insulins in polyacrylamide gel. *J. Chromatogr., 76,* 263–267.

Kosecki, R. (1988). Recirculating isoelectric focusing: A system for protein separations. *BioPharm, 1(6),* 28–31.

Koyoma, J., Nomura, J., Shiojima, Y., Ohtsu, Y., & Horii, I. (1992). Effect of column length and elution mechanism on the separation of proteins by reversed-phase high-performance liquid chromatography. *J. Chromatogr., 625,* 217–222.

Krayukhina, E., Uchiyama, S., & Fukui, K. (2012). Effects of rotational speed on the hydrodynamic properties of pharmaceutical antibodies measured by analytical ultracentrifugation sedimentation velocity. *Eur. J. Pharm. Sci., 47,* 367–374.

Krell, T., Greco, F., Nicolai, M. C., Dubayle, J., Renauld-Mongenie, G., Poisson, N. et al. (2003). The use of microcalorimetry to characterize tetanus neurotoxin, pertussis toxin and filamentous haemagglutinin. *Biotechnol. Appl. Biochem., 38,* 241–251.

Kwan, A. H., Mobli, M., Gooley, P. R., King, G. F., & Mackay, J. P. (2011). Macromolecular NMR spectroscopy for the non-spectroscopist. *FEBS J., 278,* 687–703.

Lai, C. S., Wolff, C. E., Novello, D., Griffone, L., Cuniberti, C., Molina, F. et al. (1993). Solution structure of human plasma fibronectin under different solvent conditions—Fluorescence energy transfer, circular dichroism and light-scattering studies. *J. Mol. Biol., 230,* 625–640.

Lebl, M. (1993). Observation of a conformational effect in peptide molecule by reversed-phase high-performance liquid chromatography. *J. Chromatogr., 644,* 285–287.

Li, C. H., Nguyen, X., Narhi, L., Chemmalil, L., Towers, E., Muzammil, S. et al. (2011). Applications of circular dichroism (CD) for structural analysis of proteins: Qualification of near- and far-UV CD for protein higher order structural analysis. *J. Pharm. Sci., 100,* 4642–4654.

Lin, S. Y. & Wang, S. L. (2012). Advances in simultaneous DSC-FTIR microspectroscopy for rapid solid-state chemical stability studies: Some dipeptide drugs as examples. *Adv. Drug Deliv. Rev., 64,* 461–478.

Linde, S., Welinder, B. S., & Hansen, B. (1986). Preparative reversed-phase high-performance liquid chromatography of iodinated insulin retaining full biological activity. *J. Chromatogr., 369,* 327–339.

Lloyd, L. F. & Corran, P. H. (1982). Analysis of insulin preparations by reversed-phase high-performance liquid chromatography. *J. Chromatogr., 240,* 445–454.

Maa, Y. F. & Hsu, C. C. (1996). Aggregation of recombinant human growth hormone induced by phenolic compounds. *Int. J. Pharm., 140,* 155–168.

Mach, H. & Arvinte, T. (2011). Addressing new analytical challenges in protein formulation development. *Eur. J. Pharm. Biopharm., 78,* 196–207.

Mach, H. & Middaugh, C. R. (2011). Ultraviolet spectroscopy as a tool in therapeutic protein development. *J. Pharm. Sci., 100,* 1214–1227.

Mackes, K., Itoh, M., Greene, K., & Gerich, J. (1981). Radioimmunoassay of human plasma somatostatin. *Diabetes, 30,* 728–734.

Malamud, D. & Drysdale, J. W. (1978). Isoelectric points of proteins: A table. *Anal. Biochem., 86,* 620–647.

Malik, K., Matejtschuk, P., Thelwell, C., & Burns, C. J. (2013). Differential scanning fluorimetry: Rapid screening of formulations that promote the stability of reference preparations. *J. Pharm. Biomed. Anal., 77,* 163–166.

Malzert, A., Boury, F., Renard, D., Robert, P., Proust, J. E., & Benoit, J. P. (2002). Influence of some formulation parameters on lysozyme adsorption and on its stability in solution. *Int. J. Pharm., 242,* 405–409.

Martindale, H., Marsh, J., Hallett, F. R., & Albisser, A. M. (1982). Examination of insulin formulations using quasi-elastic light scattering. *Diabetes, 31,* 364–366.

Mattison, K., Morfesis, A., & Kaszuba, M. (2003). A primer on particle sizing using dynamic light scattering. *Am. Biotechnol. Lab., 21(13)*, 20–22.

McEntire, J. (1994). Biotechnology product validation, Part 5: Selection and validation of analytical techniques. *BioPharm, 7,* 68–78.

Mellstedt, H. (2013). Clinical considerations for biosimilar antibodies. *EJC Suppl., 11,* 1–11.

Mesmin, C., Fenaille, F., Ezan, E., & Becher, F. (2011). MS-based approaches for studying the pharmacokinetics of protein drugs. *Bioanalysis, 3,* 477–480.

Micard, V. & Guilbert, S. (2000). Thermal behavior of native and hydrophobized wheat gluten, gliadin and glutenin-rich fractions by modulated DSC. *Int. J. Biol. Macromol., 27,* 229–236.

Miller, L. M., Bourassa, M. W., & Smith, R. J. (2013). FTIR spectroscopic imaging of protein aggregation in living cells. *Biochim. Biophys. Acta, 1828,* 2339–2346.

Mock, K. (1993). Routine sensitive peptide mapping using LC/MS of therapeutic proteins produced by recombinant DNA technology. *Peptide Res., 6,* 100–104.

Morgan, C. R. & Lazarow, A. (1963). Immunoassay of insulin: Two antibody system: Plasma insulin levels of normal, subdiabetic and diabetic rats. *Diabetes, 12,* 115–126.

Moriyama, Y., Sato, Y., & Takeda, K. (1993). Reformation of the helical structure of bovine serum albumin by the addition of small amounts of sodium dodecyl sulfate after the disruption of the structure by urea. *J. Colloid Interface Sci., 156,* 420–424.

Musial, S. P., Duran, M. P., & Smith, R. V. (1986). Analysis of solvent-mediated conformational changes of insulin by radioimmunoassay (RIA) techniques. *J. Pharm. Biomed. Anal., 4,* 589–600.

Nail, S. L., Jiang, S., Chongprasert, S., & Knopp, S. A. (2002). Fundamentals of freeze drying. In S. L. Nail & M. J. Akers (Eds.), *Development and manufacturing of protein pharmaceuticals* (pp. 281–360). New York: Kluwer Academic/Plenum Publishers.

Oellerich, M. (1984). Enzyme-immunoassay: A review. *J. Clin. Chem. Clin. Biochem., 22,* 895–904.

Ohno, M., Kai, M., & Ohkura, Y. (1989). High-performance liquid chromatographic determination of substance P-like arginine-containing peptide in rat brain by on-line post-column fluorescence derivatization with benzoin. *J. Chromatogr., 490,* 301–310.

Ohta, M., Tokunaga, H., Kimura, T., Satoh, H., & Kawamura, J. (1984). Analysis of insulins by high-performance liquid chromatography. III. Determination of insulins in various preparations. *Chem. Pharm. Bull., 32,* 4641–4644.

Ojeda, C. B. & Rojas, F. S. (2013). Recent applications in derivative ultraviolet/visible absorption spectroscopy: 2009–2011. *Microchem. J., 106*, 1–16.

Parsons, D. L. (1982). Fluorescence stability of human albumin solutions. *J. Pharm. Sci., 71,* 349–351.

Patapoff, T. W., Mrsny, R. J., & Lee, W. A. (1993). The application of size exclusion chromatography and computer simulation to study the thermodynamic and kinetic parameters for short-lived dissociable protein aggregates. *Anal. Biochem., 212,* 71–78.

PDR Staff. (2014). *Physicians' desk reference* (68th ed.). Montvale, NJ: PDR Network, LLC.

Pedersen, J. S., Hansen, S., & Bauer, R. (1994). The aggregation behavior of zinc-free insulin studied by small-angle neutron scattering. *Eur. Biophys. J., 22,* 379–389.

Peter, A., Szepesi, G., Balaspiri, L., & Burger, K. (1987). Coordinative interactions in the separation of insulin and its derivatives by high-performance liquid chromatography. *J. Chromatogr., 408,* 43–52.

Peters, B. H., Molnar, F., & Ketolainen, J. (2014). Structural attributes of model protein formulations prepared by rapid freeze-drying cycles in a microscale heating stage. *Eur. J. Pharm. Biopharm., 87*(2), 347–56.

Philo, J. S. (2006). Is any measurement method optimal for all aggregate sizes and types? *AAPS J., 8,* E564–E571.

Pina, A. S., Lowe, C. R., & Roque, A. C. (2014). Challenges and opportunities in the purification of recombinant tagged proteins. *Biotechnol. Adv., 32,* 366–381.

Powers, M. E., Adjei, A., Lu, M. Y. F., & Manning, M. C. (1994). Solution behavior of leuprolide acetate, an LHRH agonist, as determined by circular dichroism spectroscopy. *Int. J. Pharm., 108,* 49–55.

Pritchett, T. J. (1996). Capillary isoelectric focusing of proteins. *Electrophoresis, 17,* 1195–1201.

Privalov, P. L. & Dragan, A. I. (2007). Microcalorimetry of biological macromolecules. *Biophys. Chem., 126,* 16–24.

Rabel, S. R. & Stobaugh, J. F. (1993). Applications of capillary electrophoresis in pharmaceutical analysis. *Pharm. Res., 10,* 171–186.

Redfield, C., Smith, L. J., Boyd, J., Lawrence, G. M. P., Edwards, R. G., Gershater, C. J. et al. (1994). Analysis of the solution structure of human interleukin-4 determined by heteronuclear three-dimensional nuclear magnetic resonance techniques. *J. Mol. Biol., 238,* 23–41.

Rehfeld, J. F. (1978). Immunochemical studies on cholecystokinin II. Distribution and molecular heterogeneity in the central nervous system and small intestine of man and hog. *J. Biol. Chem., 253,* 4022–4030.

Rich, J. R. & Withers, S. G. (2009). Emerging methods for the production of homogeneous human glycoproteins. *Nat. Chem. Biol., 5,* 206–215.

Riggin, R. M., Dorulla, G. K., & Miner, D. J. (1987). A reversed-phase high-performance liquid chromatographic method for characterization of biosynthetic human growth hormone. *Anal. Biochem., 167,* 199–209.

Riggin, R. M., Shaar, C. J., Dorulla, G. K., Lefeber, D. S., & Miner, D. J. (1988). High-performance size-exclusion chromatographic determination of the potency of biosynthetic human growth hormone products. *J. Chromatogr., 435,* 307–318.

Rossmann, M. G. & Argos, P. (1981). Protein folding. *Ann. Rev. Biochem., 50,* 497–532.

Sadana, A. (1987). Enzyme folding and unfolding. *J. Biotechnol., 6,* 107–133.

Sali, A., Shakhnovich, E., & Karplus, M. (1994). Kinetics of protein folding—A lattice model study of the requirements for folding to the native state. *J. Mol. Biol., 235,* 1614–1636.

Sarciaux, J. M., Mansour, S., Hageman, M. J., & Nail, S. L. (1999). Effects of buffer composition and processing conditions on aggregation of bovine IgG during freeze-drying. *J. Pharm. Sci., 88,* 1354–1361.

Schoneich, C., Kwok, S. K., Wilson, G. S., Rabel, S. R., Stobaugh, J. F., Williams, T. D. et al. (1993). Separation and analysis of peptides and proteins. *Anal. Chem., 65,* R67–R84.

Shah, B., Jiang, X. G., Chen, L., & Zhang, Z. (2014). LC-MS/MS peptide mapping with automated data processing for routine profiling of N-glycans in immunoglobulins. *J. Am. Soc. Mass Spectrom., 25,* 999–1011.

Sharma, B. (2007). Immunogenicity of therapeutic proteins. Part 2: Impact of container closures. *Biotechnol. Adv., 25,* 318–324.

Sharma, D. K., Oma, P., & Krishnan, S. (2009, April). Silicone microdroplets in protein formulations. *Pharm. Technol., 33*(4), 74–79.

Sluzky, V., Shahrokh, Z., Stratton, P., Eberlein, G., & Wang, Y. J. (1994). Chromatographic methods for quantitative analysis of native, denatured, and aggregated basic fibroblast growth factor in solution formulations. *Pharm. Res., 11,* 485–490.

Smirnova, E., Safenkova, I., Stein-Margolina, V., Shubin, V., & Gurvits, B. (2014). Can aggregation of insulin govern its fate in the intestine? Implications for oral delivery of the drug. *Int. J. Pharm., 471,* 65–68.

Smith, D. J., Venable, R. M., & Collins, J. (1985). Separation and quantitation of insulins and related substances in bulk insulin crystals and in injectables by reversed-phase high performance liquid chromatography and the effect of temperature on the separation. *J. Chromatogr. Sci., 23,* 81–88.

Stein, S. (1989). Primary structure analysis of recombinant proteins. *BioPharm, 2(2),* 30–37.

Strege, M. A. & Lagu, A. L. (1993a). Capillary electrophoresis as a tool for the analysis of protein folding. *J. Chromatogr. A, 652,* 179–188.

Strege, M. A. & Lagu, A. L. (1993b). Capillary electrophoretic protein separations in polyacrylamide-coated silica capillaries and buffers containing ionic surfactants. *J. Chromatogr., 630,* 337–344.

Stromqvist, M. (1994). Peptide mapping using combinations of size-exclusion chromatography, reversed-phase chromatography and capillary electrophoresis. *J. Chromatogr. A, 667,* 304–310.

Surewicz, W. K., Mantsch, H. H., & Chapman, D. (1993). Determination of protein secondary structure by Fourier transform infrared spectroscopy: A critical assessment. *Biochemistry, 32,* 389–394.

Szabo, O. & Mahler, R. J. (1970). The effects of epinephrine, amino acids and albumin on the measurement of insulin by radioimmunoassay. *Horm. Metab. Res., 2,* 125–130.

Szepesi, G. & Gazdag, M. (1981). Improved high-performance liquid chromatographic method for the analysis of insulins and related compounds. *J. Chromatogr., 218,* 597–602.

Takeda, K., Hamada, S., & Wada, A. (1993). Secondary structural changes of large and small fragments of bovine serum albumin in thermal denaturation and in sodium dodecyl sulfate denaturation. *J. Protein Chem., 12,* 223–228.

Takeda, K., Sasa, K., Kawamoto, K., Wada, A., & Aoki, K. (1988). Secondary structure changes of disulfide bridge-cleaved bovine serum albumin in solutions of urea, guanidine hydrochloride, and sodium dodecyl sulfate. *J. Colloid Interface Sci., 124,* 284–289.

Takeda, K. & Wada, A. (1991). Secondary structural changes of *N*-bromosuccinimide-cleaved bovine serum albumin in solutions of sodium dodecyl sulfate, urea, and guanidine hydrochloride. *J. Colloid Interface Sci., 144,* 45–52.

Takeda, K., Wada, A., & Moriyama, Y. (1990). Secondary structural changes of chymotrypsinogen A, alpha-chymotrypsin, and the isolated polypeptides Cys(1)-Leu(13), Ile(16)-Tyr(146), and Ala(149)-Asn(245) in sodium dodecyl sulfate, urea and guanidine hydrochloride. *Colloid Polym. Sci., 268,* 612–617.

Takeda, K., Wada, A., Yamamoto, K., Moriyama, Y., & Aoki, K. (1989). Conformational change in bovine serum albumin by heat treatment. *J. Protein Chem., 8,* 653–659.

Thannhauser, T. W., Sherwood, R. W., & Scheraga, H. A. (1998). Determination of the cysteine and cystine content of proteins by amino acid analysis: Application to the characterization of disulfide-coupled folding intermediates. *J. Protein Chem., 17,* 37–43.

Thies, M. J., Kammermeier, R., Richter, K., & Buchner, J. (2001). The alternatively folded state of the antibody C(H)3 domain. *J. Mol. Biol., 309,* 1077–1085.

Trethewey, J. (1989). Bio-assays for the analysis of insulin. *J. Pharm. Biomed. Anal., 7,* 189–197.

Tsai, P. K., Volkin, D. B., Dabora, J. M., Thompson, K. C., Bruner, M. W., Gress, J. O. et al. (1993). Formulation design of acidic fibroblast growth factor. *Pharm. Res., 10,* 649–659.

United States Pharmacopoeia Convention (2013). *The United States pharmacopeia* (28th ed.). Rockville, MD: United States Pharmacopeial Convention, Inc.

van de Waterbeemd, H., Karajiannis, H., & El Tayar, N. (1994). Lipophilicity of amino acids. *Amino Acids, 7,* 129–145.

van de Weert, M., Haris, P. I., Hennink, W. E., & Crommelin, D. J. (2001). Fourier transform infrared spectrometric analysis of protein conformation: Effect of sampling method and stress factors. *Anal. Biochem., 297,* 160–169.

van den Berg, E. M. M., Jans, A. W. H., & Binst, G. V. (1986). Conformational properties of somatostatin. VI. In a methanol solution. *Biopolymers, 25,* 1895–1908.

van Veelen, P. A., Jimenez, C. R., Li, K. W., Geraerts, W. P. M., Tjaden, U. R., & van der Greef, J. (1994). New advanced analytical tools in peptide/protein research: Matrix assisted laser desorption mass spectrometry. *J. Control. Release, 29,* 223–229.

Vashist, S. K., Marion, S. E., Lam, E., Hrapovic, S., & Luong, J. H. (2014). One-step antibody immobilization-based rapid and highly-sensitive sandwich ELISA procedure for potential in vitro diagnostics. *Sci. Rep., 4,* 4407.

Verdonck, E., Schaap, K., & Thomas, L. C. (1999). A discussion of the principles and applications of modulated temperature DSC (MTDSC). *Int. J. Pharm., 192,* 3–20.

Vermeer, A. W., Bremer, M. G., & Norde, W. (1998). Structural changes of IgG induced by heat treatment and by adsorption onto a hydrophobic Teflon surface studied by circular dichroism spectroscopy. *Biochim. Biophys. Acta, 1425,* 1–12.

Vermeer, A. W. & Norde, W. (2000). The thermal stability of immunoglobulin: Unfolding and aggregation of a multi-domain protein. *Biophys. J., 78,* 394–404.

Vermeer, A. W., Norde, W., & van Amerongen, A. (2000). The unfolding/denaturation of immunogammaglobulin of isotype 2b and its F(ab) and F(c) fragments. *Biophys. J., 79,* 2150–2154.

Vigh, G., Varga-Puchony, Z., Hlavay, J., & Papp-Hites, E. (1982). Factors influencing the retention of insulins in reversed-phase high-performance liquid chromatographic systems. *J. Chromatogr., 236,* 51–59.

Vlaanderen, I., Van Grondelle, R., & Bloemendal, M. (1993). The use of fast protein liquid (size exclusion) chromatography for the fractionation of crystallins and the study of β-crystallin aggregation. *J. Liquid Chromatogr., 16,* 367–382.

Vorauer, K., Skias, M., Trkola, A., Schulz, P., & Jungbauer, A. (1992). Scale-up of recombinant protein purification by hydrophobic interaction chromatography. *J. Chromatogr., 625,* 33–39.

Wang, W. (2000). Lyophilization and development of solid protein pharmaceuticals. *Int. J. Pharm., 203,* 1–60.

Watson, E. & Kenney, W. C. (1988). High-performance size-exclusion chromatography of recombinant derived proteins and aggregated species. *J. Chromatogr., 436,* 289–298.

Watson, E., Shah, B., DePrince, R., Hendren, R. W., & Nelson, R. (1994). Matrix-assisted laser desorption mass spectrometric analysis of a pegylated recombinant protein. *Biotechniques, 16,* 278–280.

Welfle, K., Misselwitz, R., Hausdorf, G., Hohne, W., & Welfle, H. (1999). Conformation, pH-induced conformational changes, and thermal unfolding of anti-p24 (HIV-1) monoclonal antibody CB4–1 and its F_{ab} and F_c fragments. *Biochim. Biophys. Acta, 1431,* 120–131.

Welinder, B. S. (1971). Isoelectric focusing of insulin and insulin derivatives. *Acta Chem. Scand., 25,* 3737–3742.

Welinder, B. S. (1980). Gel permeation chromatography of insulin. *J. Liquid Chromatogr., 3,* 1399–1416.

Welinder, B. S., Linde, S., & Hansen, B. (1985). Reversed-phase high-performance liquid chromatography of insulin and insulin derivatives. A comparative study. *J. Chromatogr., 348,* 347–361.

Wen, J., Arthur, K., Chemmalil, L., Muzammil, S., Gabrielson, J., & Jiang, Y. (2012). Applications of differential scanning calorimetry for thermal stability analysis of proteins: Qualification of DSC. *J. Pharm. Sci., 101,* 955–964.

West, M. H. P., Wu, R. S., & Bonner, W. M. (1984). Polyacrylamide gel electrophoresis of small peptides. *Electrophoresis, 5,* 133–138.

Wetlaufer, D. B. (1962). Ultraviolet spectra of proteins and amino acids. *Adv. Protein Chem., 17,* 303–390.

Winder, A. F. & Gent, W. L. G. (1971). Correction of light scattering errors in spectrophotometric protein determinations. *Biopolymers, 10,* 1243–1252.

Wuthrich, K. (1989). Protein structure determination in solution by nuclear magnetic resonance spectroscopy. *Science, 243,* 45–50.

Yalow, R. S. (1973). Radioimmunoassay methodology: Application to problems of heterogeneity of peptide hormones. *Pharm. Rev., 25,* 161–178.

Yang, A.-S., Gunner, M. R., Sampogna, R., Sharp, K., & Honig, B. (1993). On the calculation of pK_as in proteins. *Protein Struct. Funct. Gen., 15,* 252–265.

Yonezawa, K., Yokono, K., Yaso, S., Hari, J., Amano, K., Kawase, Y. et al. (1986). Degradation of insulin by insulin-degrading enzyme and biological characteristics of its fragments. *Endocrinology, 118,* 1989–1996.

Young, L., Jernigan, R. L., & Covell, D. G. (1994). A role for surface hydrophobicity in protein-protein recognition. *Protein Sci., 3,* 717–729.

Youssef, A. M. & Winter, G. (2013). A critical evaluation of microcalorimetry as a predictive tool for long term stability of liquid protein formulations: Granulocyte colony stimulating factor (GCSF). *Eur. J. Pharm. Biopharm., 84,* 145–155.

Zhu, D., Porter, W., Long, M., Fraunhofer, W., Gleason, K., & Gao, Y. (2010). Using heat conduction microcalorimetry to study thermal aggregation kinetics of proteins. *Thermochim. Acta, 499,* 1–7.

Zink, T., Ross, A., Ambrosius, D., Rudolph, R., & Holak, T. A. (1992). Secondary structure of human granulocyte colony-stimulating factor derived from NMR spectroscopy. *FEBS Lett., 314,* 435–439.

Zolls, S., Tantipolphan, R., Wiggenhorn, M., Winter, G., Jiskoot, W., Friess, W. et al. (2012). Particles in therapeutic protein formulations, Part 1: Overview of analytical methods. *J. Pharm. Sci., 101,* 914–935.

3 Stability of Therapeutic Peptides and Proteins

3.1 INTRODUCTION

A major challenge to the formulation of peptides and proteins into efficacious dosage forms is to ensure their stability over their shelf life (Manning, Chou, Murphy, Payne, & Katayama, 2010). Typically, a shelf life of at least 1.5–2 years at room temperature storage or refrigeration storage is desirable. Storage stability in the frozen state is less relevant as commercial distribution channels are not equipped for frozen products. Most proteins are marketed as lyophilized products, and most of these need refrigerated storage (Nail, Jiang, Chongprasert, & Knopp, 2002). The safety and efficacy of the product must be maintained over the shelf life during storage and handling, and it is realized that some degradation takes place. Some degradation products may have more impact on safety than others and therefore should be treated differently. Impurities resulting from deamidation and aggregation are relatively common, and the percent impurity allowed can be discussed with the Food and Drug Administration (FDA) on a case-by-case basis at an early stage. There are currently no regulatory guidelines for acceptable levels of such impurities. Toxicity testing may be required if the impurity levels are high and/or if the degradation products may likely be toxic.

Instability of peptides and proteins may be broadly classified as physical instability or chemical instability. Physical instability refers to a change in the secondary, tertiary, or quaternary structure of a protein and includes denaturation, aggregation, precipitation, and/or adsorption to surfaces. Chemical instability involves covalent modification of the protein via bond formation or cleavage. Chemical instability is an outcome of reactions such as hydrolysis, deamidation, oxidation, disulfide exchange, β-elimination, and racemization (Manning, Patel, & Borchardt, 1989). Chemical and physical instability degradation pathways most commonly encountered are typically deamidation, oxidation, and aggregation. Physical and chemical instability reactions are tied to each other, and one may lead to the other. A peptide or protein drug comprising of several amino acid residues may possess multiple reactive sites, susceptible to chemical reactions. This amplifies any degradation that may take place, even though the degradation at any one site is relatively small. Though one or more of the aforementioned reactions can occur for a given peptide, not all of these are sufficiently rapid to compromise the shelf life of a peptide formulation at its pH of maximum stability (typically, pH 5–7) (Powell, 1994). The objective of this chapter is to discuss those reactions that are most relevant to pharmaceutical systems. Another relevant topic, accelerated stability testing, will be discussed in Chapter 4.

3.2 PHYSICAL INSTABILITY

3.2.1 Denaturation

Denaturation refers to a disruption of the higher-order structure, such as secondary and tertiary structure of a protein. Denaturation, which may be reversible or irreversible, can be caused by thermal stress, extremes of pH, or denaturing chemicals. Denaturation typically involves unfolding (see Chapter 2) of the protein. Usually, unfolding involves a sharp transition in structure from native to an unfolded state at the melting temperature. The melting temperature, T_m, is defined as the temperature that represents the midpoint of unfolding. The midpoint refers to the state where 50% of the molecules are completely unfolded. The sharp transition between native and unfolded molecules implies that denaturation is a cooperative phenomenon where disruption of any significant portion of the folded structure leads to the unfolding of the entire molecule. However, the T_m is not a fixed value as it is affected by a wide variety of environmental factors, especially pH changes.

Though unfolding involves a sharp transition from native to unfolded state, intermediate states can exist in some systems. Such systems will show a biphasic kinetics for unfolding. For example, the biphasic unfolding kinetics observed for human interferon suggests the existence of at least one unfolding intermediate in the unfolding pathway. The mechanism for reversible and irreversible thermal denaturation of interferon or other proteins may be represented as

$$\begin{array}{ccccc} N & \leftrightarrow & U^* & \leftrightarrow & U \\ & & \downarrow & & \\ & & X & & \end{array}$$

where

N and U are the native and unfolded states, respectively
U^* is an unfolding intermediate
X is a collection of inactive molecules (e.g., aggregated protein) that are kinetically and/or thermodynamically blocked from changing to N or U (Mulkerrin & Wetzel, 1989)

If the unfolded protein cannot easily recover its native state by refolding, then denaturation is considered to be irreversible. As discussed in Chapter 2, the folding is determined by the amino acid sequence or the primary structure of the protein. Failure of a protein to refold often results when the protein has undergone chemical or conformational changes. This may occur when the primary structure of a protein has been altered due to chemical instability. Sometimes, a protein may fold, but not to its correct native structure. Use of denaturants such as guanidine hydrochloride or urea, followed by dialysis, can be used to recover the native state of such a misfolded protein (Manning et al., 1989). A protein that exhibits reversible unfolding is typically small (<30 kDa), dilute (<1 mg/mL), and highly charged to inhibit aggregation (Kendrick, Tiansheng, & Chang, 2002).

3.2.2 Aggregation

Protein molecules can often undergo self-association by physical or chemical forces to form dimers, trimers, tetramers, or higher oligomers. This self-association or aggregation is a common problem during formulation development and pharmaceutical processing. Aggregation may occur due to intrinsic factors related to structure or due to extrinsic factors related to processing stress or environmental changes (Wang, Nema, & Teagarden, 2010), and it may occur at any point throughout the lifetime of a protein (Mahler, Friess, Grauschopf, & Kiese, 2009). While proteins in their folded state are only marginally stable, they are very stable in aggregated form, which partly drives the tendency toward aggregation (Roberts, 2014). Although aggregation is a physical instability, it can lead to loss of biological activity. However, some proteins may exert their biological effects as a dimer, trimer, or higher oligomeric forms and may lose biological activity upon dissociation. For instance, tumor necrosis factor exists as a compact trimer in aqueous solutions and maintenance of its trimeric structure is essential for its activity (Hlodan & Pain, 1994). In any case, a study of the aggregation phenomena is crucial to developing a successful peptide or protein formulation. Even if aggregation does not result in loss of bioactivity, formation of insoluble aggregates can cause blockage of tubing, membranes, or pumps in an infusion set, as is the case with insulin. Furthermore, since aggregation leads to an increase in effective molecular weight, the aggregated protein may be more immunogenic. The dimers, trimers, and oligomers of a protein, when freely soluble, are referred to as soluble aggregates and are typically in the 1–100 nm size range. Protein aggregates can also be insoluble. These insoluble aggregates are opalescent and can be measured by light scattering in the visible range. However, several new terms have been used in recent years. Generally, particles are defined as having a size above 0.1 μm and are further classified as subvisible (0.1–100 μm) and visible particles (above 100 μm). Submicron (0.1–1 μm) particles are a subgroup of subvisible particles (Zolls et al., 2012). These particles can include not just aggregates but any other particles in the formulation as well since all particles may influence product quality and need to be analyzed. Due to pharmacopoeial requirements of parenteral products, particles above 10 μm have received attention for a long time, but subvisible particles below 10 μm have received greater attention only in recent years. Potential immunogenicity of these particles has been the subject of debate between academic and industrial scientists and is now receiving increasing scrutiny from regulatory agencies (Carpenter et al., 2009; Ratanji, Derrick, Dearman, & Kimber, 2014; Singh et al., 2010; Wang et al., 2012). The morphology of these particles is variable, ranging from almost spherical to long, irregular fibers, and their formation is governed by slow or delayed-onset kinetics. This makes it difficult to characterize the aggregates and ensure stability of the product over the intended shelf life (Ripple & Dimitrova, 2012).

Soluble aggregates can be quantitated by various chromatographic techniques such as size-exclusion chromatography (SEC) or laser light scattering techniques or more advanced techniques such as analytical ultracentrifugation (see Chapter 2). Microflow imaging has been successfully used to detect and quantify subvisible particles (Barnard, Singh, Randolph, & Carpenter, 2011). Capillary zone electrophoresis equipped with laser-induced fluorescence detection has also been used to study aggregates (Jensen, Lee, & King, 1998). Nonreduced sodium dodecyl sulfate polyacrylamide gel electrophoresis (SDS-PAGE) can also be used to qualitatively study soluble aggregates. For example, interleukin (IL)-1 receptor type I, a 312-amino-acid recombinant glycoprotein, was run on nonreduced gels after the formulation was exposed to 30°C or 50°C for 7 days in formulation pH ranging from 3.0 to 9.0. The bands seen in various lanes (with reference to molecular weight markers) were used to determine conditions that induce aggregates (high-molecular-weight species) or breakdown (low-molecular-weight species) products. pH 6 was found to be the optimum pH for minimum aggregation and breakdown (Remmele, Nightlinger, Srinivasan, & Gombotz, 1998).

Both soluble and insoluble aggregates may involve covalent and/or noncovalent linkages (Eckhardt, Oeswein, & Bewley, 1991). Precipitation may be considered as the macroscopic equivalent of aggregation. The absence of any undefined aggregates in a protein solution is essential prior to its IV administration. A variety of techniques can be used to study the aggregation behavior of therapeutic peptides and proteins. Whether the aggregate is covalent or noncovalent will be the primary determinant of a suitable analytical technique. The use of light scattering, fluorescence spectroscopy, or SEC in the analysis of protein aggregates has been discussed in Chapter 2.

3.2.2.1 Mechanism of Aggregation

Generally, aggregation is a two-step process. The first step involves unfolding of the protein molecule, thereby exposing the buried, hydrophobic amino acid residues to the aqueous solvent. In the second step, the hydrophobic residues of the unfolded protein molecules undergo association, leading to protein aggregation. Such association takes place in order to minimize the unfavorable exposure of hydrophobic residues in the unfolded protein to water. In accordance with this mechanism, aggregation is a polymolecular, concentration-dependent process that obeys higher-order kinetics. A decrease in protein concentration lowers the rate and extent of thermally induced aggregation. Several studies have investigated the mechanisms of protein aggregation (Renthal & Velasquez, 2002; Xu, Tsai, & Nussinov, 1998). In addition to conformational stability, colloidal stability resulting from protein–protein interactions is also important for protein aggregation. If aggregation is being driven by colloidal rather than conformational stability, the protein should be stabilized against attractive intermolecular forces, possibly by adjustment of ionic strength, pH, and buffer type (Chi, Krishnan, Randolph, & Carpenter, 2003; Goldberg, Bishop, Shah, & Sathish, 2011).

In general, when processing macromolecules, energy input from any source could induce aggregation to some extent. The higher the energy input, the more the aggregation. Energy transfer is required to unfold the protein molecule. This can be induced by a temperature increase, radiation, ultrasound, or another kind of energy input (Hawe, Kasper, Friess, & Jiskoot, 2009). It can also originate from chemical changes in the environment of the protein such as changes in pH, salt concentration, or solvent composition (Persson & Gekas, 1994). Since a critical nucleation event may be required to induce precipitation, a lag time may be observed in the kinetics of protein aggregation. Proteins, which typically are 30–50 Å in diameter, coagulate into 0.1–0.2 µm primary particles, which then flocculate to form 10–20 µm precipitates. Protein aggregation has also been explained in terms of a surface free energy model that presumes that changes in surface free energy drive protein aggregation reactions. These changes can be calculated from changes in the area and charge density of the protein–water interface, with aggregation being energetically equivalent to the loss of the protein–water interface. A large positive entropy change drives aggregation and results from the release of bound water molecules when aggregation takes place (Blank, 1994). Although model systems give important information, they make several assumptions to simplify the aggregation mechanisms. This is especially true since several inactivation reactions may occur simultaneously (Lencki, Arul, & Neufeld, 1992a; Lencki, Arul, & Neufeld, 1992b). Since these mechanisms are quite complex, real systems investigated under real conditions are likely to result in more useful information (Persson & Gekas, 1994).

Aggregation can also take place within the biological cells during protein folding (King, 1989; Nguyen & Bensaude, 1994; Seckler & Jaenicke, 1992). Once the polypeptide chain is synthesized, it may interact with other proteins to form large protein aggregates called inclusion bodies, which may have diameters as high as 1 µm. These inclusion bodies can be dissociated by agents such as urea or guanidine hydrochloride. These agents unfold the proteins, and these can then be refolded to the correct native structure. Aggregation may also occur due to interaction of partially folded or misfolded proteins. The fact that hGH displays a concentration-dependent aggregation during folding may be directly related to the mechanism of inclusion body formation when this protein is overproduced in bacteria (Defelippis, Alter, Pekar, Havel, & Brems, 1993). The cytoplasm of *Escherichia coli* is very reducing and thus unsuitable for the formation of disulfide bonds. Therefore, it is likely that when large eukaryotic proteins are synthesized in *E. coli*, disulfide bonds are not formed. The resulting polypeptide may then aggregate as a result of hydrophobic or covalent interactions between protein molecules. When recombinant proteins are expressed in bacteria, inclusion bodies often form and several techniques are used to recover the native protein from these aggregates. The insoluble protein pellets are separated from other cellular components by homogenization and centrifugation. The pellets are then solubilized by denaturants such as urea or guanidine hydrochloride to unfold the protein. The denaturant is then removed to aid the refolding of the protein. Aggregation must be prevented during the refolding of the protein. Refolding often takes place through an intermediate molten globule structure that may associate because it has hydrophobic regions on its surface. Bovine and porcine growth hormones are known to fold through a stable folding intermediate that is characterized as compact and largely ∀-helical but lacking the native tertiary structure of the protein. This intermediate can undergo aggregation at high concentrations (Bastiras & Wallace, 1992; Lehrman et al., 1991).

Although stable equilibrium intermediates have been identified for nonhuman species of growth hormone, no intermediate has been identified for the folding of the human growth hormone (hGH) (Brems, Brown, & Becker, 1990). Equilibrium intermediates have also been identified in the folding pathway of human insulin in monomer-inducing solvent and in two monomeric insulin analogs. The presence of these intermediates has important stability implications as certain unfolded regions may alter the susceptibility of the insulin structure to chemical degradation. Also, the intermediate structure may play an important role in the insulin gelation/fibrillation process. Knowledge of the intermediate may allow insulin mutations to be specifically directed toward stabilizing or destabilizing these intermediates. This may result in increased shelf life of commercial insulin products, which is currently limited to approximately 2 years due to chemical and physical degradation (Millican & Brems, 1994).

Polyethylene glycol (PEG) has been used to assist the refolding because of its unique property to bind with the protein in its unfolded form and yet be preferentially excluded from a folded native protein. Thus, PEG binds to the intermediate form and prevents its association. Once folding has taken place, PEG drives the equilibrium to a compact native state as it is then excluded from the structure (Cleland & Wang, 1990).

A comprehensive review of hydrophobic interactions in protein conformation and stability and in other cellular processes has been published (Blokzijl & Engberts, 1993). Aggregation of endogenous physiological proteins can have important implications in health and disease. Protein misfolding and aggregation phenomena appear to be the underlying cause for the formation of insoluble plaques, characteristic of a number of diseases such as amyloidosis (Miller, Bourassa, & Smith, 2013). For example, a major feature of Alzheimer's disease is the presence of amyloid plaques in the brain tissue of patients. A β-amyloid peptide is the principal proteinaceous component of these amyloid deposits, and the neurotoxicity of the peptide is related to its aggregation (Barrow, Yasuda, Kenny, & Zagorski, 1992; Clements, Walsh, Williams, & Allsop, 1994; Mantyh et al., 1993).

3.2.2.2 Protein Fibrillation

Protein fibrillation refers to protein aggregates forming in the shape of long fibers. Typically, the protein molecules first form an elongated fibrous shape termed protofilament, and then two of these protofilaments twist around each other in a helical manner similar to DNA, forming the protofibril. Several protofibrils then twist around each other, forming the fibrils (van de Weert & Randolph, 2013). Fibrillation has been observed with proteins such as insulin, glucagon, and human calcitonin. Studies with glucagon, a 29-residue peptide hormone, have shown that even subtle changes in conditions can change the type of fibrils formed and that overall fibrillation is a complex phenomenon (Pedersen et al., 2006). Formation of fibrils is a major challenge to the therapeutic use of glucagon. Nile red staining has been used to detect aggregation and fibrillation of human calcitonin. Nile red samples indicated the gradual appearance of small bright objects that represent growing aggregates, even though no signal increase was detected by ultraviolet (UV) absorbance or fluorescence intensity for the first 2 h (Figure 3.1) (Mach & Arvinte, 2011). Fibrillation of insulin is discussed in detail in Section 4.5.1.

3.2.2.3 Aggregation Behavior in Liquid and Solid States

Aggregation can take place in the liquid, solid, or intermediate state. Examples of protein aggregation in the liquid state will be discussed in Chapter 4. In this section,

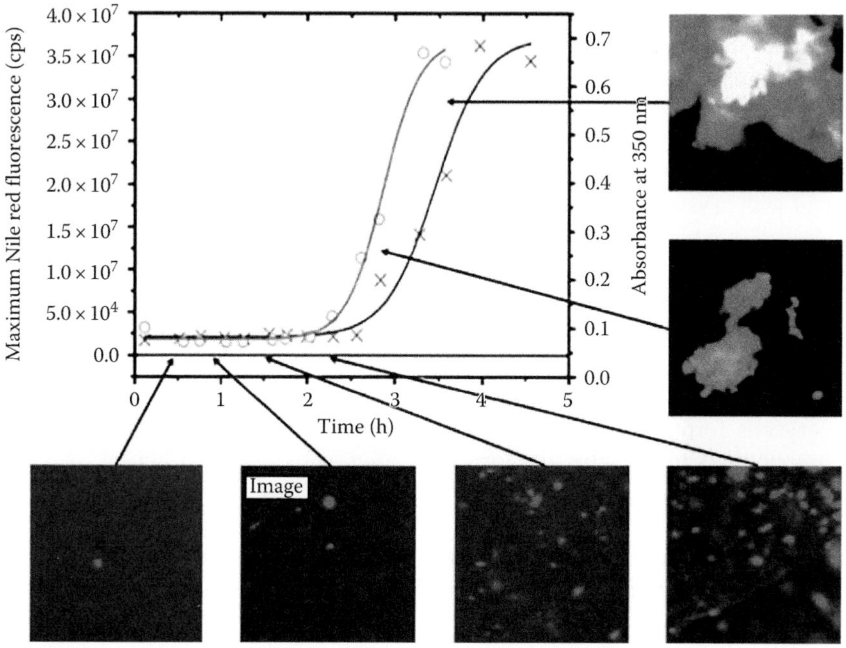

FIGURE 3.1 **(See color insert.)** Fibrillation of human calcitonin followed by microscopy, Nile red fluorescence, and UV absorbance. Nile red fluorescence (o) detected gelation phenomena earlier than UV absorbance (X), but only the microscopic study of Nile red–stained solutions revealed the presence and progression of aggregates in the early lag phase period. (Reproduced from Mach, H. and Arvinte, T., *Eur. J. Pharm. Biopharm.*, 78, 196, 2011. With permission.)

aggregation in the solid and intermediate states will be discussed. Most commercially available therapeutic peptides and proteins are available as lyophilized powder. While lyophilization may reduce instability, it does not necessarily prevent all pathways of physical and chemical degradation. Lyophilized powders may undergo aggregation during storage, especially in the presence of moisture. Bovine serum albumin (BSA) has been used as a model protein to investigate its moisture-induced aggregation in the solid state (Jordan, Yoshioka, & Terao, 1994; Katakam & Banga, 1995). As the moisture was increased in the solid state, a bell-shaped curve was obtained for the aggregation. With increasing moisture, the reaction rate increased up to monolayer coverage, but as the moisture is increased beyond monolayer coverage, the dilution of the reactants acts to inhibit the reaction. The aggregates involved covalent bonds and could be dissolved in a solution of 6 M urea and 10 mM dithiothreitol. The urea loosens the aggregates, making all covalent bonds accessible to the thiol reagent, thereby solubilizing the aggregates (Katakam & Banga, 1995). Similarly, moisture-induced aggregation has been reported in lyophilized insulin (Flores-Fernandez, Sola, & Griebenow, 2009). A rational approach to the stabilization of proteins in the solid state includes specifically targeting the mechanisms involved, maintaining optimum moisture levels, and increasing the physical stability of the lyophilized protein (Costantino, Langer, & Klibanov, 1994).

The liquid crystalline state or mesophase is an intermediate state between liquid and solid states. Two main types of liquid crystals are termed smectic (soap- or grease-like) and nematic (threadlike). Liquid crystals are mobile and have the flow properties of liquids. At the same time, they exhibit birefringence, a property associated with crystals. In birefringence, polarized light passing through the material is divided into two components with different velocities and hence different refractive indices. While the surfactant–water–oil systems commonly encountered in pharmacy form smectic structures, some hydrophobic peptides can form nematic peptide liquid crystals (PLCs). Formulations of detirelix, a hydrophobic luteinizing hormone–releasing hormone (LHRH) peptide, readily form nematic PLCs in aqueous solutions. The onset of PLC formation is concentration dependent, with no crystals forming below 4 mg/mL. Nafarelin, another LHRH analog, is less hydrophobic than detirelix and forms liquid crystals only at 8 mg/mL or above. The parent peptide, LHRH, is hydrophilic and does not exhibit birefringence even at 30 mg/mL concentration levels. In contrast, nafarelin and detirelix gel immediately at this concentration. Thus, PLC formation may be a useful early indicator of peptide aggregation and is driven by hydrophobic self-association. In order to produce stable formulations, it is desirable to prevent the formation of PLCs. It is difficult to prevent PLC formation by adding buffer counterions due to the opposing kinetic and thermodynamic forces for salts of different hydrophobicity. However, organic cosolvents such as 10%–15% of propylene glycol can suppress the formation of PLCs (Powell, Fleitman, Sanders, & Si, 1994; Powell, Sanders, Rogerson, & Si, 1991).

3.2.3 Adsorption

As discussed in Chapter 2, although the folded protein is hydrophilic, it does have hydrophobic residues that lie in the interior of the molecule. At a hydrophobic interface, the polypeptide chain can unfold to allow the hydrophobic part of the molecule to interact with the surface, leading to adsorption. The surface activity of proteins due to their being amphiphilic polyelectrolytes is largely responsible for adsorption to surfaces. Potential adsorption of therapeutic proteins to commonly encountered surfaces during processing and storage should be a part of preformulation stress screening (Bee et al., 2009; Rabe, Verdes, & Seeger, 2011). As discussed earlier, such adsorption may also be followed by aggregation. Thus, it can be expected that the primary sequence of the polypeptide chain will determine the adsorption, although some masking effect will exist due to the folded 3D structure of the protein. Adsorption is important to understand as a protein will encounter various surfaces as it is being produced, formulated, shipped, and used. These surfaces include steel tanks, glass vials or ampuls, rubber stoppers, Teflon, and IV administration set tubing.

Although the size, charge, structure, and other chemical properties of proteins will influence surface activity, they are all related to the primary sequence of the polypeptide chain. Size is considered important since larger proteins are believed to form multiple contact points upon adsorption to a surface (Horbett & Brash, 1987). This would suggest that a protein that is likely to unfold readily would be more surface active. In contrast, a disulfide cross-linked protein is less likely to unfold and will thus be less surface active. Such structural factors are likely to influence surface activity though more studies need to be done to establish such relationships. The diversity in surface activity of peptides and proteins primarily depends on the nature of the side chains of constituent amino acids. Hydrophobic analogs of LHRH, such as nafarelin and detirelix, adsorb strongly to glass and other surfaces at low concentrations (Anik & Hwang, 1983). Adsorption of the F_{ab} and F_c fragments of intact IgG molecule has been reported to occur independently, apparently due to the higher hydrophobicity of the F_{ab} fragment (Vermeer, Giacomelli, & Norde, 2001). In addition to the hydrophobicity of the protein, its polyelectrolyte behavior is also an important determinant of its adsorption tendency. Depending on the pH and ionic strength of the media, charge interactions between protein and surface will differ considerably (Hartvig, van de Weert, Ostergaard, Jorgensen, & Jensen, 2011). As expected, proteins adsorb the most to neutral or slightly charged surfaces near their isoelectric point. Based on this diversity in adsorption, it can be expected that if a mixture of proteins is allowed to adsorb onto a surface, the adsorbed layer will be richer in some proteins than others.

Horbett (1992) has provided a comprehensive discussion of the adsorption process. The adsorption process shows a saturation effect, with a plateau value close to that expected for a close-packed monolayer of protein (~0.1 to 0.5 μg/cm^2, depending on the protein diameter). Actual values may differ by a factor of two or three, based on available surface area, molecule orientation, and packing. Another important characteristic of the adsorption process is its irreversibility, which may be because of multiple contact points between protein and the surface. Simple rinsing of the surface is unlikely to remove the protein, unless a detergent is included. The ability of the detergent, sodium dodecyl sulfate, to remove albumin or fibrinogen from polymer surfaces was found to decrease as the time between adsorption and elution was increased. This was believed to result because of a molecular spreading in the postadsorption phase that allowed more contacts per molecule, making removal more difficult (Horbett & Brash, 1987). While this irreversibility of adsorption poses some problems, it does have the advantage that any denaturation upon adsorption is limited to the fraction adsorbed. In other words, the protein is not going to adsorb and then desorb so as to effectively denature the entire protein content of the solution. However, surfactants or even the presence of excess protein will allow some exchange of the adsorbed protein back into solution.

The kinetics of adsorption to solid surfaces will often consist of a rapid initial phase followed by a slower phase as adsorption approaches a steady-state value. Eventually, the solid surface may be 1000 times more concentrated than the bulk phase. A method to study the adsorption behavior of proteins at interfaces has been described by Burgess et al. (1991, 1992). This method uses an interfacial rheological technique, which permits the measurement of structural–mechanical properties of the adsorbed layers. These properties are related to the rate of interfacial adsorption, interfacial interactions, and conformational changes in the adsorbed layers and serve as an indication of denaturation of adsorbed molecules. This method has been used to study competition between proteins (such as BSA or human immunoglobulin G) and surfactants (such as Tween 80 or lecithin) for adsorption on interfaces. Denaturation is likely to be greater at the air–water interface as compared to a solid–water interface, due to the mobility of the former. For example, shaking allows a continuous creation of a new interface, thus providing a massive surface area that can lead to large-scale denaturation and/or aggregation. Some peptides, such as the neuropeptides, also have a tendency to adsorb at the oil–water interface. The antibacterial activity of the family of peptides called polymyxins is related to their affinity for the cell membrane, which leads to lysis.

Loss of insulin due to adhesion to surfaces is an obstacle to all development of insulin formulations and delivery systems. Adsorption losses are very significant for dilute solutions but become less of a concern at concentrations higher than 5 IU/mL (approximately 0.2 mg/mL, 0.03 mM) as the percentage lost becomes negligible. Insulin adsorption may be minimized by the addition of 0.1%–1% of albumin. The adsorption rate of lysozyme has been reported to increase as the pH becomes close to its isoelectric point. By monitoring charge screening effects induced by the addition of salts, it was observed that the adsorption of lysozymes at the air–water interface increased when the electrostatic interactions decreased (Malzert et al., 2002).

3.3 CHEMICAL INSTABILITY

Chemical pathways of peptide and protein degradation include hydrolysis, deamidation, oxidation, disulfide exchange, β-elimination, and racemization. While several studies have explored specific mechanisms of degradation, others have made empirical observations on degradation of polypeptides. An example of the latter is the observation that salmon calcitonin (sCT) degrades as a function of pH and temperature, with four main degradation peaks observed by reversed-phase gradient high-performance liquid chromatography (HPLC). The degradation reaction followed first-order kinetics and maximum stability was achieved at pH 3.3 (Lee, Lee, Song, Chun, & DeLuca, 1992). A subsequent mechanistic study has reported that sCT degrades via hydrolysis resulting in the cleavage of the 1–2 amide bond, deamidation of Gln^{14} and Gln^{20} residues, disulfide exchange, and dimerization (Windisch et al., 1997).

Although most studies are carried out in aqueous solutions, it should be realized that the in vivo stability of a peptide in human serum should also be evaluated. Limited stability studies in human serum and synovial fluid have been reported for the peptides that bind with the histocompatibility complex. These peptides are in development for the novel treatment of selected autoimmune disorders such as rheumatoid arthritis and insulin-dependent diabetes (Powell et al., 1993; Powell, Grey, Gaeta, Sette, & Colon, 1992). Some of the more important chemical pathways of peptide or protein degradation will be discussed in the following sections. Some of the other major protein degradation pathways or problems include N- or C-terminal heterogeneity, N-terminal modification resulting in diketopiperazine formation, desialylation, phosphorylation, or sulfation.

3.3.1 Oxidation

Oxidation is one of the major degradation pathways of peptides, both in solution and lyophilized formulations. Amino acids that may undergo oxidation include methionine, cysteine, histidine, tryptophan, and tyrosine (Ji, Zhang, Cheng, & Wang, 2009). Most oxidation reactions commonly encountered in therapeutic proteins under normal storage conditions involve methionine and/or cysteine residues. Methionine is easily oxidized, even by atmospheric oxygen, to methionine sulfoxide. Under extreme oxidative conditions, it forms methionine sulfone. Methionine can undergo auto-oxidation, chemical oxidation, and photooxidation. Auto-oxidation implies *uncatalyzed* thermal oxidation, but actually it is governed by trace amounts of peroxide, metal ions, light, base, and free radical initiators. Since these may be present as impurities in formulation excipients, prediction of auto-oxidation is not always possible (Cleland, Powell, & Shire, 1993; Manning et al., 1989; Powell, 1994). In addition to trace metals, light can also catalyze oxidation of methionine (Fransson, 1997). The presence of sodium chloride in a recombinant humanized monoclonal antibody has been reported to cause an increase in oxidation at higher temperatures, resulting from corrosion of stainless steel containers used at low pH used for the formulation that in turn generated iron ions that catalyzed the oxidation of methionine (Lam, Yang, & Cleland, 1997).

Protein oxidation by hydrogen peroxide is especially relevant to the pharmaceutical industry since several commonly used excipients such as the polysorbate surfactants often contain trace amounts of peroxides. The peroxide is partly present as an impurity from the manufacturing process and is partly generated during storage due to degradation of polysorbates. Peroxides and other chemicals may also be leached out from stoppers, IV tubings, or IV bags (Chang et al., 2010). Peroxides react with methionine residues to produce methionine sulfoxide via a nucleophilic substitution reaction (Meyer, Ho, & Manning, 2002). A low peroxide grade should be used for these excipients (see Chapter 4).

Other excipients (such as PEG) or leachable materials from the container closure system (such as gray silicon rubber stoppers) may also be a source of peroxides. The tertiary structure of a protein may protect some methionine residues from oxidation, for example, hGH undergoes oxidation at Met 14 and Met 125 positions but not at Met 170. This is because Met 170 is located in the interior of the folded native state. Separation of the oxidized variants of hGH can be done by RP-HPLC (Teshima & Canova-Davis, 1992). Alpha$_1$-antitrypsin (AAT) is known to be very unstable in solutions. Besides aggregation and thermal lability, AAT undergoes oxidation of exposed methionine residue. It has been reported that inactivated AAT could be reverted to 80% active AAT by dialysis against 0.2 M mercaptoethanol, a reducing agent (Yu, Roosdorp, & Pushpala, 1988).

Cysteine is also easily oxidized to yield cystine disulfide. During long-term storage, free sulfhydryl groups may be oxidized to intrachain or interchain disulfide linkages. Such interchain bonds lead to protein aggregation (Arakawa, Prestrelski, Kenney, & Carpenter, 1993). Once the disulfide is formed, it is much less susceptible to reaction with mild oxidants, as compared to methionine or cysteine. Thus, proteins containing unpaired cysteine are more likely to undergo oxidation and should be formulated and packaged under nitrogen to prolong their shelf life.

As in the case of oxidation of methionine, oxidation of cysteine is also affected by the conformation of the protein. The cysteine residues of recombinant human interleukin-1β (rhIL-1β) are unreactive, indicating that they are buried in the protein interior and thus inaccessible to oxidation. While oxidation of a protein does not necessarily result in loss of biological activity, there is at least a partial loss of activity in most cases. Oxidation may or may not affect protein antigenicity depending on whether conformational changes took place or not. In general, oxidation of methionine to methionine sulfoxide does not affect protein antigenicity as this reaction does not result in conformational changes. However, oxidation of cysteine residues to form disulfide bridges can lead to large changes in antigenicity, due to significant conformational changes.

Although freezing or lyophilization will generally reduce oxidation, it may sometimes increase oxidation. As freezing proceeds, the exclusion of protein from the ice matrix increases the available concentration of cysteine residues, with possible increase in oxidation. Also, the lowered temperature may increase the oxygen concentration in the solutions.

Oxidation may be minimized by the use of antioxidants such as phenolic compounds (BHT, BHA, propyl gallate, and vitamin E), reducing agents (methionine, ascorbic acid, sodium sulfite, thioglycerol, and thioglycolic acid), or chelating agents (ethylenediaminetetraacetic acid [EDTA], citric acid, and thioglycolic acid) (Cleland et al., 1993). The choice of antioxidants may actually be very limited as the phenolic compounds are not sufficiently soluble in water, while some reducing agents may induce degradation in the long run. Methionine oxidation in a recombinant monoclonal antibody was prevented by antioxidants such as methionine, sodium thiosulfate, catalase, or platinum. However, an adduct was formed when thiosulfate was used. The molar ratio of protein to antioxidant required to inhibit temperature-induced oxidation with methionine was 1:5 (Lam et al., 1997).

Other approaches to prevent oxidation include nitrogen flush, refrigeration, protection from light, and adjustment of pH. The stability of a recombinant factor VIII derivative was improved when the oxygen in the overhead space in the vial was replaced by nitrogen (Fatouros, Osterberg, & Mikaelsson, 1997). Replacement of cysteine with serine offers another possibility to prevent such oxidation reactions. Serine, a stable amino acid with a hydroxy group, mimics the size and polarity of cysteine, and the serine mutant often retains full biological activity (Wang, 1992).

3.3.2 Hydrolysis

The peptide bond can undergo hydrolysis, resulting in peptide degradation in a formulation. The Asp-Pro peptide bonds are known to be most susceptible to hydrolytic breakdown. However, the peptide bond is otherwise fairly unreactive, except at extremes of pH. Recombinant human macrophage colony-stimulating factor (rhM-CSF) was shown to degrade in acidic solutions by peptide cleavage at aspartate$_{169}$-proline$_{170}$ and aspartate$_{213}$-proline$_{214}$ positions. Interestingly, a third potential cleavage site (aspartate$_{45}$-proline$_{46}$) did not degrade under conditions that cleaved the other Asp-Pro sites. Most likely, this results from the Asp$_{45}$-Pro$_{46}$ sequence being buried in a tightly folded protein region while the other Asp-Pro pairs reside in exposed surface locations (Schrier et al., 1993). Another possibility is that the primary structure in the region adjacent to Asp-Pro residues may also influence the hydrolytic reaction. A similar observation was made with rhIL-11. The predominant degradation pathway of rhIL-11 at 50°C in an acidic solution was hydrolytic cleavage at Asp_{133}-Pro_{134} position. However, no hydrolysis was observed at the Asp_{12}-Pro_{13} position. Again, this selectivity of Asp-Pro hydrolytic cleavage may be attributed to the tertiary folded structure or the primary structure at the regional location of Asp-Pro bonds (Kenley & Warne, 1994). In some crystalline suspensions of insulin, a cleavage of the peptide bond Thr_{A8}-Ser_{A9} occurs. This peptide cleavage takes place only in preparations containing rhombohedral crystals in addition to free zinc ions and is species dependent: bovine > porcine > human insulin (Brange, Langkjaer, Havelund, & Volund, 1992). Hydrolysis can be minimized by using cosolvents, formulating at optimal pH and ionic strength, and storing at low temperature.

3.3.3 Deamidation

Another common hydrolytic reaction responsible for degradation of peptides and proteins is the nonenzymatic deamidation of Asn and Gln residues. The amide groups of asparaginyl or glutaminyl (Asn and Gln) residues are labile at extremes of pH and may be hydrolyzed easily to a free carboxylic acid (Asp and Glu, respectively). Though the resulting products are also naturally occurring amino acid residues, deamidation may have significant effects on protein bioactivity, half-life, conformation, aggregation, and/or immunogenicity. The effects have to be evaluated on a case-by-case basis as deamidation does not always affect bioactivity. However, even if the bioactivity is not affected, the deamidated protein may have a faster clearance from the body, for example, an unfolded protein that has been deamidated has lost its original primary sequence and may not be able to refold to its native state. Such damaged protein will then be effectively cleared by the body and will also be more susceptible to irreversible aggregation (Cleland et al., 1993).

The rate of deamidation of a protein depends on its primary, secondary (Xie & Schowen, 1999), and tertiary structure, in addition to other factors such as temperature, pH, buffer species, ionic strength, and special intermolecular interactions. Deamidation of Asn and Gln residues is an acid- and base-catalyzed hydrolysis so that the reaction rate is minimum at a particular pH. In general, deamidation is catalyzed by base, heat, and ionic strength and inhibited by the addition of organic solvents. The pH of maximum stability for most deamidation reactions is around pH 6. Though it is difficult to generalize, it seems that at room temperature, the half-lives for peptide deamidation would limit the shelf life of formulations to 2 years or less. If the peptide contains the hydrolytically labile Asp-Pro bond, the shelf life may be limited to about 0.4 years (Powell, 1994) though using a lower pH of about 4–5 may help to improve stability against deamidation as well as hydrolysis of Asp-Pro bonds.

Generally, Asn residues are more prone to deamidation than Gln residues. Insulin contains six residues that may undergo deamidation, Gln^{A5}, Gln^{A15}, Asn^{A18}, Asn^{A21}, Asn^{B3}, and Gln^{B4}. Of these, the three asparagine residues are the most labile, especially the C-terminal residue A^{21}. Thus, insulin undergoes rapid hydrolytic degradation in acid solutions due to extensive deamidation at residue Asn^{21}, resulting in monodesamido-(A21)-insulin. In neutral solutions, some deamidation still takes place at residue Asn^{B3}, resulting in a mixture of isoAsp and Asp derivatives (Brange et al., 1992; Darrington & Anderson, 1994; Fisher & Porter, 1981).

Adrenocorticotropic hormone (ACTH), a single-chain polypeptide containing 39 amino acids, contains a single Asn residue at position 25 that is known to deamidate under alkaline conditions. For a thorough study of the factors that influence deamidation of Asn in ACTH, investigators have used a hexapeptide (Val-Tyr-Pro-Asn-Gly-Ala). This Asn-Gly hexapeptide is based on residues 22–27 of ACTH. At low pH, the deamidation involves direct hydrolysis of the Asn residue to generate only the Asp peptide. However, in neutral or alkaline pH, both the Asp and isoAsp peptides are produced due to the formation of an intermediate cyclic imide. Depending on which bond is hydrolyzed, the imide yields either the normal α-linked (Asp) or isomerized β-linked (isoAsp) peptide. The rate of Asp-X and Y-Asp amide bond hydrolysis was not affected by the size of the amino acid on the C-terminal side of the Asp residue. This may be attributed to the reaction site of the degradation pathway being remote from the C-terminal residue. In contrast, C-terminal substitution of Gly with more bulky residues (Ser, Val) affected the rate and amount of cyclic imide being produced (Oliyai & Borchardt, 1993; Oliyai & Borchardt, 1994). Since deamidation is a hydrolysis reaction, elimination of water by lyophilization is expected to stop or decrease deamidation. However, literature has reported that the residual moisture present in the solid state will still allow deamidation reaction to take place (Patel & Borchardt, 1990).

Studies with the Asp hexapeptide (Val-Tyr-Pro-Asp-Gly-Ala) in lyophilized formulations suggest that chemical instability in the solid state follows a pseudo-first-order reversible kinetics. The pH of the starting solution did not have a statistically significant impact on the mean degradation rate constant, but stability was affected by residual moisture content and/or temperature of the lyophilized formulations. Lactose, an amorphous bulking agent, was found to be better for stability of the hexapeptide as compared to mannitol (Oliyai, Patel, Carr, & Borchardt, 1994). Pramlintide, a synthetic analog of human amylin, produced nine deamidation and four hydrolysis products when subjected to stress at 40°C for 45 days (Hekman et al., 1998). If conventional strategies such as alteration of pH, addition of excipients, or lyophilization cannot control the deamidation reaction, a last-resort strategy can be the use of genetic engineering. By recombinant DNA technology, the Asn residues can be selectively eliminated and replaced by other residues, provided that conformation and bioactivity of the protein can be maintained. Ser residue has been used for this purpose with some success because of its similar size and potential hydrogen donor functional group (Cleland et al., 1993).

3.3.4 β-Elimination and Disulfide Exchange

Under thermal stress, a protein will often undergo a destruction of the disulfide bonds by β-elimination from cystine residues, resulting in thiols that may contribute to other degradation pathways (Manning et al., 1989). Degradation of rhM-CSF under alkaline conditions is believed to take place by a β-elimination mechanism that involves parallel cleavage and intramolecular cross-linking reactions (Schrier et al., 1993). The generation of the free thiols by β-elimination may in turn catalyze disulfide interchange. Disulfide exchange may also result due to the presence of unpaired cysteine residues. These cysteine residues can react at different sites to form new disulfide bridge(s), resulting in proteins with incorrect disulfide linkages and *nonnative* conformation. The reaction is base catalyzed and promoted by mercaptoethanol, which is sometimes used as an antioxidant. It can be prevented by thiol scavengers such as p-mercuribenzoate, *N*-ethylmaleimide, or copper ions (Manning et al., 1989; Wang, 1992).

3.3.5 Racemization

All amino acids (except glycine) are chiral at the carbon bearing the side chain and may undergo base-catalyzed racemization. Racemization will generate D-enantiomers that may result in loss of bioactivity. The D-enantiomers are often more resistant to proteolytic enzymes and may improve the stability of the resulting peptide. Racemization of the Asp residue is generally much faster than that of other residues, probably due to its unique mechanism that involves the formation of a cyclic imide (Manning et al., 1989).

3.3.6 Thermal Stability

Storage at normal refrigeration temperatures (5°C) may not be enough to prevent the pathways of physical and chemical degradation of peptides and proteins. For example, storage of bovine insulin at −20°C is necessary to prevent deamidation and polymerization reactions (Fisher & Porter, 1981). Thermal stress to a protein will often induce denaturation, aggregation, or deamidation reactions, as already discussed. Freezing also produces stress to a protein by a different mechanism, as will be discussed in Chapter 5.

3.3.7 Multiple Pathways of Chemical Instability

In most cases, degradation will involve more than just one reaction. For example, deamidation at specific asparaginase residues and oxidation of a cysteine and several methionines were involved in degradation of Orthoclone OKT3, a monoclonal antibody used to prevent rejection of organ transplants (Kroon, Baldwin-Ferro, & Lalan, 1992). Furthermore, these pathways of degradation may or may not affect the bioactivity of the protein. Deamidation of OKT3 resulted in charge shifts visible by isoelectric focusing but did not affect its binding affinity, because most sites of deamidation were located in the constant regions of the immunoglobulin. In contrast, oxidative mechanisms more seriously affected the binding affinity of OKT3 (Rao & Kroon, 1993). Similarly, chimeric L6, a mouse–human monoclonal antibody, degrades by three routes: chemical degradation to smaller-molecular-weight species, irreversible aggregation, and formation of a reversible dimer (Paborji, Pochopin, Coppola, & Bogardus, 1994). Fragmentation of human therapeutic immunoglobulins (IVIGs) has also been reported to occur during manufacture due to contamination with proteases from human blood (Page et al., 1995).

Two of the most common chemical degradation pathways involve oxidation and deamidation. Becker et al. (1988) isolated the degradation products resulting from storage of biosynthetic hGH at 40°C for 2 weeks. Five degradation components were found, with the two major ones being sulfoxide and desamido derivatives. Interestingly, these products retained full biological activity, and their immunoreactivity was essentially identical to that of an hGH reference standard.

Although the overall reaction scheme is complicated, some predictions of instability may be made by the structure of the protein. The primary structure will reveal potential sites of chemical degradation. For example, a molecule of rhIL-11 includes two methionines (possible oxidation sites), a single asparaginase (possible deamidation site), seven glutamines (possible deamidation sites), and two aspartyl–prolyl couples (possible peptide bond cleavage sites) (Kenley & Warne, 1994). Peptide sequences that may be predictive of a specific degradation pathway, and its anticipated effect on the shelf life of a product, have been described in the literature (Powell, 1994).

Although proteins often undergo physical or chemical instability, it should be realized that some small peptides may have sufficient stability for formulation of a simple aqueous parenteral solution (Wang, 1992). Several problems relating to the stability of the protein could be resolved if protein fragments that retain biological activity can be used. Such smaller functional analogs may also have an improved solubility.

The degradation pathways discussed here are more likely to be encountered during formulation and pharmaceutical processing of therapeutic peptides and proteins. In addition, other degradation mechanisms will be encountered during drug delivery, such as degradation by proteolytic enzymes in stomach. Selective degradation may be encountered, for example, at a molar ratio of 172:1 (insulin/enzyme), chymotrypsin caused almost complete degradation of insulin within 40 min, while very little insulin was degraded by trypsin under similar conditions. Four of the amino acids involved in the proteolytic degradation of insulin are essential for its receptor binding, thus making it difficult to resort to chemical modification of insulin to make it resistant to such proteolytic attack (Schilling & Mitra, 1991). The implications of such degradation on the oral absorption of proteins are discussed in Chapter 7.

Furthermore, enzymatic barriers will be encountered in blood and other organs throughout the body. Measurements of peptide stability in vitro in plasma or serum can be done by adding peptide to plasma or serum, incubating at 37°C, and assaying for the amount lost. Such analysis may be a useful predictor of peptide stability under in vivo conditions. Unstable peptides, which do not circulate in the blood for more than a few minutes, are unlikely to be developed into viable therapeutic agents (Powell, 1994). It has been proposed that the chemical instability may also be reduced by stabilizing the structure of the protein via interaction with excipients (Meyer et al., 2002).

3.4 BIOSIMILARS AND COMPARABILITY PROTOCOLS

Recently, considerable attention has been focused on *biosimilars*, also known as *follow-on biologics*, or *subsequent-entry biologics*, since patent protection on many blockbuster recombinant proteins has begun to expire. Over 20 of the top-selling biopharmaceuticals will come off patent by 2020, creating significant opportunity for biosimilars, with an estimated market size for biosimilars in excess of $ 50 billion (DiPaola, 2013). Biosimilars have already been on the European market for many years, with first approval in 2006, and over a dozen approvals since then, for biosimilars of recombinant insulin, hGH, GCSF, erythropoietin, and the monoclonal antibody infliximab. Unlike generic products, a biosimilar may not be substituted for the innovator product unless it is categorized by FDA as interchangeable. Some states are already passing regulations to control such substitution. In 2012, the FDA issued three-draft guidance on biosimilar product development, based on the 2009 biologics price competition and innovation act (*Biosimilars Act*). FDA is taking a no *one-size-fits-all* approach, looking at the *totality of evidence* (Pittman, 2013; Rathore, 2009; Warner, 2003). Regulatory aspects of biosimilars in Europe are also discussed in the literature (Zuniga & Calvo, 2009).

Demonstrating bioequivalence to the reference product is difficult and expensive for various reasons. Some of these reasons include the potential need to prove that a biosimilar made by a different process is bioequivalent to the reference product in the absence of clear regulatory guidelines and the possibility of patent infringement lawsuits relating to process or other product extension type patents. A battery of physicochemical tests will be required to determine that the product has an identical folded structure as the approved product. A quality target product profile (QTPP), based on testing done on the reference product, will be required at the onset of the process. Novel analytical approaches have been used for comparability studies (Federici, Lubiniecki, Manikwar, & Volkin, 2013). One measure of protein structure for a well-characterized protein could be its resistance to thermal denaturation (Cauchy, D'Aoust, Dawson, Rode, & Hefford, 2002). In addition to having the same profile as the innovator product in terms of amino acid sequence, potency, and efficacy, the biosimilar must also have an impurity profile that does not significantly affect safety, immunogenicity, or other similar considerations. Similarly, the presence of subvisible particles or other aggregates should also be compared. While it may be difficult to have an *identical* product, it should be feasible to have a *comparable* product. FDA may have specific concerns about physicochemical and biological characterization, comparison of impurity profiles, effect on downstream processes, and effect on process controls and controls of intermediates and/or in-process materials. Comparability needs to be demonstrated using various lots of the reference product, purchased on the open market over a range of time and with different expiration dates (Li & Qiu, 2013; Pittman, 2013).

Innovator companies have filed citizen petitions with FDA with the argument that biotechnology-derived products are inseparable from the manufacturing process used to produce them. Since the follow-on manufacturer does not have access to the originator's cell bank or exact fermentation/purification process, small differences may change clinical profile. Many of the analytical tests may be using special reagents developed to look for specific impurities that are unique to the manufacturing process and may only be used to detect minor and well-controlled changes to the innovator company's own manufacturing process. A generic biopharmaceutical developed by a different manufacturing process may have a different impurity and cannot be tested by the analytical method used by the innovator company. It has been suggested that biosimilars should be assessed by globally harmonized preclinical and clinical studies and also by postmarketing pharmacovigilance programs (Owens, Landgraf, Schmidt, Bretzel, & Kuhlmann, 2012) and closely monitored for side effect profile and immunogenicity relative to the reference product (Kessler, Goldsmith, & Schellekens, 2006; Roger & Mikhail, 2007; Schellekens, 2009).

3.5 CONCLUSIONS

Therapeutic peptides and proteins can degrade by several physical and chemical pathways. Physical instability may occur by denaturation, aggregation, or adsorption. Pathways of chemical instability include oxidation, hydrolysis, deamidation, β-elimination, disulfide exchange, and racemization. In most cases, more than one pathway of physical and/or chemical instability is responsible for the degradation of peptides and proteins. The primary structure will often reveal potential sites of chemical degradation. Physical instability is more difficult to predict from primary structure. However, if most residues are hydrophobic amino acids, it does suggest a strong tendency toward adsorption and aggregation. For proteins, the secondary and tertiary structure may be more useful predictors of physical stability. Comparability protocols and regulatory guidance documents for well-characterized biologics are now allowing the introduction of biogenerics into the world market.

REFERENCES

Anik, S. T. & Hwang, J. Y. (1983). Adsorption of D-Nal(2)6 LHRH, a decapeptide, onto glass and other surfaces. *Int. J. Pharm., 16,* 181–190.

Arakawa, T., Prestrelski, S. J., Kenney, W. C., & Carpenter, J. F. (1993). Factors affecting short-term and long-term stabilities of proteins. *Adv. Drug Del. Rev., 10,* 1–28.

Barnard, J. G., Singh, S., Randolph, T. W., & Carpenter, J. F. (2011). Subvisible particle counting provides a sensitive method of detecting and quantifying aggregation of monoclonal antibody caused by freeze-thawing: Insights into the roles of particles in the protein aggregation pathway. *J. Pharm. Sci., 100,* 492–503.

Barrow, C. J., Yasuda, A., Kenny, P. T. M., & Zagorski, G. (1992). Solution conformations and aggregational properties of synthetic amyloid beta-peptides of Alzheimer's disease: Analysis of circular dichroism spectra. *J. Mol. Biol., 225,* 1075–1093.

Bastiras, S. & Wallace, J. C. (1992). Equilibrium denaturation of recombinant porcine growth hormone. *Biochemistry, 31,* 9304–9309.

Becker, G. W., Tackitt, P. M., Bromer, W. W., Lefeber, D. S., & Riggin, R. M. (1988). Isolation and characterization of a sulfoxide and a desamido derivative of biosynthetic human growth hormone. *Biotechnol. Appl. Biochem., 10,* 326–337.

Bee, J. S., Chiu, D., Sawicki, S., Stevenson, J. L., Chatterjee, K., Freund, E. et al. (2009). Monoclonal antibody interactions with micro- and nanoparticles: Adsorption, aggregation, and accelerated stress studies. *J. Pharm. Sci., 98,* 3218–3238.

Blank, M. (1994). Protein aggregation reactions: Surface free energy model. *J. Theor. Biol., 169,* 323–326.

Blokzijl, W. & Engberts, J. B. F. N. (1993). Hydrophobic effects—Opinions and facts. *Angew. Chem. Int. Ed., 32,* 1545–1579.

Brange, J., Langkjaer, L., Havelund, S., & Volund, A. (1992). Chemical stability of insulin. 1. Hydrolytic degradation during storage of pharmaceutical preparations. *Pharm. Res., 9,* 715–726.

Brems, D. N., Brown, P. L., & Becker, G. W. (1990). Equilibrium denaturation of human growth hormone and its cysteine-modified forms. *J. Biol. Chem., 265,* 5504–5511.

Burgess, D. J., Longo, L., & Yoon, J. K. (1991). A novel method of assessment of interfacial adsorption of blood proteins. *J. Parent. Sci. Tech., 45(5),* 239–245.

Burgess, D. J., Yoon, J. K., & Sahin, N. O. (1992). A novel method of determination of protein stability. *J. Parent. Sci. Technol., 46,* 150–155.

Carpenter, J. F., Randolph, T. W., Jiskoot, W., Crommelin, D. J., Middaugh, C. R., Winter, G. et al. (2009). Overlooking subvisible particles in therapeutic protein products: Gaps that may compromise product quality. *J. Pharm. Sci., 98,* 1201–1205.

Cauchy, M., D'Aoust, S., Dawson, B., Rode, H., & Hefford, M. A. (2002). Thermal stability: A means to assure tertiary structure in therapeutic proteins. *Biologicals, 30,* 175–185.

Chang, J. Y., Xiao, N. J., Zhu, M., Zhang, J., Hoff, E., Russell, S. J. et al. (2010). Leachables from saline-containing IV bags can alter therapeutic protein properties. *Pharm. Res., 27,* 2402–2413.

Chi, E. Y., Krishnan, S., Randolph, T. W., & Carpenter, J. F. (2003). Physical stability of proteins in aqueous solution: Mechanism and driving forces in nonnative protein aggregation. *Pharm. Res., 20,* 1325–1336.

Cleland, J. L., Powell, M. F., & Shire, S. J. (1993). The development of stable protein formulations—A close look at protein aggregation, deamidation, and oxidation. *Crit. Rev. Ther. Drug Carr. Syst., 10,* 307–377.

Cleland, J. L. & Wang, D. I. C. (1990). Cosolvent assisted protein refolding. *Bio/Technology, 8,* 1274–1278.

Clements, A., Walsh, D. M., Williams, C. H., & Allsop, D. (1994). Aggregation of Alzheimer's peptides. *Biochem. Soc. Trans., 22,* S16.

Costantino, H. R., Langer, R., & Klibanov, A. M. (1994). Solid-phase aggregation of proteins under pharmaceutically relevant conditions. *J. Pharm. Sci., 83,* 1662–1669.

Darrington, R. T. & Anderson, B. D. (1994). The role of intramolecular nucleophilic catalysis and the effects of self-association on the deamidation of human insulin at low pH. *Pharm. Res., 11,* 784–793.

Defelippis, M. R., Alter, L. A., Pekar, A. H., Havel, H. A., & Brems, D. N. (1993). Evidence for a self-associating equilibrium intermediate during folding of human growth hormone. *Biochemistry, 32,* 1555–1562.

DiPaola, M. (2013, May). Biosimilar approaches: How to best outsource the development of biosimilars. *Contract Pharma,* 5, 68–69.

Eckhardt, B. M., Oeswein, J. Q., & Bewley, T. A. (1991). Effect of freezing on aggregation of human growth hormone. *Pharm. Res., 8,* 1360–1364.

Fatouros, A., Osterberg, T., & Mikaelsson, M. (1997). Recombinant factor VIII SQ—Influence of oxygen, metal ions, pH and ionic strength on its stability in aqueous solution. *Int. J. Pharm., 155,* 121–131.

Federici, M., Lubiniecki, A., Manikwar, P., & Volkin, D. B. (2013). Analytical lessons learned from selected therapeutic protein drug comparability studies. *Biologicals, 41,* 131–147.

Fisher, B. V. & Porter, P. B. (1981). Stability of bovine insulin. *J. Pharm. Pharmacol., 33,* 203–206.

Flores-Fernandez, G. M., Sola, R. J., & Griebenow, K. (2009). The relation between moisture-induced aggregation and structural changes in lyophilized insulin. *J. Pharm. Pharmacol., 61,* 1555–1561.

Fransson, J. R. (1997). Oxidation of human insulin-like growth factor I in formulation studies. 3. Factorial experiments of the effects of ferric ions, EDTA, and visible light on methionine oxidation and covalent aggregation in aqueous solution. *J. Pharm. Sci., 86,* 1046–1050.

Goldberg, D. S., Bishop, S. M., Shah, A. U., & Sathish, H. A. (2011). Formulation development of therapeutic monoclonal antibodies using high-throughput fluorescence and static light scattering techniques: Role of conformational and colloidal stability. *J. Pharm. Sci., 100,* 1306–1315.

Hartvig, R. A., van de Weert, M., Ostergaard, J., Jorgensen, L., & Jensen, H. (2011). Protein adsorption at charged surfaces: The role of electrostatic interactions and interfacial charge regulation. *Langmuir, 27,* 2634–2643.

Hawe, A., Kasper, J. C., Friess, W., & Jiskoot, W. (2009). Structural properties of monoclonal antibody aggregates induced by freeze-thawing and thermal stress. *Eur. J. Pharm. Sci., 38,* 79–87.

Hekman, C., DeMond, W., Dixit, T., Mauch, S., Nuechterlein, M., Stepanenko, A. et al. (1998). Isolation and identification of peptide degradation products of heat stressed pramlintide injection drug product. *Pharm. Res., 15,* 650–659.

Hlodan, R. & Pain, R. H. (1994). Tumour necrosis factor is in equilibrium with a trimeric molten globule at low pH. *FEBS Lett., 343,* 256–260.

Horbett, T. A. (1992). Adsorption of proteins and peptides at interfaces. In T. J. Ahern & M. C. Manning (Eds.), *Pharmaceutical biotechnology. Volume 2. Stability of protein pharmaceuticals. Part A: Chemical and physical pathways of protein degradation* (pp. 195–214). New York: Plenum Publishers.

Horbett, T. A. & Brash, J. L. (1987). Proteins at interfaces: Current issues and future prospects. In J. L. Brash & T. A. Horbett (Eds.), *Proteins at interfaces: Physicochemical and biochemical studies* (pp. 1–33). Washington, DC: American Chemical Society.

Jensen, P. K., Lee, C. S., & King, J. A. (1998). Temperature effects on refolding and aggregation of a large multimeric protein using capillary zone electrophoresis. *Anal. Chem., 70,* 730–736.

Ji, J. A., Zhang, B., Cheng, W., & Wang, Y. J. (2009). Methionine, tryptophan, and histidine oxidation in a model protein, PTH: Mechanisms and stabilization. *J. Pharm. Sci., 98,* 4485–4500.

Jordan, G. M., Yoshioka, S., & Terao, T. (1994). The aggregation of bovine serum albumin in solution and in the solid state. *J. Pharm. Pharmacol., 46,* 182–185.

Katakam, M. & Banga, A. K. (1995). Aggregation of proteins and its prevention by carbohydrate excipients: Albumins and gamma-globulin. *J. Pharm. Pharmacol., 47,* 103–107.

Kendrick, B. S., Tiansheng, L., & Chang, B. S. (2002). Physical stabilization of proteins in aqueous solution. In J. F. Carpenter & M. C. Manning (Eds.), *Rational design of stable protein formulations: Theory and practice* (pp. 61–84). New York: Kluwer Academic Press/Plenum Publishers.

Kenley, R. A. & Warne, N. W. (1994). Acid-catalyzed peptide bond hydrolysis of recombinant human interleukin 11. *Pharm. Res., 11(1),* 72–76.

Kessler, M., Goldsmith, D., & Schellekens, H. (2006). Immunogenicity of biopharmaceuticals. *Nephrol. Dial. Transplant., 21(Suppl. 5),* v9–v12.

King, J. (1989, April). Deciphering the rules of protein folding. *Chem. Eng. News., 10,* 32–54.

Kroon, D. J., Baldwin-Ferro, A., & Lalan, P. (1992). Identification of sites of degradation in a therapeutic monoclonal antibody by peptide mapping. *Pharm. Res., 9,* 1386–1393.

Lam, X. M., Yang, J. Y., & Cleland, J. L. (1997). Antioxidants for prevention of methionine oxidation in recombinant monoclonal antibody HER2. *J. Pharm. Sci., 86,* 1250–1255.

Lee, K. C., Lee, Y. J., Song, H. M., Chun, C. J., & DeLuca, P. P. (1992). Degradation of synthetic *Salmon* calcitonin in aqueous solution. *Pharm. Res., 9,* 1521–1523.

Lehrman, S. R., Tuls, J. L., Havel, H. A., Haskell, R. J., Putnam, S. D., & Tomich, C. C. (1991). Site-directed mutagenesis to probe protein folding: Evidence that the formation and aggregation of a bovine growth hormone folding intermediate are dissociable processes. *Biochemistry, 30,* 5777–5784.

Lencki, R. W., Arul, J., & Neufeld, R. J. (1992a). Effect of subunit dissociation, denaturation, aggregation, coagulation, and decomposition on enzyme inactivation kinetics: I. First-order behavior. *Biotechnol. Bioeng., 40,* 1421–1426.

Lencki, R. W., Arul, J., & Neufeld, R. J. (1992b). Effect of subunit dissociation, denaturation, aggregation, coagulation, and decomposition on enzyme inactivation kinetics: II. Biphasic and grace period behavior. *Biotechnol. Bioeng., 40,* 1427–1434.

Li, M. & Qiu, Y. X. (2013). A review on current downstream bio-processing technology of vaccine products. *Vaccine, 31,* 1264–1267.

Mach, H. & Arvinte, T. (2011). Addressing new analytical challenges in protein formulation development. *Eur. J. Pharm. Biopharm., 78,* 196–207.

Mahler, H. C., Friess, W., Grauschopf, U., & Kiese, S. (2009). Protein aggregation: Pathways, induction factors and analysis. *J. Pharm. Sci., 98,* 2909–2934.

Malzert, A., Boury, F., Renard, D., Robert, P., Proust, J. E., & Benoit, J. P. (2002). Influence of some formulation parameters on lysozyme adsorption and on its stability in solution. *Int. J. Pharm., 242,* 405–409.

Manning, M. C., Chou, D. K., Murphy, B. M., Payne, R. W., & Katayama, D. S. (2010). Stability of protein pharmaceuticals: An update. *Pharm. Res., 27,* 544–575.

Manning, M. C., Patel, K., & Borchardt, R. T. (1989). Stability of protein pharmaceuticals. *Pharm. Res., 6,* 903–918.

Mantyh, P. W., Ghilardi, J. R., Rogers, S., Demaster, E., Allen, C. J., Stimson, E. R. et al. (1993). Aluminum, iron, and zinc ions promote aggregation of physiological concentrations of beta-amyloid peptide. *J. Neurochem., 61,* 1171–1174.

Meyer, J. D., Ho, B., & Manning, M. C. (2002). Effects of conformation on the chemical stability of pharmaceutically relevant polypeptides. In J. F. Carpenter & M. C. Manning (Eds.), *Rational design of stable protein formulations: Theory and practice* (pp. 85–107). New York: Kluwer Academic Press/Plenum Publishers.

Miller, L. M., Bourassa, M. W., & Smith, R. J. (2013). FTIR spectroscopic imaging of protein aggregation in living cells. *Biochica et Biophysica Acta, 1828,* 2339–2346.

Millican, R. L. & Brems, D. N. (1994). Equilibrium intermediates in the denaturation of human insulin and two monomeric insulin analogs. *Biochemistry, 33,* 1116–1124.

Mulkerrin, M. G. & Wetzel, R. (1989). pH dependence of the reversible and irreversible thermal denaturation of gamma interferons. *Biochemistry, 28,* 6556–6561.

Nail, S. L., Jiang, S., Chongprasert, S., & Knopp, S. A. (2002). Fundamentals of freeze drying. In S. L. Nail & M. J. Akers (Eds.), *Development and manufacturing of protein pharmaceuticals* (pp. 281–360). New York: Kluwer Academic Press/Plenum Publishers.

Nguyen, V. T. & Bensaude, O. (1994). Increased thermal aggregation of proteins in ATP-depleted mammalian cells. *Eur. J. Biochem., 220,* 239–246.

Oliyai, C. & Borchardt, R. T. (1993). Chemical pathways of peptide degradation. IV. Pathways, kinetics, and mechanism of degradation of an aspartyl residue in a model hexapeptide. *Pharm. Res., 10,* 95–102.

Oliyai, C. & Borchardt, R. T. (1994). Chemical pathways of peptide degradation. VI. Effect of the primary sequence on the pathways of degradation of aspartyl residues in model hexapeptides. *Pharm. Res., 11,* 751–758.

Oliyai, C., Patel, J. P., Carr, L., & Borchardt, R. T. (1994). Chemical pathways of peptide degradation. VII. Solid state chemical instability of an aspartyl residue in a model hexapeptides. *Pharm. Res., 11,* 901–908.

Owens, D. R., Landgraf, W., Schmidt, A., Bretzel, R. G., & Kuhlmann, M. K. (2012). The emergence of biosimilar insulin preparations—A cause for concern? *Diabetes Technol. Ther., 14,* 989–996.

Paborji, M., Pochopin, N. L., Coppola, W. P., & Bogardus, J. B. (1994). Chemical and physical stability of chimeric L6, a mouse-human monoclonal antibody. *Pharm. Res., 11,* 764–771.

Page, M., Ling, C., Dilger, P., Bentley, M., Forsey, T., Longstaff, C. et al. (1995). Fragmentation of therapeutic human immunoglobulin preparations. *Vox Sang., 69,* 183–194.

Patel, K. & Borchardt, R. T. (1990). Deamidation of asparaginyl residues in proteins: A potential pathway for chemical degradation of proteins in lyophilized dosage forms. *J. Parent. Sci. Technol., 44,* 300–301.

Pedersen, J. S., Dikov, D., Flink, J. L., Hjuler, H. A., Christiansen, G., & Otzen, D. E. (2006). The changing face of glucagon fibrillation: Structural polymorphism and conformational imprinting. *J. Mol. Biol., 355,* 501–523.

Persson, K. M. & Gekas, V. (1994). Factors influencing aggregation of macromolecules in solution. *Process Biochem., 29,* 89–98.

Pittman, D. (2013, June). No one-size-fits-all approach in FDA approving biosimilars. AAPS Newsmagazine, pp. 30–31.

Powell, M. F. (1994). Peptide stability in aqueous parenteral formulations. In J. L. Cleland & R. Langer (Eds.), *Formulation and delivery of proteins and peptides* (pp. 100–117). Washington, DC: American Chemical Society.

Powell, M. F., Fleitman, J., Sanders, L. M., & Si, V. C. (1994). Peptide liquid crystals: Inverse correlation of kinetic formation and thermodynamic stability in aqueous solution. *Pharm. Res., 11,* 1352–1354.

Powell, M. F., Grey, H., Gaeta, F., Sette, A., & Colon, S. (1992). Peptide stability in drug development: A comparison of peptide reactivity in different biological media. *J. Pharm. Sci., 81,* 731–735.

Powell, M. F., Sanders, L. M., Rogerson, A., & Si, V. (1991). Parenteral peptide formulations: Chemical and physical properties of native luteinizing hormone-releasing hormone (LHRH) and hydrophobic analogues in aqueous solution. *Pharm. Res., 8,* 1258–1263.

Powell, M. F., Stewart, T., Otvos, L., Urge, L., Gaeta, F. C. A., Sette, A. et al. (1993). Peptide stability in drug development. II. Effect of single amino acid substitution and glycosylation on peptide reactivity in human serum. *Pharm. Res., 10,* 1268–1273.

Rabe, M., Verdes, D., & Seeger, S. (2011). Understanding protein adsorption phenomena at solid surfaces. *Adv. Colloid Interface Sci., 162,* 87–106.

Rao, P. E. & Kroon, D. J. (1993). Orthoclone OKT3: Chemical mechanisms and functional effects of degradation of a therapeutic monoclonal antibody. In Y. J. Wang & R. Pearlman (Eds.), *Pharmaceutical biotechnology. Vol. 5. Stability and characterization of proteins and peptide drugs: Case histories* (pp. 135–158). New York: Plenum Publishers.

Ratanji, K. D., Derrick, J. P., Dearman, R. J., & Kimber, I. (2014). Immunogenicity of therapeutic proteins: Influence of aggregation. *J. Immunotoxicol., 11,* 99–109.

Rathore, A. S. (2009). Follow-on protein products: Scientific issues, developments and challenges. *Trends Biotechnol., 27,* 698–705.

Remmele, R. L., Nightlinger, N. S., Srinivasan, S., & Gombotz, W. R. (1998). Interleukin-1 receptor (IL-1R) liquid formulation development using differential scanning calorimetry. *Pharm. Res., 15,* 200–208.

Renthal, R. & Velasquez, D. (2002). Self-association of helical peptides in a lipid environment. *J. Protein Chem., 21,* 255–264.

Ripple, D. C. & Dimitrova, M. N. (2012). Protein particles: What we know and what we do not know. *J. Pharm. Sci., 101,* 3568–3579.

Roberts, C. J. (2014). Therapeutic protein aggregation: Mechanisms, design, and control. *Trends Biotechnol., 32,* 372–380.

Roger, S. D. & Mikhail, A. (2007). Biosimilars: Opportunity or cause for concern? *J. Pharm. Pharm. Sci., 10,* 405–410.

Schellekens, H. (2009). Biosimilar therapeutics-what do we need to consider? *NDT Plus., 2,* i27–i36.

Schilling, R. J. & Mitra, A. K. (1991). Degradation of insulin by trypsin and alpha-chymotrypsin. *Pharm. Res., 8,* 721–727.

Schrier, J. A., Kenley, R. A., Williams, R., Corcoran, R. J., Kim, Y. K., Northey, R. P. et al. (1993). Degradation pathways for recombinant human macrophage colony-stimulating factor in aqueous solution. *Pharm. Res., 10,* 933–944.

Seckler, R. & Jaenicke, R. (1992). Protein folding and protein refolding. *FASEB J., 6,* 2545–2552.

Singh, S. K., Afonina, N., Awwad, M., Bechtold-Peters, K., Blue, J. T., Chou, D. et al. (2010). An industry perspective on the monitoring of subvisible particles as a quality attribute for protein therapeutics. *J. Pharm. Sci., 99,* 3302–3321.

Teshima, G. & Canova-Davis, E. (1992). Separation of oxidized human growth hormone variants by reversed- phase high-performance liquid chromatography: Effect of mobile phase pH and organic modifier. *J. Chromatogr., 625,* 207–215.

van de Weert, M. & Randolph, T. W. (2013). Physical instability of peptides and proteins. In L. Hovgaard, S. Frokjaer, & M. van de Weert (Eds.), *Pharmaceutical formulation development of peptides and proteins* (2nd ed., pp. 107–129). Boca Raton, FL: CRC Press.

Vermeer, A. W., Giacomelli, C. E., & Norde, W. (2001). Adsorption of IgG onto hydrophobic teflon. Differences between the F(ab) and F(c) domains. *Biochica et Biophysica Acta, 1526,* 61–69.

Wang, W., Nema, S., & Teagarden, D. (2010). Protein aggregation—Pathways and influencing factors. *Int. J. Pharm., 390,* 89–99.

Wang, W., Singh, S. K., Li, N., Toler, M. R., King, K. R., & Nema, S. (2012). Immunogenicity of protein aggregates—Concerns and realities. *Int. J. Pharm., 431,* 1–11.

Wang, Y. J. (1992). Parenteral products of peptides and proteins. In K. E. Avis, H. A. Lieberman, & L. Lachman (Eds.), *Pharmaceutical dosage forms: Parenteral medications* (1st ed., pp. 283–319). New York: Marcel Dekker, Inc.

Warner, S. (2003, October). FDA caution tempers race for generic biologics. *TheScientist, 27,* 44–45.

Windisch, V., DeLuccia, F., Duhau, L., Herman, F., Mencel, J. J., Tang, S. Y. et al. (1997). Degradation pathways of salmon calcitonin in aqueous solution. *J. Pharm. Sci., 86,* 359–364.

Xie, M. & Schowen, R. L. (1999). Secondary structure and protein deamidation. *J. Pharm. Sci., 88,* 8–13.

Xu, D., Tsai, C. J., & Nussinov, R. (1998). Mechanism and evolution of protein dimerization. *Protein Sci., 7,* 533–544.

Yu, C., Roosdorp, N., & Pushpala, S. (1988). Physical stability of a recombinant alpha-antitrypsin injection. *Pharm. Res., 5,* 800–802.

Zolls, S., Tantipolphan, R., Wiggenhorn, M., Winter, G., Jiskoot, W., Friess, W. et al. (2012). Particles in therapeutic protein formulations, Part 1: Overview of analytical methods. *J. Pharm. Sci., 101,* 914–935.

Zuniga, L. & Calvo, B. (2009). Regulatory aspects of biosimilars in Europe. *Trends Biotechnol., 27,* 385–387.

4 Preformulation and Formulation of Therapeutic Peptides and Proteins

4.1 INTRODUCTION

Formulation of peptide and protein drugs is very different from the formulation of traditional drugs in many ways. Unlike traditional drugs, proteins have primary, secondary, tertiary, and quaternary structures, and thus, degradation of proteins is not a simple one-step reaction. As discussed in Chapter 3, multiple inactivation pathways exist for the degradation of peptides and proteins. Furthermore, the products of these degradation reactions may not be easily detected by any one analytical technique. Also, accelerated stability studies for shelf-life determination may be more difficult to apply for peptide and protein drugs (Nail, Jiang, Chongprasert, & Knopp, 2002). Formulation development will include a consideration of several factors such as structure of the protein, factors affecting its physical and chemical stability, and means of stabilization (Wang, 1999). A formulation scientist involved in formulation or product development for peptide and protein drugs must have a strong background in protein chemistry in addition to physical pharmacy and formulation principles. Also, support of a strong bioanalytical group is a must for successful formulation development. Table 4.1 provides a listing of some peptide and protein products that are currently on the market for parenteral administration and are obtained from natural or synthetic sources, and Table 4.2 provides a listing of some recombinant proteins on the market. Only a few products are listed in these tables, but a more complete list is available in other publications (PDR Staff, 2014; Spada & Walsh, 2005), and some of the other recombinant proteins approved more recently are discussed in Chapter 1. A list of excipients used in these products is also included. The monoclonal antibodies on the market and the excipients included in their formulation are listed separately in Table 1.1.

4.2 PREFORMULATION STUDIES

Product development of a peptide or protein drug starts with preformulation studies on the bulk material. These studies include physicochemical characterization of the bulk material and evaluation of its solubility and stability. The primary sequence and carbohydrate profile of the protein will be established, and its extinction coefficient

TABLE 4.1
Some Therapeutic Peptides and Proteins on Market from Natural or Synthetic Sources for Parenteral Administration

Peptide/ Protein Drug	Common Abb.	No. of Amino Acids	Mol. Wt.	Brand	Origin	Physical Form/ Diluent	Formulation[a]	Storage/Stability
$Alpha_1$-proteinase inhibitor (or—antitrypsin)	AAT	394	52 kDa (glycoprotein)	Prolastin (Grifols)	Pooled human plasma	Powder/sterile WFI USP	Reconstituted product contains AAT, Na^+ (100–210 mEq/L), Cl^- (60–180 mEq/L), PEG (NMT 5 ppm), and sucrose (NMT 0.1%).	2°C–8°C; avoid freezing.
Asparaginase (L-)	—	NA	135 kDa (tetramer)	Elspar (Merck)	Derived from *E. coli*	Lyophilized powder/ sterile WFI USP	Each vial contains 10,000 IU of protein with 80 mg of mannitol.	2°C–8°C; discard reconst. soln. after 8 h or sooner, if cloudy.
Calcitonin	CT	32	3418 Da	Miacalcin (Novartis)	Synthetic	Solution	Each mL contains 200 IU of protein with acetic acid 2.25 mg, sodium acetate 2 mg, NaCl 7.5 mg, and phenol 5 mg.	2°C–8°C.
Chorionic gonadotropin (human)	HCG	NA	NA	Pregnyl (Merck) Novarel (Ferring)	Human placenta/ pregnancy urine	Lyophilized powder/ bacteriostatic WFI USP	Each vial of Pregnyl contains 10,000 USP units of protein with s.p dibasic 4.4 mg, s.p. monobasic 5 mg.	15°C–30°C; after reconstitution, refrigerate and use within 30 days.
Desmopressin acetate	DDAVP	9	1069.2 Da (free base)	DDAVP® (Sanofi-Aventis)	Synthetic	Solution	Each mL contains Desmopressin acetate 4 mcg, chlorobutanol 0.5%, sodium chloride 0.9%.	Keep refrigerated at about 4°C.

(Continued)

TABLE 4.1 (*Continued*)
Some Therapeutic Peptides and Proteins on Market from Natural or Synthetic Sources for Parenteral Administration

Peptide/ Protein Drug	Common Abb.	No. of Amino Acids	Mol. Wt.	Brand	Origin	Physical Form/ Diluent	Formulation[a]	Storage/Stability
Darbepoetin alfa (synthetic erythropoietin)		165	37.1 kDa	Aranesp (Amgen)	Synthetic	Solution	25–500 mcg/mL in vials or prefilled syringes; formulated with polysorbate 80, phosphate buffer, and sod. chloride in WFI.	2°C–8°C; avoid freezing.
Glucagon	—	29	3483 Da	Glucagon for Injection USP (Lilly)	Beef and pork pancreas	Lyophilized powder/ diluting solution or sterile WFI USP.	Each 1 mg vial contains 49 mg of lactose. The diluting soln. contains glycerin 1.6% and phenol 0.2%.	15°C–30°C; after reconstitution, 5°C up to 48 h if diluting soln. used; use immediately if sterile WFI is used.
Interferon alfa-n3	IF-α-n3	166 (approx)	16–27 kDa	Alferon N (Hemispherx Biopharma)	Pooled human leukocytes	Solution	Each 1 mL vial contains 5 million IU protein with 3.3 mg phenol, and 1 mg HSA in a pH 7.4 phosphate-buffered saline solution (8 mg/mL NaCl, 1.74 mg/mL s.p. dibasic, 0.2 mg/mL pot. phosphate monobasic, and 0.2 mg/mL pot. chloride.)	2°C–8°C; do not freeze or shake.

(Continued)

TABLE 4.1 (*Continued*)
Some Therapeutic Peptides and Proteins on Market from Natural or Synthetic Sources for Parenteral Administration

Peptide/ Protein Drug	Common Abb.	No. of Amino Acids	Mol. Wt.	Brand	Origin	Physical Form/ Diluent	Formulation[a]	Storage/Stability
Leuprolide acetate	—	9	1209 Da	Lupron (AbbVie)	Synthetic	Solution	Each multiple-dose vial contains 5 mg/mL of protein with 0.9% of benzyl alcohol and sodium chloride for tonicity adjustment.	Refrigerate until dispensed; patient may store it unrefrigerated below 86°F; avoid freezing and protect from light.
Urokinase	—	267.00	NA	Abbokinase (Abbott)	Tissue culture of human kidney cells	Lyophilized powder/ sterile WFI USP	Each vial contains 250,000 IU of protein with 25 mg of mannitol, 250 mg of albumin (human), and 50 mg of NaCl.	2°C–8°C

Source: Compiled from physician desk reference and literature/websites.
Notes: All solutions are in water for injection USP; bacteriostatic water for injection contains benzyl alcohol as a preservative.
Abbreviations: s.p., sodium phosphate; HSA, human serum albumin; NaCl, sodium chloride; NMT, not more than.
NA: Information not available.

TABLE 4.2
Some Recombinant Therapeutic Proteins on Market for Parenteral Administration

Peptide/Protein Drug	Common Abb.	No. of Amino Acids	Mol. Wt.	Brand	Physical Form/ Diluent	Formulation[a]	Storage/Stability
Filgrastim (granulocyte colony-stimulating factor)	G-CSF	175	19 kDa	Neupogen (Amgen)	Solution	Each single-use vial or prefilled syringe contains 300 or 480 mcg protein in a sod. acetate buffer containing sorbitol and 0.004% Tween® 80.	2°C–8°C; do not freeze or shake.
Granulocyte-macrophage colony-stimulating factor (Sargramostim)	GM-CSF	127	15.5–19.5 kDa (glycoprotein)	Leukine (Sanofi-Aventis)	Lyophilized powder/ sterile WFI USP	Each vial contains 250 or 500 mcg protein with 40 mg mannitol, 10 mg sucrose, and 1.2 mg tromethamine USP.	2°C–8°C; do not freeze or shake.
Erythropoietin	Epoetin Alfa	165	30.4 kDa (glycoprotein)	Epogen (Amgen) Procrit7 (Janssen Biotech)	Solution	Single-dose vial containing 2–10,000 U of protein with citrate buffer (pH 6.9) sodium chloride 0.58% HSA 0.25%. Also available as multidose vial preserved with 1% benzyl alcohol (pH 6.1).	2°C–8°C; do not freeze or shake.
Factor VIII (antihemophilic factor)	rFVIII	—	80–90 kDa subunits (glycoprotein)	Kogenate FS (Bayer Healthcare)	Lyophilized powder/ sterile WFI USP	Final reconstituted product contains 100–400 IU/mL of protein and is stabilized with sucrose, glycine, and histidine in place of HSA as used in Kogenate.	2°C–8°C; avoid freezing; lyophilized powder may be stored up to 25°C for 3 months without loss of activity.

(*Continued*)

TABLE 4.2 (*Continued*)
Some Recombinant Therapeutic Proteins on Market for Parenteral Administration

Peptide/Protein Drug	Common Abb.	No. of Amino Acids	Mol. Wt.	Brand	Physical Form/ Diluent	Formulation[a]	Storage/Stability
Human growth hormone (Somatotropin)	hGH	191	22.125 kDa	Humatrope (Lilly)	Lyophilized powder/ dilution solution	5 mg vial with mannitol 25 mg, glycine 5 mg. Diluting solution contains m-cresol 0.3% and glycerin 1.7%. Also available in cartridges with diluting soln. in accompanying syringes and HumatroPen for injection.	2°C–8°C; use within 14 days after reconstitution.
Interferon alfa-2b	IF-α-2b	—	19.27 kDa	Intron A (Merck)	Solution or lyophilized powder/ bacteriostatic WFI (benzyl alcohol preserved)	*Powder*: 3–50 million IU protein with 1.8 mg s.p. dibasic, 1.3 mg s.p. monobasic, 7.5 mg NaCl, 0.1 mg EDTA, 0.1 mg polysorbate 80, and 1.5 mg m-cresol. *Solution in vials or multidose pens*: 3–25 million IU protein with all inactive ingredients as in powder.	2°C–8°C.
Interferon beta-1b	IF-β-1b	165	18.5 kDa	Betaseron (Bayer Healthcare)	Lyophilized powder	Each vial contains 0.3 mg (9.6 million IU) protein, 15 mg HSA, and 15 mg mannitol.	2°C–8°C; use within 3 h of reconstitution.

(*Continued*)

TABLE 4.2 (*Continued*)
Some Recombinant Therapeutic Proteins on Market for Parenteral Administration

Peptide/Protein Drug	Common Abb.	No. of Amino Acids	Mol. Wt.	Brand	Physical Form/ Diluent	Formulation[a]	Storage/Stability
Interferon gamma-1b	IF-γ-1b	—	2 × 16.46 kDa (dimer)	Actimmune (Vidara Therapeutics)	Solution	Each vial contains 100 mcg of protein with 20 mg mannitol, 0.36 mg sodium succinate, and 0.05 mg polysorbate 20.	2°C–8°C; do not freeze or shake.
Interleukin-2	IL-2	133	15.3 kDa	Proleukin (Prometheus/ Bayer/ Novartis)	Lyophilized powder/ sterile WFI USP	Each reconstituted vial contains 18 million IU protein with 50 mg mannitol, 0.18 mg SDS, 0.17 mg monobasic s.p., and 0.89 mg dibasic s.p. (pH 7.2–7.8)	2°C–25°C; stable for 48 h after reconstitution.
Tissue plasminogen activator (alteplase)	tPA	527	59.04/63.0 kDa	Activase (Genentech, CA)	Lyophilized powder/ sterile WFI	Each vial contains 20, 50, or 100 mg protein. The 100 mg vial contains L-arginine 3.5 g, phosphoric acid 1.0 g, and less than 11 mg of polysorbate 80.	

Source: Compiled from physician desk reference, websites, and literature references.
Notes: All solutions are in water for injection USP; bacteriostatic water for injection contains benzyl alcohol as a preservative.
Abbreviations: s.p., sodium phosphate; HSA, human serum albumin; NaCl, sodium chloride; NMT, not more than.

will be determined. The solubility profile of the protein, its stability, and isoelectric point (pI) will determine what pH should be used for formulation development. For example, a pH range of 3–8 may be investigated at a constant ionic strength of 100–150 mM. Then using the pH of maximum stability, effect of ionic strength on stability may be investigated using an NaCl concentration of 0–1 M. Using optimized pH and ionic strength, effect of various buffer species on stability can be screened. The buffer species used has been shown to affect the aggregation tendency of a humanized antibody following heat treatment (Kameoka, Masuzaki, Ueda, & Imoto, 2007). Using the optimized buffer system, the effect of protein concentration itself on protein stability can be screened. Protein–excipient screening studies will also be run using the optimized buffer system, and other studies such as freeze–thaw stability, agitation stability, and UV light stability will be run (Sarabia, 2003). The peptide content of the bulk powder should also be determined and stated for other users such as for the formulation scientist who will take over after the preformulation studies are completed. Most chloride and acetate salts of peptides are hygroscopic and may contain nonstoichiometric amount of acid, and thus, obtaining required amounts by just weighing without adjustment for peptide content can lead to error (Fang, Fernando, Ugwu, & Blanchard, 1995). Relevant stability-indicating assays are developed, and any potential stability issues or formulation challenges are identified. Forced degradation studies are carried out to understand degradation pathways, and this understanding in turn helps to develop strategies to stabilize the protein. Such studies can include exposure to extremes of temperature, pH, and light, as well as repeated freezing and thawing.

Preformulation studies should also characterize the impurities in the bulk protein as these will affect the stability of the product. For example, the presence of a protease, even at very low concentrations, can destabilize the product. However, variants (which may result from the manufacturing process) with similar biological activity as the product are considered to be product-related substances rather than impurities. Impurities, when present, may be product related or process related. Process-related impurities may include host cell proteins, culture media components, lipids, polysaccharides, and viruses. These impurities can affect the immunogenicity of the protein as discussed in Chapter 1. Trace levels of DNA from host cells may also be present and can be quantitated using a DNA hybridization test. The process history and purification history of a protein such as its exposure to different pH, buffers, and contaminating proteases may also affect its potency, stability, safety, pharmacokinetics, or immunogenicity and thus should be considered. However, a purified protein is not necessarily stable. While the advances in biotechnology can now provide us with highly purified products, the purified protein may actually be more susceptible to processes such as shear and agitation. This is because the purified protein does not have the natural environment that normally contributes to its stability. This environment may sometimes include other proteins, carbohydrates, lipids, or salts that help to stabilize the structure (Hanson & Rouan, 1992).

The analytical methods most useful for degradation or purity tests are listed in Table 4.3. Based on analytical testing and all the preformulation work, specifications will be developed for the bulk protein. These will typically include appearance, color, bioassay (potency), and purity specifications based on UV, sodium

TABLE 4.3
Analytical Methods for Some Degradation/Purity Tests

Degradation Reaction	Analytical Methods
Purity testing	SDS-PAGE (reducing and nonreducing), SE-HPLC, RP-HPLC, ion-exchange HPLC, LAL test, ECP ELISA, Western blot, affinity chromatography, immunoaffinity chromatography, immunoprecipitation
Deamidation	Isoelectric focusing, ion-exchange chromatography, RP-HPLC
Oxidation	RP-HPLC, peptide analysis, fluorescence, HPLC-MS, HPLC-MS/MS
Hydrolysis	HPLC, HPLC-MS, HPLC-MS/MS
Aggregation	SE-HPLC, SDS-PAGE, dynamic light scattering, analytical ultracentrifugation, microflow imaging

dodecyl sulfate polyacrylamide gel electrophoresis (SDS-PAGE), reversed-phase high-performance liquid chromatography (RP-HPLC), and/or ion-exchange HPLC assays. Limits for specific degradation products can be based on the analytical methods listed in Table 4.3. Limits for endotoxin and residual DNA can also be determined by Limulus amoebocyte lysate (LAL) test or DNA hybridization test, respectively. Generally, the final product should contain no more than 100 pg of cellular DNA per dose. The rabbit pyrogen test will detect all pyrogens and is subject to false positives and high variability. Thus, the LAL test for pyrogenic lipopolysaccharides from gram-negative bacteria such as *Escherichia coli* (endotoxins) is also used (DiPaolo, Pennetti, Nugent, & Venkat, 1999). Identification test can include analytical methods such as amino acid analysis, tryptic mapping, circular dichroism, N- and C-terminal sequencing, isoelectric focusing, and/or mass spectrometry. See Chapter 2 for a detailed discussion of these analytical methods. As would be clear from this discussion, preformulation is a broad term, and many of the topics discussed elsewhere in this book would also be relevant for an understanding of preformulation activities. Once a final formula is developed, additional specifications such as those relating to pH, bioburden, osmolarity, and particulate matter will be developed, and sterility testing will be performed.

4.3 FORMULATION DEVELOPMENT

Preformulation development is followed by formulation of the product. During formulation development and stability testing, the focus is on product-related impurities such as the degradation products. However, process-related impurities can also have a significant impact on the overall stability of the molecule as well as lot-to-lot variation. Appropriate packaging is also very important and may sometimes minimize the instability of a peptide or protein formulation. For example, protein can be filled in ampuls under nitrogen flushing to avoid oxidative degradation. A systematic investigation of the potential of extractables and leachables from the container that may affect protein stability is also needed, based on risk assessment, review of literature, and use of sensitive analytical methods (Wakankar et al., 2010). Formulation challenges are different for solution and lyophilized formulations.

Lyophilized formulations are discussed in more details in Chapter 5. If a liquid formulation can have adequate shelf life, then it may be preferred as it would reduce the developmental time as well as operational costs. Formulation challenges may also be more unique for high-concentration protein formulations. The need to frequently administer high doses of monoclonal antibodies in the limited volumes allowable for subcutaneous administration, on an outpatient basis, will drive more research into these unique challenges such as those related to concentration-dependent aggregation or the high viscosity (Shire, Shahrokh, & Liu, 2004). This can pose solubility as well as syringeability problems (Bai et al., 2003). Injection forces will depend on syringe properties such as needle diameter and also on injection speed and rheological properties of the formulation (Allmendinger et al., 2014). The viscosity of high-concentration protein solutions has been shown to depend not only on concentration but also on the electrostatic intermolecular interactions, and therefore, the rheological profile of the formulations may be pH dependent (Yadav, Liu, Shire, & Kalonia, 2010; Yadav, Shire, & Kalonia, 2010). The amount of fluid that can be injected subcutaneously into skin can be increased by the use of the enzyme hyaluronidase that can temporarily degrade hyaluronan, which is a major component of skin. Halozyme Therapeutics, Inc. makes a recombinant human version of the enzyme for use as an adjuvant to increase the absorption and dispersion of other injected drugs. Halozyme's Enhanze technology is being used to reformulate many biologics and monoclonal antibodies (Banga, 2011; Narasimhan, Mach, & Shameem, 2012).

Most recombinant proteins are now available in prefilled syringes, and these may pose some unique challenges as well. For example, Filgrastim (Neupogen®) is available in vials or as single-dose preservative-free prefilled syringes (SingleJect®). Siliconized syringes may result in generation of subvisible particles in the formulation (Badkar, Wolf, Bohack, & Kolhe, 2011; Demeule, Messick, Shire, & Liu, 2010; Majumdar et al., 2011; Sharma, Oma, & Krishnan, 2009). Some of the factors important to formulation development of peptides and proteins are discussed in the following subsections. The use of pharmaceutical excipients in formulations will be discussed in Section 4.4.

4.3.1 Buffer System

Buffers prevent small changes in pH of formulation that may adversely affect stability. Buffer systems that may be acceptable for some peptides and proteins include phosphate (pH 6.2–8.2), acetate (pH 3.8–5.8), citrate (pH 2.1–6.2; pK of 3.15, 4.8, and 6.4), succinate (pH 3.2–6.6; pK of 4.2 and 5.6), histidine (pK of 1.8, 6.0, and 9.0), glycine (pK of 2.35 and 9.8), arginine (pK of 2.18 and 9.1), triethanolamine (pH 7.0–9.0), tris-hydroxymethylaminomethane (THAM, pK 8.1), and maleate buffer (Goldstein & Audhya, 1992; Nail et al., 2002). Glycine and arginine are not effective at neutral pH. Citrate buffer can contribute to a high ionic strength due to its multiple pK values and, in rare cases, may bind to proteins.

Choice of appropriate buffer systems is very important. For example, zinc insulin prepared in phosphate buffer may lead to precipitation of zinc phosphate from solution (Lougheed, Albisser, Martindale, Chow, & Clement, 1983; Lougheed,

Woulfe-Flanagan, Clement, & Albisser, 1980). Also, phosphate buffer may react with calcium leached from glass vials and zinc leached from rubber stoppers and may generate a haze in the solution due to phosphate-forming insoluble salts of divalent metal cations. Several factors can contribute to delamination or generation of glass flakes in vials (Iacocca et al., 2010). For stabilization of γ-interferon, a low pH (about 5.0) was desirable, and thus, inorganic buffers such as phosphate buffer were undesirable. Instead, organic acid buffers such as lactate, succinate, or acetate buffers were found to be suitable (Hwang-Felgner, Jones, & Maher, 1992). However, generally inorganic buffers are more commonly used. Buffers can have significant effects on the tertiary and quaternary structures of proteins, but the mechanisms by which buffers may stabilize or destabilize proteins are complex, and thus, it may not be possible to predict what buffer may be the best for any specific protein. However, in general, desirable attributes for an ideal buffer for protein stabilization would include being zwitterionic, preferentially excluded from the protein domain, and be able to scavenge free radicals. Also, it should not make the protein more susceptible to mechanical stress and should not undergo or catalyze complexation with the carbohydrate part of glycoproteins (Ugwu & Apte, 2004). Histidine buffer has the physiological pH (7.4) within its buffering range and has been frequently used to buffer monoclonal antibodies (see Table 1.1). High-concentration antibody formulations may have significant self-buffering ability. It has been reported that at concentrations exceeding 60–80 mg/mL, buffer capacities of monoclonal antibodies may exceed that of 10 mM acetate buffer, which is commonly used in the pH range of 4–6 (Gokarn et al., 2008). In these cases, the addition of a buffer is not required if the formulation can maintain pH in the desired range.

Besides pH control, the buffer system used can also affect the solubility of the protein. A drastic example is the solubility of tissue plasminogen activator (tPA). At pH 6.0 and a buffer concentration of 20 mM, the rank order for the solubility of tPA was observed to be imidazole < phosphate < histidine < succinate < ethylenediaminetetraacetic acid (EDTA) < citrate, with a 50-fold difference in solubility from imidazole to citrate. Furthermore, the solubility was dependent on ionic strength, increasing about 15-fold as the amount of sodium chloride in a succinate buffer was increased from 0 to 0.3 M (Nguyen & Ward, 1993). When the salt concentration is much greater than that of the buffer, the salt may become the effective buffer in the reaction (Ugwu & Apte, 2004). A very high ionic strength should not be used as it will increase osmolarity and may cause pain upon injection.

4.3.2 pH of the Vehicle

The pH of the protein formulation can be critical to its stability and bioactivity. The net charge on a protein is zero at its pI, positive at pH below pI, and negative at pH above pI. The solubility is the lowest at pI, and thus, buffering at pH very close to pI is not desirable. The choice of pH is made difficult by the fact that solubility is just one consideration. The stability of the protein is also dictated by pH. For example, deamidation reaction may be pH dependent. Furthermore, the best choice of pH to avoid deamidation may be dictated by the mechanism of deamidation. If the mechanism is general acid/base catalyzed, then a pH of 6.0 will minimize deamidation.

Deamidation that proceeds through cyclic imide intermediate is base catalyzed, and acidic pH would thus be desirable in this case (Cleland, Powell, & Shire, 1993). If low pH is avoided to minimize deamidation, high pH may induce oxidation. Thus, a compromise pH for maximum solubility and stability should be chosen. Extremes of pH may also cause irritation at the injection site, though other factors such as the type of buffer species and buffer strength will also be involved.

4.3.3 Protein Solubility

Successful formulation of an aqueous solution of a protein formulation requires careful consideration of the solubility of the protein. The solubility of a protein is affected by several factors such as solution pH and excipients, for example, protamine has been reported to increase the solubility of human growth hormone (hGH) at a molar ratio of 1:23 (hGH/protamine) (Ablinger, Wegscheider, Keller, Prassl, & Zimmer, 2012). Typically, it may be considered as the maximum amount that can be dissolved under defined conditions so that the solution is clear and does not gel or precipitate. While most proteins are very soluble, several are insoluble or slightly soluble. Individual amino acids vary greatly in their aqueous solubility. The aqueous solubility of an amino acid is inversely related to the size of the nonpolar portion of the molecule (Needham, Paruta, & Gerraughty, 1971). Protein solubility is based on the ability of polar residues to interact with water in such a way that the rest of the protein (such as hydrophobic residues buried in the interior) can maintain an active structure (Schein, 1990). However, protein solubility cannot be clearly predicted from a consideration of the structure of the protein. Proteins that exist more in disordered or helical conformations are generally more soluble than those that form β-sheets. It is often believed that covalent attachment of carbohydrates increases the solubility of proteins. While this is true for several glycoproteins, exceptions exist and no generalization can be made. Once the formulation is designed, solubility must be maintained during the shelf life of the product. This can be a challenge since physical denaturation, such as by aggregation, can lead to a loss of solubility. Measurement of protein solubility will often need some innovative approach. Traditional techniques of preparing a saturated solution containing excess solid may not be viable due to the cost of the protein. Also, addition of excess protein may form a gel rather than the creation of a solid phase. Precipitation of proteins by polyethylene glycol (PEG) 8000 can be used as a measure of solubility since plots of logarithm of protein solubility versus PEG concentration are often linear. Extrapolation of these plots back to zero PEG concentration provides a reasonable estimate of solubility. PEG molecules are excluded from the surface of the protein, and thus, the protein gets concentrated in a smaller volume and precipitates as it exceeds its solubility or due to interaction with neighboring protein molecules (Bai et al., 2003; Schein, 1990). However, the PEG method is time consuming, and better methods are needed for a high throughput in the postgenomic age. One solubility assay based on light scattering has been developed. It can be performed in a microplate and is based on an increase in light scattering as the protein is becoming less soluble (Bai et al., 2003).

Several environmental factors such as pH can affect the solubility of a protein (Middaugh & Volkin, 1992). Protein solubility is generally minimum at its pI and increases substantially as the pH is changed to the acidic or basic side of pI. This is because the net charge on the protein is zero at its pI, resulting in maximum interaction between protein molecules. The pH–solubility profile of tPA exhibits a typical V-shape with a minimum around pH 6.0 (Nguyen & Ward, 1993). If a protein is resolubilized by adjusting pH upward or downward from its pI, there is a possibility that residues may change orientation and may not be able to regain their former position (Schein, 1990). Temperature increases also increase the solubility of most proteins, but beyond a certain point, the protein will unfold and denature (Middaugh & Volkin, 1992). As is well known, organic solvents reduce protein solubility and can cause precipitation. This is because they lower the dielectric constant of the mixture, leading to an enhancement of electrostatic attractions, resulting in reduced solubility. Dielectric constants of proteins and relevant models have been described in the literature (Schutz & Warshel, 2001). Considerations relating to protein solubility will become even more important when working with high-concentration protein formulations (Shire et al., 2004).

4.3.4 Selection of Solvent System

Protein–solvent interactions in pharmaceutical formulations have been discussed by Arakawa, Kita, and Carpenter (1991). Polyhydric alcohols such as glycerol, erythritol, and sorbitol are likely to stabilize the protein, and further support can be obtained by ensuring that they are increasing the transition temperature of the protein. The stabilizing effects of cosolvents on proteins can be described by the preferential interaction of the protein with solvent components. The cosolvent increases the chemical potential of the protein, and as a result, the free energy of the system is increased. This creates a thermodynamically unfavorable situation whereby a protein would not unfold since that would increase the contact surface area between protein and solvent. Thus, the native structure of the protein is favored, resulting in stabilization. Proteins that can bind to ligands may also have increased thermodynamic activity, as it takes energy to remove the ligand. Preferentially excluded compounds are likely to increase the surface tension of water. It should be realized that these compounds can also bind to the protein, such as by hydrophobic interactions or hydrogen bonding, and the net results will relate to the balance between such binding and preferential exclusion.

4.3.5 Preservation of a Formulation

Peptide and protein drugs, being biological products, are susceptible to microbial contamination and growth. Therefore, preservatives can be an important formulation component. This is especially true for multiple-dose vials. Even lyophilized powders need to be reconstituted, and the reconstituted solution is required to be stable for about 2 weeks. The preservative in this case can be added to the powder formulation, or it can be a part of the solution used for reconstitution. The choice of an appropriate preservative is very critical since several preservatives may cause precipitation or turbidity in the reconstituted solutions (Kwan, 1985, 1989). Regulatory agencies generally do not encourage addition of preservatives to single-dose injectables as the manufacturer may be doing that to cover up for inadequate aseptic processing. Sodium bisulfite, a commonly used stabilizer in injectable preparations, has been reported to cause degradation of human insulin in the pH range of 4.0–7.0. It should, however, be noted that many infusion solutions contain glucose that may help to stabilize the protein in the presence of a bisulfite. This is due to the presence of an aldehyde group in glucose that can form a bisulfite–glucose adduct (Asahara, Yamada, & Yoshida, 1991). If the formulation contains a surfactant, incompatibilities may also exist between surfactants and preservatives. Of the many preservatives tested to stabilize interferon in one particular formulation, it was claimed that only thimerosal (0.02 mg/mL) was found to be effective. Reconstituted solutions with this formulation were reported to be stable for at least 4 weeks (Kwan, 1989). However, other preservatives may be effective in a different formulation. The clinical need for a multiple-dose preserved formulation of a protein needs to be balanced against the difficulties of developing such a formulation. Preservatives and other excipients/solvents that reduce the polarity of the formulation may make the protein more susceptible to unfolding. In a study with humanized monoclonal antibody, benzyl alcohol was found to be the most effective preservative, but it induced significant aggregation at high concentrations. In this study, the protein was most stable in the presence of methyl and propylparaben, while phenol and m-cresol were not compatible with the protein (Gupta & Kaisheva, 2003). In a recent study, antimicrobial preservatives were reported to induce protein unfolding and aggregation in the order m-cresol > phenol > benzyl alcohol > phenoxyethanol > chlorobutanol (Hutchings, Singh, Cabello-Villegas, & Mallela, 2013). When antibodies are used at high concentrations, an increase in opalescence may be observed, and it can be characterized by using light-scattering-derived second virial coefficient, which is indicative of attractive or repulsive forces between molecules in solution (MacMillan & Kantor, 2003).

4.3.6 Choice of Container

Only type I glass should be used for proteins due to its inertness. When vials are used, proper selection of elastomeric closure is very important as it can interact with the formulation. This is because rubber is a very complex material composed of multiple ingredients in addition to the basic polymers such as vulcanizing agents, accelerators, activators, antioxidants, fillers, lubricating agents, and pigments. These ingredients can leach out and cause stability or other problems with the formulation. Also, the preservative efficacy of the formulation may be reduced due to sorption of the preservative into the rubber. The choice of the elastomeric closure will be influenced by several factors such as the drug itself, the buffer and pH used, preservative, desired method of sterilization, and moisture vapor/gas protection. Extractables such as phthalates from plastic and rubber can leach into formulation and cause allergic reactions or increase immunogenicity (see also Section 1.4) (Sharma, 2007). The presence of polysorbate 80 in formulations may result in higher leaching from containers and closures. During stability testing, some vials can be stored inverted to detect possible problems resulting from interaction of the formulation with the stopper. Siliconization of vials to achieve complete drainage of formulation from the container and siliconization of elastomeric closures for easier insertion into the vials are industry practices that may pose unique problems when working with protein drugs, primarily due to adsorption problems but possibly also due to generation of particulates. This should be carefully considered, and alternative methods are now generally used such as rubber stoppers coated with polytetrafluoroethylene (PTFE; Teflon®), ethylene tetrafluoroethylene (ETFE; FluroTec®), or a fluorinated polymer (Helvoet®). Other types of coatings have been used as well (Hoger et al., 2013).

4.4 PHARMACEUTICAL EXCIPIENTS IN FORMULATIONS

This section discusses the individual excipients useful to prevent or minimize the various physical and chemical degradation pathways. Excipients used are generally those that are listed as *generally recognized as safe* (GRAS) in the FDA inactive ingredients list. A list of all inactive ingredients present in approved products can be found at the FDA's website (2014). Excipients derived from human (e.g., human serum albumin [HSA]) or animal (e.g., Tweens) sources should be avoided due to the risk associated with transmissible diseases such as Creutzfeldt–Jakob disease, bovine spongiform encephalopathy, hepatitis, and HIV. The purity of the excipients used is very important and should be carefully considered. A purity grade even higher than the pharmacopeial monograph may be required if some specific impurity may be implicated in a degradation reaction. If the protein concentration is low, even trace amounts of impurities such as metal ions, peroxides, proteases, or reducing sugars will result in a high impurity/protein ratio (Chang & Hershenson, 2002). A detailed study of peroxide formation in polysorbate 80 has been reported. It was observed that peroxide formation in a 20% solution was much faster than in neat polysorbate 80, suggesting that the raw material should preferably be stored neat or as a concentrated solution. Peroxide formation was induced in polysorbate 80 by autoxidation and could be controlled by an antioxidant or by reducing contact with molecular oxygen in other ways. Light was found to accelerate peroxide formation but only in the presence of air (Ha, Wang, & Wang, 2002). Storage conditions may also affect stability, for example, long-term storage of bulk formulations having sodium chloride in stainless steel tanks may cause well-known interactions of stainless steel with halides, which can result in metal-catalyzed oxidation of proteins (Clark, 2000).

Different excipients may exert their stabilizing effect by different mechanisms. While specific mechanisms will be discussed under the various subsections for individual excipients, a few of the more common mechanisms with wide application are discussed here. Several of the cosolvents stabilize proteins by being preferentially excluded from the surface of the proteins (Figure 4.1) (Arakawa & Timasheff, 1982; Back, Oakenfull, & Smith, 1979; Gekko & Timasheff, 1981; Lee & Timasheff, 1981; Ohtake, Kita, & Arakawa, 2011; Timasheff, 1992). Since these cosolvents are required at high concentrations (>0.25 M), it seems logical that their stabilizing action is not due to specific binding with the protein. Timasheff (1992) has shown that all of these cosolvents give negative binding stoichiometry to proteins.

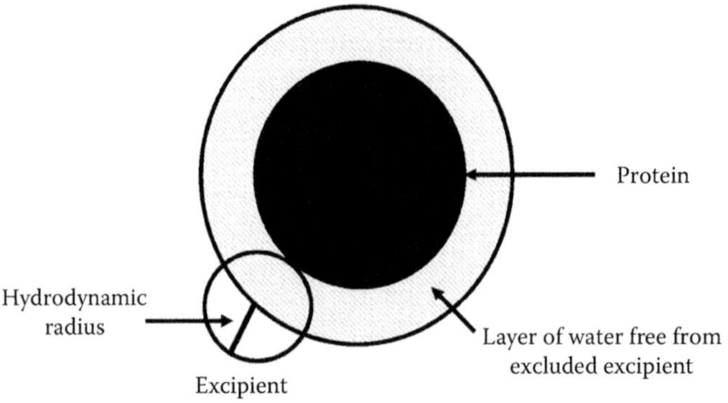

FIGURE 4.1 Schematic illustration of the excluded volume effect where the protein is surrounded by a layer of water that is free from the excluded excipient. (Reproduced with permission from Ohtake, S. et al., *Adv. Drug Deliv. Rev.*, 63, 1053, 2011.)

Negative binding means that the cosolvent is depleted (excluded) in the domain of the proteins, which signifies the enhancement of the solvent with respect to water in that domain (hydration). Thus, stabilization of a protein may result due to preferential exclusion of an added excipient (Figure 4.1). Since preferential exclusion from the unfolded (D) form must be greater than that from the native (N) form, by Le Chatelier's principle, the equilibrium, N ∍ D, must shift toward the native (N) form. However, if the excipient binds to the many newly exposed hydrophobic sites of the unfolded protein, it may denature the protein. Many of these excipients/cosolvents have a significant hydrophobic character and will bind to denatured protein. Thus, while they act as protein stabilizers at low temperatures, they may facilitate denaturation at higher temperatures (Kendrick, Tiansheng, & Chang, 2002). Preferential exclusion of cosolvents may be controlled either by the solvent or by the chemical nature of the protein surface. When controlled by solvent, steric exclusion or change in the surface tension of water may be involved. Steric exclusion results because of size differences between cosolvent and water molecules. The smaller water molecules preferentially hydrate the protein, while the larger cosolvent molecules are unable to penetrate this shell. Cosolvents can also be depleted in surface layer because they may increase the surface tension of water. When controlled by the protein surface, a solvophobic effect may be involved. If the contact between nonpolar regions of a protein and the cosolvent is entropically even more unfavorable than contact with water, the cosolvent molecule will move away into bulk. Glycerol and some other polyols belong to this category.

Some excipients stabilize proteins by specifically binding to them. Acidic fibroblast growth factor (aFGF) is stabilized by several specific ligands that bind to the aFGF polyanion binding site. However, other commonly used excipients were also able to stabilize aFGF by preferential exclusion (Tsai et al., 1993). Similarly, basic fibroblast growth factor is also stabilized by heparin (a sulfated polysaccharide) or other sulfated ligands (Vemuri et al., 1994). Several amorphous excipients stabilize a protein in the solid state by altering the glass transition temperature (T_g) and, therefore, the reactant mobility and protein flexibility. Additionally, the excipient reduces the effective concentration of the protein in the solid state, thereby minimizing the rate of intermolecular decomposition pathways in the presence of moisture. Furthermore, the amorphous excipients are often hygroscopic and may act to *buffer* any water that is absorbed in the solid state (Hageman, 1992).

4.4.1 Albumin

Human albumin is a protein (mol. wt. 66.4 kDa) that consists of 585 amino acids in a single polypeptide chain folded into nine loops, which are further organized into three homologous domains of three loops each. These loops are stabilized by disulfide bridges. Albumin can bind with a wide variety of biological materials. Circulating albumin normally contains 1–2 fatty acids per molecule and is the main transport vehicle for long-chain free fatty acids in the circulation (Spector & Fletcher, 1978). Albumin constitutes about 60% of the total plasma proteins and contributes about 80% of the colloid osmotic pressure that keeps fluid within the blood vessels (Peters, 1985). The plasma concentration of albumin is high (3–5 mg/mL), and half-life is about 20 days. Albumin molecule is highly charged, and this makes it very soluble. Also, the disulfide bridges in the structure make the molecule very stable (Foster, 1977). Because of its excellent solubility and stability, albumin has gained widespread use as an excipient. HSA, 0.1–1.0%, is often used to stabilize and to prevent adsorption of therapeutic peptides and proteins to various surfaces. Some of the marketed peptide or protein formulations (Tables 4.1 and 4.2) that contain HSA include Epogen® (erythropoietin), Alferon® N (interferon α-n3), and Abbokinase® (urokinase). Some of the interferon and other products have been reformulated to remove HSA. Albumin prevents surface adsorption of proteins by preferentially adsorbing to surfaces. Indirectly, this may stabilize the protein molecule since less adsorption would lead to higher recovery of the protein drug. One disadvantage of using HSA as an excipient in formulation is that it may interfere with many of the analytical techniques for quantitation of proteins. Also, as discussed earlier, there are concerns about potential contamination of serum albumin with blood-borne pathogens, and thus, recent trend is to avoid using albumin as an excipient (Osterberg, Fatouros, & Mikaelsson, 1997). Many of the products that already have albumin are being reformulated to remove albumin. Recombinant albumin is now available and can be used to avoid the potential contamination issues. It seems that there are no functional differences between recombinant albumin and HSA, drug binding is similar, and crystal structures are identical (Chen, He, Shi, & Yang, 2013; Chuang, Kragh-Hansen, & Otagiri, 2002). However, production of sufficient quantities still presents challenges to meet the needs (Chen et al., 2013). Also, it will still interfere with analytical methods, and it may be best to avoid albumin altogether. This may be especially relevant now since developing a well-characterized product (see Chapter 3) will need reliance on biophysical analytical methods that may be confounded by the presence of albumin. However, any reformulation should be done with caution.

4.4.2 Amino Acids

Amino acids have been used to reduce surface adsorption, to inhibit aggregate formation, and to stabilize proteins against heat denaturation (Arakawa, Tsumoto, Kita, Chang, & Ejima, 2007; Wang & Hanson, 1988). Several empirical studies have reported these beneficial effects of using amino acids as excipients, even though the mechanism of such protection is not very clear. The dicarboxylic amino acids, aspartic acid, and glutamic acid, at their pI, have been reported to reduce the aggregation of insulin (Bringer, Heldt, & Grodsky, 1981). Although the mechanism of this protection is not clear, it could be as simple as establishment of an acidic pH due to these amino acids. Alternatively, the carboxyl groups of the amino acids could chelate the zinc in insulin, thereby reducing aggregation. Amino acids may also increase protein solubility. Small neutral amino acids (e.g., Gly) or those containing charged side chains (e.g., Asp, Glu, Arg, Lys, His) are most effective to increase solubility. In contrast, large amino acids with apolar side chains may have either no effect or may even decrease solubility. Again, the mechanism is not clear, but they may bind directly to protein or may exert an indirect effect by chelation of metal ions (Middaugh & Volkin, 1992; Quinn & Andrade, 1983). Amino acids can also stabilize proteins due to preferential exclusion (Timasheff, 1992). In a patent awarded to Goldstein and Audhya (1992), it was reported that while glycine (2% w/w) was found to stabilize the pentapeptide, thymopentin, other amino acids did not exert any stabilizing effect. The mechanism of stabilization was not clear. Similarly, another patent reported that addition of glycine or alanine to α-interferon solutions prior to lyophilization provided a product with improved stability. The reconstituted solutions contained about 20 mg/mL of the amino acid in a phosphate buffer (pH 7.2) system containing α-interferon (Kwan, 1985). Amino acids have been used as excipients in marketed products. The most commonly used amino acid in marketed products is glycine, which is present in products such as Kogenate® FS (Factor VIII), Humatrope® (hGH), and many monoclonal antibody formulations. Arginine is also present in Activase® (tPA), where it helps to stabilize and solubilize tPA. Arginine has a dramatic effect on the solubility of tPA. In an arginine phosphate buffer at pH 7.2, type I tPA was almost insoluble at concentrations below 75 mM. As arginine concentration was increased to 100 mM, the solubility increased to 30 mg/mL. At 175 mM, it reached 90 mg/mL, while at 200 mM, 150 mg/mL could be dissolved and the solution became gelatinous (Nguyen & Ward, 1993). Arginine can assist in the refolding of proteins and prevent aggregation (Arakawa et al., 2007).

4.4.3 Carbohydrates

Sugars have been reported to stabilize proteins against a variety of stress situations, such as heating and lyophilization, or simply to improve stability in solution (Back et al., 1979; Carpenter, Crowe, & Crowe, 1987; Carpenter, Martin, Crowe, & Crowe, 1987; Lee & Timasheff, 1981). The mechanism of stabilization of proteins by sugars is not well understood but is usually explained by the effect of sugars on the structure of water. Sucrose may be preferentially excluded from the protein domain by increasing the surface tension of water and due to higher cohesive forces of the sucrose–water solvent system. This leads to protein stabilization since the unfolded state of the protein is not favored thermodynamically in the presence of sucrose (Lee & Timasheff, 1981). Sucrose is one of the most effective agents to protect hemoglobin from spontaneous oxidation to methemoglobin during lyophilization and subsequent storage (Pristoupil, Kramlova, Fortova, & Ulrych, 1985). However, sucrose may also directly interact with the protein as was shown for a PEGylated protein (Mosharraf, Malmberg, & Fransson, 2007). Carbohydrate excipients have also been found to reduce the moisture-induced aggregation of proteins in the solid state (Katakam & Banga, 1995). Carbohydrates may also increase the solubility of proteins. When using reducing sugars such as maltose or lactose as excipients for peptides and proteins, it must be realized that Maillard or Browning reaction must be avoided. This reaction is initiated by the attack of amino nucleophiles such as lysine on reducing sugars and the formation of glycosylamines. The glycosylamines undergo further rearrangement and degradation to unsaturated carbonyls. Since the reaction takes place at high relative humidity, the presence of solid humectant such as sorbitol will reduce the rate of browning by decreasing the free water available for mobilization of reactants (Hageman, 1988). Also, short-term use, under defined conditions, may not pose a problem. For example, the reducing sugar maltose was found to have no detrimental effect on an enzyme's catalytic activity as a consequence of protein browning (Carpenter et al., 1987). Since the reaction may occur above the T_m of the protein (Back et al., 1979), these conditions may not be encountered under normal processing and storage. However, all precautions must be taken to avoid the reaction, and protein should not be stored in dextrose solutions for prolonged periods. Nonreducing sugars such as sucrose or trehalose should be used where possible. Trehalose is generally considered to be a promising excipient for stabilization of proteins during lyophilization (Roser, 1991) and is GRAS listed. Carbohydrates can be excipients of choice as they are natural compounds for ingestion. Thus, large quantities can be used, and they will usually not disrupt protein structure or function when so used.

4.4.4 Chelating and Reducing Agents

Some anions and cations will directly bind to the protein. For example, divalent cations will often bind to certain proteins and lower their solubility. Removal of these ions by a chelating agent such as EDTA may help to maintain protein solubility. Also, EDTA will inhibit the metal-catalyzed oxidation of sulfhydryl groups. aFGF contains three cysteine residues, and thus, small amounts of EDTA have been reported to stabilize aFGF against heat-induced aggregation. For the same reason, reducing agents such as dithiothreitol and β-mercaptoethanol can stabilize aFGF (Tsai et al., 1993). While disodium EDTA has acceptability as a pharmaceutical excipient, the reducing agents may not be acceptable. Such agents are more likely to be used to dissociate covalent aggregates in basic research investigations. Citrate buffer also has some capability of binding trace metal contaminants in solutions.

4.4.5 Cyclodextrins

Cyclodextrins are carbohydrates but are being discussed separately as their mechanism of stabilization of proteins is unique and different from that of other carbohydrates. Cyclodextrins have been investigated to solubilize, stabilize, and promote the delivery of peptide and protein drugs (Brewster, Hora, Simpkins, & Bodor, 1991; Brewster, Simpkins, Hora, Stern, & Bodor, 1989; Johnson, Hoesterey, & Anderson, 1994; Serno, Geidobler, & Winter, 2011; Simpkins, 1991; Szejtli, 1994). The natural cyclodextrins, α, β, and γ are cyclic oligosaccharides of 6, 7, and 8 glucopyranose units, respectively. The ring structure resembles a truncated core, and the fundamental basis of their pharmaceutical applications is the capability to form inclusion complexes due to the hydrophobic property of the cavity (Duchene & Wouessidjewe, 1990a, 1990b; Harada, Li, & Kamachi, 1993). The cavity size is the smallest (≈ 5 Å) for α-cyclodextrin, followed by that of β- (≈ 6 Å) and γ-cyclodextrin (≈ 8 Å). For conventional drugs, the application of cyclodextrins is a result of formation of inclusion complexes with the cyclodextrin cavity. For protein drugs, the mechanism is somewhat different as will be discussed shortly. In this case, complexation does not involve encapsulation of the entire macromolecule. The most widely used cyclodextrin is β-cyclodextrin. This is because the cavity of α-cyclodextrin is too small to accommodate most drugs and γ-cyclodextrin is too expensive to produce. Unfortunately, β-cyclodextrin has a very low water solubility. Cyclodextrin derivatives (as compared to natural cyclodextrins) exhibit higher solubility and lower toxicity by the parenteral route while retaining their efficacy of molecular encapsulation. Hydroxypropyl-β-cyclodextrin (HPβ-CD) is a derivative with high solubility (>50%) and less toxicity (Duchene & Wouessidjewe, 1990c; Szejtli, 1991a). An intravenous (IV) injection of HPβ-CD, given as a single dose as high as 3.0 g, is safe and well tolerated by the human body. When administered orally, only insignificant amounts are absorbed. The necessary toxicological and human clinical data on HPβ-CD are available. Cyclodextrin–drug complexes are on the market in several countries (Strattan, 1992a, 1992b; Szejtli, 1991b). Two types of cyclodextrins used in parenterals are HPβ-CD (Encapsin™) and a sulfobutyl ether β-cyclodextrin (Captisol™). Several captisol-enabled FDA-approved products are on the market.

The use of cyclodextrins to stabilize proteins can be illustrated by considering protein aggregation. Aggregation could result because a protein adsorbs and then unfolds at the air/water interfaces generated by shaking or shear, thereby exposing the hydrophobic amino acids, which are normally located in the interior. The exposed hydrophobic amino acid side chains of one molecule interact with those of another to form aggregates. The mechanism by which cyclodextrins are likely to minimize such aggregation is by molecular encapsulation of these side chains, thereby preventing these hydrophobic interactions (Strattan, 1991). Aromatic amino acids are known to form an inclusion complex with cyclodextrins (Matsuyama, El-Gizawy, & Perrin, 1987). Brewster et al. (1991) have reported that 20–40 molecules of hydroxypropyl-β-cyclodextrin interact with one molecule of interleukin-2 (IL-2), a protein with 133 amino acids. In this study, the cyclodextrin was found to inhibit the aggregation of IL-2 and insulin. The bioactivity of IL-2 was not affected by complexation with cyclodextrin. In general, changes in bioactivity are not expected if the protein site encapsulated by the cyclodextrin is not essential to bioactivity (Strattan, 1991). In this case, the protein–cyclodextrin complex can still bind to receptor as the region affected by the complexation is redundant to bioactivity. HPβ-CD is also reported to solubilize ovine growth hormone (Simpkins, 1991) and to prevent aggregation of porcine growth hormone (Charman, Mason, & Charman, 1993).

4.4.6 Polyhydric Alcohols

These include molecules that are trihydric or higher, such as glycerol, erythritol, arabitol, xylitol, sorbitol, or mannitol. Mannitol is often used as a bulking agent in lyophilization, which will be discussed in Chapter 4. Polyhydric alcohols, in a concentration of 2%–5% by weight, have been reported to stabilize gamma-interferon (Hwang-Felgner et al., 1992). Glycerol causes preferential hydration of proteins as discussed earlier. The thermodynamically unfavorable interaction between glycerol and protein tends to minimize the contact surface, thus leading to stabilization of the native structure (Gekko & Timasheff, 1981). Although glycerol also has an affinity for polar groups of protein, preferential exclusion predominates. Glycerol has also been reported to suppress the aggregation of proteins (Kim & Lee, 1993). Though glycerol stabilizes a protein in solution, it may act as a humectant on the powder and cause decomposition by the Maillard or Browning reaction. Also, glycerol may be a good substrate for bacteria (Schein, 1990).

4.4.7 Polyethylene Glycol

PEG is unique in that even though it is preferentially excluded from the protein due to steric exclusion, it may denature or destabilize the protein. This is especially true at high temperatures. This is most likely because PEG can bind to the denatured protein by hydrophobic interactions. However, PEG may be used to reduce surface adsorption of proteins since it has a binding affinity for surfaces. Also, it can stabilize proteins during freeze–thawing. This is because at subzero temperatures, the hydrophobic characters of the compound become weak, and hence, preferential exclusion effect predominates (Arakawa et al., 1991; Timasheff, 1992). Covalent attachment of PEG to proteins is a different application and is discussed in Section 6.5.1.

4.4.8 Salts

Salts are excluded from the protein–solvent interface due to increased surface tension of water, and thus, buffers and/or salts generally stabilize the protein. The ranking in effectiveness of stabilization follows the Hofmeister series for anions (Kendrick et al., 2002). However, salts can also bind weakly to charges on the protein surface. The balance between these two forces varies with the nature of ions. The contribution of each ion to exclusion and to binding must be considered. For example, while NaCl and Na_2SO_4 are strongly excluded, $CaCl_2$ and $MgCl_2$ show considerable binding. Since the last two salts also increase surface tension, it follows that a surface tension increase does not necessarily lead to preferential exclusion

(Arakawa & Timasheff, 1982). Similarly, while guanidinium sulfate and guanidinium chloride both raise surface tension of water, guanidinium chloride binds to protein and is a denaturant, while guanidinium sulfate is excluded and is a weak stabilizer (Timasheff, 1992). At low concentrations of salts, proteins are surrounded by counterions that decrease their electrostatic free energy, resulting in increased solubility. This increased solubility at low ionic strength is known as *salting in*. At higher salt concentrations, the ions compete with the protein for water molecules, resulting in *salting out* or decreased solubility (Middaugh & Volkin, 1992). Thus, salts should be used with caution. The effect of salts on the stability of a protein is related to the water mobility. The inactivation rate of β-galactosidase in all salt solutions increased with increasing salt concentration up to 200 mM. At higher concentrations, salts that decreased water mobility (KF, phosphate, and Na_2SO_4) resulted in stabilization of β-galactosidase, while others (KI, KBr, and KCl) continued to increase its inactivation rate (Yoshioka, Aso, Izutsu, & Terao, 1993). Aluminum salts are also used to adsorb proteins and function as an adjuvant (see Section 10.1.1).

4.4.9 Surfactants

Surfactants are known to stabilize proteins, if optimal conditions exist. Polysorbate 80 is one of the most widely used surfactants and is present in the marketed products such as Neupogen (granulocyte colony-stimulating factor), Kogenate (Factor VIII), Orthoclone® OKT3 (muromonab-CD3), and Activase (tPA). Other surfactants present in marketed products include polysorbate 20 (Interferon gamma-lb; Actimmune®) and sodium dodecyl sulfate (SDS) (IL-2; Proleukin®). However, as discussed earlier, polysorbate 80 may contain trace amounts of peroxide that can accelerate the degradation of the protein drug. A low peroxide grade made from vegetable sources is now commercially available and should be used. Peroxide formation is also a function of storage conditions, so more peroxide may be generated in a low peroxide starting material. However, since the level of total polysorbate 80 in a formulation is typically less than 1%, this peroxide may or may not affect stability and should be investigated on a case-by-case basis (Ha et al., 2002).

The mechanism of protein stabilization by surfactants is not very clear. One possible mechanism is believed to be the preferential adsorption of the surfactant at the interface. Thus, the protein cannot adsorb at the interface to unfold and undergo aggregation (see Figure 4.2). Since a surfactant will completely cover the surface as a monolayer at its critical micelle concentration (CMC), the CMC is most likely the optimum concentration. However, this has not been proven conclusively. Several measurements of surfactant–protein stoichiometry have shown that a direct interaction is involved and protein–surfactant aggregates form well below the CMC for the surfactant. The surfactant may bind to the protein and reduce its available hydrophobic surface area, thus reducing protein's self-association and interactions with hydrophobic surfaces. The hydrophobic portion of nonionic surfactants can bind to the hydrophobic patches on the proteins so that the protein–surfactant complex is more hydrophilic than either the surfactant or the protein and thereby increasing the protein solubility and reducing its tendency to aggregate (Bummer & Koppenol, 2000; Randolph & Jones, 2002). It has been reported that polysorbate-induced

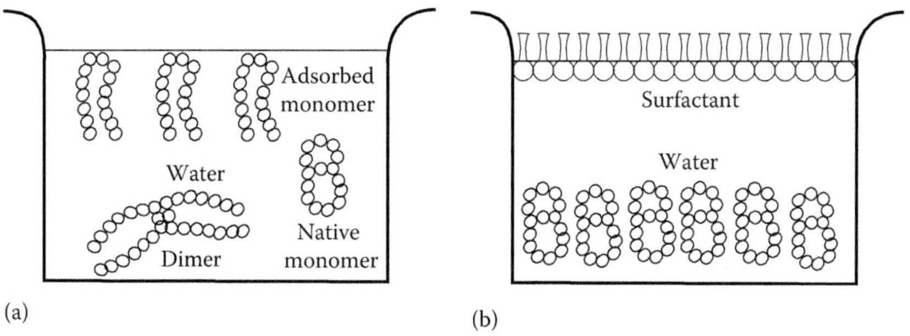

FIGURE 4.2 Schematic for the mechanism by which a surfactant may stabilize protein by preferential adsorption at the air/water interface: (a) in the absence of surfactant, protein adsorbs at surface, unfolds, and undergoes aggregation; (b) in the presence of surfactant at its CMC, the surface is covered with a monolayer of surfactant molecules, which stabilize the protein by preventing its adsorption to the surface.

protection of growth hormone does not correlate with the CMC of the surfactant but rather with the amount of surfactant needed to saturate the hydrophobic patches (Bam et al., 1998). It has also been suggested that the biophysical effects of polysorbates depend on the structure of the polysorbate used and the polysorbate to protein ratio (Deechongkit et al., 2009). The interaction of excipients with proteins in solution and in dry state has been reviewed (Kamerzell, Esfandiary, Joshi, Middaugh, & Volkin, 2011; Ohtake et al., 2011). The protective effect of polysorbate 80 on stability of lactate dehydrogenase upon freeze–thawing was observed to be related to the surface area of the ice crystals. It was hypothesized that the surfactant molecules compete with the protein for sites on the ice surface (Hillgren, Lindgren, & Alden, 2002).

Also, it should be realized that if the denaturation of proteins is caused by its binding to electrically charged surfaces, then the surfactant will not stabilize the protein unless a different mechanism is involved. Surfactants can also be used to solubilize proteins. For example, a nonionic surfactant, Laureth-12, has been reported to solubilize recombinant human interferon-β (Hershenson, Stewart, Carroll, & Shaked, 1989). Higher concentrations of surfactants can also denature proteins. The shorter the hydrocarbon chain of the surfactant, the higher the concentration that may be required to disrupt the secondary structure of a protein (Moriyama, Sasaoka, Ichiyanagi, & Takeda, 1992).

Nonionic and ionic surfactants containing the hydrophobic group, $CH_3(CH_2)_N$, with $N = 7–16$, have been reported to stabilize crystalline zinc insulin formulations. The authors proposed that the long hydrophobic groups reduced the effective polarity of the solvent (Lougheed et al., 1983). Use of surfactants in a formulation must be restricted to the lowest levels due to potential toxicity, especially in pediatric formulations (Hanson & Rouan, 1992). Surfactants may sometimes be present even when not added in the formulation. For example, interferon-β produced by recombinant DNA may contain residual amounts of SDS or other surfactants, used in the extraction and purification steps. The presence of such chemicals may make the material unacceptable for clinical use. In contrast, low concentrations of SDS in IL-2 have been used to ensure the solubility of IL-2 upon reconstitution with water (Geigert, Solli, Woehleke, & Vemuri, 1993). Another surfactant, sodium laurate, has been used to stabilize and solubilize IL-2 or interferon-β for use as a parenteral injection (Thomson, 1989). Nonionic surfactants such as polysorbates and poloxamers can reduce aggregation of proteins. Three nonionic surfactants, polysorbate 80, Pluronic F-68, and Brij 700, were found to surface stabilize monoclonal antibodies during a shipping simulation (Levine, Ransohoff, Kawahata, & Mcgregor, 1991). Polyols of the pluronic type are polypropylene glycol (PPG)/PEG block polymers, and they have also been reported to prevent both the adsorption and aggregation of insulin. The PPG chain of the molecule is believed to be responsible for the stabilizing effect (Thurow & Geisen, 1984). Polysorbate 80, in concentrations above 0.01% (w/v), was found to reduce the aggregation of hGH upon lyophilization and also when reconstituted formulation is shaken (Pearlman & Oeswein, 1992).

4.4.10 Miscellaneous

Excipients that increase the solubility of a protein may also prevent self-association. Sodium salicylate, for example, has been reported to markedly increase the solubility of porcine insulin and thereby reduce its tendency toward self-association (Touitou et al., 1987). The addition of urea in a concentration range of 1–3 mg/mL was found to inhibit self-association of insulin as well as prevent its surface adsorption. Since urea breaks down the water structure, it presumably acts by decreasing the interactions between dimers. Higher concentrations of urea, however, were found to denature insulin and lead to accelerated self-association (Sato, Ebert, & Kim, 1983). Polymers have also been used as excipients to stabilize proteins. Insulin formulated to contain an additive, polypropylene polyethylene glycol, demonstrated a reduced propensity toward aggregation (Feingold, Jenkins, & Kraegen, 1984; Lougheed et al., 1983). Low concentrations (<1%) of polyvinylpyrrolidone (PVP) have also been found to stabilize an antibody against heat-induced aggregation, but slightly higher concentrations can induce protein destabilization (Gombotz et al., 1994). Apparently, serum also contains a factor that prevents the aggregation of insulin (Albisser, Lougheed, Perlman, & Bahoric, 1980). Ammonium tartrate has been suggested as a stabilizing agent for liquid nasal formulations of calcitonin, luteinizing hormone releasing factor, vasopressin, and other peptides (Ceschel, Segu, & Ronchi, 1992).

4.4.11 Selection of Excipient

It is evident from the discussion in this chapter that while several excipients can be used, each has its own advantages and disadvantages. If an excipient can provide the desired buffering capacity, while enhancing protein solubility and stabilizing the formulation at the same time, that would be close to an ideal excipient. However, usually more than one excipient may be required in the formulation to provide all these desirable attributes. The best excipients must be found for each individual protein, and it is hard to make generalizations. This is not to say that guiding principles cannot be used but only to say that these principles provide only a starting point. Even a simple formulation component such as sodium chloride has been reported to cause extensive aggregation of hGH upon lyophilization (Pikal, Dellerman, Roy, & Riggin, 1991). Also, the choice of stabilizer depends on the goals, such as whether the primary goal is to prevent aggregation or to stabilize against hydrolysis. Different stabilizers working by different mechanisms may be required. Sugars are good stabilizers in both solution and in dry state. Glycerol and other polyols are very good stabilizers but may induce self-association. Amino acids and salts may be unpredictable due to their ionic nature and consequent interactions with charged proteins. In general, a strongly excluded excipient leaves the protein in an aqueous environment, which is conducive to the stability of the native state. The quantity of the excipient used is also important. As an example, while carbohydrates can stabilize proteins in the dried state, excess carbohydrates may eliminate the interaction responsible for such stabilization (Arakawa et al., 1991). Also, it should be recognized that each excipient in the formulation has the potential to contribute to impurities. A rigorous quality control is required to ensure that the excipient is 100% pure or that any impurity present would not compromise the stability or efficacy of the formulation. For example, the contamination of clinical albumin solutions with adventitious metal ions can damage the metal-binding sites of albumin, which in turn will affect its ability to mediate drug metabolism. Also, hemodialysis patients receiving vanadium-contaminated albumin will not be able to excrete vanadium, resulting in serious body accumulation of the metal (Quinlan, Coudray, Hubbard, & Gutteridge, 1992).

Differential scanning calorimetry (DSC) (see Chapter 2 for a discussion of the DSC technique) is gaining increasing use as a tool to screen potential excipients that can stabilize the desired protein (Chang, Randall, & Lee, 1993). The thermal transition in this case represents the unfolding of the native structure, which results in an endothermic peak. Stabilizing excipients, sugars such as glucose and sucrose, and polyols such as sorbitol and glycerol were found to increase the denaturation temperature (T_m) of the proteins, ovalbumin, lysozyme, and α-chymotrypsinogen (Back et al., 1979). For tPA, the melting temperature (T_m) in phosphate buffer was found to be about 66°C. In the presence of arginine, a stabilizer used in the formulation, the T_m shifted up to 71°C (Pearlman & Nguyen, 1992). The thermogram for this observation is shown in Figure 4.3. Similarly, the transition temperature of aFGF was found to rise from 63°C to 67°C to 74°C as the concentration of trehalose was increased from 0 to 0.5 to 1.5 M, respectively (Tsai et al., 1993). In studies with a humanized monoclonal antibody, preservatives such as benzyl alcohol or chlorobutanol were found to destabilize the protein, with denaturation temperatures decreasing by 2°C–8°C in a concentration-dependent manner. However, the addition of methyl or propylparaben did not affect protein stability (Gupta & Kaisheva, 2003). In contrast, sorbitol has been reported to be a stabilizer for antibodies, with T_m shifting to higher temperatures (Gonzalez, Murature, & Fidelio, 1995). It should also be realized that T_m is pH dependent. A combination of excipients will often be required to stabilize a peptide or protein formulation. A combination of mannitol and amorphous glycine was found to provide the greatest protection against degradation and aggregation of hGH (Pikal et al., 1991). The utility of DSC to predict protein stability can be confirmed by verifying the predictions by a more direct measure. In a study with IL-1

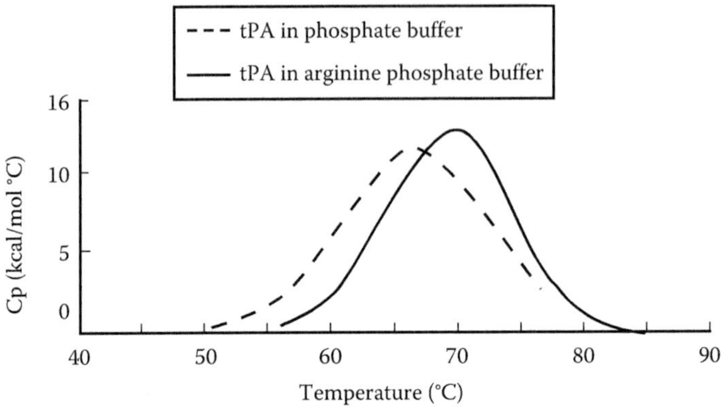

FIGURE 4.3 Plot of heat capacity (Cp) versus temperature for a solution of tPA in either phosphate buffer or arginine phosphate buffer. (Replotted with permission from Pearlman, R. and Nguyen, T., *J. Pharm. Pharmacol.*, 44, 178, 1992.)

receptor type I, direct measurement by size exclusion chromatography was used to confirm that DSC data correctly predicted the rank and order of stability in the presence of preservatives (Remmele, Nightlinger, Srinivasan, & Gombotz, 1998).

4.5 AGGREGATION IN PROTEIN FORMULATIONS

A peptide or protein formulation must be stable against the various physical and chemical pathways of degradation discussed in Chapter 3. The aggregation behavior of proteins and the underlying mechanisms have been discussed in Chapter 3. Aggregation may be introduced during pharmaceutical processing by shaking, exposure to hydrophobic surfaces, heating, shear, or other miscellaneous processes such as sonication. While these general factors will be discussed in Chapter 5, some examples of the aggregation of proteins and ways to prevent these are discussed in the following subsections. Aggregation may also be influenced by the route of administration. A 31-amino-acid acylated peptide has been used as a model system to show that its aggregation status can have a dramatic effect on its subcutaneous bioavailability and pharmacokinetics (Clodfelter et al., 1998).

4.5.1 Aggregation Behavior of Insulin

Insulin is an example of a protein that has a quaternary structure, that is, it normally exists in a self-associated form rather than as a monomer. Insulin exists as a monomer only at a very low concentration (<0.1 μM, ~0.6 μg/mL). At higher concentrations, insulin exists as a dimer. The dimers are believed to result from the hydrophobic association of the B23–B28 regions on insulin monomers. In the presence of zinc ions and in the pH range of 4–8, three dimers come together to form a hexamer. At concentrations ≥2 mM, the hexamer is formed at neutral pH without the assistance of zinc ions. The insulin formulations on the market predominantly are either neutral solutions or suspensions of zinc insulin and exist primarily in hexameric form

(Brange & Langkjoer, 1993), while insulin concentrations in blood are sufficiently low so that insulin circulates and brings about its biological effects as a monomer (Bi et al., 1984). However, there is no anomaly here since at low concentrations, insulin dissociates into monomers. Therefore, in solution, insulin may exist as an equilibrium mixture of monomers, dimers, tetramers, hexamers, or higher associated states, the relative amounts being dependent on factors such as concentration of insulin, pH, nature and concentration of metal ions, purity, processing methods, storage temperature, or ionic strength (Brange, 1987; Lougheed et al., 1980). Many of the recently introduced monomeric insulin analogs also exist in hexameric form in the formulation but dissociate into monomers more easily upon administration. The diameters of the insulin monomer, dimer, tetramer, and hexamer are approximately 30, 39, 50, and 59 Å, respectively. Chelation of zinc ions by EDTA has been reported to cause hexamers to deaggregate to dimers. Since three dimers result from dissociation of hexamer, the enzymatic degradation of insulin by alpha-chymotrypsin was enhanced threefold in the presence of EDTA (Liu, Kildsig, & Mitra, 1991). Similarly, sodium glycocholate, a bile salt, may be capable of dissociating insulin oligomers to monomers, and this may partly explain the role of bile salts as enhancers of insulin bioavailability across mucosal barriers. In studies with zinc insulin (hexamers) and sodium insulin (dimers), the rate of degradation by chymotrypsin in the presence of bile salts was increased by a factor of 5.4- and 2.1-fold, respectively. These values are close to six- and twofold increase that would be expected by the complete dissociation of hexamers and dimers to monomers (Li, Shao, & Mitra, 1992). It may, however, be noted that covalent, higher-molecular-weight transformation products can result upon long-term storage of commercial insulin preparations. Accelerated stability studies indicate that the main product is covalent insulin dimer, but in protamine-containing products, the formation of covalent insulin–protamine also takes place. At temperatures greater than 25°C, covalent oligo- and polymers can also form (Brange, Havelund, & Hougaard, 1992). Although the formation of such covalent aggregates is slower than the chemical decomposition of insulin, their presence may lead to immunological side effects. Aggregation of zinc-free insulin as a function of concentration of sodium chloride (10–100 mM), pH value (7.5–10.5), and insulin concentration (1.8–13.4 mg/mL) has been investigated. At the lowest pH and the highest salt concentration (pH 7.5, 100 mM NaCl, 12 mg/mL insulin), the weight average molar mass was found to be close to that of a hexamer. In contrast, the weight average molar mass was close to that of monomer at the highest pH and the lowest salt concentration (pH 10.5, 10 mM NaCl, 1.9 mg/mL). Since the monomer carries two negative charges at pH 7.5 and six negative charges at pH 10.5, the results can be explained in terms of electrostatic repulsion between insulin monomers. As the pH is reduced, protein charge–charge repulsions are reduced, shifting the equilibrium toward oligomers. Similarly, increased ionic strength screens the charge repulsions, favoring the formation of oligomers. However, it may be noted that in all cases, a distribution of oligomers is present with a relative Gaussian width of about 30% (Kadima et al., 1993).

Problems relating to aggregation of insulin result not from the dimers or hexamers (provided that they are noncovalent) but due to the formation of insoluble precipitates, often referred to as fibrillation of insulin. For a more general discussion of protein fibrillation, see Section 3.2.2.2. Formation of such insoluble aggregates is a major obstacle to the development of insulin infusion systems since the aggregates result in blockage of tubing, membranes, and pumps. Insulin aggregates can develop in both implantable and portable systems, and the reservoirs may have to be flushed or changed every 3–4 days (Lougheed et al., 1980). Obviously, this makes it difficult to develop implantable systems. Any irregularities in the tubings such as the microscopic barbs and distortions produced while sectioning capillary bore PTFE tubing have been reported to induce aggregation of insulin. Surface irregularities such as these may also attract platelets from the blood stream to the roughened regions, which may contribute to insulin aggregation (Lougheed & Albisser, 1980).

The mechanism of formation of insoluble aggregates in lyophilized insulin at an elevated temperature and a high humidity has been investigated. The aggregates were formed by noncovalent interactions and β-elimination of cystines, followed by thiol-catalyzed disulfide interchange in the solid state. Insulin can thus be stabilized by lowering the formation of thiols or transforming them chemically to nonreactive species. Controlling the humidity will also help to stabilize insulin in the solid state (Costantino, Langer, & Klibanov, 1994). Bovine insulin is more susceptible to fibrillation as compared to porcine or human insulin.

In the electron microscope, insulin fibrils are seen as long fibers with a diameter of 10–50 nm. Although the exact mechanism of fibril formation is not known, the main driving force appears to be shielding of hydrophobic surfaces and also possible formation of a β-sheet to further stabilize the fibrillar structure. It is believed that formation of fibril nuclei first requires the monomerization of oligomeric insulin. This is because the hydrophobic interfaces of insulin monomers are normally buried in the dimer or the hexamer. By monomerization, these hydrophobic surfaces are exposed, allowing monomers to associate. Therefore, it seems logical that prevention of insulin dissociation will stabilize the molecule against fibrillation. As an example, this could be achieved by addition of surplus zinc ions that will further stabilize the hexameric structure. Also, blockage of hydrophobic surfaces by addition of surfactant can also counter insulin fibrillation (Brange, Andersen, Laursen, Meyn, & Rasmussen, 1997; Brange & Langkjoer, 1993). Another study has proposed that a partially folded intermediate is the precursor for association and eventually fibrillation (Nielsen, Frokjaer, Brange, Uversky, & Fink, 2001).

Marketed insulin preparations often show frosting, which is the formation of a finely divided precipitate on the walls of the containers. The process can be accelerated by the presence of a large headspace within the vial, suggesting that denaturation at the air–water interface is involved. Other factors that may contributed to frosting include zinc concentration, pH, and the presence or absence of additives (Manning, Patel, & Borchardt, 1989). Frosting is most likely a macroscopic manifestation of insulin fibrillation and is observed more with human NPH insulin. Since human insulin contains more monomeric insulin than animal insulins, fibrillation is more likely in accordance with the mechanism of fibrillation discussed earlier (Brange & Langkjoer, 1993).

Chemical modification of insulin to produce sulfated insulin has also been reported to produce a nonaggregating insulin. The introduction of charged sulfate groups in the molecule disrupts the interaction between hydrophobic regions of insulin molecules, thereby preventing aggregation (Pongor, Brownlee, & Cerami, 1983). Another approach to reducing the self-association of insulin is the development of a Lys, Pro analog of human insulin (Galloway, 1993), which is resistant to self-association. Simulation studies can help with the rational design of insulin analogs that have reduced tendency for aggregation (Berhanu & Masunov, 2012). The development of this analog and its pharmacokinetics, as well as other monomeric insulin analogs that have recently become available, is discussed in Sections 1.2.1.1.2 and 6.6.3.

4.5.2 Aggregation Behavior of Human Growth Hormone

hGH can undergo aggregation to form a dimer or higher-molecular-weight oligomers (Becker et al., 1987). Such aggregation can take place during formulation, processing, storage, and reconstitution (Pearlman & Oeswein, 1992). Phenolic compounds also induce aggregation of hGH (Maa & Hsu, 1996). The most common aggregation product is a stable, noncovalent dimer, which is chemically identical to hGH but is essentially inactive in bioassay. Zinc ions (Zn^{2+}) can also induce dimerization of hGH. Two Zn^{2+} ions associate per dimer of hGH, and the formation of the Zn^{2+}–hGH dimeric complex may be important for the storage of hGH in secretory granules. Replacement of potential Zn^{2+} ligands such as His^{18}, His^{21}, and Glu^{174} in hGH with alanine has been reported to weaken Zn^{2+} binding, thereby preventing formation of hGH dimer (Cunningham, Mulkerrin, & Wells, 1991). Aggregation of hGH was known to be a problem during its preparation from pituitary gland, before the availability of the recombinant protein. A process of ultrafiltration (Lewis, Cheever, & Seavey, 1969) was shown to result in an aggregate-free preparation (Lewis et al., 1969). Aggregation of hGH can be induced by freezing and thawing (Schwartz & Batt, 1973), agitation (Bam et al., 1998), or oxidation (Mulinacci, Poirier, Capelle, Gurny, & Arvinte, 2013). Product literature of recombinant hGH from Lilly (Humatrope) instructs to reconstitute the lyophilized powder by injecting the diluent against the glass wall and mixing by gentle swirling, without any shaking. As currently formulated, hGH readily precipitates out of neutral solution with vigorous mixing. Besides stability and bioactivity considerations, any aggregation of hGH may increase the antigenicity of the product.

4.5.3 Aggregation Behavior of Alpha$_1$-Antitrypsin

A congenital deficiency of alpha$_1$-antitrypsin (AAT) is known to be a major cause of chronic emphysema. A product made from pooled human plasma of normal donors is commercially available (Prolastin®, Grifols Therapeutics, Inc.) for chronic replacement therapy of individuals having congenital deficiency of AAT with clinically demonstrable panacinar emphysema (PDR Staff, 2014). AAT is known to undergo physical stability problems, in addition to loss of activity in solutions. Particulate formation at different storage temperatures was found to follow a first-order loss. Shelf life (t_{90}) at 25°C was only 8.7 days. For rAAT injection, extrapolation of the data (25°C–90°C range) to 4°C gave a predicted shelf life of 5.1 months. However, the activity loss is much faster than the rate of loss of rAAT to particulate formation

(Yu, Roosdorp, & Pushpala, 1988). When the degradation rate constant (k) of rAAT was measured at various pH values, a plot of log k versus pH gave a V-shaped stability profile. The amount of monomeric rAAT decreased at both low and high pH. Also, rAAT was unstable in phosphate-buffered solutions. Aggregation was minimized in citrate buffer and also with sucrose and higher salt concentrations (Vemuri, Yu, & Roosdorp, 1993).

4.5.4 Aggregation Behavior of Human Interferon

Recombinant human interferon brings about its physiological effects as a dimer (Brochon, Tauc, Merola, & Schoot, 1993). Recombinant human interferon is known to aggregate at neutral to slightly alkaline pH, and the resulting solutions may form visible precipitates. Such solutions may have decreased potency and also may cause thrombosis, if injected. Formulation of a liquid dosage form buffered around pH 5.0 was found to prevent such aggregation (Hwang-Felgner et al., 1992). This pH dependence of the aggregation of human interferon has been exhaustively analyzed by Mulkerrin and Wetzel (1989). They found that the pH dependence occurs over a narrow range between pH 5.0 and 6.0 and is controlled by the effect of pH on the solution properties of the unfolded protein. The rate of thermally induced aggregation was found to change little from pH 8.5 to 6.0 but then decreased to near zero at pH 5.0. The molecule was actually found to unfold more readily at pH 5.0, but aggregation resulted only at pH 6.0. Most likely, this resulted due to a loss of positive charge during this pH transition, due to deprotonation of one or both of the histidines, producing a dramatic change in the solubility of the thermally unfolded state, rendering it more susceptible to aggregation. Aggregates of interferon-alpha have been reported to play a key role in its immunogenicity (Braun, Kwee, Labow, & Alsenz, 1997). Antimicrobial preservatives added to formulations may also induce aggregation of interferon (Bis & Mallela, 2014).

4.5.5 Aggregation of Monoclonal Antibodies

IV immune globulin products need to be characterized for IgG multimers as these aggregates may result from the pasteurization process and can cause adverse reactions in patients (Nadler, Paliwal, & Regnier, 1994). Similarly, aggregation can occur in monoclonal antibodies as well. As discussed in Chapter 1, monoclonal antibodies are one of the largest categories of proteins in development. Thus, an understanding of aggregation of immunoglobulins becomes very important. Monoclonal antibodies are typically formulated at high protein concentrations (10–50 mg/mL or higher), and the potential of aggregation or other physical stability problems becomes even greater. The denaturation of multidomain proteins can typically be described by the denaturation of individual domains. However, the unfolding of IgG is still a complex process, and it seems that F_{ab} fragment is the most sensitive to heat treatment. Unfolding is typically followed quickly by an irreversible aggregation step (Vermeer & Norde, 2000). Polysorbate 80 has been shown to have a stabilization effect on IgG in solution during mechanical agitation (Vidanovic, Milic, Stankovic, & Poprzen, 2003), and agitation-induced aggregation of IgG may also be dependent on binding of ions

(Fesinmeyer et al., 2009). A related problem when handling monoclonal antibodies can be the high viscosity of the solutions due to the high concentrations used (Narasimhan et al., 2012). Highly viscous formulations are difficult to withdraw into a syringe and inject, and also during processing, they have undesirable properties like increased back pressure during filtration or excessive frothing. Innovative methods to reduce viscosity by using certain salts or control of pH or ionic strength of the formulation have been attempted (Liu & Shire, 2002). Also, it is desirable to develop analytical methods to analyze high-concentration solutions without dilution since extrapolation of stability indicating parameters from work done with diluted solutions may or may not be valid (Harn, Allan, Oliver, & Middaugh, 2007).

Administration instructions for Orthoclone OKT3 instruct that the protein solution may develop fine translucent particles that are shown not to affect potency, and the solution can be drawn into a syringe through a low-protein-binding 0.2-micron filter (PDR Staff, 2014). Loss of stability by hydrolysis and aggregation was accelerated at 30°C at all the pH ranges investigated as compared to 5°C for a murine IgG2a monoclonal antibody (Riggin, Clodfelter, Maloney, Rickard, & Massey, 1991).

4.5.6 Aggregation Behavior of Some Other Proteins

Recombinant human keratinocyte growth factor, a 163 residue protein, unfolds at relatively low temperatures, and the unfolded protein aggregates rapidly to produce large visible precipitates. Heparin and high-molecular-weight dextran were able to minimize aggregation. Among the small ions tested, sodium citrate was the most effective stabilizer (Chen et al., 1994). Other proteins for which investigations of their aggregation have been reported in the literature include lysozyme (Georgalis, Umbach, Raptis, & Saenger, 1997) and recombinant human Factor VIII (Cho, Garanchon, Kashi, Wong, & Besman, 1996). Recombinant human IL-1β also undergoes aggregation at or above 39°C and upon exposure to repeated freeze–thaw cycles in the absence of stabilizing excipients. The resulting precipitates are soluble in 2.5% SDS solution, and SDS-PAGE analysis of dissolved precipitates shows the presence of monomers (≤30%) and oligomers (≥70%). All oligomers convert into monomers in the presence of β-mercaptoethanol, suggesting that the aggregation is covalent in nature. At 60°C, the aggregation/precipitation followed apparent first-order kinetics and is apparently due to autoxidation of cysteine residues (Gu & Fausnaugh, 1993).

4.6 NOVEL FORMULATION APPROACHES

While a conventional formulation involves the use of pharmaceutical excipients and selection of appropriate vehicle and pH, several novel approaches to formulation development are also feasible. Currently, proteins are formulated almost exclusively as parenterals. However, with increasing knowledge and research, alternate routes such as nasal, pulmonary, oral, transdermal, or ophthalmic are likely to become feasible as various barriers are overcome. This will create a need to formulate products in unique dosage forms for its intended use. Several additional factors such as the delivery device will need to be considered (Hanson & Rouan, 1992). Also, since nasal or ophthalmic solutions are not used all at once, the use of preservatives becomes more important. We will discuss here the potential use of liposomes, reverse micelles, emulsions, and the use of genetic engineering or chemical modification to formulate a therapeutic peptide or protein. Other approaches such as computer modeling, lyophilization, spray drying, aerosols, microspheres, nonaqueous formulations, and protein crystals are also feasible and have been discussed in Chapters 5 and 6. With the completion of the human genome project and the current focus on proteomics (see Chapter 1), the importance of screening *high-throughput formulations* has increased. This would involve the use of experimental design approaches (see Section 4.8) and possible use of databases and other programs that can assess the physical and chemical properties of proteins based on their primary sequence. These programs could include predictive protein algorithms based on primary sequence, sequence homology analysis software, or those that can calculate important properties such as pI, hydrophobicity profile, positioning of secondary structure elements, or conformational flexibility. This information will help the formulation scientist to anticipate the behavior of the protein in development (Nayar & Manning, 2002).

4.6.1 Liposomes

A liposomal formulation of a protein may offer several advantages for drug delivery. The aqueous interior of the liposome will preserve the structure and conformation of the protein, while the lipid exterior may help to improve absorption across biological barriers. The use of a liposomal formulation may also alter the biodistribution of the drug, prolonging the circulation time or help in site-specific targeting (Allen & Cullis, 2013; Sharma & Sharma, 1997; Weiner, 1989, 1990a, 1990b). Liposomes can also be used for the targeted and controlled delivery of peptides and proteins (Crommelin et al., 1997). Recent advances include development of *second-generation liposomes* in which the presence of PEG or other functional moieties on surface can be utilized to develop stealth liposomes to reduce clearance by the RES, resulting in long circulating liposomes. Also, polymeric scaffolds that have entrapped drug-loaded liposomes can be used as depots (Allen & Cullis, 2013; Immordino, Dosio, & Cattel, 2006; Mufamadi et al., 2011; Rai, Vance, Poon, Mogridge, & Kane, 2008). Figure 4.4 shows the schematic for a stealth liposome. A quality by design (QbD) (see Section 4.8) approach has been used to study the factors affecting encapsulation efficiency of liposomes (Xu, Costa, Khan, & Burgess, 2012).

Generally, a peptide or protein is expected to reside in the aqueous compartment of the liposome. However, it should be recognized that the hydrophobic part of the protein may interact with the lipid membrane. Such protein–lipid interactions may or may not affect the bioactivity of the protein. Proteins that can transform into a *molten globule* state appear to exist in either a water-soluble or a membrane-bound form. In the *molten globule* state, the protein exists in an unfolded intermediate conformation. This conformation makes it easier for the hydrophobic part of the protein to partition into the lipid bilayers. This mechanism has been proposed for the interaction of recombinant granulocyte colony-stimulating factor (rhG-CSF) and tumor necrosis factor-α (TNF-α) with lipid vesicles. However, charge interactions can be equally important and must be considered (Collins & Cha, 1994; Hlodan & Pain, 1994).

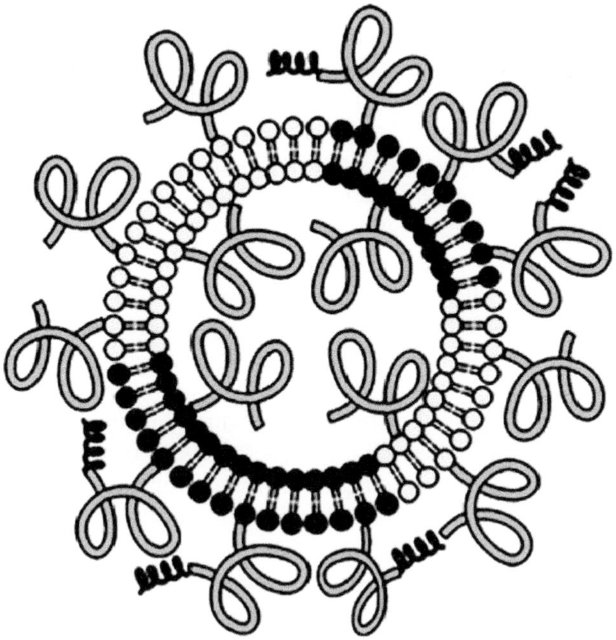

FIGURE 4.4 **(See color insert.)** Schematic depicting a stealth PEGylated liposome. (Reproduced from open access article Mufamadi, M.S. et al., *J. Drug Deliv.*, 2011, 939851, 2011, and with permission from Rai, P. et al., *Chemistry*, 14, 7748, 2008.)

The incorporation of IL-2 into liposomes was strongly dependent on the charge of the liposomes and the pH and ionic strength of the hydration medium. The highest incorporation efficiency (81%) was achieved with negatively charged liposomes composed of phosphatidylcholine/phosphatidylglycerol (9:1). Coinjection of IL-2 containing liposomes resulted in an enhancement of the immune response (Bergers, Otter, Dullens, Kerkvliet, & Crommelin, 1993). The binding of recombinant human epidermal growth factor (hEGF) to membrane phospholipids also occurs primarily by electrostatic interactions. The presence of Trp–Trp sequence in hEGF allows fluorescence studies to be used as a spectroscopic probe of environmental changes. The binding of hEGF to SUV made of negatively charged phospholipids modifies the excitation and emission spectra, and thus, fluorescence studies can be used to study the interaction (Sierra, Vincent, Padron, & Gallay, 1992). The liposome-inserted form of rhG-CSF was found to retain its biological activity and was more stable than the free form. The mechanism of stabilization is not clear, but the insertion of hydrophobic regions into lipid membranes may present protein–protein contacts at these regions, thereby preventing aggregation. Therefore, it seems that if a protein can enter an intermediate conformation such as the molten globular state, it may be possible to stabilize it by liposomes or micelles (Collins & Cha, 1994). In cases where such interaction takes place, liposomes may also sustain the release of the drug even after they have leaked their entire aqueous compartment.

Surfactant protein-A (SP-A), which is the most abundant protein of the pulmonary surfactant, is also known to bind phospholipids. The preferential interaction of SP-A with dipalmitoylphosphatidylcholine (DPPC) may play an important role in preventing alveolar collapse (Casals, Miguel, & Perezgil, 1993).

Liposomes are not stable for long-term use and are thus often lyophilized into a dry preparation. During lyophilization, liposomes can undergo a change in size and can leach their contents due to freezing injury. Cryoprotectants such as saccharides can help to prevent leakage. Proteins such as albumin or gelatin have also been reported to act as cryoprotectants when added to the inside of the liposomal membrane. In a multilamellar vesicle (MLV), there are several lipid bilayers on the inside of the aqueous layer. These can interact with the hydrophobic region of a protein that can thus cause a lipid–protein–lipid bridging among the membranes of the MLV liposome (Yasui, Fujioka, & Nakamura, 1993). Similarly, carbohydrates have been used as free and membrane-bound cryoprotectants during lyophilization of liposomes. These cryoprotectants can prevent fusion at low concentrations and aggregation at higher concentrations. Also, membrane stabilization results which prevents leakage (Engel et al., 1994). The patent literature also describes a method for administering dehydrated liposomes by inhalation. Spray-dried liposome particles containing the entrapped drug are released into the air or propellant stream, or by release from a pressurized can containing a liposome suspension in the propellant. Aerosols of liposome–oxytocin are claimed to provide sustained delivery similar to that provided by an IV infusion (Radhakrishnan, Mihalko, & Abra, 1990).

4.6.2 Reverse Micelles

Reverse micelles can provide a unique opportunity to put a peptide or protein in an organic solvent while still maintaining an aqueous environment in the immediate vicinity of the molecule. Normally, micelles form in water whereby the inside of the micelle provides a lipophilic environment. However, if the surfactant is added to an organic solvent, the molecules can form a reverse micelle where the interior of the micelle would be hydrophilic. Such reverse micelles can be used to entrap enzymes in reverse micelles of surfactant in organic solvents. Such a formulation of a protein drug could be useful in drug delivery since the lipophilic exterior of the micelle may be able to better partition into some mucosae. Nasal delivery of peptides may be enhanced from such reverse micelles (Su, 1991).

4.6.3 Emulsions

While the emulsion dosage form is normally associated with oral delivery, submicron injectable emulsions have been used as a vehicle for IV administration of lipophilic drugs. For injectable emulsions, the particle size of the droplets must be <1 μm, and a usual size range is 100–500 nm. In addition, the emulsion must be sterile, isotonic, nonpyrogenic, nontoxic, biodegradable, and stable (Benita & Levy, 1993). While peptide and protein drugs are currently not administered orally, an emulsion formulation is also sometimes used in research investigations to enhance oral delivery.

For the formulation of a peptide–lipid emulsion, the shearing force required for mixing should be minimized so that it does not denature the protein. Surfactants or a mixture of surfactants with hydrophile–lipophile balance (HLB) matching the lipid phase will reduce the need for high-shear mixing (Weiner, 1990b). Using albumin as a model protein, the secondary structure was found to change to some extent as determined by FTIR when it was incorporated into water-in-oil emulsions (Jorgensen, van de, Vermehren, Bjerregaard, & Frokjaer, 2004). Microemulsions, which can form spontaneously, may also offer a good means to avoid such shear (Bhargava, Narurkar, & Lieb, 1987) and should be investigated for the incorporation of peptides and proteins. A multiple water-in-oil-in-water (w/o/w) emulsion containing bovine growth hormone for sustained release has also been described in the patent literature (Tyle, 1989). Sterilization of a parenteral emulsion poses problems since conventional sterilization techniques such as autoclaving, radiation, or filtration are not viable. The only viable method is aseptic processing. Unfortunately, aseptic processing would involve numerous steps, thereby increasing the risk of contamination. Lidgate, Trattner, Shultz, and Maskiewicz (1992) have described a unique approach in which it became possible to carry out sterile filtration of a parenteral emulsion containing [thr^1]-muramyldipeptide in an oil-in-water emulsion. This peptide product could be an effective adjuvant eliciting both cell-mediated and humoral immune response. Using a microfluidizer, the emulsion was produced at an internal equipment pressure of greater than 16,000 psi. After at least five cycles through the interaction chamber, the resulting emulsion had small droplets and a narrow size range distribution, to allow sterile filtration through a 0.22 μm cartridge filter.

4.6.4 Genetic Engineering or Chemical Modification

The formulation scientist trying to stabilize a peptide or protein formulation will often have an impossible situation at hand. Several degradation routes may be involved. For example, a pH that prevents aggregation may enhance deamidation. A last resort strategy when everything else fails is to substitute some amino acids by recombinant DNA technology or by chemical modification. Site-directed mutagenesis can be used to make amino acid substitutions at specific sites in a protein. This can be done only if the protein conformation and activity are not affected by the substitution. If site-directed mutagenesis is not feasible to replace reactive groups, it may also be possible to block such groups with chemical agents. Using molecular dynamics simulations, sites with a propensity toward aggregation can be identified. Mutations at these sites can then be done to develop more stable therapeutic proteins (Chennamsetty, Voynov, Kayser, Helk, & Trout, 2009; Voynov, Chennamsetty, Kayser, Helk, & Trout, 2009). Similarly, structure-based protein design can be done to identify sites for mutations to reduce immunogenicity (Parker, Choi, Griswold, & Bailey-Kellogg, 2013). Coupling of a peptide to PEG or lipid groups may result in stabilization. Such PEGylated proteins are discussed in Chapter 6. Cleland et al. (1993) have discussed an approach to substitute the deamidating residue Asn by Ser. Ser is an obvious choice for Asn because it has similar size and potential hydrogen donor functional group. For proteins containing Cys residues, aggregation may take place by the formation of interchain disulfide bonds. For this reason, any free thiol groups in the protein may sometimes be mutated to a less reactive amino acyl residue through the use of recombinant techniques, as in the case of IL-2 (Arakawa, Prestrelski, Kenney, & Carpenter, 1993). Development of prodrugs offers another approach for the improved formulation, stability, or delivery of peptide and protein drugs (Bundgaard & Moss, 1990; Bundgaard & Rasmussen, 1991a, 1991b; Delie et al., 1994; Kahns & Bundgaard, 1991; Kahns, Buur, & Bundgaard, 1993). A prodrug converts to the parent drug once in the body or once the delivery barrier has been crossed. Prodrug approach has been used to improve the aqueous solubility of cyclic peptides. Cyclic peptides have lower aqueous solubility as their ionizable C- and N-terminus participate in cyclization. By a prodrug approach, ionizable moieties can also be introduced into the polypeptide to increase its aqueous solubility (Oliyai & Stella, 1993). Proteins may also be stabilized by decreasing their conformational flexibility. Due to its lack of a side chain, Gly has the most conformational freedom, and its replacement with Ala should restrict this freedom. Stabilization may also be achieved by introduction of a disulfide bond or an ion binding site (Manning et al., 1989).

4.7 ACCELERATED STABILITY TESTING

For conventional drugs, rate constants determined at higher temperatures are routinely used to predict the shelf life at lower storage temperatures, by the use of the Arrhenius equation. This allows for the determination of shelf life without a real-time analysis. Until recently, many believed that such accelerated stability studies cannot be done for peptide or protein drugs since these would degrade at the higher temperatures used. While there is some truth to these observations, it is now believed that such generalization cannot be made and limited accelerated stability studies are feasible. Such studies are performed at temperatures well below those at which a protein would unfold. This is because one of the key requirements to use the Arrhenius equation is that the same reaction is responsible for degradation over the entire temperature range being used. As a protein unfolds, the reaction pathways will change. Generally, proposed ICH storage conditions (25C/60% RH, 30C/60% RH, and 40C/75% RH) are still applicable for protein drugs, but more than one stability indicating assay method should be used. Also, stability in refrigerator (2°C–8°C) should also be monitored. Such stability studies are useful to screen formulations, but real-time stability data are still required for product approval. When the same strength and exact container/closure is used for three or more fill contents, manufacturer may elect to place only the smallest and largest container size into the stability program, that is, bracketing.

From the thermogram for tPA (Figure 4.3), it can be seen that tPA undergoes a change in heat capacity at around 50°C. Since this indicates that structural change starts around this temperature, accelerated stability studies should not exceed 50°C. It was thus found that by working at 40°C and below, accelerated stability data for tPA can be extrapolated down to 5°C. Thus, accelerated studies can be done for proteins if an assay specifically monitors one degradation event over the temperature range used. For example, accelerated studies can be used if the loss of activity is followed and is due to one reaction, say, unfolding of protein to denatured state, rather than multiple covalent chemical degradation reactions. Use of several assay methods, rather than one to measure final activity loss, can preclude the use of accelerated studies. For example, below 30°C, IL-1β degraded by deamidation, as measured by isoelectric focusing and SDS-PAGE. However, at 39°C or above, the degradation mechanism was aggregation, and different analytical techniques are required. More importantly, the mechanism of degradation is changing over the temperature range used, and this may preclude the prediction of shelf life from these data (Gu et al., 1991). In a study done by Yoshioka, Aso, Izutsu, and Terao (1994), the data for activity remaining as a function of time for several model protein formulations were fitted into equations by nonlinear regression analysis. It was found that the reciprocal of t_{90}, a measure of inactivation rate, exhibited an approximate linear Arrhenius relationship. Thus, even though the proteins show complex kinetics, it may be possible, in some cases, to extrapolate results at higher temperatures to predict the shelf life of protein preparations. At this time, several concerns remain, and usually real-time stability data are supplied to regulatory agencies for shelf-life determination of therapeutic peptides and proteins (Cleland et al., 1993). The presence of certain peptide sequences in the primary structure of a protein may also predict the probability of a specific degradation pathway, which may compromise the shelf life of the product.

Accelerated stability testing is not limited to temperature and should include other conditions/processing used to accelerate protein degradation. These include exposure to light, humidity, oxygen, or multiple freeze–thaw cycles and exposure to other stresses, for example, agitation to simulate stress likely to be experienced during shipping. While it may be possible to extrapolate these types of stress to predict stability, no quantitative relationships are currently and reliably known. For agitation, even the type of agitation used may result in different types of aggregation, both qualitatively and quantitatively (Kiese, Papppenberger, Friess, & Mahler, 2008). Stability testing requirements are well defined in regulatory guidelines, but standard procedures for forced degradation testing of proteins are not defined, with the exception of photostability (Hawe et al., 2012). The types of stress testing to be conducted will also depend on the type of formulation to be designed, for example, resistance to agitation or accidental freezing would be more relevant for a liquid formulation rather than a lyophilized formulation (Chang & Hershenson, 2002). Bioburden and endotoxin testing should also be performed at the beginning and the end of the stability testing.

4.8 QUALITY BY DESIGN FOR FORMULATION DEVELOPMENT

QbD concepts were introduced into the FDAs CMC review process in 2004 as a result of cGMP for the twenty-first century initiative and have been used for biotechnology products and processes as well (Rathore, 2009; Rouiller et al., 2012), in addition to formulation development (Xu et al., 2012). In essence, the concepts state that quality should be achieved by rational design of the product and process and should not rely on testing representative samples. A systematic approach is taken to meet predefined objectives by understanding the product and the process, as well as process control, based on sound science and quality risk management (Rathore, 2014). The concepts are defined in the ICH guidelines Q8, Q9, and Q10 and require a complete understanding of the relationship between critical process parameters and the product's critical quality attributes. To facilitate the implementation of QbD, FDA encourages the use of process analytical technology (PAT), for example, during downstream bioprocessing of vaccine products (Li & Qiu, 2013) or during manufacturing and development of monoclonal antibodies (Kozlowski & Swann, 2006).

High-throughput formulation screening can be used to find *aggregation hot spots* within a formulation space (Li, Mach, & Blue, 2011). A design of experiment approach coupled with high-throughput screening has been utilized to find the range of optimal buffer compositions that can maximize thermal stability and minimize the viscosity of monoclonal antibody formulations (He et al., 2011). Statistical design techniques could be very useful for preformulation and formulation development of biopharmaceuticals. These designs substantially reduce the large number of studies that would need to be conducted by using the more traditional approach of investigating one factor at a time since a multitude of factors exist, which can affect protein stability. Furthermore, they yield an understanding of the interactions between the various factors. In a preformulation study with recombinant PEGylated staphylokinase mutant, the effects of buffer strength, NaCl concentration, and pH on the conformation and stability of the protein were monitored by evaluating four responses. The responses were protein unfolding transition temperature (T_m), ellipticity of the protein at 220 nm, and monitoring stability (% loss) and % dePEGylation following 2-week storage at 40°C. A central composite design using two-level full factorial study was performed. The shapes of the CD spectra for different runs were similar, suggesting that different formulations had similar secondary structures. T_m was found to be predictive of stability, and pH was the most significant factor affecting T_m (Bedu-Addo, Moreadith, & Advant, 2002). An experimental design technique has also been used to investigate the effect of preservatives on stability of a humanized monoclonal antibody. Stability was assessed using SEC-HPLC, DSC, UV spectroscopy, light scattering, and potency testing (Gupta & Kaisheva, 2003). Using response surface methodology, the thermal stability of lysozyme, as determined by microcalorimetry, was found to be pH dependent. The unfolding transition temperature was observed to be around 75°C and was reversible as observed by rescanning the samples. Sucrose was found to stabilize the protein, with each percentage increase in sucrose concentration resulting in a +0.2°C increase in unfolding temperature (Branchu, Forbes, York, & Nyqvist, 1999). Thus, a few percentage increases in sucrose concentration would lead to a significant change in unfolding temperature. A response surface plot has also been used to understand the oxidation of methionine in the presence of ferric ions, EDTA, and light (Fransson, 1997).

4.9 CONCLUSIONS

Development of a stable and efficacious formulation of a peptide or protein product requires a careful consideration of the pH, buffer system, cosolvents, and drug solubility. Several excipients such as amino acids, carbohydrates, salts, surfactants, and polyhydric alcohols are commonly used in marketed formulations of therapeutic peptides and proteins to stabilize them into efficacious products. Selection of appropriate excipients to stabilize the formulation and considerations related to impurities in excipients is very critical to the stability of the formulation. The formulation must be stable against various physical and chemical pathways of degradation. While most therapeutic proteins are currently formulated for parenteral administration, this trend will change as novel delivery routes are developed. This will require the use of novel formulation approaches such as the use of liposomes, microspheres, protein crystals, nonaqueous formulations, or emulsions. Use of genetic engineering or chemical modification may be required if conventional approaches fail to stabilize a peptide or protein in the formulation. Accelerated stability testing is more difficult for protein formulations as compared to that for traditional drugs. However, limited studies are feasible if the underlying mechanisms are understood.

REFERENCES

Ablinger, E., Wegscheider, S., Keller, W., Prassl, R., & Zimmer, A. (2012). Effect of protamine on the solubility and deamidation of human growth hormone. *Int. J. Pharm., 427,* 209–216.

Albisser, A. M., Lougheed, W., Perlman, K., & Bahoric, A. (1980). Nonaggregating insulin solutions for long-term glucose control in experimental and human diabetes. *Diabetes, 29,* 241–243.

Allen, T. M. & Cullis, P. R. (2013). Liposomal drug delivery systems: From concept to clinical applications. *Adv. Drug Deliv. Rev., 65,* 36–48.

Allmendinger, A., Fischer, S., Huwyler, J., Mahler, H. C., Schwarb, E., Zarraga, I. E. et al. (2014). Rheological characterization and injection forces of concentrated protein formulations: An alternative predictive model for non-Newtonian solutions. *Eur. J. Pharm. Biopharm., 87(2),* 318–328.

Arakawa, T., Kita, Y., & Carpenter, J. F. (1991). Protein-solvent interactions in pharmaceutical formulations. *Pharmaceut. Res., 8,* 285–291.

Arakawa, T., Prestrelski, S. J., Kenney, W. C., & Carpenter, J. F. (1993). Factors affecting short-term and long-term stabilities of proteins. *Adv. Drug Deliv. Rev., 10,* 1–28.

Arakawa, T. & Timasheff, S. N. (1982). Preferential interactions of proteins with salts in concentrated solutions. *Biochemistry, 21,* 6545–6552.

Arakawa, T., Tsumoto, K., Kita, Y., Chang, B., & Ejima, D. (2007). Biotechnology applications of amino acids in protein purification and formulations. *Amino Acids, 33,* 587–605.

Asahara, K., Yamada, H., & Yoshida, S. (1991). Stability of human insulin in solutions containing sodium bisulfite. *Chem. Pharm. Bull., 39,* 2662–2666.

Back, J. F., Oakenfull, D., & Smith, M. B. (1979). Increased thermal stability of proteins in the presence of sugars and polyols. *Biochemistry, 18,* 5191–5196.

Badkar, A., Wolf, A., Bohack, L., & Kolhe, P. (2011). Development of biotechnology products in pre-filled syringes: Technical considerations and approaches. *AAPS PharmSciTech., 12,* 564–572.

Bai, S. J., Katayama, D. S., Chou, D. K. C., Anchordoquy, T. J., Nayar, R., & Manning, M. C. (2003). High-throughput formulation of biopharmaceutical products in the postgenomic age. *Genomic/Proteomic Technol., October,* 28F–28H.

Bam, N. B., Cleland, J. L., Yang, J., Manning, M. C., Carpenter, J. F., Kelley, R. F. et al. (1998). Tween protects recombinant human growth hormone against agitation-induced damage via hydrophobic interactions. *J. Pharmaceut. Sci., 87,* 1554–1559.

Banga, A. K. (2011). *Transdermal and intradermal delivery of therapeutic agents: Application of physical technologies*. Boca Raton, FL: CRC Press, Taylor & Francis.

Becker, G. W., Bowsher, R. R., Mackellar, W. C., Poor, M. L., Tackitt, P. M., & Riggin, R. M. (1987). Chemical, physical, and biological characterization of a dimeric form of biosynthetic human growth hormone. *Biotechnol. Appl. Biochem., 9,* 478–487.

Bedu-Addo, F., Moreadith, R., & Advant, S. J. (2002). Preformulation development of recombinant pegylated staphylokinase SY161 using statistical design. *AAPS PharmSciTech., 4,* 1–10.

Benita, S. & Levy, M. Y. (1993). Submicron emulsions as colloidal drug carriers for intravenous administration—Comprehensive physicochemical characterization. *J. Pharmaceut. Sci., 82,* 1069–1079.

Bergers, J. J., Otter, W. D., Dullens, H. F. J., Kerkvliet, C. T. M., & Crommelin, D. J. A. (1993). Interleukin-2-containing liposomes: Interaction of interleukin-2- with liposomal bilayers and preliminary studies on application in cancer vaccines. *Pharmaceut. Res., 10,* 1715–1721.

Berhanu, W. M. & Masunov, A. E. (2012). Controlling the aggregation and rate of release in order to improve insulin formulation: Molecular dynamics study of full-length insulin amyloid oligomer models. *J. Mol. Model., 18,* 1129–1142.

Bhargava, H. N., Narurkar, A., & Lieb, L. M. (1987). Using microemulsions for drug delivery. *Pharm. Technol., 11(3),* 46–54.

Bi, R. C., Dauter, Z., Dodson, E., Dodson, G., Giordano, F., & Reynolds, C. (1984). Insulin's structure as a modified and monomeric molecule. *Biopolymers, 23,* 391–395.

Bis, R. L. & Mallela, K. M. (2014). Antimicrobial preservatives induce aggregation of interferon alpha-2a: The order in which preservatives induce protein aggregation is independent of the protein. *Int. J. Pharmaceut., 472,* 356–361.

Branchu, S., Forbes, R. T., York, P., & Nyqvist, H. (1999). A central composite design to investigate the thermal stabilization of lysozyme. *Pharmaceut. Res., 16,* 702–708.

Brange, J. (1987). *Galenics of insulin*. Berlin, Germany: Springer-Verlag.

Brange, J., Andersen, L., Laursen, E. D., Meyn, G., & Rasmussen, E. (1997). Toward understanding insulin fibrillation. *J. Pharmaceut. Sci., 86,* 517–525.

Brange, J., Havelund, S., & Hougaard, P. (1992). Chemical stability of insulin. 2. Formation of higher molecular weight transformation products during storage of pharmaceutical preparations. *Pharmaceut. Res., 9,* 727–734.

Brange, J. & Langkjoer, L. (1993). Insulin structure and stability. In Y. J. Wang & R. Pearlman (Eds.), *Pharmaceutical biotechnology. Vol. 5: Stability and characterization of proteins and peptide drugs: Case histories* (pp. 315–350). New York: Plenum Press.

Braun, A., Kwee, L., Labow, M. A., & Alsenz, J. (1997). Protein aggregates seem to play a key role among the parameters influencing the antigenicity of interferon alpha (IFN-alpha) in normal and transgenic mice. *Pharmaceut. Res., 14,* 1472–1478.

Brewster, M. E., Hora, M. S., Simpkins, J. W., & Bodor, N. (1991). Use of 2-hydroxypropyl-β-cyclodextrin as a solubilizing and stabilizing excipient for protein drugs. *Pharm. Res., 8,* 792–795.

Brewster, M. E., Simpkins, J. W., Hora, M. S., Stern, W. C., & Bodor, N. (1989). The potential use of cyclodextrins in parenteral formulations. *J. Parent. Sci. Tech., 43(5),* 231–240.

Bringer, J., Heldt, A., & Grodsky, G. M. (1981). Prevention of insulin aggregation by dicarboxylic amino acids during prolonged infusion. *Diabetes, 30,* 83–85.

Brochon, J. C., Tauc, P., Merola, F., & Schoot, B. M. (1993). Analysis of a recombinant protein preparation on physical homogeneity and state of aggregation. *Anal. Chem., 65,* 1028–1034.

Bummer, P. M. & Koppenol, S. (2000). Chemical and physical considerations in protein and peptide stability. In E. L. McNally (Ed.), *Protein formulation and delivery* (pp. 5–69). New York: Marcel Dekker, Inc.

Bundgaard, H. & Moss, J. (1990). Prodrugs of peptides. 6. Bioreversible derivatives of thyrotropin-releasing hormone (TRH) with increased lipophilicity and resistance to cleavage by the TRH-specific serum enzyme. *Pharmaceut. Res., 7,* 885–892.

Bundgaard, H. & Rasmussen, G. J. (1991a). Prodrugs of peptides. 11. Chemical and enzymatic hydrolysis kinetics of N-acyloxymethyl derivatives of a peptide-like bond. *Pharmaceut. Res., 8,* 1238–1242.

Bundgaard, H. & Rasmussen, G. J. (1991b). Prodrugs of peptides. 9. Bioreversible N-α-hydroxyalkylation of the peptide bond to effect protection against carboxypeptidase or other proteolytic enzymes. *Pharmaceut. Res., 8,* 313–322.

Carpenter, J. F., Crowe, L. M., & Crowe, J. H. (1987). Stabilization of phosphofructokinase with sugars during freeze-drying: Characterization of enhanced protection in the presence of divalent cations. *Biochim. Biophys. Acta, 923,* 109–115.

Carpenter, J. F., Martin, B., Crowe, L. M., & Crowe, J. H. (1987). Stabilization of phosphofructokinase during air-drying with sugars and sugar/transition metal mixtures. *Cryobiology, 24,* 455–464.

Casals, C., Miguel, E., & Perezgil, J. (1993). Tryptophan fluorescence study on the interaction of pulmonary surfactant protein-a with phospholipid vesicles. *Biochem. J., 296,* 585–593.

Ceschel, G., Segu, A. M., & Ronchi, C. Liquid pharmaceutical composition for nasal administration containing a polypeptide as active ingredient. U.S. Patent 5,124,315, June 23, 1992.

Chang, B. S. & Hershenson, S. (2002). Practical approaches to protein development. In J. F. Carpenter & M. C. Manning (Eds.), *Rational design of stable protein formulations: Theory and practice* (pp. 1–25). New York: Kluwer Academic/Plenum Publishers.

Chang, B. S., Randall, C. S., & Lee, Y. S. (1993). Stabilization of lyophilized porcine pancreatic elastase. *Pharmaceut. Res., 10,* 1478–1483.

Charman, S. A., Mason, K. L., & Charman, W. N. (1993). Techniques for assessing the effects of pharmaceutical excipients on the aggregation of porcine growth hormone. *Pharmaceut. Res., 10,* 954–962.

Chen, B. L., Arakawa, T., Hsu, E., Narhi, L. O., Tressel, T. J., & Chien, S. L. (1994). Strategies to suppress aggregation of recombinant keratinocyte growth factor during liquid formulation development. *J. Pharm. Sci., 83,* 1657–1661.

Chen, Z., He, Y., Shi, B., & Yang, D. (2013). Human serum albumin from recombinant DNA technology: Challenges and strategies. *Biochim. Biophys. Acta, 1830,* 5515–5525.

Chennamsetty, N., Voynov, V., Kayser, V., Helk, B., & Trout, B. L. (2009). Design of therapeutic proteins with enhanced stability. *Proc. Natl. Acad. Sci. USA, 106,* 11937–11942.

Cho, Y., Garanchon, C., Kashi, R., Wong, C., & Besman, M. J. (1996). Characterization of aggregates of recombinant human factor VIII by size-exclusion chromatography and immunoassay. *Biotechnol. Appl. Biochem., 24,* 55–59.

Chuang, V. T., Kragh-Hansen, U., & Otagiri, M. (2002). Pharmaceutical strategies utilizing recombinant human serum albumin. *Pharmaceut. Res., 19,* 569–577.

Clark, A. R. S. S. J. (2000). Formulation of proteins for pulmonary delivery. In E. L. McNally (Ed.), *Protein formulation and delivery* (pp. 201–234). New York: Marcel Dekker, Inc.

Cleland, J. L., Powell, M. F., & Shire, S. J. (1993). The development of stable protein formulations—A close look at protein aggregation, deamidation, and oxidation. *Crit. Rev. Ther. Drug Carrier Syst., 10,* 307–377.

Clodfelter, D. K., Pekar, A. H., Rebhun, D. M., Destrampe, K. A., Havel, H. A., Myers, S. R. et al. (1998). Effects of non-covalent self-association on the subcutaneous absorption of a therapeutic peptide. *Pharmaceut. Res., 15,* 254–262.

Collins, D. & Cha, Y. S. (1994). Interaction of recombinant granulocyte colony stimulating factor with lipid membranes—Enhanced stability of a water-soluble protein after membrane insertion. *Biochemistry, 33,* 4521–4526.

Costantino, H. R., Langer, R., & Klibanov, A. M. (1994). Moisture-induced aggregation of lyophilized insulin. *Pharmaceut. Res., 11,* 21–29.

Crommelin, D. J. A., Daemen, T., Scherphof, G. L., Vingerhoeds, M. H., Heeremans, J. L. M., Kluft, C. et al. (1997). Liposomes: Vehicles for the targeted and controlled delivery of peptides and proteins. *J. Contr. Rel., 46,* 165–175.

Cunningham, B. C., Mulkerrin, M. G., & Wells, J. A. (1991). Dimerization of human growth hormone by zinc. *Science, 253,* 545–548.

Deechongkit, S., Wen, J., Narhi, L. O., Jiang, Y., Park, S. S., Kim, J. et al. (2009). Physical and biophysical effects of polysorbate 20 and 80 on darbepoetin alfa. *J. Pharmaceut. Sci., 98,* 3200–3217.

Delie, F., Couvreur, P., Nisato, D., Michel, J. B., Puisieux, F., & Letourneux, Y. (1994). Synthesis and in vitro study of a diglyceride prodrug of a peptide. *Pharmaceut. Res., 11,* 1082–1087.

Demeule, B., Messick, S., Shire, S. J., & Liu, J. (2010). Characterization of particles in protein solutions: Reaching the limits of current technologies. *AAPS J., 12,* 708–715.

DiPaolo, B., Pennetti, A., Nugent, L., & Venkat, K. (1999). Monitoring impurities in biopharmaceuticals produced by recombinant technology. *Pharm. Sci. Technol. Today, 2,* 70–82.

Duchene, D. & Wouessidjewe, D. (1990a). Pharmaceutical uses of cyclodextrins and derivatives. *Drug Dev. Ind. Pharm., 16,* 2487–2499.

Duchene, D. & Wouessidjewe, D. (1990b). Physicochemical characteristics and pharmaceutical uses of cyclodextrin derivatives, part I. *Pharm. Technol., 14(6),* 26–34.

Duchene, D. & Wouessidjewe, D. (1990c). Physicochemical characteristics and pharmaceutical uses of cyclodextrin derivatives, part II. *Pharm. Technol., 14(8),* 22–30.

Engel, A., Bendas, G., Wilhelm, F., Mannova, M., Ausborn, M., & Nuhn, P. (1994). Freeze drying of liposomes with free and membrane-bound cryoprotectants—The background of protection and damaging processes. *Int. J. Pharmaceut., 107,* 99–110.

Fang, X. J., Fernando, Q., Ugwu, S. O., & Blanchard, J. (1995). An improved method for determination of acid dissociation constants of peptides. *Pharmaceut. Res., 12,* 1423–1429.

FDA. (2014). (http://www.accessdata.fda.gov/scripts/cder/iig/index.Cfm). (Accessed September 27, 2014.)

Feingold, V., Jenkins, A. B., & Kraegen, E. W. (1984). Effect of contact material on vibration-induced insulin aggregation. *Diabetologia, 27,* 373–378.

Fesinmeyer, R. M., Hogan, S., Saluja, A., Brych, S. R., Kras, E., Narhi, L. O. et al. (2009). Effect of ions on agitation- and temperature-induced aggregation reactions of antibodies. *Pharmaceut. Res., 26,* 903–913.

Foster, J. F. (1977). Some aspects of the structure and conformational properties of serum albumin. In V. M. Rosenoer, M. Oratz, & M. A. Rothschild (Eds.), *Albumin structure, function & uses* (pp. 53–84). Oxford, U.K.: Pergamon Press, Inc.

Fransson, J. R. (1997). Oxidation of human insulin-like growth factor I in formulation studies. 3. Factorial experiments of the effects of ferric ions, EDTA, and visible light on methionine oxidation and covalent aggregation in aqueous solution. *J. Pharmaceut. Sci., 86,* 1046–1050.

Galloway, J. A. (1993). New directions in drug development—Mixtures, analogues, and modeling. *Diabetes Care, 16,* 16–23.

Geigert, J., Solli, N., Woehleke, P., & Vemuri, S. (1993). Development and shelf-life determination of recombinant interleukin-2 (proleukin). In Y. J. Wang & R. Pearlman (Eds.), *Pharmaceutical biotechnology. Vol. 5: Stability and characterization of proteins and peptide drugs: Case histories* (pp. 249–262). New York: Plenum Press.

Gekko, K. & Timasheff, S. N. (1981). Mechanism of protein stabilization by glycerol: Preferential hydration in glycerol-water mixtures. *Biochemistry, 20,* 4667–4676.

Georgalis, Y., Umbach, P., Raptis, J., & Saenger, W. (1997). Lysozyme aggregation studied by light scattering. II. Variations of protein concentration. *Acta Cryst., D53,* 703–712.

Gokarn, Y. R., Kras, E., Nodgaard, C., Dharmavaram, V., Fesinmeyer, R. M., Hultgen, H. et al. (2008). Self-buffering antibody formulations. *J. Pharmaceut. Sci., 97,* 3051–3066.

Goldstein, G. & Audhya, T. Stabilized aqueous formulations of thymopentin. U.S. Patent 5,140,010, August 18, 1992.

Gombotz, W. R., Pankey, S. C., Phan, D., Drager, R., Donaldson, K., Antonsen, K. P. et al. (1994). The stabilization of a human IgM monoclonal antibody with poly(vinylpyrrolidone). *Pharmaceut. Res., 11,* 624–632.

Gonzalez, M., Murature, D. A., & Fidelio, G. D. (1995). Thermal stability of human immunoglobulins with sorbitol. A critical evaluation. *Vox Sang., 68,* 1–4.

Gu, L. & Fausnaugh, J. (1993). Stability and characterization of human interleukin-1β. In Y. J. Wang & R. Pearlman (Eds.), *Pharmaceutical biotechnology. Vol. 5: Stability and characterization of proteins and peptide drugs: Case histories* (pp. 221–248). New York: Plenum Press.

Gu, L. C., Erdos, E. A., Chiang, H., Calderwood, T., Tsai, K., Visor, G. C. et al. (1991). Stability of interleukin 1β (IL-1b) in aqueous solution: Analytical methods, kinetics, products, and solution formulation implications. *Pharmaceut. Res., 8,* 485–490.

Gupta, S. & Kaisheva, E. (2003). Development of a multidose formulation for a humanized monoclonal antibody using experimental design techniques. *AAPS PharmSciTech., 5,* 74–82.

Ha, E., Wang, W., & Wang, Y. J. (2002). Peroxide formation in polysorbate 80 and protein stability. *J. Pharmaceut. Sci., 91,* 2252–2264.

Hageman, M. J. (1988). The role of moisture in protein stability. *Drug Dev. Ind. Pharm., 14(14),* 2047–2070.

Hageman, M. J. (1992). Water sorption and solid-state stability of proteins. In T. J. Ahern & M. C. Manning (Eds.), *Pharmaceutical biotechnology. Vol. 2: Stability of protein pharmaceuticals. Part A: Chemical and physical pathways of protein degradation* (pp. 273–309). New York: Plenum Press.

Hanson, M. A. & Rouan, S. K. E. (1992). Introduction to formulation of protein pharmaceuticals. In T. J. Ahern & M. C. Manning (Eds.), *Pharmaceutical biotechnology. Vol. 3: Stability of protein pharmaceuticals: Part B: In vivo pathways of degradation and strategies for protein stabilization* (pp. 209–233). New York: Plenum Press.

Harada, A., Li, J., & Kamachi, M. (1993). Macromolecular recognition—Formation of inclusion complexes of polymers with cyclodextrins. *Proc. Jpn. Acad. B, 69,* 39–44.

Harn, N., Allan, C., Oliver, C., & Middaugh, C. R. (2007). Highly concentrated monoclonal antibody solutions: Direct analysis of physical structure and thermal stability. *J. Pharmaceut. Sci., 96,* 532–546.

Hawe, A., Wiggenhorn, M., van de Weert, M., Garbe, J. H., Mahler, H. C., & Jiskoot, W. (2012). Forced degradation of therapeutic proteins. *J. Pharmaceut. Sci., 101,* 895–913.

He, F., Woods, C. E., Trilisky, E., Bower, K. M., Litowski, J. R., Kerwin, B. A. et al. (2011). Screening of monoclonal antibody formulations based on high-throughput thermostability and viscosity measurements: Design of experiment and statistical analysis. *J. Pharmaceut. Sci., 100,* 1330–1340.

Hershenson, S., Stewart, T., Carroll, C., & Shaked, Z. (1989). Formulation of recombinant interferon-β using laureth-12, a novel nonionic surfactant. In D. Marshak & D. Liu (Eds.), *Therapeutic peptides and proteins: Formulations, delivery and targeting* (pp. 31–36). Cold Spring Harbor, NY: Cold Spring Harbor Laboratory.

Hillgren, A., Lindgren, J., & Alden, M. (2002). Protection mechanism of Tween 80 during freeze-thawing of a model protein, LDH. *Int. J. Pharmaceut., 237,* 57–69.

Hlodan, R. & Pain, R. H. (1994). Tumour necrosis factor is in equilibrium with a trimeric molten globule at low pH. *FEBS Lett., 343,* 256–260.

Hoger, K., Becherer, T., Qiang, W., Haag, R., Friess, W., & Kuchler, S. (2013). Polyglycerol coatings of glass vials for protein resistance. *Eur. J. Pharm. Biopharm., 85,* 756–764.

Hutchings, R. L., Singh, S. M., Cabello-Villegas, J., & Mallela, K. M. (2013). Effect of antimicrobial preservatives on partial protein unfolding and aggregation. *J. Pharmaceut. Sci., 102,* 365–376.

Hwang-Felgner, J., Jones, R. E., & Maher, J. F. Gamma interferon formulation. U.S. Patent 5,151,265, September 29, 1992.

Iacocca, R. G., Toltl, N., Allgeier, M., Bustard, B., Dong, X., Foubert, M. et al. (2010). Factors affecting the chemical durability of glass used in the pharmaceutical industry. *AAPS PharmSciTech., 11,* 1340–1349.

Immordino, M. L., Dosio, F., & Cattel, L. (2006). Stealth liposomes: Review of the basic science, rationale, and clinical applications, existing and potential. *Int. J. Nanomedicine, 1,* 297–315.

Johnson, M. D., Hoesterey, B. L., & Anderson, B. D. (1994). Solubilization of a tripeptide HIV protease inhibitor using a combination of ionization and complexation with chemically modified cyclodextrins. *J. Pharmaceut. Sci., 83,* 1142–1146.

Jorgensen, L., van de, W. M., Vermehren, C., Bjerregaard, S., & Frokjaer, S. (2004). Probing structural changes of proteins incorporated into water-in-oil emulsions. *J. Pharmaceut. Sci., 93,* 1847–1859.

Kadima, W., Ogendal, L., Bauer, R., Kaarsholm, N., Brodersen, K., Hansen, J. F. et al. (1993). The influence of ionic strength and ph on the aggregation properties of zinc-free insulin studied by static and dynamic laser light scattering. *Biopolymers, 33,* 1643–1657.

Kahns, A. H. & Bundgaard, H. (1991). Prodrugs of peptides. 13. Stabilization of peptide amides against α-chymotrypsin by the prodrug approach. *Pharmaceut. Res., 8,* 1533–1538.

Kahns, A. H., Buur, A., & Bundgaard, H. (1993). Prodrugs of peptides. 18. Synthesis and evaluation of various esters of desmopressin (dDAVP). *Pharmaceut. Res., 10,* 68–74.

Kameoka, D., Masuzaki, E., Ueda, T., & Imoto, T. (2007). Effect of buffer species on the unfolding and the aggregation of humanized IgG. *J. Biochem., 142,* 383–391.

Kamerzell, T. J., Esfandiary, R., Joshi, S. B., Middaugh, C. R., & Volkin, D. B. (2011). Protein-excipient interactions: Mechanisms and biophysical characterization applied to protein formulation development. *Adv. Drug Deliv. Rev., 63,* 1118–1159.

Katakam, M. & Banga, A. K. (1995). Aggregation of proteins and its prevention by carbohydrate excipients: Albumins and gamma-globulin. *J. Pharm. Pharmacol., 47,* 103–107.

Kendrick, B. S., Tiansheng, L., & Chang, B. S. (2002). Physical stabilization of proteins in aqueous solution. In J. F. Carpenter & M. C. Manning (Eds.), *Rational design of stable protein formulations: Theory and practice* (pp. 61–84). New York: Kluwer Academic/Plenum Publishers.

Kiese, S., Papppenberger, A., Friess, W., & Mahler, H. C. (2008). Shaken, not stirred: Mechanical stress testing of an IgG1 antibody. *J. Pharmaceut. Sci., 97,* 4347–4366.

Kim, D. & Lee, Y. J. (1993). Effect of glycerol on protein aggregation—Quantitation of thermal aggregation of proteins from CHO cells and analysis of aggregated proteins. *J. Therm. Biol., 18,* 41–48.

Kozlowski, S. & Swann, P. (2006). Current and future issues in the manufacturing and development of monoclonal antibodies. *Adv. Drug Deliv. Rev., 58,* 707–722.

Kwan, H. K. Biologically stable alpha-interferon formulations. U.S. Patent 4,496,537, January 29, 1985.

Kwan, H. K. H. Biologically stable interferon compositions comprising thimerosal. U.S. Patent 4,847,079, July 11, 1989.

Lee, J. C. & Timasheff, S. N. (1981). The stabilization of proteins by sucrose. *J. Biol. Chem., 256,* 7193–7201.

Levine, H. L., Ransohoff, T. C., Kawahata, R. T., & Mcgregor, W. C. (1991). The use of surface tension measurements in the design of antibody-based product formulations. *J. Parent. Sci. Technol., 45,* 160–165.

Lewis, U. J., Cheever, E. V., & Seavey, B. K. (1969). Aggregate-free human growth hormone. I. Isolation by ultrafiltration. *Endocrinology, 84,* 325–331.

Lewis, U. J., Parker, D. C., Okerlund, M. D., Boyar, R. M., Litteria, M., & Vanderlaan, W. P. (1969). Aggregate-free human growth hormone. II. Physicochemical and biological properties. *Endocrinology, 84,* 332–339.

Li, M. & Qiu, Y. X. (2013). A review on current downstream bio-processing technology of vaccine products. *Vaccine, 31,* 1264–1267.

Li, Y., Mach, H., & Blue, J. T. (2011). High throughput formulation screening for global aggregation behaviors of three monoclonal antibodies. *J. Pharmaceut. Sci., 100,* 2120–2135.

Li, Y., Shao, Z., & Mitra, A. K. (1992). Dissociation of insulin oligomers by bile salt micelles and its effect on alpha-chymotrypsin-mediated proteolytic degradation. *Pharmaceut. Res., 9,* 864–869.

Lidgate, D. M., Trattner, T., Shultz, R. M., & Maskiewicz, R. (1992). Sterile filtration of a parenteral emulsion. *Pharmaceut. Res., 9,* 860–863.

Liu, F., Kildsig, D. O., & Mitra, A. K. (1991). Insulin aggregation in aqueous media and its effect on alpha-chymotrypsin-mediated proteolytic degradation. *Pharmaceut. Res., 8,* 925–929.

Liu, J. & Shire, S. J. (2002). Reduced-viscosity concentrated protein formulations. [WO 02/30463 A2], pp. 1–50.

Lougheed, W. & Albisser, A. M. (1980). Insulin delivery and the artificial beta cell: Luminal obstructions in capillary conduits. *Int. J. Art. Organs, 3,* 50–56.

Lougheed, W. D., Albisser, A. M., Martindale, H. M., Chow, J. C., & Clement, J. R. (1983). Physical stability of insulin formulations. *Diabetes, 32,* 424–432.

Lougheed, W. D., Woulfe-Flanagan, H., Clement, J. R., & Albisser, A. M. (1980). Insulin aggregation in artificial delivery systems. *Diabetologia, 19,* 1–9.

Maa, Y. F. & Hsu, C. C. (1996). Aggregation of recombinant human growth hormone induced by phenolic compounds. *Int. J. Pharmaceut., 140,* 155–168.

MacMillan, S. & Kantor, A. (2003). Using B22 values to understand opalescence and predict the stability of concentrated liquid rhMAbs. *IBC Third International Conference*, Philadelphia, PA, September 22–24.

Majumdar, S., Ford, B. M., Mar, K. D., Sullivan, V. J., Ulrich, R. G., & D'Souza, A. J. (2011). Evaluation of the effect of syringe surfaces on protein formulations. *J. Pharmaceut. Sci., 100,* 2563–2573.

Manning, M. C., Patel, K., & Borchardt, R. T. (1989). Stability of protein pharmaceuticals. *Pharmaceut. Res., 6,* 903–918.

Matsuyama, K., El-Gizawy, S., & Perrin, J. H. (1987). Thermodynamics of binding of aromatic amino acids to alpha, beta and gamma-cyclodextrins. *Drug Dev. Ind. Pharm., 13,* 2687–2691.

Middaugh, C. R. & Volkin, D. B. (1992). Protein solubility. In T. J. Ahern & M. C. Manning (Eds.), *Pharmaceutical biotechnology. Vol. 2: Stability of protein pharmaceuticals. Part A: Chemical and physical pathways of protein degradation* (pp. 109–134). New York: Plenum Press.

Moriyama, Y., Sasaoka, H., Ichiyanagi, T., & Takeda, K. (1992). Secondary structural changes of metmyoglobin and apomyoglobin in anionic and cationic surfactant solutions: Effect of the hydrophobic chain length of the surfactants on the structural changes. *J. Protein Chem., 11,* 583–588.

Mosharraf, M., Malmberg, M., & Fransson, J. (2007). Formulation, lyophilization and solid-state properties of a pegylated protein. *Int. J. Pharmaceut., 336,* 215–232.

Mufamadi, M. S., Pillay, V., Choonara, Y. E., Du Toit, L. C., Modi, G., Naidoo, D. et al. (2011). A review on composite liposomal technologies for specialized drug delivery. *J. Drug Deliv., 2011,* 939851.

Mulinacci, F., Poirier, E., Capelle, M. A., Gurny, R., & Arvinte, T. (2013). Influence of methionine oxidation on the aggregation of recombinant human growth hormone. *Eur. J. Pharm. Biopharm., 85,* 42–52.

Mulkerrin, M. G. & Wetzel, R. (1989). pH dependence of the reversible and irreversible thermal denaturation of gamma interferons. *Biochemistry, 28,* 6556–6561.

Nadler, T. K., Paliwal, S. K., & Regnier, F. E. (1994). Rapid, automated, two-dimensional high-performance liquid chromatographic analysis of immunoglobulin G and its multimers. *J. Chromatogr. A, 676,* 331–335.

Nail, S. L., Jiang, S., Chongprasert, S., & Knopp, S. A. (2002). Fundamentals of freeze drying. In S. L. Nail & M. J. Akers (Eds.), *Development and manufacturing of protein pharmaceuticals* (pp. 281–360). New York: Kluwer Academic/Plenum Publishers.

Narasimhan, C., Mach, H., & Shameem, M. (2012). High-dose monoclonal antibodies via the subcutaneous route: Challenges and technical solutions, an industry perspective. *Ther. Deliv., 3,* 889–900.

Nayar, R. & Manning, M. C. (2002). High throughput formulation: Strategies for rapid development of stable protein products. In J. F. Carpenter & M. C. Manning (Eds.), *Rational design of stable protein formulations: Theory and practice* (pp. 177–198). New York: Kluwer Academic/Plenum Publishers.

Needham, T. E., Paruta, A. N., & Gerraughty, R. J. (1971). Solubility of amino acids in pure solvent systems. *J. Pharmaceut. Sci., 60,* 565–567.

Nguyen, T. H. & Ward, C. (1993). Stability characterization and formulation development of alteplase, a recombinant tissue plasminogen activator. In Y. J. Wang & R. Pearlman (Eds.), *Pharmaceutical biotechnology. Vol. 5: Stability and characterization of proteins and peptide drugs: Case histories* (pp. 91–134). New York: Plenum Press.

Nielsen, L., Frokjaer, S., Brange, J., Uversky, V. N., & Fink, A. L. (2001). Probing the mechanism of insulin fibril formation with insulin mutants. *Biochemistry, 40,* 8397–8409.

Ohtake, S., Kita, Y., & Arakawa, T. (2011). Interactions of formulation excipients with proteins in solution and in the dried state. *Adv. Drug Deliv. Rev., 63,* 1053–1073.

Oliyai, R. & Stella, V. J. (1993). Prodrugs of peptides and proteins for improved formulation and delivery. *Annu. Rev. Pharmacol. Toxicol., 32,* 521–544.

Osterberg, T., Fatouros, A., & Mikaelsson, M. (1997). Development of a freeze-dried albumin-free formulation of recombinant factor VIII SQ. *Pharmaceut. Res., 14,* 892–898.

Parker, A. S., Choi, Y., Griswold, K. E., & Bailey-Kellogg, C. (2013). Structure-guided deimmunization of therapeutic proteins. *J. Comput. Biol., 20,* 152–165.

PDR Staff. *Physicians' Desk Reference* (2014). (68th ed.) Montvale, NJ: PDR Network, LLC.

Pearlman, R. & Nguyen, T. (1992). Pharmaceutics of protein drugs. *J. Pharm. Pharmacol., 44,* 178–185.

Pearlman, R. & Oeswein, J. Q. Human growth hormone formulation. U.S. Patent 5,096,885, March 17, 1992.

Peters, T. (1985). Serum albumin. *Adv. Protein Chem., 37,* 161–245.

Pikal, M. J., Dellerman, K. M., Roy, M. L., & Riggin, R. M. (1991). The effects of formulation variables on the stability of freeze-dried human growth hormone. *Pharmaceut. Res., 8,* 427–436.

Pongor, S., Brownlee, M., & Cerami, A. (1983). Preparation of high-potency, non-aggregating insulins using a novel sulfation procedure. *Diabetes, 32,* 1087–1091.

Pristoupil, T. I., Kramlova, M., Fortova, H., & Ulrych, S. (1985). Haemoglobin lyophilized with sucrose: The effect of residual-moisture on storage. *Haematologia, 18,* 45–52.

Quinlan, G. J., Coudray, C., Hubbard, A., & Gutteridge, J. M. C. (1992). Vanadium and copper in clinical solutions of albumin and their potential to damage protein structure. *J. Pharmaceut. Sci., 81,* 611–614.

Quinn, R. & Andrade, J. D. (1983). Minimizing the aggregation of neutral insulin solutions. *J. Pharm. Sci., 72,* 1472–1473.

Radhakrishnan, R., Mihalko, P. J., & Abra, R. M. Method and apparatus for administering dehydrated liposomes by inhalation. U.S. Patent 4,895,719, January 23, 1990.

Rai, P., Vance, D., Poon, V., Mogridge, J., & Kane, R. S. (2008). Stable and potent polyvalent anthrax toxin inhibitors: Raft-inspired domain formation in liposomes that contain PEGylated lipids. *Chemistry, 14,* 7748–7751.

Randolph, T. W. & Jones, L. S. (2002). Surfactant-protein interactions. In J. F. Carpenter & M. C. Manning (Eds.), *Rational design of stable protein formulations: Theory and practice* (pp. 159–175). New York: Kluwer Academic/Plenum Publishers.

Rathore, A. S. (2009). Roadmap for implementation of quality by design (QbD) for biotechnology products. *Trends Biotechnol., 27,* 546–553.

Rathore, A. S. (2014). QbD/PAT for bioprocessing: Moving from theory to implementation. *Curr. Opin. Chem. Eng., 6,* 1–8.

Remmele, R. L., Nightlinger, N. S., Srinivasan, S., & Gombotz, W. R. (1998). Interleukin-1 receptor (IL-1R) liquid formulation development using differential scanning calorimetry. *Pharmaceut. Res., 15,* 200–208.

Riggin, A., Clodfelter, D., Maloney, A., Rickard, E., & Massey, E. (1991). Solution stability of the monoclonal antibody-vinca alkaloid conjugate, KS1/4-DAVLB. *Pharmaceut. Res., 8,* 1264–1269.

Roser, B. (1991). Trehalose drying: A novel replacement for freeze-drying. *BioPHARM, 4,* 47–53.

Rouiller, Y., Solacroup, T., Deparis, V., Barbafieri, M., Gleixner, R., Broly, H. et al. (2012). Application of quality by design to the characterization of the cell culture process of an Fc-Fusion protein. *Eur. J. Pharm. Biopharm., 81,* 426–437.

Sarabia, R. E. (2003). Preformulation of protein therapeutics. *IBC Third International Conference*, Philadelphia, PA, September 22–24, 2003.

Sato, S., Ebert, C. D., & Kim, S. W. (1983). Prevention of insulin self-association and surface adsorption. *J. Pharmaceut. Sci., 72,* 228–232.

Schein, C. H. (1990). Solubility as a function of protein structure and solvent components. *Bio/Technology, 8,* 308–317.

Schutz, C. N. & Warshel, A. (2001). What are the dielectric "constants" of proteins and how to validate electrostatic models? *Proteins, 44,* 400–417.

Schwartz, P. L. & Batt, M. (1973). The aggregation of [^{125}I] human growth hormone in response to freezing and thawing. *Endocrinology, 92,* 1795–1798.

Serno, T., Geidobler, R., & Winter, G. (2011). Protein stabilization by cyclodextrins in the liquid and dried state. *Adv. Drug Deliv. Rev., 63,* 1086–1106.

Sharma, A. & Sharma, U. S. (1997). Liposomes in drug delivery: Progress and limitations. *Int. J. Pharmaceut., 154,* 123–140.

Sharma, B. (2007). Immunogenicity of therapeutic proteins. Part 2: Impact of container closures. *Biotechnol. Adv., 25,* 318–324.

Sharma, D. K., Oma, P., & Krishnan, S. (2009). Silicone microdoplets in protein formulations. *Pharm. Technol., April,* 74–79.

Shire, S. J., Shahrokh, Z., & Liu, J. (2004). Challenges in the development of high protein concentration formulations. *J. Pharmaceut. Sci., 93,* 1390–1402.

Sierra, I. M., Vincent, M., Padron, G., & Gallay, J. (1992). Interaction of recombinant human epidermal growth factor with phospholipid vesicles. A steady state and time-resolved fluorescence study of the bis-tryptophan sequence (TRP49–TRP50). *Eur. Biophys. J., 21,* 337–344.

Simpkins, J. W. (1991). Solubilization of ovine growth hormone with 2-hydroxypropyl-beta-cyclodextrin. *J. Parent. Sci. Technol., 45,* 266–269.

Spada, S. & Walsh, G. (2005). *Directory of approved biopharmaceutical products.* Boca Raton, FL: CRC Press.

Spector, A. A. & Fletcher, J. E. (1978). Fatty acid binding by serum albumin. In T. Peters & I. Sjoholm (Eds.), *Albumin: Structure, biosynthesis and function* (pp. 51–60). Oxford, U.K.: Pergamon Press.

Strattan, C. E. (1991). Cyclodextrins and biological macromolecules. *BioPHARM, 4,* 44–51.

Strattan, C. E. (1992a). 2-Hydroxypropy1-beta-cyclodextrin, Part I: Patents and regulatory issues. *Pharm. Technol., 16(1),* 69–74.

Strattan, C. E. (1992b). 2-Hydroxypropyl-beta-cyclodextrin, Part II: Safety and manufacturing issues. *Pharm. Technol., 16(2),* 52–58.

Su, K. S. E. (1991). Nasal route of peptide and protein drug delivery. In V. H. L. Lee (Ed.), *Peptide and protein drug delivery* (pp. 595–631). New York: Marcel Dekker, Inc.

Szejtli, J. (1991a). Cyclodextrins in drug formulations: Part I. *Pharm. Technol., 15,* 36–44.

Szejtli, J. (1991b). Cyclodextrins in drug formulations: Part II. *Pharm. Technol., 15,* 24–38.

Szejtli, J. (1994). Medicinal applications of cyclodextrins. *Med. Res. Rev., 14,* 353–386.

Thomson, J. W. Stable formulation of biologically active proteins for parenteral injection. U.S. Patent 4,816,440, March 28, 1989.

Thurow, H. & Geisen, K. (1984). Stabilization of dissolved proteins against denaturation at hydrophobic interfaces. *Diabetologia, 27,* 212–218.

Timasheff, S. N. (1992). Stabilization of protein structure by solvent additives. In T. J. Ahern & M. C. Manning (Eds.), *Pharmaceutical biotechnology. Vol. 3: Stability of protein pharmaceuticals. Part B: In vivo pathways of degradation and strategies for protein stabilization* (pp. 265–285). New York: Plenum Press.

Touitou, E., Alhaique, F., Fisher, P., Memoli, A., Riccieri, F. M., & Santucci, E. (1987). Prevention of molecular self-association by sodium salicylate: Effect on insulin and 6-carboxyfluorescein. *J. Pharm. Sci., 76,* 791–793.

Tsai, P. K., Volkin, D. B., Dabora, J. M., Thompson, K. C., Bruner, M. W., Gress, J. O. et al. (1993). Formulation design of acidic fibroblast growth factor. *Pharmaceut. Res., 10,* 649–659.

Tyle, P. Sustained release growth hormone compositions for parenteral administration and their use. U.S. Patent, August 15, 1989.

Ugwu, S. O. & Apte, S. P. (2004). The effects of buffers on protein conformational stability. *Pharm. Technol., 28,* 86–108.

Vemuri, S., Beylin, I., Sluzky, V., Stratton, P., Eberlein, G., & Wang, Y. J. (1994). The stability of bFGF against thermal denaturation. *J. Pharm. Pharmacol., 46,* 481–486.

Vemuri, S., Yu, C. T., & Roosdorp, N. (1993). Formulation and stability of recombinant α1-antitrypsin. In Y. J. Wang & R. Pearlman (Eds.), *Pharmaceutical biotechnology. Vol. 5: Stability and characterization of protein and peptide drugs: Case histories* (pp. 263–286). New York: Plenum Press.

Vermeer, A. W. & Norde, W. (2000). The thermal stability of immunoglobulin: Unfolding and aggregation of a multi-domain protein. *Biophys. J., 78,* 394–404.

Vidanovic, D., Milic, A. J., Stankovic, M., & Poprzen, V. (2003). Effects of nonionic surfactants on the physical stability of immunoglobulin G in aqueous solution during mechanical agitation. *Pharmazie, 58,* 399–404.

Voynov, V., Chennamsetty, N., Kayser, V., Helk, B., & Trout, B. L. (2009). Predictive tools for stabilization of therapeutic proteins. *MAbs, 1,* 580–582.

Wakankar, A. A., Wang, Y. J., Canova-Davis, E., Ma, S., Schmalzing, D., Grieco, J. et al. (2010). On developing a process for conducting extractable-leachable assessment of components used for storage of biopharmaceuticals. *J. Pharmaceut. Sci., 99,* 2209–2218.

Wang, W. (1999). Instability, stabilization, and formulation of liquid protein pharmaceuticals. *Int. J. Pharmaceut., 185,* 129–188.

Wang, Y. J. & Hanson, M. A. (1988). Parenteral formulations of proteins and peptides: Stability and stabilizers. *J. Parent. Sci. Technol., 42,* S3–S26.

Weiner, A. L. (1989). Liposomes as carriers for polypeptides. *Adv. Drug Deliv. Rev., 3,* 307–341.

Weiner, A. L. (1990a). Developing lipid-based vehicles for peptide and protein drugs. Part I: Selection and analysis issues. *BioPHARM, 3(3),* 27–32.

Weiner, A. L. (1990b). Lipid-based vehicles for peptide and protein drugs: Part II: Manufacturing variables. *BioPHARM, 3(4),* 16–21.

Xu, X., Costa, A. P., Khan, M. A., & Burgess, D. J. (2012). Application of quality by design to formulation and processing of protein liposomes. *Int. J. Pharmaceut., 434,* 349–359.

Yadav, S., Liu, J., Shire, S. J., & Kalonia, D. S. (2010). Specific interactions in high concentration antibody solutions resulting in high viscosity. *J. Pharmaceut. Sci., 99,* 1152–1168.

Yadav, S., Shire, S. J., & Kalonia, D. S. (2010). Factors affecting the viscosity in high concentration solutions of different monoclonal antibodies. *J. Pharmaceut. Sci., 99,* 4812–4829.

Yasui, K., Fujioka, H., & Nakamura, Y. (1993). Cryoprotective effect of gelatin and albumin on recombinant human tumor necrosis factor liposome. *Chem. Pharm. Bull., 41(12),* 2138–2140.

Yoshioka, S., Aso, Y., Izutsu, K., & Terao, T. (1993). The effect of salts on the stability of β-galactosidase in aqueous solution, as related to the water mobility. *Pharmaceut. Res., 10,* 1484–1487.

Yoshioka, S., Aso, Y., Izutsu, K., & Terao, T. (1994). Application of accelerated testing to shelf-life prediction of commercial protein preparations. *J. Pharmaceut. Sci., 83,* 454–456.

Yu, C., Roosdorp, N., & Pushpala, S. (1988). Physical stability of a recombinant alpha-antitrypsin injection. *Pharmaceut. Res., 5,* 800–802.

5 Lyophilization, Pharmaceutical Processing, and Handling of Therapeutic Peptides and Proteins

5.1 INTRODUCTION

As discussed in Chapter 4, the formulation of peptide and protein drugs is a difficult task and the formulation scientist has to overcome several challenges to develop a stable and efficacious formulation. Once a formulation is identified, it needs to be scaled up for large-scale manufacturing. During such scale-up and subsequent manufacture, the peptide or protein is exposed to several types of stresses. Also, production of the pure protein itself prior to its formulation also exposes the protein to several stress situations. These stress situations can be loosely defined as *pharmaceutical processing*. These include the generation of extensive air–water interfaces due to the turbulence in mixing tanks, foaming, adsorption to filters or tubing, freezing and drying stress during lyophilization, and other unique situations such as exposure to light, organic solvents, or heavy metals (Akers & Nail, 1994). This chapter discusses the stress imposed on peptide and protein drugs during pharmaceutical processing and means to minimize the resulting deleterious effects on protein stability. Emphasis will be placed on protein lyophilization. In addition, handling of therapeutic peptides and proteins in other settings such as in a hospital will also be discussed. A quality by design approach (see Section 4.8) has been described to understand the effect of various variables and their interactions for a lyophilized monoclonal antibody (Awotwe-Otoo et al., 2012).

5.2 PROTEIN DESTABILIZATION INDUCED BY PHARMACEUTICAL PROCESSING

5.2.1 Adsorption Induced by Pharmaceutical Processing

As discussed in Chapter 3, peptides and proteins undergo adsorption to air–water and solid–water interfaces. Such interfaces are readily generated during processing. For example, mixing of solutions will generate extensive air–water interfaces due to the resulting turbulence, while filtration of solutions will generate a solid–water interface. Even after the final product is marketed, the container headspace still provides an air–water interface, while the container walls provide a solid–water interface. Important determinants of adsorption on a solid–water interface will be surface-to-volume ratio, chemical nature of the solid, and the solution/ionic properties of the protein in the formulation. Shear may not directly denature proteins but has indirect effect of creating air–water interfaces, including those created by shear flow during processes such as pumping, filtration, and mixing (Ashton, Dusting, Imomoh, Balabani, & Blanch, 2009; Thomas & Geer, 2011). The adsorption may be minimized by reducing the surface area or solid exposed to solution, selecting a surface with minimum adsorption or controlling the solution properties of the protein. The solution properties of the protein can be controlled by changing the pH, ionic strength, or by addition of cosolvents. For adsorption on air–water interfaces, shaking must be minimized. The other approach to prevent adsorption, obviously, is the addition of appropriate pharmaceutical excipients to the formulation.

While a general discussion of adsorption was provided in Chapter 3, adsorption to filters during pharmaceutical processing of peptide or protein drugs is discussed here. Since proteins are heat labile, they are often sterilized by filtration. Obviously, loss of protein on the filter by adsorption is not desirable for both quality and economic reasons. An understanding of protein adsorption to filters and strategies to overcome such adsorption are, therefore, very important. In studies done by Sarry and Sucker (1992, 1993), the adsorption of several proteins on a variety of microporous membrane filters was investigated. The proteins used included the enzymes, malate dehydrogenase (MDH), lactate dehydrogenase (LDH), catalase (CAT), acid phosphatase (AP), and a lipophilic polypeptide, dihydrocyclosporin A (DH-CyA). The filters studied could be divided into two categories. Polymers with hydrophilic groups included cellulose nitrate (CN), polysulfone (PS-1), nylon 66 (N-1, N-2), regenerated cellulose (RC), and mixed esters (CA/CN). The second category, polymers with nonpolar groups, included cellulose triacetate (CA-1, CA-2), polyvinylidene difluoride (PVDF), polycarbonate (PC), and polysulfone (PS-2). It was found that filtration conditions such as pressure and pore size had no effect on protein adsorption. The filter type was found to have the greatest influence on the extent of adsorption. The most adsorbent filters were CN and CA/CN, followed by N-1, N-2, PS-1, CA-1, and CA-2. Finally, the filters that adsorbed very little protein were PVDF, PC, RC, and PS-2 (Sarry & Sucker, 1993). The adsorption could not be correlated with a measurable property of the polymers. One of the reasons could be that the hydrophobic filters are treated in order to make them wettable. Of the filters tested, the PS-2 and PVDF filters, with their surface made hydrophilic, were found to be particularly suitable for sterilizing dilute protein solutions. The compatibility of the filter with the solvent system used must also be considered before filter selection is made. In the case of MDH, LDH, and CAT, pH of the solution was also found to have a significant effect on adsorption of enzymes. Adsorption was stronger from acidic solutions than from alkaline solutions. In a study by Martin and Manteuffel (1988), it was found that microporous membrane filters adsorb only small quantities of protein nonspecifically before achieving continuous steady-state equilibrium of quantitative protein recovery during the remaining filtration process. Individual membrane types were not distinguishable by their *protein-binding* characteristics for high concentrations (≥1 mg/mL) protein solutions. The *low-protein-binding* membrane filters are useful only with very dilute (about 10 μg/mL or less) protein solutions. Another situation where these filters would be useful is when the protein drug is very expensive and loss of even a milligram quantity is of concern.

5.2.2 Aggregation Induced by Pharmaceutical Processing

The process of protein aggregation, its mechanism, and several individual examples have been discussed in Chapter 3. In this section, several processing factors that may induce such aggregation are discussed. Aggregation may be induced during pharmaceutical processing by several stress situations such as shaking, heating, exposure to hydrophobic surfaces, freeze-thaw cycles, and lyophilization or upon long-term storage of proteins in solid state, especially in the presence of moisture. As mentioned earlier, shear may not directly denature proteins but has indirect effect of creating air–water interfaces. Bee et al. (2009) have reported that shear alone did not cause aggregation of proteins when exposed to stress exceeding that expected during normal processing, using a rheometer. Instead, air-bubble entrainment, pump cavitational stress, adsorption, and contamination by particulates are more likely to result in protein aggregation. Aerosolization of proteins also induces aggregation and this will be discussed in Section 5.4.4.

5.2.2.1 Shaking

Shaking may result during manufacture, shipping, or handling (e.g., reconstitution of lyophilized product) of the product. Aggregation could result during shaking because a protein adsorbs and then unfolds at the air–water interfaces generated by shaking or shear, thereby exposing the hydrophobic amino acids, which are normally located in the interior. The exposed hydrophobic amino acid side chains of one molecule interact with those of another to form aggregates. As much as 70% of porcine growth hormone was lost by aggregation when a 0.5 mg/mL solution was vortexed for just 1.0 min (Charman, Mason, & Charman, 1993). Since shaking results in rapid change of surface, it is possible that some protein may first denature at surface, enter the bulk, and then aggregate in bulk. Aggregation of gamma globulin on shaking resulted in two types of aggregates. Under electron microscope, these appeared as either long fibers or as small particles. It is believed that the fibers formed at the interface, while the particles formed in the bulk based on the hypothesis that proteins coagulated in bulk will show less orientation in structure than that coagulated at surface (Henson, Mitchell, & Mussellwhite, 1970).

5.2.2.2 Hydrophobic Surfaces

As discussed in Section 5.2.2.1, shaking a solution generates air–water interfaces, inducing adsorption of protein on the hydrophobic air interface. Therefore, it seems logical that other hydrophobic surfaces such as Teflon will also induce adsorption and aggregation. This was demonstrated elegantly in a study by Sluzky, Tamada, Klibanov, and Langer (1991). In this study, when insulin solution was shaken in a glass tube in the absence of any headspace, no aggregation occurred, as expected. However, aggregation took place when the solution was shaken either with headspace or with the introduction of Teflon spheres in the glass tube. When the kinetics of insulin aggregation was followed, it was seen that as the number of Teflon spheres was doubled, the induction period was shorter and the slope, that is, rate of aggregation, was steeper. Aggregation of neutral regular porcine insulin was found to be influenced by the contact material. Of the medical grade materials likely to be used in an insulin pump, silicone rubber may be one of the most active in promoting insulin aggregation (Lougheed, Albisser, Martindale, Chow, & Clement, 1983). Processing aids such as silicone oil or extrinsic particles shed by some pumps can lead to nucleation-induced aggregation (Nayak, Colandene, Bradford, & Perkins, 2011). Polysorbate 20 has been reported to inhibit silicone oil–induced aggregation during agitation (Thirumangalathu et al., 2009). Relatively hydrophilic materials such as polyamide, cellulose butyrate, and titanium produced the least (<3%) aggregation under the conditions used in a study by Feingold, Jenkins, and Kraegen (1984). This was followed by materials that induced moderate (7%–16%) aggregation. These included silicone elastomer, polypropylene, and polytetrafluoroethylene. Finally, polyvinyl chloride (37%) and polypropylene syringes (100%) induced the most aggregation.

5.2.2.3 Heating

Aggregation or other denaturation reactions may be induced by heating a protein solution. Heating alters the interactions of the solute and solvent and changes the Gibb's free energy (ΔG) of the system. The protein thus transforms to a physical form for which the ΔG is minimized. Dynamic light scattering has been used to monitor the heat-induced aggregation, and the self-cooling dissociation for insulin and immunoglobulins (Singh, Bohidar, & Chopra, 1991). Porcine growth hormone was also reported to form insoluble aggregates between 50°C and 70°C, as seen by the increase in optical density at 450 nm (Charman et al., 1993).

5.2.2.4 Shear-Induced Aggregation

Shear can induce denaturation or aggregation of peptides and proteins, though the exact mechanism is not known and may involve a complex process (Cartwright, Senussi, & Grady, 1977; Charm & Lai, 1971; Charm & Wong, 1970, 1981; Joubert, Luo, Nashed-Samuel, Wypych, & Narhi, 2011; Lencki, Tecante, & Choplin, 1993; Thomas & Dunnill, 1979; Virkar, Narendranathan, Hoare, & Dunnill, 1981). Some of the earlier literature has assumed that shear could directly disrupt or distort the molecules. More recent literature, however, seems to indicate that shear does not directly denature the molecules. Instead, shear accelerates the turnover of air–liquid interfaces, which causes denaturation and/or aggregation, as discussed earlier. Increased hydrostatic pressure has been reported to cause structural changes in protein molecules due to compression. These changes may then lead to dissociation, aggregation, or gelation of proteins (Yamamoto, Hayashi, & Yasui, 1993). Aggregation can also result during ultrafiltration of proteins. Aggregation of albumin during ultrafiltration was found to be primarily due to a high shear rate combined with high protein concentration. This could be prevented by using low transmembrane pressure so as to minimize the concentration differential and to reduce the shear (Kim, Chen, & Fane, 1993). Shear and surface stress can also occur in aerosol devices used for pulmonary dosing and in needleless jet injector guns. Such stress can also induce aggregation, and use of excipients may be required to stabilize the drug under these situations (Pearlman & Oeswein, 1992). In carefully controlled experiments, Pikal, Dellerman, Roy, and Riggin (1991) have shown that human growth hormone (hGH) does not aggregate even at the highest shear rates. However, it will aggregate readily if a hydrophobic interface is generated.

5.2.2.5 Miscellaneous Factors

Other processes that lead to an input of energy such as radiation, sonication, or ultrasound can also lead to aggregation of proteins. Literature in the food science area suggests that proteolytic enzymes can also induce aggregation in concentrated protein solutions. This is because proteolytic enzymes degrade proteins, resulting in free polypeptides with changed size and conformation, which can undergo aggregation (Persson & Gekas, 1994). Certain metal ions can also induce the aggregation of proteins. High concentrations of certain metals may have a role in the pathogenesis of Alzheimer's disease as they promote the aggregation of β-amyloid peptide (Mantyh et al., 1993). Aggregation of beta-amyloid peptide was also found to depend on the solution conditions such as pH and peptide concentration. The solution conformation of the peptide was reported to be a mixture of beta-sheet, alpha-helix, and random coils, with relative ratios being dependant on the solution conditions (Barrow, Yasuda, Kenny, & Zagorski, 1992). Polymerization of lysozyme is also pH dependent, with dimer being the predominant species between pH 5 and 9 (Branchu, Forbes, York, & Nyqvist, 1999).

5.3 LYOPHILIZATION AS A PHARMACEUTICAL PROCESS

Only a few proteins have sufficient stability to be marketed as solutions with a shelf life of a year or more. Therefore, it becomes important to be able to dry the protein in order to prevent degradation. A dry product also weighs less, thus cutting down transportation costs. Since proteins are heat sensitive, such drying must be done without using elevated temperatures. The most common method to accomplish this is lyophilization or freeze-drying. Over half the protein products on the market are freeze-dried formulations. In addition to preventing degradation, freeze-drying also results in a product with a very high specific surface area. These physical attributes result in fast and complete dissolution during product reconstitution. Also, freeze-drying process can be carried out aseptically to result in a sterile product (Gatlin & Nail, 1994). This allows the formulation of injectable products, the dosage form most commonly used for peptide and protein drugs. Lyophilization is currently the primary practical approach for product development, in spite of its various limitations that will be discussed. However, a solution formulation can be investigated in parallel and may be preferred, if successful. Alternatively, a company may first market a lyophilized product and then work or continue to work on a solution formulation.

5.3.1 Process Details

Freeze-drying is a process where water or other solvent is removed from a frozen solution by sublimation. The liquid formulation is aseptically placed in vials that are partially stoppered with slotted rubber closures that allow the escape of water vapor. The vials are then aseptically transferred to the freeze-dryer, which is equipped with a chamber containing shelves, a vacuum pump, a condenser, and a refrigeration system. The freeze-dryer should be capable of being sterilized by steam prior to use. In addition, many systems are equipped with a mechanism to stopper the vials within the chamber at the end of secondary drying, automatic clean-in-place equipment, automatic loading and unloading of vials, and computerized monitoring and control (Gatlin & Nail, 1994). While chamber-type freeze-dryers with heat-controlled shelves are typically used for production, laboratory-scale work can be done by using manifold systems where frozen containers are attached to an evacuated manifold having a chilled condenser (Cappola, 2000).

Lyophilization process consists of three steps: (1) freezing, (2) primary drying, and (3) secondary drying. To monitor the process, the product temperature, shelf temperature, and chamber pressure are continually recorded during the process (Figure 5.1). In the first step, the water and solutes are completely frozen. The microstructure formed during freezing must be retained during drying to get the desired physical attributes of a freeze-dried product.

After freezing is accomplished, the chamber pressure is reduced and the temperature of the shelves is increased to provide energy for primary drying. Primary drying removes frozen water under vacuum through sublimation. The driving force for sublimation is the differences in vapor pressure of water within the system. The process is limited by the rate of transfer of heat from the shelf to the sublimation front and rate of transfer of water vapor from the product to the condenser. The predominant mechanism of heat transfer is conduction, and thus, good thermal contact between the shelf and the product is essential. Typically, the liquids to be

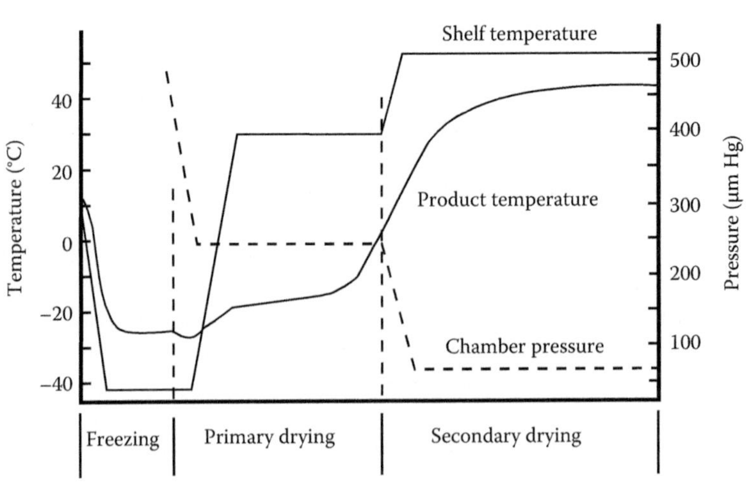

FIGURE 5.1 Process variables during the freeze-dry cycle. (Replotted from Gatlin, L.A. and Nail, S.L., Freeze drying: A practical overview, in R.G. Harrison (Ed.), *Protein Purification Process Engineering*, Marcel Dekker, Inc., New York, 1994, pp. 317–367. With permission from Taylor & Francis.)

freeze-dried are filled into containers to a thickness of 10–20 mm or less. Higher thickness would limit drying and prolong lyophilization times. The desired thickness may be achieved by a simple fill or the liquid may be spin frozen on the walls of a vial (Cappola, 2000; Nail, Jiang, Chongprasert, & Knopp, 2002).

During primary drying, a receding boundary is observed in the vial as the thickness of dry solids increases as the sublimation interface and the thickness of frozen solution decreases. Due to the small dimensions of the pores in a typical dried product, flow in the dried product contributes to most of the total product resistance, which in turn determines the economics of the freeze-drying process (Pikal, Shah, Senior, & Lang, 1983). If the formulation has a high solids content (above 30%), the rate of water removal may be slowed. Once all water is removed, the apparently *dry* product still contains some nonfrozen *bound* water, about 0.3–0.35 g/g of globular protein that constitutes the hydration shell of the protein (Wang, 2000). Secondary drying removes a part of this unfrozen *bound* water. The shelf temperature is further increased, and the chamber pressure is further decreased for this stage of lyophilization. During this secondary drying, heat of sublimation is no longer required and the applied heat thus causes an increase in the product temperature. This serves as an indicator that secondary drying has begun (Figure 5.1). The product temperature can be measured by using thermocouples placed in the center of select vials located in the center of the shelf. Generally, the water content of a lyophilized protein product is less than 10%. Thus, part of the hydration shell is removed, which may disrupt the native state of the protein. Excipients that help to stabilize the protein are discussed in Section 5.3.6.3.

5.3.2 Maximum Allowable Product Temperature

Primary drying must be carried out below the eutectic temperature for a crystalline solute. As an example, the dissolution of sodium chloride (salt) in water can be considered. This addition of solute to water causes a depression in the freezing point of water. As this system is cooled below the new freezing point, pure water begins to crystallize out as ice. As cooling continues, the interstitial fluid becomes more concentrated with salt. Finally, crystallization of the concentrated interstitial fluid occurs as a eutectic mixture at the eutectic temperature. A eutectic mixture is an intimate physical mixture of two or more crystalline solids that melt as they were a single compound. Many solutes, including peptides and proteins, do not crystallize during freezing. For such amorphous materials, eutectic crystallization does not occur. Instead, the system freezes by a change in viscosity of the system from a viscous liquid to a glass at a temperature called the glass transition temperature. This is typically written as T_g', to differentiate it from the softening point or a true glass transition, T_g of protein/excipients in the solid state or of a pure polymer. The T_g' can be detected by differential scanning calorimetry (DSC) as a second-order increase in the baseline of the thermogram and typically occurs just prior to the onset of the endotherm for the melting of ice. The T_g' for pure sucrose is –32°C, while that of pure trehalose is –30°C. In the solid state with water contents of 1%, the T_g for pure sucrose and trehalose are about 100°C and 65°C, respectively (Carpenter, Chang, Garzon-Rodriquez, & Randolph, 2002). The water content should be specified since water has a very low T_g (−135°C) and thus increasing water content will greatly reduce T_g. Compared to excipients, dried proteins have a relatively high T_g. The overall T_g for the formulation must be higher than the storage temperature (e.g., room temperature) to maintain protein stability. However, methionine oxidation can still occur rapidly in a dried protein formulation and should be prevented or minimized by sealing vials under nitrogen or using additives such as free methionine to compete with protein residues for reactive oxygen species (Carpenter et al., 2002). The glass transition temperature should be considered when planning accelerated stability studies (Duddu & Dal Monte, 1997). Molecular mobility is reduced below the glass transition temperature. The viscosity changes by 3–4 orders of magnitude in the glass transition region. The glass transition temperature can be seen on a DSC scan as a shift in the baseline toward higher heat capacity. For an amorphous solute, primary drying must be carried out below the glass transition temperature. The eutectic temperature or the glass transition temperature of the formulation represents the maximum allowable product temperature during lyophilization. The consequences of lyophilizing above these temperatures will be discussed under *Lyophilization Cycle*. Temperature dependence of molecular mobility below T_g and its possible affect on long-term stability of lyophilized proteins have been investigated (Duddu, Zhang, & DalMonte, 1997).

5.3.3 Measurement of Lyophilization-Induced Thermal Changes

The changes in transition states induced by lyophilization may be measured by DSC. See Chapter 2 for a discussion of DSC as an analytical tool and the use of microcalorimeter and modulated DSC for characterization of proteins. A freeze-drying microscope can also be utilized to directly visualize the ice crystals since it uses a low-temperature stage that can be evacuated (Nail et al., 2002). As complete solidification of a formulation takes place during freezing, the resistance increases and thus resistivity measurements can also be used to monitor the process. X-ray powder diffraction can be used as reliable method for characterization of the physical state of a freeze-dried solid and should be used periodically during the stability studies as well. NMR technique can be used to measure molecular mobility of frozen systems through the spin–lattice (T_1) and spin–spin (T_2) relaxation times. Another powerful technique, infrared spectroscopy, discussed in Chapter 2 can be used to monitor changes in secondary structure of a protein following lyophilization. Proteins undergo reversible rearrangement of their secondary structure upon removal of water. Using FTIR spectroscopy, the α-helix content of recombinant human albumin was found to drop from 58% in solution to 25%–30% in the dehydrated state. Also, the β-sheet content increased from 0% to 10%–20% and unordered structures increased from 40% to 50%–60% (Costantino, Griebenow, Mishra, Langer, & Klibanov, 1995).

5.3.4 Protein Denaturation during Lyophilization

Since lyophilization can dry a protein solution without heat, it may seem that this is the perfect solution to all the formulation problems relating to peptide and protein drugs. Unfortunately, this is not the case. Lyophilization process induces conformational instability in many proteins, and the freeze-dried protein also is subject to degradation in the solid state. Two different types of stress can be encountered during lyophilization. The mechanism of protein denaturation is different for these two situations. The first stress is due to the freezing itself, while the second stress is due to the drying process. Different proteins tolerate these stresses to various degrees. Protein denaturation during lyophilization may be monitored by infrared spectroscopy as discussed in Chapter 2. A common denaturation reaction during lyophilization is the aggregation of protein drugs. Aggregates may form during freezing or drying or during reconstitution of structurally perturbed protein molecules.

Proteins that normally exist as an oligomer may dissociate during lyophilization. For example, L-asparaginase normally exists as a tetramer in the active form. Upon lyophilization, it dissociates into inactive monomer subunits. These monomers can reassociate into active tetramer if reconstituted at neutral pH and low ionic strength. However, if reconstituted at high pH and high ionic strength, the monomers are not able to reassociate into a tetramer (Hellman, Miller, & Cammack, 1983). Aggregation during lyophilization may also be dependent on formulation pH. Interleukin-2 has been reported to unfold extensively upon lyophilization at pH 7 but was an order of magnitude more stable at pH 5 with respect to formation of aggregates during lyophilization (Prestrelski, Pikal, & Arakawa, 1995). Structural descriptors of proteins that may predict aggregation during lyophilization have been investigated (Roughton, Iyer, Bertelsen, Topp, & Camarda, 2013).

5.3.5 Freeze-Thaw Stability

Since freezing of a protein solution may result in situations other than preparation for freeze-drying, the freeze-thaw stability also becomes important. During the processing, storage, and shipping of proteins, freezing and freeze-thaw stress may arise, either as a desired manipulation or even by accident. Some proteins may be stored frozen for long-term storage and then thawed before use. A reconstituted lyophilized protein is also sometimes subdivided into aliquots and stored frozen. Finally, there is a need to be prepared for a situation where accidental freezing takes place during shipping and handling (Arakawa, Prestrelski, Kenney, & Carpenter, 1993). Several mechanisms may be involved in the denaturation of protein during freezing. Several empirical studies have reported the denaturation and aggregation of proteins during freezing or thawing (Barnard, Singh, Randolph, & Carpenter, 2011; Hawe, Kasper, Friess, & Jiskoot, 2009; Koseki, Kitabatake, & Doi, 1990; Shikama & Yamazaki, 1961; Soliman & Berg, 1971; Zhang, Singh, Shirts, Kumar, & Fernandez, 2012). A common mechanism of denaturation relates to the changes in the microenvironment of the protein during freezing. As the solute starts to freeze, pure water first starts separating as ice, while the protein and salts are excluded from the ice crystals. Thus, the concentration of all solutes increases dramatically in the non-ice phase, and this localized high salt or solute concentration may denature the protein. Such denaturation may be due to a direct effect of salt on the protein or due to indirect effects such as a dramatic change in pH in the microenvironment of the protein. For example, a pH 7.5 sodium phosphate buffer may drop to a pH as low as 4.5 during freezing, due to the selective crystallization of Na_2HPO_4. This happens because the solubility of the disodium salt is lower than that of the monosodium salt. On the other hand, a potassium phosphate buffer may increase in pH since the monobasic salt of potassium phosphate, being less soluble than the dibasic salt, tends to precipitate during freezing. However, such significant pH shifts may only occur if eutectic crystallization takes place. A more typical shift seen is about 0.5 units due to freeze concentration of solutes. A sodium phosphate buffer has been reported to give more turbid solutions upon reconstitution of lyophilized bovine IgG as compared to potassium phosphate buffer, and this effect was not attributed to pH shifts during freezing (Sarciaux, Mansour, Hageman, & Nail, 1999).

Protein may also adsorb to the ice–water interface. Rapid or quench cooling may generate small ice crystals or a large ice–water interface as compared to slow cooling and may thus have a greater denaturation effect. However, slower cooling may generate large ice crystals in which secondary drying may slow down due to smaller surface area. A moderate degree of supercooling (10°C–15°C) may thus be required as a compromise (Nail et al., 2002; Wang, 2000). Also, the protein will increase in concentration in the non-ice phase, and this may also lead to aggregation of individual molecules. Though the freezing point of water is 0°C, it may not actually freeze at 0°C unless seeded with crystalline ice. Instead, it may remain liquid at subzero temperatures. Such supercooling is common in lyophilization cycles before freezing takes place. As ice crystals form, there is an exothermic release of energy of crystallization. The ideal freezing methodology for proteins produces moderate supercooling and rapid ice growth so that the protein is not exposed for too long to the hostile environment of the concentrated aqueous system formed during freezing. This supercooling depends on factors such as volume of water, its purity, and the cooling rate. Such supercooling or undercooling *without freezing* may be desirable to stabilize some proteins. This is because while cooling retards all kinetic rate processes, freezing may actually damage the proteins as we discussed in Section 5.3.4. The nucleation probability at subzero temperatures can be reduced significantly by dispersing an aqueous solution into an immiscible phase as microscopic droplets. Protein stabilization by undercooling has been done for laboratory storage of small quantities of intermediates or chromatographic fractions. This process may not be viable for therapeutic proteins due to safety and regulatory reasons (Franks, 1993).

5.3.6 Formulation Components for Lyophilization

A typical lyophilized protein formulation may contain a bulking agent, buffer, tonicity modifier, and other excipients that may be classified as cryoprotectants and lyoprotectants (Hanson & Rouan, 1992). Nonionic surfactants can be used to reduce protein aggregation. Since phosphate buffer often causes problems during lyophilization, alternate buffers such as tris, histidine, or citrate may be used. Trehalose or sucrose are often used to stabilize the protein as they are both nonreducing sugars that can act as both cryo- and lyoprotectants and provide a glassy matrix by remaining amorphous during lyophilization. Bulking agents typically used are discussed in Section 5.3.6.1. Some excipients may also be added to the reconstituting solution rather than to the formulation directly. For example, the presence of nonionic surfactant such as Tweens during lyophilization has been reported to adversely affect the stability of a protein antibody but helps to optimize protein recovery if added to the reconstituting solution (Jones et al., 2001). We will briefly consider the other excipients first and then discuss the cryoprotectants and lyoprotectants in further details.

5.3.6.1 Bulking Agents

Since protein drugs are very potent, very small quantities are required in the product. In the absence of a bulking agent, it will be hard to process the product and the resulting lyophilized vial will look almost empty. Bulking agents, therefore, allow pharmaceutical processing and allow the production of a presentable lyophilized product. Mannitol is a commonly used bulking agent. Other possible bulking agents include glycine or hydroxyethyl starch. The latter is an amorphous polymer with a high collapse temperature (around −10°C) and high T_g. However, it does not stabilize the protein and, if used, will still need the concomitant use of some stabilizing disaccharides.

5.3.6.2 Tonicity Modifiers

Mannitol, in addition to being a bulking agent, can also adjust tonicity if present in the right concentration. Dextrose or sodium chloride can also be used to control tonicity. However, sodium chloride can create problems during lyophilization. It can reduce the collapse temperature of the formulation even if a fraction of the salt does not crystallize. It has been reported to enhance aggregation during lyophilization, as in the case of hGH. In situations where it is not possible to adjust tonicity prior to lyophilization, it may be possible to reconstitute the product with an isotonic diluent before use (Hanson & Rouan, 1992). Also, excipients such as mannitol or glycine, used as crystalline bulking agents, can also render the formulation isotonic without the need for addition of salts. The concentration of any buffers used should be kept low to avoid chances of precipitation during freezing.

5.3.6.3 Cryoprotectants and Lyoprotectants

Since the mechanism of protein denaturation during freezing is different from that during drying, it follows that some excipients may stabilize the protein during freezing, while others may stabilize the protein during drying. The excipients that help against freezing are called cryoprotectants, while those that help against drying are called lyoprotectants. This is because the stabilization mechanism of excipients depends on whether the protein molecules are in the dried, solution, or frozen state.

5.3.6.3.1 Cryoprotectants

Stabilization of proteins in the frozen state is based on the same mechanisms as stabilization in aqueous solutions. Therefore, the preferential exclusion mechanism discussed in Chapter 4 to explain the protective effect of excipients in formulation applies here as well. Carpenter and Crowe (1988) investigated over two dozen different compounds that protected LDH from damage during freeze-thawing. These solutes came from different classes such as sugars, polyols, and amino acids, and they all stabilize the enzyme. The only property common to this diverse group of compounds was that they all were shown to be preferentially excluded from the surface of the protein in aqueous solutions (Arakawa et al., 1993).

5.3.6.3.2 Lyoprotectants

Thermodynamic considerations of preferential exclusion of structure-stabilizing compounds may not be applicable when water is removed from the system. Such excipients may stabilize proteins by binding to the dried protein, thus serving as *water substitute*, when the hydration shell of the protein is removed. An alternate mechanism that has been proposed involves vitrification, which depends on the immobilization of protein molecules accompanied by glass formation (Arakawa, Kita, & Carpenter, 1991; Carpenter, Prestrelski, & Arakawa, 1993; Liao, Brown, Quader, & Martin, 2002; Prestrelski, Arakawa, & Carpenter, 1993). Therefore, such a mechanism could explain the stabilization of a lyophilized protein by excipients. Carbohydrates such as trehalose or maltose were found to stabilize phosphofructokinase during freeze-drying. In the absence of these excipients, the enzyme was completely inactivated, while in the presence of these excipients, it retained up to 80% of original activity after freeze-drying (Carpenter, Crowe, & Crowe, 1987). Lyoprotectants may also stabilize the protein during storage subsequent to freeze-drying. The addition of lyoprotectants had no measurable effect on the in-process loss of a mouse IgG monoclonal antibody during freeze-drying. However, they had a dramatic influence on stability of the antibody during storage. The best lyoprotectant in this study was HPβ-CD, and this was ascribed to a relatively high collapse temperature (≈−9°C), amorphous nature, good water solubility, and ability of complexation (Ressing et al., 1992). Similarly, a 360:1 molar ratio of lyoprotectant to protein has been reported to be required for storage stability of a model recombinant humanized monoclonal antibody (Gregoriadis et al., 1999). Since high concentrations may be required for monoclonal antibody formulations, such excipient-to-protein molar ratios may result in hypertonic solutions and a compromise may sometimes be required to lower the stability requirements to achieve appropriate tonicity (Shire, Shahrokh, & Liu, 2004).

Based on the mechanism of hydrogen bonding as discussed, it may be expected that all sugars should act as lyoprotectants. However, this is not the case. In order to effectively hydrogen bond with the protein, the excipient must be in an amorphous form. Furthermore, amorphous excipients may increase protein–protein distance in the solid state and thus prevent aggregation (Costantino et al., 1998). A concentration of 0.3 M or higher is also typically needed for stabilization. Disaccharides such as trehalose, sucrose, maltose, and lactose seem to be more effective than monosaccharides like glucose or larger saccharides like maltohexaose. Sucrose and trehalose are commonly used as lyoprotectants for biomolecules (Wang, 2000). DSC and infrared spectroscopic techniques have been used to show that sugars that acted as lyoprotectants for freeze-drying of enzymes were in the amorphous form. Excipients such as polyethylene glycol (PEG) that existed in crystalline form were good cryoprotectants (by preferential exclusion) but not good lyoprotectants. It follows from this discussion that PEG/sugar mixtures were highly effective in stabilizing enzymes during the overall freeze-drying process (Prestrelski et al., 1993).

Similarly, it has been reported that when freeze-drying enzymes from a phosphate buffer, mannitol must be in an amorphous state for stabilization of these enzymes. Presumably, crystallization leads to a loss of the interaction between the protein and mannitol as the mannitol molecules interact among themselves to form the crystal (Izutsu, Yoshioka, & Terao, 1994). Such crystallization can take place at temperature as low as –25°C, and thus, the lyophilization cycle used is very important, as will be discussed in the next section. The state of the drug itself is also very important to its stability. Generally, crystalline form is presumed to be more stable, but amorphous insulin has been reported to be more stable than crystalline insulin (Pikal & Rigsbee, 1997).

5.3.7 Lyophilization Cycle

The first step to develop an optimum lyophilization cycle is to characterize the thermal properties of the formulation. During freezing, eutectic crystallization or glass formation (vitrification) would typically occur. For a formulation that forms a eutectic mixture upon cooling, lyophilization should begin below the melting point of the eutectic to avoid melt back. If product temperature exceeds the eutectic temperature, ice can melt and drying will take place from the liquid state instead of solid state. Thus, desirable properties of the freeze-dried product may be lost. For formulations that do not crystallize but form an amorphous glass upon cooling, lyophilization must begin below the glass transition temperature to avoid collapse of the cake structure. The collapse temperature can be directly and visually determined using a freeze-drying microscope, or an indirect approach using several runs in a freeze-dryer or by using DSC can be used for its determination. If lyophilization is performed above the glass transition temperature, the glassy interstitial material will lose viscosity and may not have enough rigidity to support its own weight, resulting in collapse of the product (Gatlin & Nail, 1994; Hanson & Rouan, 1992). The collapse temperature is the temperature above which the *dried* region adjacent to the ice–vapor interface attains sufficient fluidity to flow, thus destroying the cake structure. The collapse temperature (T_{col}) would be higher than T_g', though the value would differ for each excipient. For example, T_g' and T_{col} for citric acid is –54°C and –53°C, respectively, whereas those for mannitol are –33°C and –27°C, respectively (Wang, 2000). Collapse temperatures observed with amorphous solutes are generally much lower than the eutectic melting temperature of ice and most organic compounds. Formulations with collapse temperatures of –40°C or lower should be avoided since most commercial-scale lyophilizers cannot consistently maintain such low product temperatures (Nail et al., 2002). Excipients such as mannitol, glycine, lactose, and sucrose could be useful as they have a relatively high collapse temperature and produce a cosmetically elegant product. Collapse of the formulation can generally be observed visually, and any such vials in a batch should be rejected. Protein stability in collapsed formulations is expected to be low, partially due to the high residual moisture resulting from the lower specific surface area of the collapsed product (Nail et al., 2002). Polymers such as dextran and hydroxyethyl starch also have relatively high collapse temperatures (Carpenter, Pikal, Chang, & Randolph, 1997).

A formulation with amorphous components may have *unfrozen* water as high as 20% following primary drying. This is because not all the water initially present in the formulation was converted to ice during freezing. The formulation thus typically needs a prolonged secondary drying to remove this high concentration of *unfrozen* water. In contrast, almost all water is present as ice for solutes that crystallize and no secondary drying may be needed. A typical residual moisture content following secondary drying may be about 0.5%–4% w/w. Some residual moisture is required to maintain protein stability (Nail et al., 2002), but it should typically be less than 3%. During secondary drying, the temperature should be raised to a point that removes water but does not cause degradation of the product. The temperature and duration of secondary drying will control the final moisture content of the product, a critical factor that in turn may determine the long-term stability of the protein. The importance of freeze-drying cycle development to control moisture content in a monoclonal antibody formulation as a function of shelf temperature, chamber pressure, and drying time has been reported (Ma, Wang, Bouffard, & MacKenzie, 2001), and effect of residual moisture on stability of a lyophilized monoclonal antibody has been investigated (Breen, Curley, Overcashier, Hsu, & Shire, 2001). An efficient single-step lyophilization cycle has also been reported where the shelf temperature was set for secondary drying and product temperature during primary drying was maintained below T_g' by adjusting the chamber pressure (Chang & Fischer, 1995).

When formulating freeze-dried pharmaceuticals, crystalline drug or excipient is sometimes desired. It might be possible to induce some amorphous materials to crystallize by freezing the product and then cycling it above and below its T_g. A technique called aging, or annealing, can be used to induce crystallization of the drug or excipient by holding the system for 2–6 h above the glass transition temperature of the metastable glass, while avoiding melting the product (Izutsu, Yoshioka, & Terao, 1993; Nail et al., 2002). A crystalline drug may sometimes be desired for improved shelf life, especially with antibiotics. Since proteins are amorphous, these characteristics do not apply to proteins. However, a mixture of a crystalline and amorphous excipient is sometimes desirable. The amorphous excipient stabilizes the protein, while the crystalline excipient such as glycine yields good cake properties upon lyophilization. The presence of a crystalline sugar (mannitol) and an amorphous protectant (dextran, sucrose, trehalose, or 2-hydroxypropyl-β-cyclodextrin) was reported to reduce the formation of dimeric and oligomeric species during lyophilization (Hora, Rana, & Smith, 1992). Mannitol–sucrose mixtures have also been successfully used for protein lyophilization (Johnson, Kirchhoff, & Gaud, 2002). An annealing step is often required when using mannitol or glycine to assure maximum crystallization and a good cake structure. DSC can be utilized to design an annealing protocol to complete mannitol crystallization. Problems have been reported where mannitol was not completely crystallized and rather formed a metastable glass that under certain shipping conditions altered the water content of the formulation, resulting in increased degradation of the protein in the formulation (Carpenter et al., 1997). Mannitol hemihydrate can crystallize during freeze-drying, and this should be avoided since it can release water into the formulation during storage (Larsen, Trnka, & Grohganz, 2014). The pH and ionic strength of the formulation and the salt form of glycine have been reported to influence glycine crystallization during freezing (Akers, Milton, Byrn, & Nail, 1995).

Data collected during the development of a lyophilization cycle are generally used to scale up the process on a trial-and-error basis. The freezing rate is expected to affect the stability of the protein during lyophilization (e.g., due to the size of the ice crystals formed) and possibly even storage stability and should be investigated. A concentration effect may also be investigated since it has been reported, at least for antibodies, that an increase in concentration reduced the level of aggregates upon lyophilization. While residual moisture may play a role, this increased stabilization could not be completely attributed to change in residual moisture (Sarciaux et al., 1999). Most likely, a relatively high protein concentration increases the *intrinsic* resistance of a protein to denaturation. Also, an increased protein-to-disaccharide mass ratio will increase the formulation collapse temperature and the T_g of the final dried product as proteins tend to have a high T_g. Furthermore, a concentrated solution will reduce the total volume to be lyophilized, thereby reducing the duration and cost of the process (Carpenter et al., 2002).

Validation must also be done and must include aseptic processing and filling validation. The filling process can introduce physical stress as shear, friction, cavitation, foaming, contact with filling materials, and possibly an increase in temperature (Nayak et al., 2011). Since peptide and protein drugs can be very expensive, a placebo formulation that contains all components except the drug may be used for scale-up studies. However, it must be ensured that the drug is not determining the freezing characteristics of the formulation. Once the process enters full-scale manufacturing, the cost of one freeze-dryer load of a recombinant protein can be as high as $1 million, and thus strict control of equipment and lyophilization cycle is a must (Gatlin & Nail, 1994). In addition to cost of the drug, freeze-drying is also an expensive step in terms of both operative and equipment costs. Since vapor pressure of ice increases with increased product temperature, it is desirable to maintain product at a relatively high temperature to increase process efficiency. However, the product must be at least 3°C–4°C below the collapse temperature to avoid any potential loss of the product. The use of an inappropriate lyophilization cycle may also lead to aggregation of the protein, which may manifest itself as a haze upon reconstitution. Thus, a simple formulation with a high T_g, no (or minimum) buffers or salts, and a combination of an amorphous excipient such as sucrose and a crystallizing excipient such as glycine may be ideal. While much attention has been given to the drying process, a good understanding of the freezing step is also required to develop an efficient lyophilization cycle (Kasper & Friess, 2011).

5.4 ALTERNATIVE PHARMACEUTICAL PROCESSES

The lyophilization of peptides and proteins, though very useful and common, suffers from several disadvantages. A protein can denature or aggregate during the freezing step, as already discussed. Also, lyophilization needs substantial capital investment, high energy input, and long process times. A crucial factor that determines the shelf life of a product is its final physical state, irrespective of the method used to remove the water. This offers alternatives to lyophilization, such as controlled evaporative drying. Being an endothermic process, such a technique will allow control of the temperature of drying. In contrast, freezing is an exothermic process and thus latent heat of crystallization needs to be removed quickly to prevent ice from melting back into the product. Also, evaporation is faster and more energy efficient and requires less capital investment (Franks, Hatley, & Mathias, 1991). Coprecipitation of enzymes with a water-soluble starch in an organic solvent has also been proposed as an alternative to freeze-drying. Using this technique, 100% of proteolytic enzyme activity was recovered for krill proteases. Coprecipitation with starch reduces the denaturation of the protein by physically entrapping the enzyme into a support matrix (Randen, Nilson, & Edman, 1988). A granulation technique, wet spherical agglomeration, has also been explored for peptides and proteins as a low-cost alternative to lyophilization. The procedure would involve suspending the protein in a suspension liquid and then agglomerating it by adding a small amount of a second immiscible liquid called bridging liquid. Bausch and Leuenberger (1994) have reported that for water-soluble proteins such as bovine serum albumin or recombinant bovine α-interferon, *n*-hexane was an ideal suspension liquid. Agglomeration was achieved by adding an aqueous buffer as the bridging liquid. Before agglomeration, the protein was mixed with an excipient such as mannitol so that a fast dissolving product is obtained. The procedure of agglomeration using *n*-hexane had no influence on the biological activity of α-interferon, and the residual organic solvent could be reduced to less than 0.01%.

While these techniques may be of more academic interest, spray drying or use of nonaqueous solvents offers a more feasible alternative to lyophilization and will be discussed in greater details. Some of the other processing techniques not discussed so far are also discussed in this section.

5.4.1 Spray Drying

Spray drying, a well-established conventional drying process, has relatively recently found some acceptance as a processing technique for protein formulations, especially to develop products with uniform particle size for systemic delivery via pulmonary route. Spray drying has been explored for its potential use as an alternative to lyophilization and air-jet milling of peptide and protein drugs. Limited early studies in this area were most likely due to concerns that proteins would not be able to withstand the heat during the spray drying process. More success has been received based on a judicious choice of stabilizing excipients and their concentration for the specific protein being used and consideration of the complexities of spray drying conditions, T_g, particle size and morphology, and residual water content. Sucrose and trehalose have been found to be effective stabilizers for spray drying of proteins (Lee, 2002). In one relatively early study, the feasibility of spray drying solutions of recombinant methionyl hGH and tissue-type plasminogen activator (t-PA) was investigated. While hGH underwent extensive aggregation during atomization, t-PA remained intact upon atomization. Degradation of hGH during processing was due to surface denaturation at the air–liquid interface of the droplets in the spray and could be reduced by the addition of polysorbate 20 into the formulation.

Since polysorbate 20 is a surfactant, it is likely that it acts by replacing hGH molecules at the air–liquid interface. This study found that it was feasible to spray dry t-PA in the current marketed formulation. Although high temperatures are used in spray drying, the key to avoid degradation is that the contact time between droplets and hot air in the spray drying chamber be very small. A 40 s exposure of a t-PA solution to 50°C fails to degrade the protein and even a 20 s exposure to 80°C results in only negligible degradation. However, a 40 s exposure at 80°C results in 16% degradation, thus emphasizing the importance of contact time. A free-flowing powder with residual moisture contents similar to those obtained by lyophilization can thus be obtained by spray drying (Mumenthaler, Hsu, & Pearlman, 1994). Another study has also shown the protective effect of surfactants to reduce protein instability during spray drying (Adler, Unger, & Lee, 2000). Aggregation of spray-dried protein powders upon storage has been shown to be dependent on the excipient-to-protein ratio and residual moisture, but a compromise may be required between optimal stability and the need to maintain the powder characteristics that are optimal for aerosol administration (Andya et al., 1999; Maa et al., 1998).

Unlike lyophilization, the spray drying process can also be used to control the shape and size of the dried product (Hanson & Rouan, 1992; Lee, Heng, Ng, Chan, & Tan, 2011; Mobus, Siepmann, & Bodmeier, 2012). Spray drying has also been used to produce stable dried liposomes that could be administered by inhalation. The preparation of the dry product will solve the stability problems. Once administered, the inhaled particles will rehydrate in contact with the moist surface of the airways. Using 10% lactose as a bulking excipient, particles ranging from 1 to 20 μm (mean diameter 7 μm) were generated (Goldbach, Brochart, & Stamm, 1993a, 1993b). A related technology, spray coating, uses seed particles as inert carriers and can be applied to biopharmaceuticals. A recent report used spray coating to produce microsized vaccine powder formulation (40–60 μm) for epidermal powder immunization (Maa, Ameri, Rigney, Payne, & Chen, 2004).

5.4.2 Use of Nonaqueous Solvents

The use of nonaqueous solvents will allow a *dry* state for proteins as an alternative to lyophilization. Such use may be feasible though the implications for protein stability are likely to vary on a case-by-case basis. The solubility of the protein in a nonaqueous solvent may be enhanced by first lyophilizing the protein from water at a pH away from its pI to alter the ionization state of its functional groups. The residual water content can also affect solubility (Stevenson, 2000). Many nonaqueous formulations for recombinant proteins such as leuprolide acetate, rhGH, GLP-1, rh α-IFN, and rhRelaxin have been developed, and these have been reported to be stable for 3 months to more than 1 year at 37°C (3914). Leuprolide formulated in DMSO was on the market in Viadur™ (Bayer Healthcare), a once-yearly implant for the palliative treatment of advanced prostate cancer. This was the first product to use the former Alza's DUROS® implant technology and provided continuous, 12-month suppression of testosterone with a single treatment. The product was phased out due to less market demand and growing manufacturing cost. The pH of a nonaqueous solution may be measurable but will have little theoretical meaning and should be interpreted carefully. Glass electrodes may show discrepancy of as much as one pH unit due to liquid junction potential, and also pH scale will vary since dissociation constants of many organic solvents are different from that of water. Proteins solubilized in nonaqueous solvents might unfold, which may in turn be reversible or irreversible. Such unfolding can be prevented by using protein suspensions in these solvents.

5.4.3 Compaction and Tablets

In general, peptide and protein drugs are not compacted into tablets as these drugs are not active orally. However, research into oral delivery is ongoing and there could be situations where such compaction will be required. In addition, compaction of peptide or protein drugs may be required in other instances, such as the manufacture of implants for parenteral administration. Also, compacts of proteinaceous enzymes have applications in the pharmaceutical and other industries. Theoretically, such compacts may help to maintain the stability of a protein since they can be manufactured without the use of aqueous or organic solvents. In practice, however, some degradation may occur upon compaction. Degradation of the enzymes urease, lipase, α-amylase, and β-glucuronidase, under compaction, has been investigated. Compactional pressure was found to have an effect on the catalytic activity of these four enzymes. The enzymatic activity decreased with increasing pressure up to a point that represented the maximum density that a compact can achieve. The exact mechanism for the loss of activity is not known but appears to be a reduction in volume between molecules or particles. Upon compaction, the secondary structure was affected for α-amylase and β-glucuronidase but not for urease or lipase. For urease, some effect on the tertiary and quaternary structure resulted in aggregation (Groves & Teng, 1992). For wheat germ lipase, no appreciable degradation was observed at pressures up to 175 MPa. Beyond this pressure, about 30% activity was lost even though circular dichroism and SDS-PAGE analysis did not reveal any changes in protein structure. Thus, it appears that structural changes, if any, are too subtle to be detected by these analytical techniques (Zarrintan, Teng, & Groves, 1990).

A small number of tablets for laboratory experimentation may also be made by other means. Takada et al. (1994) have made small tablets of recombinant human granulocyte colony-stimulating factor (rhG-CSF) to evaluate its pharmacological activity in in vivo rat experiments. In addition to rhG-CSF, the tablets contained protease inhibitors and other pharmaceutical additives. These components were added to a solution of rhG-CSF, dissolved, and stirred in a mortar at room temperature to evaporate all the solvent. The resulting mixture was dried under vacuum overnight and then pulverized in a mortar with a pestle. After pulverization, $NaHCO_3$ was added to make an effervescent tablet. A few drops of ethanol-to-water (1:1) mixture were added to form the mass into small tablets by hand. The tablets were then enteric coated using hydroxypropyl-methylcellulose phthalate in a methylene chloride-to-methanol (7:3) mixture. The % stability of rhG-CSF following all this treatment was not reported in this study, but the tablets were pharmacologically active.

5.4.4 Aerosolization

The use of aerosols for pulmonary delivery of peptides and proteins will be discussed in Chapter 9. In this section, the stability issues for peptide and protein drugs related to the unique operational mechanism of aerosols are discussed. An aerosol may be generated by a nebulizer, a metered dose inhaler (MDI), or a dry-powder inhaler. The MDI uses chlorofluorocarbons as propellants and is currently the most common form of inhalation aerosol. With each actuation of the MDI, a precise volume/dose is delivered through a metered valve. Conventional drugs currently marketed as MDIs include metaproterenol, triamcinolone, beclomethasone, terbutaline, dexamethasone, isoproterenol, phenylephrine, cromolyn, epinephrine, ergotamine, and albuterol, among others. Research on the pulmonary delivery of peptides and proteins is currently underway, and initial results are very promising (Byron, 1987; Lee & Sciarra, 1976). Thus, it is expected that more and more peptide and protein drugs will be commercialized as aerosols in the near future. One protein that has already been marketed for pulmonary delivery is recombinant human deoxyribonuclease I (rhDNase) (Pulmozyme®, Genentech). A pulmonary delivery system for insulin is also currently under development.

Excipients commonly used in an MDI include the propellant, dispersing agents, cosolvents, antioxidants, flavors, and antimicrobials. Commonly used dispersing agents are sorbitan trioleate, oleic acid, and soya lecithin. Dispersibility may not be a concern with protein drugs if a nebulizer is used, since aqueous solutions can be used in this case. Aerosolization of a protein may generate extensive air–water interfaces, which may cause the protein to undergo aggregation or other denaturation reactions. Also, the respiratory tract exposes the airborne aerosol particles to almost 100% relative humidity, which may accelerate the rate of aggregation of hygroscopic particles during their transit to the lung (Adjei & Garren, 1990). Two other less likely possibilities of protein destabilization during aerosolization are the direct effect of shear forces and the evaporation of droplets. Evaporation of droplets is known to increase the protein concentration within the nebulizer reservoir fluid, and this may also destabilize the protein in some cases.

hGH has been reported to undergo aggregation when aerosolized from a nebulizer. A 10 min aerosolization resulted in more extensive aggregation of hGH than 2 h of vigorous shaking. Most of the aggregates were soluble oligomers and were quantitated by size exclusion chromatography. An inverse relationship of concentration and percent aggregate formed was observed in these aerosolization studies. Thus, aggregation could be reduced by using higher concentrations of the protein. The addition of a nonionic surfactant also reduced aggregation (Oeswein et al., 1991). LDH and recombinant human granulocyte colony-stimulating factor (rhG-CSF) also undergo inactivation upon aerosolization. Aerosolization of LDH resulted in an irreversible time-dependent loss of enzymatic activity in the reservoir solution. A change in the fluid volume in the reservoir of the nebulizer was found to affect the rate at which the enzyme is denatured and inactivated. After 10 min of nebulization, about 20% of enzyme is lost when a 40 mL volume is used at a pressure of 40 psig. The activity loss increases to 65% at 60 min of nebulization. Nebulization of rhG-CSF resulted in a new peak in the SEC chromatogram, eluting

3 min before the peak corresponding to the unnebulized sample. This peak was identified as a dimer. In the case of rhG-CSF, aggregate formation reached a plateau of about 35%–40%, following 5–10 min of nebulization. The reasons for such plateau formation are now clear but may involve reaching an equilibrium state or the aggregates might be hindering further molecules from reaching the air interface. PEG was found to stabilize both LDH and rhG-CSF (Niven et al., 1994). In contrast to these studies, aerosolization of rhDNase indicated that there was no loss in enzymatic activity and no increase in aggregation. Furthermore, circular dichroism spectrophotometry was used to show that there was no change in the secondary or the tertiary structure of rhDNase following nebulization. A slight drop in pH was observed for the collected nebulized samples. It was shown that this pH change is not due to any chemical changes in the protein molecule. Instead, it results from the absorption of CO_2 from the air by the fine aerosol mist, resulting in the formation of carbonic acid (Cipolla, Gonda, & Shire, 1994).

5.5 HANDLING OF THERAPEUTIC PEPTIDES AND PROTEINS

5.5.1 Sterilization

Peptide and protein drugs are currently typically formulated as parenterals, and these are required to be sterile. Most commonly used terminal sterilization techniques such as autoclaving are too harsh for the delicate structure of these drugs (Banga & Reddy, 1994). Similarly, radiation will also have harmful effects on protein stability due to free radical formation. The direct effects of radiation on protein in solid state are much less than the indirect effect of free radicals in an aqueous solution. Radiation affects the sulfur-containing and aromatic residues by causing cross-linking, which results in protein aggregation. However, radiation sterilization may be feasible for some polymeric controlled release systems, such as microspheres (Merkli, Heller, Tabatabay, & Gurny, 1994). The only viable means of sterilization of protein solutions seem to be filter sterilization. Thus, aseptic processing becomes essential and meticulous care must be taken to produce a high-quality product. New technology, such as high-speed continuous lines, form fill seal, and capabilities of clean-in-place and steam-in-place, and robotics, are now giving higher sterility assurance of aseptic processes. These new technologies will allow sterility assurance levels greater than 10^{-3}, with the goal of matching the 10^{-6} levels provided by terminal sterilization (Rech, Panaggio, & Bontempo, 1991; White, 1992). Adequate control of endotoxin levels and compliance with all other regulatory issues is required for these products (Avallone, 1990; Donroe et al., 1990). One advantage of sterile filtration would be that it will remove subvisible particles, at least those larger than 0.2 μm (Nayak et al., 2011).

5.5.2 Sanitizing Agents and Protein Stability

Sanitizing agents of diverse chemical compositions are used in the pharmaceutical industry to prepare equipment and areas for aseptic processing of parenterals.

Trace amounts of these chemicals may stay in equipment as residues and may cause degradation of the protein drug. Such a scenario becomes even more plausible during lyophilization since a concentration effect will occur during the freezing step. As discussed earlier, pure water crystallizes first as ice, leaving behind the solute in a more concentrated form. This increases the concentration of any residual sanitizing agent, thus causing deleterious effects on protein stability. Kirsch, Riggin, Gearhart, Lefeber, and Lytle (1993) have studied the effect of several commercial sanitizing agents on biosynthetic hGH obtained from several raw material lots. While isopropanol and methanol did not affect hGH chemically or physically, several other agents had a significant effect on stability of hGH. Glutaraldehyde (36 μmol/0.27 μmol hGH) completely destroyed the protein while formaldehyde also induced hGH degradation. Hydrogen peroxide also caused extensive covalent modifications. Phenol and phenolic compounds contained in commercial sanitizing agents induced aggregation of hGH during the freezing step of lyophilization. The more hydrophobic phenol compounds such as phenylphenol and 4-*tert*-amyl phenol were more harmful, presumably because they can interact with the hydrophobic regions of the hGH molecule. In order to avoid the presence of trace amounts of sanitizing agents in the pharmaceutical industry, as well as to avoid the presence of potential residues of cellular components in the biotechnology industry, the validation of cleaning between product lots becomes very important. In the manufacture of biopharmaceuticals by recombinant DNA technology, potential residues can include proteins, nucleic acids, carbohydrates, endotoxins, detergents, and the drug itself. Baffi et al. (1991) have proposed a total organic carbon (TOC) analysis method for validating cleaning between products in biopharmaceutical manufacturing. The accuracy of the TOC method for measuring trace proteins was evaluated using solutions of bovine serum albumin, tumor necrosis factor, and hGH. An advantage of TOC analysis over ELISA methodology is that it responds equally well to both native and denatured proteins. Since protein residuals may denature as a result of high temperatures or due to exposure to cleaning agents, TOC analysis would be highly complementary to the use of ELISA and Lowry protein analysis.

5.5.3 Preparation of a Typical Batch

Based on our discussion of pharmaceutical processing of peptides and proteins in this chapter, the preparation of a typical batch of a peptide or protein formulation can now be discussed. Strictly speaking, there is no typical batch or formulation or procedure when dealing with peptide and protein drugs. Each drug presents unique opportunities and problems, and successful formulation and processing have to be tailored to the unique situation for that particular peptide or protein. However, an attempt is made to present some general procedures that may be applicable in many situations. A discussion of the manufacturing methods may be found in the patent literature, and the reader may refer to it for additional information (Goldstein & Audhya, 1992; Kwan, 1985). Also, process should be designed taking into consideration the principles of QbD (see Section 4.8).

For preparation of a typical batch, a stainless steel or glass vessel is filled with Water for Injection (WFI) USP to about 75% of the batch volume. The salts are then added and dissolved to constitute the buffer system. Then, any solubilizers and/or stabilizers are added and dissolved. The pH is then adjusted before the addition of the peptide or protein. A pH optimal for the stability of the peptide or protein should be used to prevent any shock to the peptide or protein. This can then be followed by the addition of peptide or protein (adjusted for peptide or protein content) with slow stirring until it is dissolved. The pH of the resulting solution is then readjusted to the required value (typically, about 7.2), and additional water is added after in-process release to adjust to volume. The solution is then mixed and aseptically filtered through a sterilized 0.22 μm filter into a sterilized vessel. The integrity of the filter is tested after filtration. The initial steps for manufacturing of a formulation as a lyophilized powder are the same as those for an aqueous solution. Once the sterile solution is ready, it is aseptically loaded into a sterilized lyophilizer. After lyophilization, the vials are aseptically stoppered in the chambers. Finally, the vials are removed from the chamber and sealed. Manufacturing should be completed in as little time as possible, and time outside refrigeration (TOR) should be monitored to ensure that the stability of the formulation is not compromised.

A few observations can be made with regard to this procedure. The peptide or protein is dissolved by very slow stirring to prevent any foaming or aggregation. Filtration must be done through a filter that does not adsorb the peptide or protein being filtered. Also, the bulk active protein may bind to the vessel, and studies should be done to ensure that any such binding is within tolerable limits. If the bulk active was frozen and is thawed before use, the solution should be mildly stirred. This is because when the bulk solution was frozen, pure water probably froze first, so that the active is nonuniformly dispersed in the frozen material. The hold time of the thawed bulk active and the batch should be minimized, especially if the solutions are unpreserved. For filtration of protein solutions, pressure filtration is preferred to vacuum filtration as foaming may result if vacuum filtration is used (Richmond, 1990).

In the coming years, changes may be anticipated in manufacturing and testing methods used by the pharmaceutical industry, and many of these may apply to manufacture of protein-based pharmaceuticals as well. As a part of the FDA's *Pharmaceutical cGMPs for the 21st Century: A risk-based approach*, FDA has initiated process analytical technologies (PAT) (U.S. Food and Drug Administration, 2003), which essentially proposes to use on-line, in-line, or at-line measurements to streamline pharmaceutical manufacturing and improve product quality. Using spectroscopy techniques such as near-IR (NIR) or Raman spectroscopy or by using physical (e.g., pH, temperature) or other methods of in-line process monitoring, the product quality is expected to improve and the need for some finished product testing may be reduced. The technique of NIR is particularly relevant to PAT (Anderson, 2003; Dziki & Novak, 2003). A possible example of application of PAT to protein-based products could be the use of light scattering techniques to monitor the aggregation status of a protein through various stages of manufacture. Similarly, other high-resolution analytical techniques can perhaps be used for monitor specific attributes of the molecule or process. The bulk manufacture of the protein can be monitored by a combination of chromatographic techniques and optical sensors or other means.

5.5.4 Storage in Solid State

Lyophilized powders can be quite stable as long as they are not reconstituted. For example, Activase® lyophilized powder shows no significant loss of bioactivity after storage for more than 4 years at controlled room temperature (Nguyen & Ward, 1993). Degradation in the solid state often takes place by aggregation. The stability of a protein in the solid state is dependent on the moisture content of the solid, temperature, and composition of the formulation. Stability of proteins in the solid state upon exposure to moisture and/or elevated temperatures is very important since such conditions are frequently encountered in pharmaceutical systems. Examples would include the storage stability of lyophilized formulations or the stability of proteins in a controlled release system under in vivo conditions, that is, upon exposure to an aqueous environment at 37°C. As we freeze-dry a protein, the protein still retains some residual moisture in the solid state. Such residual moisture is actually critical to the stability of the protein. There is an optimum level of residual moisture required to maintain the stability of a protein formulation (Chang, Randall, & Lee, 1993; Greiff, 1971). Moisture-induced aggregation of lyophilized proteins in the solid state has been investigated for several proteins such as albumin (Liu, Langer, & Klibanov, 1991), insulin (Costantino, Langer, & Klibanov, 1994), β-galactosidase (Yoshioka, Aso, Izutsu, & Terao, 1993), and ribonuclease A (Townsend & DeLuca, 1991). While aggregation of these proteins in solution may often involve noncovalent forces, aggregation in the solid state generally involves covalent forces. While albumin is very stable as an aqueous solution, it readily aggregated when wetted with moisture and incubated at 37°C. Within just 24 h of incubation, as much as 97% of the protein became insoluble due to aggregation. It was found that exposure to both moisture and elevated temperature was required to induce aggregation. Strong denaturants such as 6 M urea, 6 M guanidine hydrochloride, and 0.1% sodium dodecyl sulfate were unable to dissolve the aggregates, indicating that aggregation involved intermolecular covalent reactions. Thiol reagents such as 10 mM dithiothreitol dissolved the aggregates, suggesting that the covalent linkages involved were disulfide bonds. Interestingly, the effect of increasing moisture on percent aggregation of bovine serum albumin exhibited a bell-shaped dependence (Liu et al., 1991). An explanation of this effect may be found in a review by Hageman (1988) on the role of moisture in protein stability. As moisture is introduced in a solid-state system to about 6%–8%, it is believed to form a monolayer on the protein surface. At this level, very little change is observed. As moisture is increased above monolayer levels, loosely bound water becomes available on the protein surface to mobilize reactants leading to increased aggregation, especially at higher temperatures. As the water content continues to increase beyond about 20%–25%, the reactants or functional groups of the protein get solubilized and the dilution effect comes into play, to decrease the proximity of reactants. At this point, the reactivity decreases again, thus resulting in the bell-shaped dependence, as partly discussed in Chapter 3.

Another source of residual moisture remaining in the product is the container itself. Lyophilization closures absorb moisture during steam sterilization and, if not adequately dried, can release this moisture into the product. This problem can become especially critical for peptide and protein formulations as these formulations have a low cake weight as compared to conventional formulations. To avoid these problems, a stopper formulation that absorbs very little moisture during autoclaving, such as a butyl formulation, should be selected (DeGrazio & Flynn, 1992). Thus, butyl and halobutyl rubber is generally used in stoppers due to its low-moisture permeability. It has been suggested that the amount of residual moisture in a vial may also be affected by its position or location in the freeze-drying equipment. In a study by Greiff (1990), vials were positioned on a base defined by Cartesian coordinates (x,y) and the residual moisture was measured as an axis perpendicular to the base (z). The resulting distribution of residual moisture was described best by an equation for a second-order polynomial. In addition to position of vial, other factors that were found to affect residual moisture were the orientation of openings in the split stoppers on the vials, the location of the shelf on which the vials were placed, and the elapsed times and shelf temperatures. Thus, in order to optimize levels of residual moisture, the microenvironment of each vial must be considered during validation studies. Storage of lyophilized cakes at high humidity and/or high temperature can result in collapse of the cake structure. Such collapse may be accompanied by crystallization of the excipients. In studies using β-galactosidase as a model protein and inositol as an excipient, it was shown that inositol had a protein-stabilizing effect when in its amorphous form. The protective effect was maintained even in the event of structural collapse, as long as inositol retained its amorphous form. However, crystallization of inositol during storage removed its stabilizing effect (Izutsu, Yoshioka, & Kojima, 1994).

5.5.5 Handling of Recombinant Proteins in a Hospital Setting

Handling of therapeutic peptides and proteins in a hospital setting requires knowledge of the unique properties of these drugs in order to avoid potential problems that may develop. For example, the protein may aggregate if reconstituted with vigorous shaking or may adsorb to the huge surface area of an intravenous (IV) administration set during delivery. Also, incompatibilities may develop during preparation of admixtures or simple step such as dilution may cause precipitation of the protein. These technical problems are discussed in this section. In addition, pharmacoeconomic issues also become important in a hospital setting. The impact of pharmacoeconomics issues on the regulatory process was discussed in Chapter 1. Pharmacoeconomics is also very important to the hospital formulary as biotechnology-derived proteins can be very expensive (Sapienza, 1994).

5.5.5.1 Reconstitution

Currently, most recombinant proteins are developed as parenteral dosage forms and this trend is likely to continue, at least in the immediate future. The hospital pharmacist will be closely associated with such therapy (Banga & Reddy, 1994). About half of the parenteral products are marketed as lyophilized powders, which need to be reconstituted just before use. The diluent is usually WFI USP and may be provided with the dry powder as a convenience. Products such as Humatrope®

(hGH) supply their own diluting solution that in this case contains 0.3% *m*-cresol as a preservative and 1.7% glycerin as a stabilizer and tonicity agent. Some products are to be reconstituted with Bacteriostatic Water for Injection that contains benzyl alcohol as a preservative. These products can generally be stored for about 14 days after reconstitution. For products that do not contain any antimicrobial preservatives, the manufacturer will generally specify that the product be used immediately after reconstitution to minimize the possibility of compromised sterility. Some hospitals wish to extend these expiration dates due to cost considerations. In a study by Pfeifer, Siegel, and Ayers (1994), the authors investigated the ability of IV immune globulin (IVIG) products to support bacterial and yeast growth following direct inoculation. The IVIG preparations are generally believed to be ideal media for microbial growth as they are essentially solutions of protein stabilized in a sugar base without antimicrobial preservatives. The results of the study showed that contrary to expectation, none of the IVIG preparations supported bacterial growth at any temperature over a 7-day period. Yeast growth was not seen at 3°C, but most preparations did support yeast growth at 25°C and 37°C. Since the cost of one dose of an IVIG product is high, the authors suggest that a reconsideration of recommended expiration dates for IVIG products may be warranted. However, such decisions must be taken very carefully and after exhaustive studies, as the consequences of a wrong decision can be disastrous. Also, the complexity of proteins makes it easy to overlook certain unforeseen situations in a limited study. In contrast, the manufacturer has generally done exhaustive studies over a prolonged period of time under the scrutiny of the FDA, before it came up with its recommendations. Another problem that may be encountered during reconstitution of therapeutic proteins is their aggregation. Proteins such as hGH, urokinase, asparagine, and several others are known to readily aggregate. Following reconstitution, thin translucent filaments may form in urokinase formulations but these do not indicate any loss of potency and no clinical problems have been associated with these filaments. Similarly, a small number of gelatinous fiber-like particles may develop occasionally in asparaginase formulations on shaking. Filtration through a 5 µm filter during administration will remove these particles with no loss in potency (PDR Staff, 2014). In contrast, aggregation of hGH can lead to loss of potency (Chapter 4). Similarly, trastuzumab (Herceptin) should be diluted in 0.9% sodium chloride rather than 5% dextrose to avoid aggregation (Demeule, Palais, Machaidze, Gurny, & Arvinte, 2009). A more detailed discussion of protein aggregation can be found in Chapters 3 and 4, in addition to this chapter.

While diluents are typically required for reconstitution, they may sometimes also be used to dilute a more concentrated product. For example, the treatment of children with diabetes may need the dilution of the commercially available insulin products, which are typically formulated at 100 units/mL. A smaller volume of the formulation may not always be practical as the volume needed may be in microliters and may not be accurately measured with commonly available syringes. In such cases, sterile diluents may possibly be used but stability of the dilution should be confirmed. For example, U-10 and U-50 dilutions of lispro insulin were found to be stable when stored at 5°C and 30°C for 32 days (Stickelmeyer, Graf, Frank, Ballard, & Storms, 2000).

5.5.5.2 Incompatibilities

The compatibility of the drug and excipients must be ensured. Reconstituted solutions of interleukin-2 (IL-2) were found to be stable for 48 h at room temperature but advised to be used within 4 h due to potential incompatibility with bacteriostatic agents. Compatibility problems may also arise when a patient receiving an IV infusion of a protein drug is given other drugs via Y-site administration. In such studies, granulocyte macrophage colony-stimulating factor (GM-CSF) at 6 mcg/mL was found to be visually incompatible with vancomycin hydrochloride. The reason for the incompatibility is not clear but may involve pH differences (Matsuura, 1992). In another study, Trissel and Martinez (1994) studied the compatibility of filgrastim (recombinant human granulocyte colony-stimulating factor, Amgen) with 97 selected drugs by visual observation, turbidity measurement, and particle sizing and counting. For these studies, 5 mL samples of filgrastim (30 μg/mL) in 5% dextrose injection were combined with 5 mL samples of solutions of other drugs in 5% dextrose injection at clinically used concentrations at 22°C. Incompatibility was defined as haze, visible particulate matter or turbidity exceeding that in controls. Of the 97 drugs tested, 22 exhibited various incompatibilities. A common incompatibility was the formation of particles or filaments, possibly due to protein aggregation. Compatibility problems may also result when different types of insulin are mixed together (Adams, Haines-Nutt, & Town, 1987). Admixtures of the inhalation solutions Pulmozyme (dornase alfa) and tobramycin products were found to be stable for simultaneous inhalation (Klemmer, Kramer, & Kamin, 2014).

In a hospital setting, a conflict often exists between what a physician wants to use and what the manufacturer wants to recommend. For peptide and protein drugs, it would be dangerous to deviate from what the manufacturer has recommended (Geigert, 1989). This is because the stability of a protein is very formulation dependent and even minor changes in excipients can cause drastic changes in the stability of the formulation. Therefore, formulated products should be used only in accordance with the instructions on the package insert. A situation may sometimes arise where a physician wishes to use a protein drug for a nonapproved indication, which may need a different dose or a different route of administration. Particular caution is required if the dilutions or other handling required is different from handling involved in normal usage. As an example, consider the formulation of Activase® (tissue plasminogen activator). The product is meant for parenteral administration and contains arginine as a solubilizer and stabilizer. However, the product has been used for some investigative studies in ophthalmic research. The manufacturer does not recommend dilution below 0.5 mg/mL, but these studies diluted the product much below this level as the required dose for the nonapproved indication was only 25 μg. Upon dilution, solubilizing excipients may lose their effectiveness. Furthermore, the dilutions were done with balanced salt solution that contains calcium and magnesium salts. These salts will react with phosphate anion present in the formulation and cause precipitation of insoluble calcium and magnesium phosphate salts. Also, since Activase does not contain a preservative, it is intended only for single use. Some of the studies in question stored unused portions in freezer for reuse, which is not acceptable (Ward & Weck, 1990).

5.5.5.3 Adsorption

Peptide and proteins have a tendency to adsorb to surfaces, and this can pose a significant problem in their delivery. This is especially true with an IV infusion since the administration set provides a huge surface area on which adsorption may take place. Furthermore, most biotechnology-derived proteins are very potent, so that the doses are very low. Thus, even small amounts of adsorption may lead to a significant loss of drug. Adsorption may also occur on an in-line filter in an administration set. When an in-line filter was added to an IV administration set for the delivery of IL-2, almost all the drug was lost (Geigert, 1989). Adsorption of insulin to glass and other materials has been investigated in several studies (McElnay, Elliott, & D'Arcy, 1987; Mizutani, 1980; Mizutani & Mizutani, 1978). Adsorption depends on the concentration of insulin, contact time with the surface, and the formulation. If insulin is being administered from a syringe as a concentrated solution, the loss due to adsorption may not be significant. However, at low concentrations, the loss due to adsorption can be significant. Such low concentrations will often be encountered with infusions. In this case, the flow rate of the infusion will also influence the extent of adsorption. Petty and Cunningham (1974) found that when they added 30 units of insulin to 1.0 L of Ringer's lactate solution, the patient received only 6.4 units of insulin due to losses by adsorption. More than 50% of insulin was adsorbed by glass and polyvinylchloride containers within 15 s of injection. Of the remaining amount, another 50% was lost by adsorption to the IV infusion set. Adsorption of insulin to glass was reduced but not eliminated by the addition of albumin. In practice, therefore, addition of albumin may or may not be effective since insulin may quickly adsorb to the infusion set and reach an equilibrium. Also, it should be realized that the rate of insulin infusion may be adjusted to the blood glucose response, and thus insulin lost through binding may become less important. A recent study has suggested that following a 20 mL flush, insulin infusions can be started without any dwell time in the IV tubing to decrease adsorption (Thompson, Vital-Carona, & Faustino, 2012).

Colony-stimulating factors, filgrastim (G-CSF) and sargramostim (GM-CSF), are also known to bind to infusion containers. It has been reported that if the final concentration of filgrastim is between 2 and 15 mcg/mL, 0.2% of human serum albumin must be added to prevent adsorption to infusion container. Since these colony-stimulating factors are very expensive, the cost of albumin in comparison is justified (Wordell, 1992).

Interleukin-1β (IL-1β) has also been reported to adsorb to plastics at low (100 ng/mL) concentrations. The addition of 1% human serum albumin was found to prevent such adsorption. However, at higher concentrations (1 μg/mL), IL-1β could be stored and delivered from either polypropylene-based syringe pump delivery systems or PVC infusion bags for up to 24 h (Visor et al., 1990). Adsorption of D-Nal $(2)^6$ LHRH on glass, plastic, tubings, syringes, and filters was found to reach a steady state within 2 h. Prevention of this adsorption by ionic compounds, inert proteins, or amino acids was investigated. Of these, phosphate and acetate ions were found to be most effective (Anik & Hwang, 1983). Plasma proteins are also known to adsorb to polymeric surfaces (Baszkin & Lyman, 1980; Chuang, Mohammad, Sharma, & Mason, 1980; Ihlenfeld & Cooper, 1979;

Soderquist & Walton, 1980), and this may have relevance to the use of PVC and silastic surfaces in vivo or in any situation where a polymer is in contact with blood. Other aspects of adsorption, such as principles and mechanisms, have been discussed in Chapter 3, while adsorption to filter during pharmaceutical processing was discussed in Section 5.2.1.

5.5.5.4 Other Considerations

Considerations of potency and viral infectivity of proteins are also of interest in a hospital setting. Potency of therapeutic peptides and proteins is often based on an international standard. This standard is expressed as international units per milliliter (IU/mL) and is prepared by the World Health Organization (WHO). If an international standard is not available, the company develops its own standard, which stays in use till uniform international standards are developed. However, the use of such in-house standards leads to confusion among pharmacists and physicians (Geigert, 1989). This is likely to be less of a problem in the future as FDA, USP, and WHO resolve these problems. For protein products prepared from pooled human plasma or other body fluids or tissues, there is a risk that causative agents of hepatitis, AIDS, or other viral diseases may be present. Such products are often heat treated in order to reduce the potential for transmission of infectious agents. A typical pasteurization treatment may consist of heating the protein in solution at about 60°C for about 10 h. Such treatment is known to destroy hepatitis viruses in human albumin (Peters, 1985) and antithrombin III (Busby, Atha, & Ingham, 1981). Examples of proteins that are or were exposed to such heat treatment or other treatment to reduce the risk of viral contamination are alpha-1 antitrypsin, antihemophilic factor, and alglucerase (modified β-glucocerebrosidase) (PDR Staff, 2014). Such treatment cannot totally eliminate the risk of viral infectivity for these products. However, pasteurization increases the quality assurance of the product and has been shown to inactivate several DNA viruses (cytomegalovirus, herpes, and hepatitis B) and RNA viruses (rubella, mumps, measles, and poliomyelitis). The risk of protein aggregation or other instability as a result of the long-term heating process required by pasteurization should be considered (Gonzalez, Murature, & Fidelio, 1995).

5.6 CONCLUSIONS

The manipulation of peptides and proteins in the biotechnology and pharmaceutical industry for their manufacture leads to several stress situations that can destabilize the drug. For example, adsorption during filtration or aggregation during shaking may occur. Another process that exposes protein to stress is lyophilization. Lyophilization, despite its unfavorable economics, has become a well-accepted and widely used method for the manufacture of stable protein formulations. Lyophilization is a complex process, and thus, it may seem that a trial-and-error approach can be taken to formulation development. However, a proper understanding of the interplay of the physics, chemistry, and engineering aspects will allow a scientific approach to develop the formulation. Appropriate excipients must be chosen carefully, and the lyophilization cycle should be developed systematically. Traditionally, lyophilization chambers were sanitized with liquid detergents such as hydrogen peroxide. However, these sanitizing agents may affect the chemical and physical stability of proteins and caution must be observed. Alternatives to lyophilization such as spray drying or use of nonaqueous solvents are being investigated. Manufacturing processes are being improved by process analytical technologies and quality by design approaches. Handling of therapeutic peptides and proteins in a hospital setting also needs careful consideration to avoid potential problems such as aggregation during reconstitution of a lyophilized protein, potential incompatibilities, or adsorption to an infusion set.

REFERENCES

Adams, P. S., Haines-Nutt, R. F., & Town, R. (1987). Stability of insulin mixtures in disposable plastic insulin syringes. *J. Pharm. Pharmacol., 39,* 158–163.

Adjei, A. & Garren, J. (1990). Pulmonary delivery of peptide drugs: Effect of particle size on bioavailability of leuprolide acetate in healthy male volunteers. *Pharm. Res., 7,* 565–569.

Adler, M., Unger, M., & Lee, G. (2000). Surface composition of spray-dried particles of bovine serum albumin/trehalose/surfactant. *Pharm. Res., 17,* 863–870.

Akers, M. J., Milton, N., Byrn, S. R., & Nail, S. L. (1995). Glycine crystallization during freezing: The effects of salt form, pH, and ionic strength. *Pharm. Res., 12,* 1457–1461.

Akers, M. J. & Nail, S. L. (1994). Top 10 current technical issues in parenteral science. *Pharm. Technol., 18,* 26–36.

Anderson, C. A. (2003). PAT applications: Qualification of equipment and validation of analytical methods. *Am. Pharm. Rev., 6,* 64–68.

Andya, J. D., Maa, Y. F., Costantino, H. R., Nguyen, P. A., Dasovich, N., Sweeney, T. D. et al. (1999). The effect of formulation excipients on protein stability and aerosol performance of spray-dried powders of a recombinant humanized anti-IgE monoclonal antibody. *Pharm. Res., 16,* 350–358.

Anik, S. T. & Hwang, J. Y. (1983). Adsorption of D-Nal(2)6 LHRH, a decapeptide, onto glass and other surfaces. *Int. J. Pharm., 16,* 181–190.

Arakawa, T., Kita, Y., & Carpenter, J. F. (1991). Protein-solvent interactions in pharmaceutical formulations. *Pharm. Res., 8,* 285–291.

Arakawa, T., Prestrelski, S. J., Kenney, W. C., & Carpenter, J. F. (1993). Factors affecting short-term and long-term stabilities of proteins. *Adv. Drug Deliv. Rev., 10,* 1–28.

Ashton, L., Dusting, J., Imomoh, E., Balabani, S., & Blanch, E. W. (2009). Shear-induced unfolding of lysozyme monitored in situ. *Biophys. J., 96,* 4231–4236.

Avallone, H. (1990). Current regulatory issues regarding sterile products. *J. Parenter. Sci. Technol., 44,* 228–230.

Awotwe-Otoo, D., Agarabi, C., Wu, G. K., Casey, E., Read, E., Lute, S. et al. (2012). Quality by design: Impact of formulation variables and their interactions on quality attributes of a lyophilized monoclonal antibody. *Int. J. Pharm., 438,* 167–175.

Baffi, R., Dolch, G., Garnick, R., Huang, Y. F., Mar, B., Matsuhiro, D. et al. (1991). A total organic carbon analysis method for validating cleaning between products in biopharmaceutical manufacturing. *J. Parenter. Sci. Technol., 45,* 13–19.

Banga, A. K. & Reddy, I. K. (1994). Biotechnology drugs: Pharmaceutical issues. *Pharm. Times, 60(3),* 68–76.

Barnard, J. G., Singh, S., Randolph, T. W., & Carpenter, J. F. (2011). Subvisible particle counting provides a sensitive method of detecting and quantifying aggregation of monoclonal antibody caused by freeze-thawing: Insights into the roles of particles in the protein aggregation pathway. *J. Pharm. Sci., 100,* 492–503.

Barrow, C. J., Yasuda, A., Kenny, P. T. M., & Zagorski, G. (1992). Solution conformations and aggregational properties of synthetic amyloid beta-peptides of Alzheimer's disease: Analysis of circular dichroism spectra. *J. Mol. Biol., 225,* 1075–1093.

Baszkin, A. & Lyman, D. J. (1980). The interaction of plasma proteins with polymers. I. Relationship between polymer surface energy and protein adsorption/desorption. *J. Biomed. Mater. Res., 14,* 393–403.

Bausch, A. & Leuenberger, H. (1994). Wet spherical agglomeration of proteins as a new method to prepare parenteral fast soluble dosage forms. *Int. J. Pharm., 101,* 63–70.

Bee, J. S., Stevenson, J. L., Mehta, B., Svitel, J., Pollastrini, J., Platz, R. et al. (2009). Response of a concentrated monoclonal antibody formulation to high shear. *Biotechnol. Bioeng., 103,* 936–943.

Branchu, S., Forbes, R. T., York, P., & Nyqvist, H. (1999). A central composite design to investigate the thermal stabilization of lysozyme. *Pharm. Res., 16,* 702–708.

Breen, E. D., Curley, J. G., Overcashier, D. E., Hsu, C. C., & Shire, S. J. (2001). Effect of moisture on the stability of a lyophilized humanized monoclonal antibody formulation. *Pharm. Res., 18,* 1345–1353.

Busby, T. F., Atha, D. H., & Ingham, K. C. (1981). Thermal denaturation of antithrombin III. *J. Biol. Chem., 256,* 12140–12147.

Byron, P. R. (1987). Pulmonary targeting with aerosols. *Pharm. Technol., 11(5),* 42–56.

Cappola, M. L. (2000). Freeze-drying concepts: The basics. In E. L. McNally (Ed.), *Protein formulation and delivery* (pp. 159–199). New York: Marcel Dekker, Inc.

Carpenter, J. F., Chang, B. S., Garzon-Rodriquez, W., & Randolph, T. W. (2002). Rational design of stable lyophilized protein formulations: Theory and practice. In J. F. Carpenter & M. C. Manning (Eds.), *Rational design of stable protein formulations: Theory and practice* (pp. 109–133). New York: Kluwer Academic/Plenum Publishers.

Carpenter, J. F. & Crowe, J. H. (1988). The mechanism of cryoprotection of proteins by solutes. *Cryobiology, 25,* 244–255.

Carpenter, J. F., Crowe, L. M., & Crowe, J. H. (1987). Stabilization of phosphofructokinase with sugars during freeze-drying: Characterization of enhanced protection in the presence of divalent cations. *Biochim. Biophy. Acta, 923,* 109–115.

Carpenter, J. F., Pikal, M. J., Chang, B. S., & Randolph, T. W. (1997). Rational design of stable lyophilized protein formulations: Some practical advice. *Pharm. Res., 14,* 969–975.

Carpenter, J. F., Prestrelski, S. J., & Arakawa, T. (1993). Separation of freezing-and drying induced denaturation of lyophilized proteins using stress-specific stabilization. *Arch. Biochem. Biophys., 303,* 456–464.

Cartwright, T., Senussi, O., & Grady, M. D. (1977). The mechanism of the inactivation of human fibroblast interferon by mechanical stress. *J. Gen. Virol., 36,* 317–321.

Chang, B. S. & Fischer, N. L. (1995). Development of an efficient single-step freeze-drying cycle for protein formulations. *Pharm. Res., 12,* 831–837.

Chang, B. S., Randall, C. S., & Lee, Y. S. (1993). Stabilization of lyophilized porcine pancreatic elastase. *Pharm. Res., 10,* 1478–1483.

Charm, S. E. & Lai, C. J. (1971). Comparison of ultrafiltration systems for concentration of biologicals. *Biotechnol. Bioeng., 13,* 185–202.

Charm, S. E. & Wong, B. L. (1970). Enzyme inactivation with shearing. *Biotechnol. Bioeng., 12,* 1103–1109.

Charm, S. E. & Wong, B. L. (1981). Shear effects on enzymes. *Enzyme Microb. Technol., 3,* 111–118.

Charman, S. A., Mason, K. L., & Charman, W. N. (1993). Techniques for assessing the effects of pharmaceutical excipients on the aggregation of porcine growth hormone. *Pharm. Res., 10,* 954–962.

Chuang, H. Y. K., Mohammad, S. F., Sharma, N. C., & Mason, R. G. (1980). Interaction of human a-thrombin with artificial surfaces and reactivity of adsorbed a-thrombin. *J. Biomed. Mater. Res., 14,* 467–476.

Cipolla, D., Gonda, I., & Shire, S. J. (1994). Characterization of aerosols of human recombinant deoxyribonuclease I (rhDNase) generated by jet nebulizers. *Pharm. Res., 11,* 491–498.

Costantino, H. R., Carrasquillo, K. G., Cordero, R. A., Mumenthaler, M., Hsu, C. C., & Griebenow, K. (1998). Effect of excipients on the stability and structure of lyophilized recombinant human growth hormone. *J. Pharm. Sci., 87,* 1412–1420.

Costantino, H. R., Griebenow, K., Mishra, P., Langer, R., & Klibanov, A. (1995). Fourier-transform infrared spectroscopic investigation of protein stability in the lyophilized form. *Biochim. Biophys. Acta, 1253,* 69–74.

Costantino, H. R., Langer, R., & Klibanov, A. M. (1994). Moisture-induced aggregation of lyophilized insulin. *Pharm. Res., 11,* 21–29.

DeGrazio, F. & Flynn, K. (1992). Lyophilization closures for protein based drugs. *J. Parenter. Sci. Technol., 46,* 54–61.

Demeule, B., Palais, C., Machaidze, G., Gurny, R., & Arvinte, T. (2009). New methods allowing the detection of protein aggregates: A case study on trastuzumab. *MAbs, 1,* 142–150.

Donroe, A., Gabler, R., Winter, P., Christiansen, G., DeRosa, D., Hill, D. et al. (1990). Current practices in endotoxin and pyrogen testing in biotechnology. *J. Parenter. Sci. Technol., 44,* 39–45.

Duddu, S. P. & Dal Monte, P. R. (1997). Effect of glass transition temperature on the stability of lyophilized formulations containing a chimeric therapeutic monoclonal antibody. *Pharm. Res., 14,* 591–595.

Duddu, S. P., Zhang, G. Z., & DalMonte, P. R. (1997). The relationship between protein aggregation and molecular mobility below the glass transition temperature of lyophilized formulations containing a monoclonal antibody. *Pharm. Res., 14,* 596–600.

Dziki, W. & Novak, K. J. (2003). Strategies for successful implementation of PAT in pharmaceutical manufacturing. *Am. Pharm. Rev., 5,* 54–60.

Feingold, V., Jenkins, A. B., & Kraegen, E. W. (1984). Effect of contact material on vibration-induced insulin aggregation. *Diabetologia, 27,* 373–378.

Franks, F. (1993). Storage stabilization of proteins. In F. Franks (Ed.), *Protein biotechnology: Isolation, characterization and stabilization* (pp. 489–531). Totowa, NJ: Humana Press.

Franks, F., Hatley, R. H. M., & Mathias, S. F. (1991). Materials science and the production of shelf stable biologicals. *BioPHARM, 4,* 38–55.

Gatlin, L. A. & Nail, S. L. (1994). Freeze drying: A practical overview. In R.G. Harrison (Ed.), *Protein purification process engineering* (pp. 317–367). New York: Marcel Dekker, Inc.

Geigert, J. (1989). Overview of the stability and handling of recombinant protein drugs. *J. Parenter. Sci. Technol., 43,* 220–224.

Goldbach, P., Brochart, H., & Stamm, A. (1993a). Spray-drying of liposomes for a pulmonary administration. I. Chemical stability of phospholipids. *Drug Dev. Ind. Pharm., 19(19),* 2611–2622.

Goldbach, P., Brochart, H., & Stamm, A. (1993b). Spray-drying of liposomes for a pulmonary administration. II. Retention of encapsulated materials. *Drug Dev. Ind. Pharm., 19(19),* 2623–2636.

Goldstein, G. & Audhya, T. (1992, August 18). Stabilized aqueous formulations of thymopentin. U.S. Patent 5,140,010.

Gonzalez, M., Murature, D. A., & Fidelio, G. D. (1995). Thermal stability of human immunoglobulins with sorbitol. A critical evaluation. *Vox Sang., 68,* 1–4.

Gregoriadis, G., McCormack, B., Obrenovic, M., Saffie, R., Zadi, B., & Perrie, Y. (1999). Vaccine entrapment in liposomes. *Methods, 19,* 156–162.

Greiff, D. (1971). Protein structure and freeze-drying: The effects of residual moisture and gases. *Cryobiology, 8,* 145–152.

Greiff, D. (1990). Factors affecting the statistical parameters and patterns of distribution of residual moistures in arrays of samples following lyophilization. *J. Parenter. Sci. Technol., 44(3),* 118–129.

Groves, M. J. & Teng, C. D. (1992). The effect of compaction and moisture on some physical and biological properties of proteins. In T.J. Ahern & M. C. Manning (Eds.), *Pharmaceutical biotechnology: Stability of protein pharmaceuticals. Part A: Chemical and physical pathways of protein degradation* (Vol. 2, pp. 311–359). New York: Plenum Press.

Hageman, M. J. (1988). The role of moisture in protein stability. *Drug Dev. Ind. Pharm., 14(14),* 2047–2070.

Hanson, M. A. & Rouan, S. K. E. (1992). Introduction to formulation of protein pharmaceuticals. In T.J. Ahern & M. C. Manning (Eds.), *Pharmaceutical biotechnology: Stability of protein pharmaceuticals: Part B: In vivo pathways of degradation and strategies for protein stabilization* (Vol. 3, pp. 209–233). New York: Plenum Press.

Hawe, A., Kasper, J. C., Friess, W., & Jiskoot, W. (2009). Structural properties of monoclonal antibody aggregates induced by freeze-thawing and thermal stress. *Eur. J. Pharm. Sci., 38,* 79–87.

Hellman, K., Miller, D., & Cammack, K. A. (1983). The effect of freeze-drying on the quaternary structure of L-asparaginase from *Erwinia carotovora. Biochim. Biophys. Acta, 749,* 133–142.

Henson, A. F., Mitchell, J. R., & Mussellwhite, P. R. (1970). The surface coagulation of proteins during shaking. *J. Colloid Interf. Sci., 32,* 162–165.

Hora, M. S., Rana, R. K., & Smith, F. W. (1992). Lyophilized formulations of recombinant tumor necrosis factor. *Pharm. Res., 9,* 33–36.

Ihlenfeld, J. V. & Cooper, S. L. (1979). Transient in vivo protein adsorption onto polymeric biomaterials. *J. Biomed. Mater. Res., 13,* 577–591.

Izutsu, K., Yoshioka, S., & Terao, T. (1993). Decreased protein-stabilizing effects of cryoprotectants due to crystallization. *Pharm. Res., 10,* 1232–1237.

Izutsu, K., Yoshioka, S., & Terao, T. (1994). Effect of mannitol crystallinity on the stabilization of enzymes during freeze-drying. *Chem. Pharm. Bull., 42(1),* 5–8.

Izutsu, K. I., Yoshioka, S., & Kojima, S. (1994). Physical stability and protein stability of freeze-dried cakes during storage at elevated temperatures. *Pharm. Res., 11,* 995–999.

Johnson, R. E., Kirchhoff, C. F., & Gaud, H. T. (2002). Mannitol-sucrose mixtures—Versatile formulations for protein lyophilization. *J. Pharm. Sci., 91,* 914–922.

Jones, L. S., Randolph, T. W., Kohnert, U., Papadimitriou, A., Winter, G., Hagmann, M. L. et al. (2001). The effects of Tween 20 and sucrose on the stability of anti-L-selectin during lyophilization and reconstitution. *J. Pharm. Sci., 90,* 1466–1477.

Joubert, M. K., Luo, Q., Nashed-Samuel, Y., Wypych, J., & Narhi, L. O. (2011). Classification and characterization of therapeutic antibody aggregates. *J. Biol. Chem., 286,* 25118–25133.

Kasper, J. C. & Friess, W. (2011). The freezing step in lyophilization: Physico-chemical fundamentals, freezing methods and consequences on process performance and quality attributes of biopharmaceuticals. *Eur. J. Pharm. Biopharm., 78,* 248–263.

Kim, K. J., Chen, V., & Fane, A. G. (1993). Some factors determining protein aggregation during ultrafiltration. *Biotechnol. Bioeng., 42,* 260–265.

Kirsch, L. E., Riggin, R. M., Gearhart, D. A., Lefeber, D. S., & Lytle, D. L. (1993). In-process protein degradation by exposure to trace amounts of sanitizing agents. *J. Parenter. Sci. Technol., 47,* 155–160.

Klemmer, A., Kramer, I., & Kamin, W. (2014). Physicochemical compatibility and stability of nebulizable drug admixtures containing dornase alfa and tobramycin. *Pulm. Pharmacol. Ther., 28,* 53–59.

Koseki, T., Kitabatake, N., & Doi, E. (1990). Freezing denaturation of ovalbumin at acid pH. *J. Biochem., 107,* 389–394.

Kwan, H. K. (1985, January 29). Biologically stable alpha-interferon formulations. U.S. Patent 4,496,537.

Larsen, H. M., Trnka, H., & Grohganz, H. (2014). Formation of mannitol hemihydrate in freeze-dried protein formulations—A design of experiment approach. *Int. J. Pharm., 460,* 45–52.

Lee, G. (2002). Spray drying of proteins. In J.F. Carpenter & M. C. Manning (Eds.), *Rational design of stable protein formulations: Theory and practice* (pp. 135–157). New York: Kluwer Academic/Plenum Publishers.

Lee, S. H., Heng, D., Ng, W. K., Chan, H. K., & Tan, R. B. (2011). Nano spray drying: A novel method for preparing protein nanoparticles for protein therapy. *Int. J. Pharm., 403,* 192–200.

Lee, S. W. & Sciarra, J. J. (1976). Development of an aerosol dosage form containing insulin. *J. Pharm. Sci., 65,* 567–572.

Lencki, R. W., Tecante, A., & Choplin, L. (1993). Effect of sear on the inactivation kinetics of the enzyme dextransucrase. *Biotechnol. Bioeng., 42,* 1061–1067.

Liao, Y. H., Brown, M. B., Quader, A., & Martin, G. P. (2002). Protective mechanism of stabilizing excipients against dehydration in the freeze-drying of proteins. *Pharm. Res., 19,* 1854–1861.

Liu, W. R., Langer, R., & Klibanov, A. M. (1991). Moisture-induced aggregation of lyophilized proteins in the solid state. *Biotechnol. Bioeng., 37,* 177–184.

Lougheed, W. D., Albisser, A. M., Martindale, H. M., Chow, J. C., & Clement, J. R. (1983). Physical stability of insulin formulations. *Diabetes, 32,* 424–432.

Ma, X., Wang, D. Q., Bouffard, R., & MacKenzie, A. (2001). Characterization of murine monoclonal antibody to tumor necrosis factor (TNF-MAb) formulation for freeze-drying cycle development. *Pharm. Res., 18,* 196–202.

Maa, Y. F., Ameri, M., Rigney, R., Payne, L. G., & Chen, D. (2004). Spray-coating for biopharmaceutical powder formulations: Beyond the conventional scale and its application. *Pharm. Res, 21,* 515–523.

Maa, Y. F., Nguyen, P. A., Andya, J. D., Dasovich, N., Sweeney, T. D., Shire, S. J. et al. (1998). Effect of spray drying and subsequent processing conditions on residual moisture content and physical/biochemical stability of protein inhalation powders. *Pharm. Res., 15,* 768–775.

Mantyh, P. W., Ghilardi, J. R., Rogers, S., Demaster, E., Allen, C. J., Stimson, E. R. et al. (1993). Aluminum, iron, and zinc ions promote aggregation of physiological concentrations of beta-amyloid peptide. *J. Neurochem., 61,* 1171–1174.

Martin, J. M. & Manteuffel, R. L. (1988). Protein recovery from effluents of microporous membranes. *BioPHARM, 1(10),* 20–27.

Matsuura, G. (1992). Visual compatibility of sargramostim (GM-CSF) during simulated Y-site administration with selected agents. *Hosp. Pharm., 27,* 200–209.

McElnay, J. C., Elliott, D. S., & D'Arcy, P. F. (1987). Binding of human insulin to burette administration sets. *Int. J. Pharm., 36,* 199–203.

Merkli, A., Heller, J., Tabatabay, C., & Gurny, R. (1994). Gamma sterilization of a semi-solid poly(ortho ester) designed for controlled drug delivery—Validation and radiation effects. *Pharm. Res., 11,* 1485–1491.

Mizutani, T. (1980). Decreased activity of proteins adsorbed onto glass surfaces with porous glass as a reference. *J. Pharm. Sci., 69,* 279–281.

Mizutani, T. & Mizutani, A. (1978). Estimation of adsorption of drugs and proteins on glass surfaces with controlled pore glass as a reference. *J. Pharm. Sci., 67,* 1102–1105.

Mobus, K., Siepmann, J., & Bodmeier, R. (2012). Zinc-alginate microparticles for controlled pulmonary delivery of proteins prepared by spray-drying. *Eur. J. Pharm. Biopharm., 81,* 121–130.

Mumenthaler, M., Hsu, C. C., & Pearlman, R. (1994). Feasibility study on spray-drying protein pharmaceuticals: Recombinant human growth hormone and tissue-type plasminogen activator. *Pharm. Res., 11(1),* 12–20.

Nail, S. L., Jiang, S., Chongprasert, S., & Knopp, S. A. (2002). Fundamentals of freeze drying. In S.L. Nail & M. J. Akers (Eds.), *Development and manufacturing of protein pharmaceuticals* (pp. 281–360). New York: Kluwer Academic/Plenum Publishers.

Nayak, A., Colandene, J., Bradford, V., & Perkins, M. (2011). Characterization of subvisible particle formation during the filling pump operation of a monoclonal antibody solution. *J. Pharm. Sci., 100(10),* 4198–4204.

Nguyen, T. H. & Ward, C. (1993). Stability characterization and formulation development of alteplase, a recombinant tissue plasminogen activator. In Y.J. Wang & R. Pearlman (Eds.), *Pharmaceutical biotechnology: Stability and characterization of proteins and peptide drugs: Case histories* (Vol. 5, pp. 91–134). New York: Plenum Press.

Niven, R. W., Ip, A. Y., Mittelman, S. D., Farrar, C., Arakawa, T., & Prestrelski, S. J. (1994). Protein nebulization. I. Stability of lactate dehydrogenase and recombinant granulocyte-colony stimulating factor to air-jet nebulization. *Int. J. Pharm., 109,* 17–26.

Oeswein, J. Q., Daugherty, A. L., Cornavaca, E. J., Moore, J. A., Gustafson, H. M., Eckhardt, B. M. et al. (1991). Aerosolization of protein pharmaceuticals. In R.N. Dalby & R. Evans (Eds.), *Proceedings of the Second Respiratory Drug Delivery Symposium* (pp. 14–49). Lexington, KY: University of Kentucky/Continuing Pharmacy Education.

Pearlman, R. & Oeswein, J. Q. (1992, March 17). Human growth hormone formulation. U.S. Patent 5,096,885.

Persson, K. M. & Gekas, V. (1994). Factors influencing aggregation of macromolecules in solution. *Proc. Biochem., 29,* 89–98.

Peters, T. (1985). Serum albumin. *Adv. Protein Chem., 37,* 161–245.

Petty, C. & Cunningham, N. L. (1974). Insulin adsorption by glass infusion bottles, polyvinyl chloride infusion containers and intravenous tubing. *Anesthesiology, 40,* 400–404.

PDR Staff. (2014). *Physicians' desk reference* (68th ed.). Montvale, NJ: PDR Network, LLC.

Pfeifer, R. W., Siegel, J., & Ayers, L. W. (1994). Assessment of microbial growth in intravenous immune globulin preparations. *Am. J. Hosp. Pharm., 51,* 1676–1679.

Pikal, M. J., Dellerman, K. M., Roy, M. L., & Riggin, R. M. (1991). The effects of formulation variables on the stability of freeze-dried human growth hormone. *Pharm. Res., 8,* 427–436.

Pikal, M. J. & Rigsbee, D. R. (1997). The stability of insulin in crystalline and amorphous solids: Observation of greater stability for the amorphous form. *Pharm. Res., 14,* 1379–1387.

Pikal, M. J., Shah, S., Senior, D., & Lang, J. E. (1983). Physical chemistry of freeze-drying: Measurement of sublimation rates for frozen aqueous solutions by a microbalance technique. *J. Pharm. Sci., 72,* 635–650.

Prestrelski, S. J., Arakawa, T., & Carpenter, J. F. (1993). Separation of freezing-and drying-induced denaturation of lyophilized proteins using stress-specific stabilization. II. Structural studies using infrared spectroscopy. *Arch. Biochem. Biophys., 303,* 465–473.

Prestrelski, S. J., Pikal, K. A., & Arakawa, T. (1995). Optimization of lyophilization conditions for recombinant human interleukin-2 by dried-state conformational analysis using Fourier-transform infrared spectroscopy. *Pharm. Res., 12,* 1250–1259.

Randen, L., Nilson, J., & Edman, P. (1988). Coprecipitation of enzymes with water soluble starch—An alternative to freeze-drying. *J. Pharm. Pharmacol., 40,* 763–766.

Rech, M., Panaggio, A., & Bontempo, J. A. (1991). Current trends in facilities and equipment for aseptic processing. *Pharm. Technol., 15,* 54–60.

Ressing, M. E., Jiskoot, W., Talsma, H., van Ingen, C. W., Beuvery, E. C., & Crommelin, D. J. A. (1992). The influence of sucrose, dextran, and hydroxypropyl-beta-cyclodextrin as lyoprotectants for a freeze-dried mouse IgG(2a) monoclonal antibody (MN12). *Pharm. Res., 9,* 266–270.

Richmond, G. W. (1990). Challenges in large scale product manufacturing. In *32nd Annual International Industrial Pharmaceutical Research Conference: Challenges in the Development of Protein Drug Products*, Land O'Lake, WI.

Roughton, B. C., Iyer, L. K., Bertelsen, E., Topp, E. M., & Camarda, K. V. (2013). Protein aggregation and lyophilization: Protein structural descriptors as predictors of aggregation propensity. *Comput. Chem. Eng., 58,* 369–377.

Sapienza, A. (1994). Biotechnology and the formulary. *Pharm. News, 1,* 19–20.

Sarciaux, J. M., Mansour, S., Hageman, M. J., & Nail, S. L. (1999). Effects of buffer composition and processing conditions on aggregation of bovine IgG during freeze-drying. *J. Pharm. Sci., 88,* 1354–1361.

Sarry, C. & Sucker, H. (1992). Adsorption of proteins on microporous membrane filters: Part I. *Pharm. Technol., 16(10),* 72–79.

Sarry, C. & Sucker, H. (1993). Adsorption of proteins on microporous membrane filters: Part II. *Pharm. Technol., 17(1),* 60–70.

Shikama, K. & Yamazaki, I. (1961). Denaturation of catalase by freezing and thawing. *Nature, 190,* 83–84.

Shire, S. J., Shahrokh, Z., & Liu, J. (2004). Challenges in the development of high protein concentration formulations. *J. Pharm. Sci., 93,* 1390–1402.

Singh, B. P., Bohidar, H. B., & Chopra, S. (1991). Heat aggregation studies of phycobilisomes, ferritin, insulin, and immunoglobulin by dynamic light scattering. *Biopolymers, 31,* 1387–1396.

Sluzky, V., Tamada, J. A., Klibanov, A. M., & Langer, R. (1991). Kinetics of insulin aggregation in aqueous solutions upon agitation in the presence of hydrophobic surfaces. *Proc. Natl. Acad. Sci., 88,* 9377–9381.

Soderquist, M. E. & Walton, A. G. (1980). Structural changes in proteins adsorbed on polymer surfaces. *J. Colloid Interface Sci., 75,* 386–397.

Soliman, F. S. & Berg, L. V. D. (1971). Factors affecting freezing damage of lactic dehydrogenase. *Cryobiology, 8,* 73–78.

Stevenson, C. L. (2000). Characterization of protein and peptide stability and solubility in non-aqueous solvents. *Curr. Pharm. Biotechnol., 1,* 165–182.

Stickelmeyer, M. P., Graf, C. J., Frank, B. H., Ballard, R. L., & Storms, S. M. (2000). Stability of U-10 and U-50 dilutions of insulin lispro. *Diabetes Technol. Ther., 2,* 61–66.

Takada, K., Nakahigashi, Y., Tanaka, T., Kishimoto, S., Ushirogawa, Y., & Takaya, T. (1994). Pharmacological activity of tablets containing recombinant human granulocyte colony-stimulating factor (rhG-CSF) in rats. *Int. J. Pharm., 101,* 89–96.

Thirumangalathu, R., Krishnan, S., Ricci, M. S., Brems, D. N., Randolph, T. W., & Carpenter, J. F. (2009). Silicone oil- and agitation-induced aggregation of a monoclonal antibody in aqueous solution. *J. Pharm. Sci., 98,* 3167–3181.

Thomas, C. R. & Dunnill, P. (1979). Action of shear on enzymes: Studies with catalase and urease. *Biotechnol. Bioeng., 21,* 2279–2302.

Thomas, C. R. & Geer, D. (2011). Effects of shear on proteins in solution. *Biotechnol. Lett., 33,* 443–456.

Thompson, C. D., Vital-Carona, J., & Faustino, E. V. (2012). The effect of tubing dwell time on insulin adsorption during intravenous insulin infusions. *Diabetes Technol. Ther., 14,* 912–916.

Townsend, M. W. & DeLuca, P. P. (1991). Nature of aggregates formed during storage of freeze-dried ribonuclease A. *J. Pharm. Sci., 80,* 63–66.

Trissel, L. A. & Martinez, J. F. (1994). Compatibility of filgrastim with selected drugs during simulated Y-site administration. *Am. J. Hosp. Pharm., 51,* 1907–1913.

U.S. Food and Drug Administration. *Pharmaceutical cGMPs for the 21st Century: A risk-based approach*; http://www.fda.gov/downloads/Drugs/DevelopmentApprovalProcess/Manufacturing/QuestionsandAnswersonCurrentGoodManufacturingPracticescGMPforDrugs/UCM176374.pdf (accessed January 17, 2015).

Virkar, P. D., Narendranathan, T. J., Hoare, M., & Dunnill, P. (1981). Studies of the effects of shear on globular proteins: Extension to high shear fields and to pumps. *Biotechnol. Bioeng., 23,* 425–429.

Visor, G. C., Tsai, K. P., Duffy, J., Miller, M. D., Calderwood, T., & Knepp, V. M. (1990). Quantitative evaluation of the stability and delivery of interleukin-1B by infusion. *J. Parenter. Sci. Technol., 44(3),* 130–132.

Wang, W. (2000). Lyophilization and development of solid protein pharmaceuticals. *Int. J. Pharm., 203,* 1–60.

Ward, C. & Weck, S. (1990). Dilution and storage of recombinant tissue plasminogen activator (Activase) in balanced salt solutions. *Am. J. Ophthalmol., 109,* 98–99.

White, T. X. (1992). Terminal sterilization and aseptic processing. *Pharm. Technol., 16,* 52–58.

Wordell, C. J. (1992). Addition of albumin to colony stimulating factor infusions. *Hosp. Pharm., 27,* 421–424.

Yamamoto, K., Hayashi, S., & Yasui, T. (1993). Hydrostatic pressure-induced aggregation of myosin molecules in 0.5 M KCl at pH 6.0. *Biosci. Biotechnol. Biochem., 57,* 383–389.

Yoshioka, S., Aso, Y., Izutsu, K., & Terao, T. (1993). Aggregates formed during storage of beta-galactosidase in solution and in the freeze-dried state. *Pharm. Res., 10,* 687–691.

Zarrintan, M. H., Teng, C. D., & Groves, M. J. (1990). The effect of compactional pressure on a wheat germ lipase preparation. *Pharm. Res., 7,* 247–250.

Zhang, A., Singh, S. K., Shirts, M. R., Kumar, S., & Fernandez, E. J. (2012). Distinct aggregation mechanisms of monoclonal antibody under thermal and freeze-thaw stresses revealed by hydrogen exchange. *Pharm. Res., 29,* 236–250.

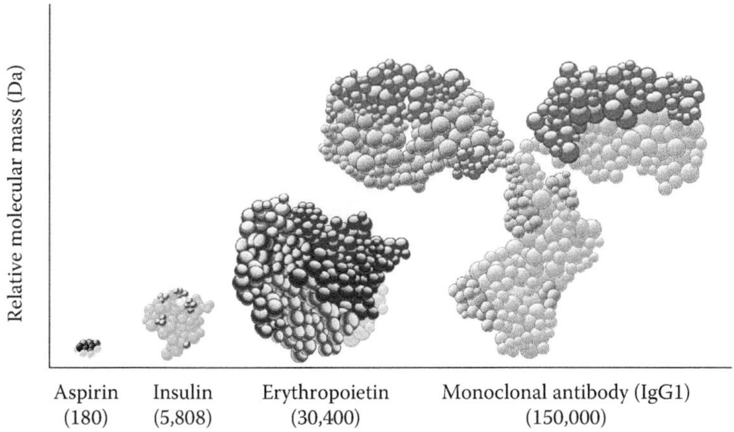

FIGURE 2.1 Structural complexity of monoclonal antibodies relative to small-molecule agents and lower-molecular-weight biologics. (Reproduced from Mellstedt, H., *EJC Suppl.*, 11, 1, 2013. With permission.)

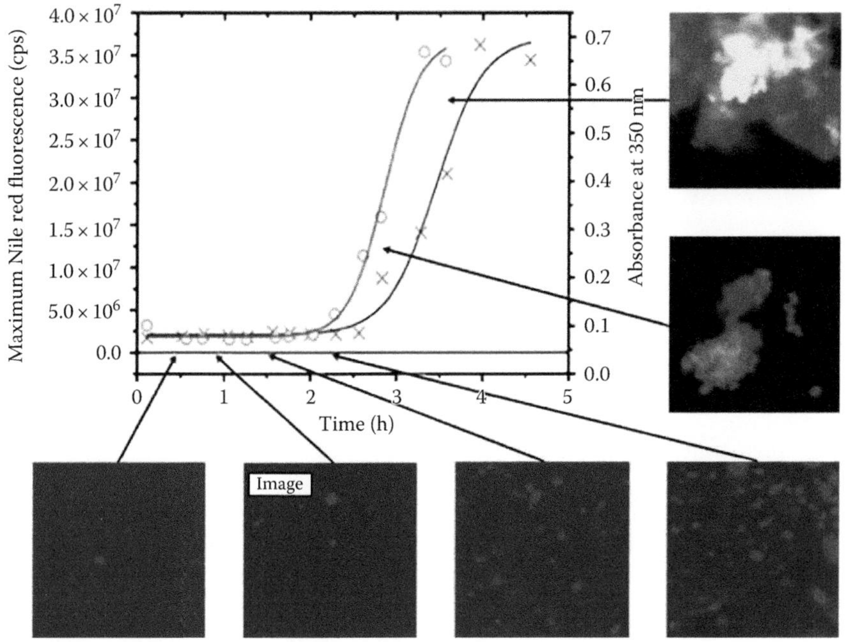

FIGURE 3.1 Fibrillation of human calcitonin followed by microscopy, Nile red fluorescence, and UV absorbance. Nile red fluorescence (o) detected gelation phenomena earlier than UV absorbance (X), but only the microscopic study of Nile red–stained solutions revealed the presence and progression of aggregates in the early lag phase period. (Reproduced from Mach, H. and Arvinte, T., *Eur. J. Pharm. Biopharm.*, 78, 196, 2011. With permission.)

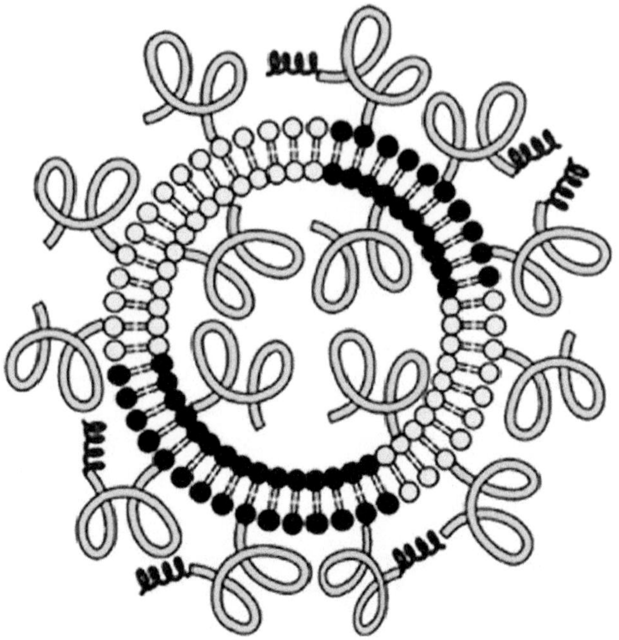

FIGURE 4.4 Schematic depicting a stealth PEGylated liposome. (Reproduced from open access article Mufamadi, M.S. et al., *J. Drug Deliv.*, 2011, 939851, 2011, and with permission from Rai, P. et al., *Chemistry*, 14, 7748, 2008.)

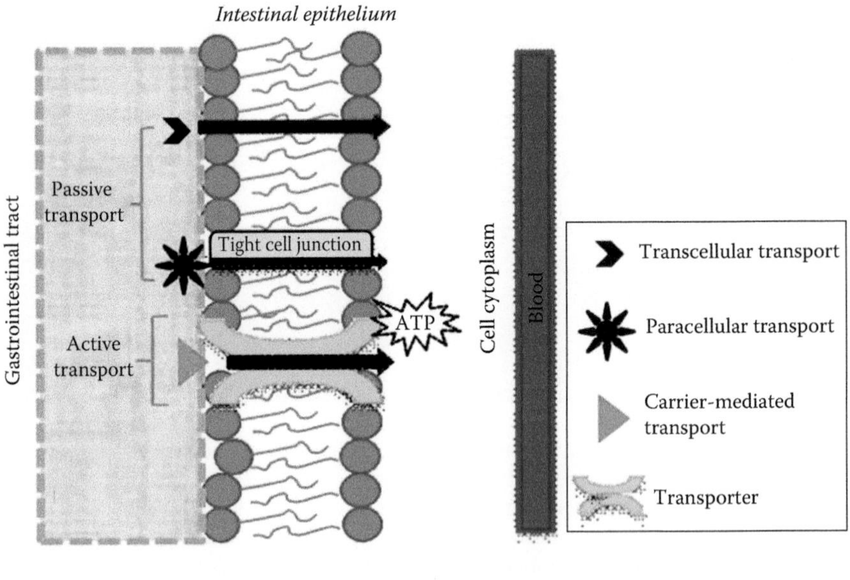

FIGURE 7.1 Intestinal drug transport mechanisms. (Reproduced from Renukuntla, J. et al., *Int. J. Pharm.*, 447, 75, 2013. With permission.)

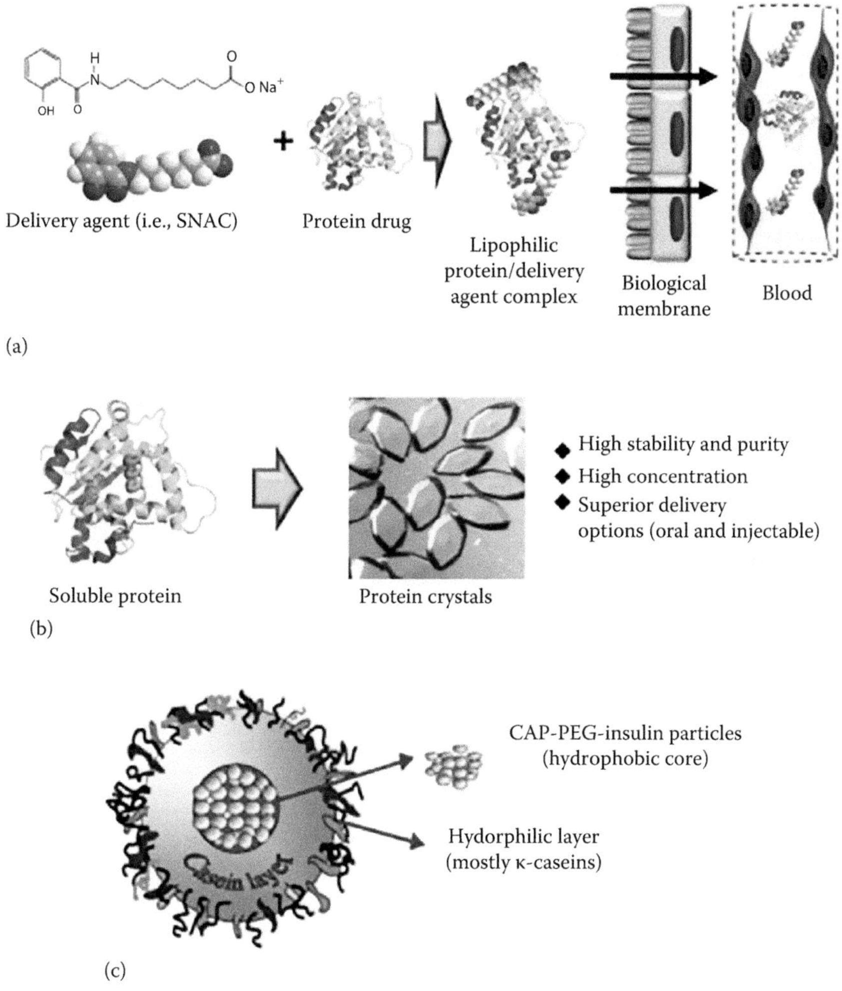

FIGURE 7.2 Examples of oral protein delivery technologies developed for clinical applications. (a) Emisphere: Eligen™ technology. (b) Altus: Cross-linked enzyme crystal (CLEC®) technology. (c) BioSante: BioOral system (calcium phosphate nanoparticles). *(Continued)*

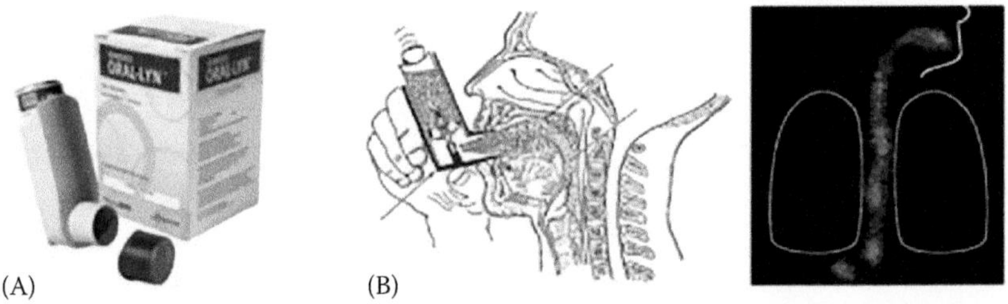

(A) (B)

(d)

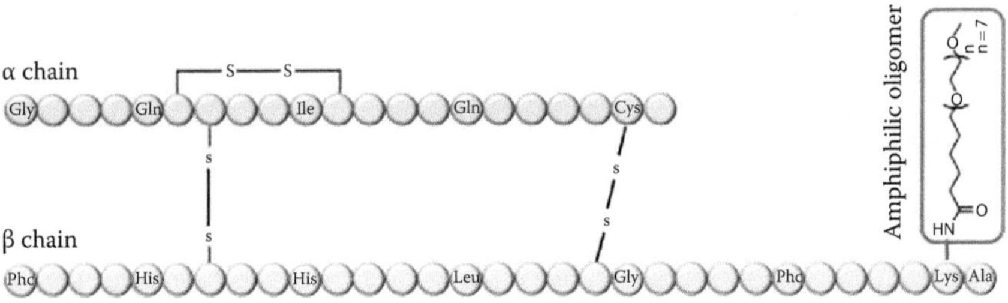

(e)

FIGURE 7.2 (Continued) Examples of oral protein delivery technologies developed for clinical applications. (d) Generex: Oral-lyn™ (A) the RapidMist™ device and (B) the formulation is delivered to the oral cavity for absorption via the oral mucosa. (e) NOBEX and Biocon: Amphiphilic oligomers. (Reproduced from Park, K. et al., *React. Funct. Polym.*, 71, 280, 2011. With permission.)

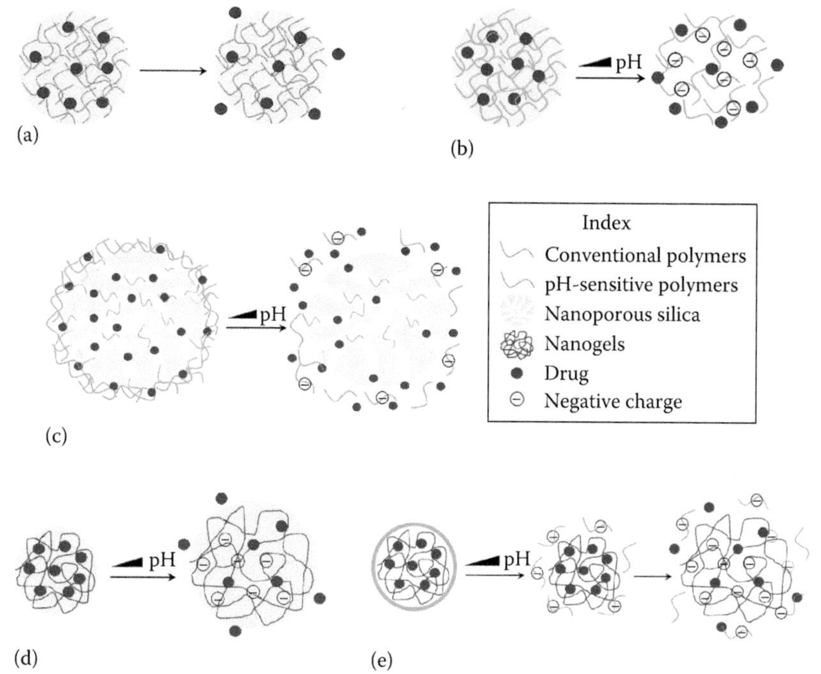

FIGURE 7.3 Drug release mechanism from carrier systems. (a) Release from conventional nanoparticles by diffusion, (b) release from pH-sensitive nanoparticles after the material dissolution at a specific pH, (c) release from pH-sensitive nanomatrix, (d) release from pH-sensitive nanoparticles after the material swelling at a certain pH, and (e) release from pH-sensitive nanoparticles as the coeffect of dissolution and swelling of the materials at specific pH. (Reproduced from Wang, X.Q. and Zhang, Q., *Eur. J. Pharm. Biopharm.*, 82, 219, 2012. With permission.)

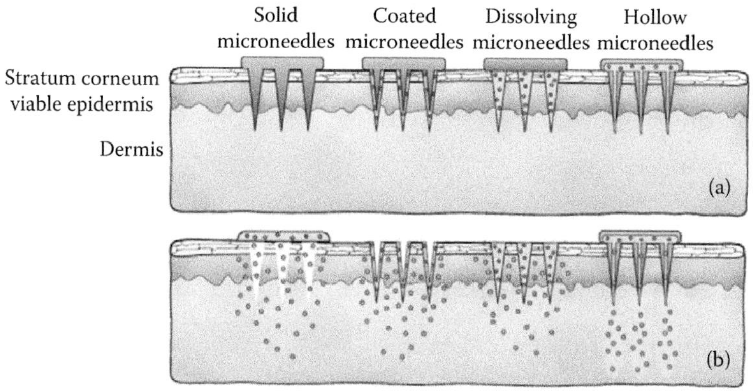

FIGURE 8.1 Microneedles (MN) applied on the skin (a) to achieve drug delivery (b). The skin can be pretreated with solid microneedles followed by drug application, or the drug may be coated on microneedles or incorporated within dissolving microneedles or injected into the skin using hollow microneedles. (Reproduced from Kim, Y.C. et al., *Adv. Drug Deliv. Rev.*, 64, 1547, 2012. With permission.)

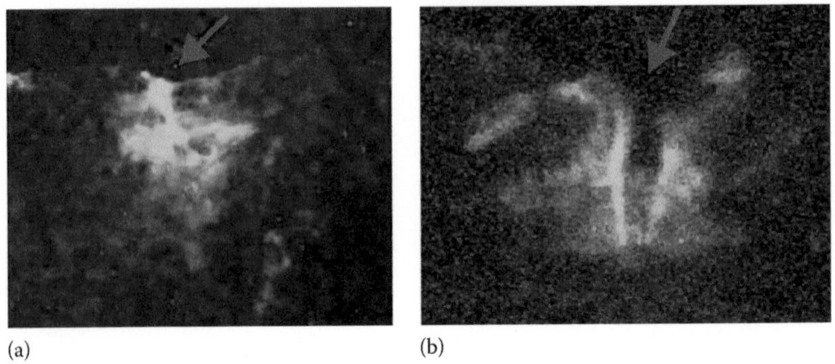

FIGURE 8.2 Fluorescent micrographs of IgG transport down through the microchannels (a) and radial diffusion (b) into the surrounding tissue in hairless rat skin following treatment with 500 μm long maltose microneedles. (Reproduced from Li, G. et al., *J. Pharm. Sci.*, 99, 1931, 2009. With permission.)

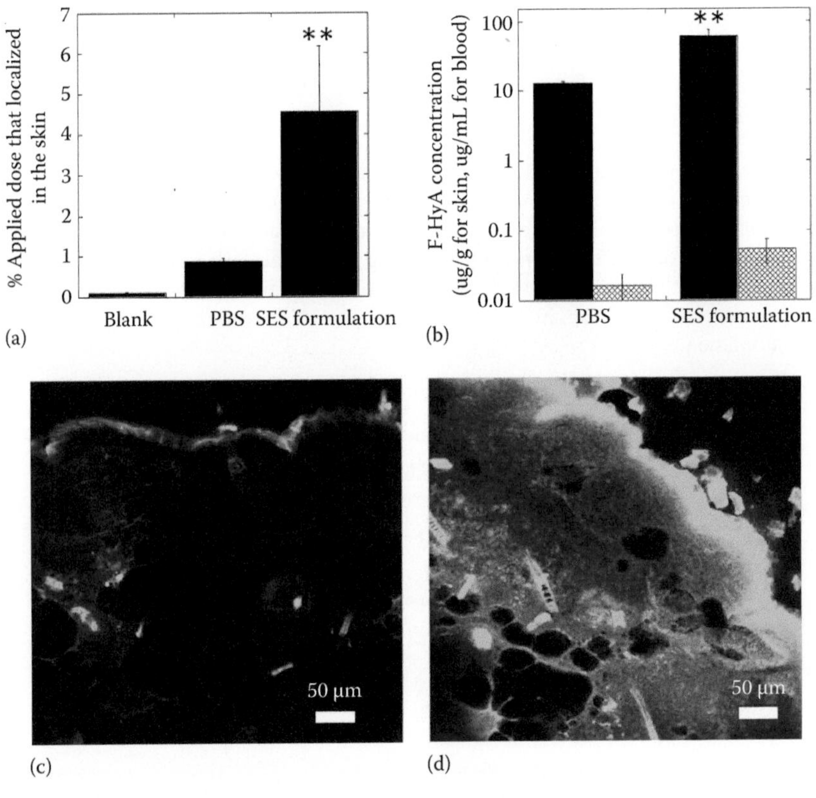

FIGURE 8.3 SPACE-peptide ethosomal carrier (SES) enhances penetration of high-MW HA into hairless mice. The figure shows delivery into the skin as percentage of the applied dose (a) or as an amount in the skin and blood (b). Confocal microscopy shows that penetration was weaker from PBS buffer (c) and much stronger from SES formula (d). (Reproduced from Chen, M. et al., *J. Control. Release*, 173, 67, 2014. With permission.)

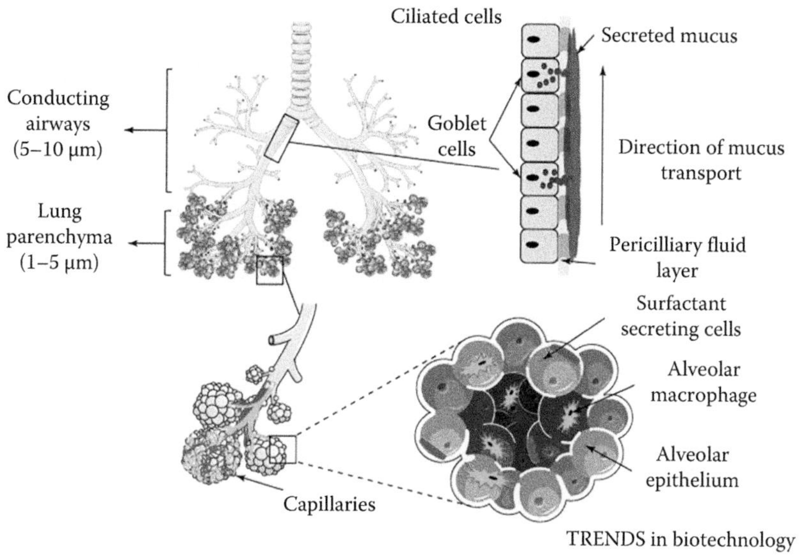

FIGURE 9.1 Diagram of the lung and particle size requirements based on intended deposition region in the respiratory tract. (Reproduced from Sou, T. et al., *Trends Biotechnol.*, 29, 191, 2011. With permission.)

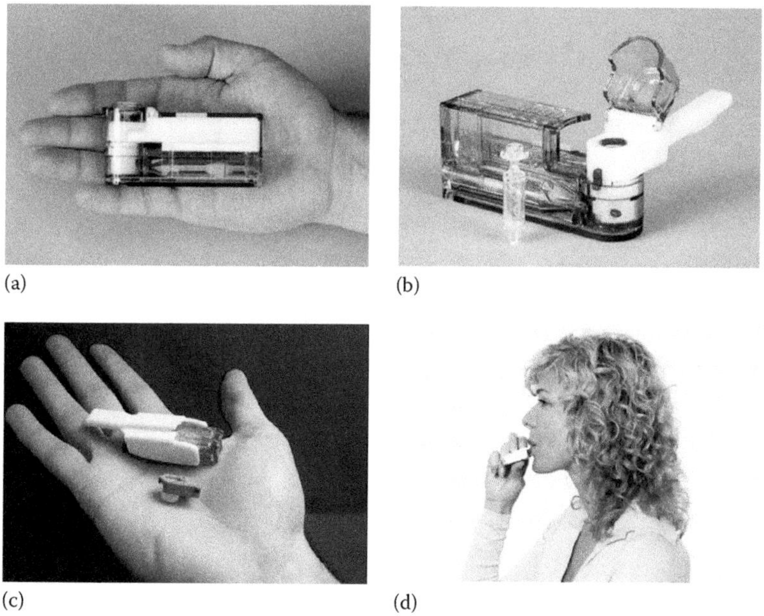

FIGURE 9.2 AFREZZA insulin inhalation device (Mannkind Corporation, Valencia, CA) approved by FDA. It consists of Technosphere® insulin inhalation powder premetered into single-use dose cartridges for administration with an AFREZZA™ inhaler. A palm-sized (a and b) or thumb-sized (c and d) device is shown. (Reproduced from Neumiller, J.J. and Campbell, R.K., *BioDrugs*, 24, 165, 2010. With permission.)

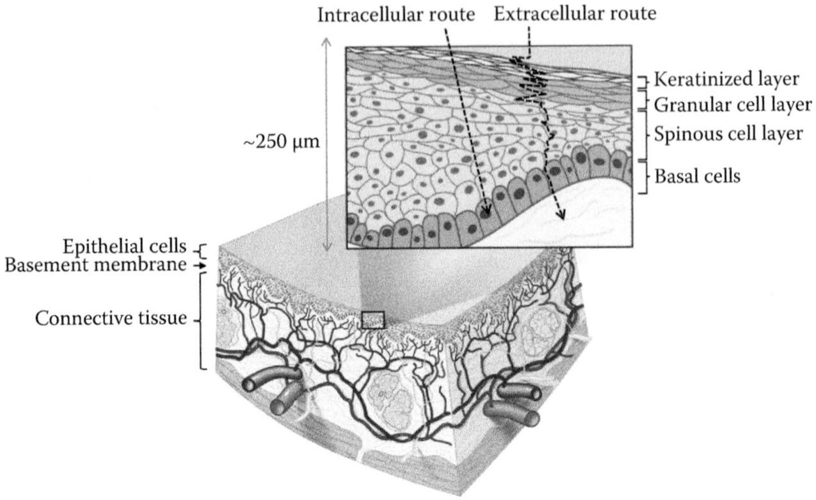

FIGURE 9.4 Cartoon of the structure of the oral mucosa. Insert shows different routes by which drugs can cross the oral mucosa. (Reproduced from Hearnden, V. et al., *Adv. Drug Deliv. Rev.*, 64, 16, 2012. With permission.)

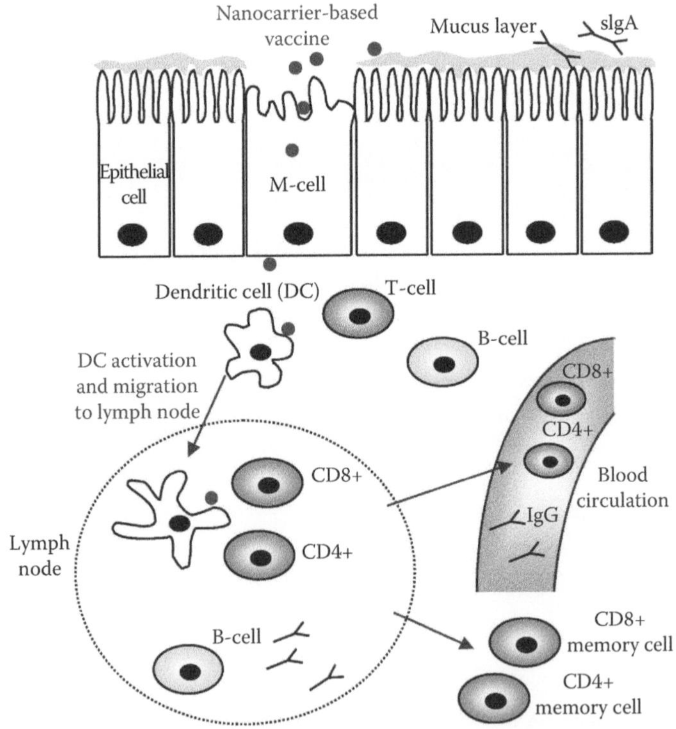

FIGURE 10.1 Development of cellular and humoral immune responses following mucosal administration of nanocarrier-based vaccines. (Reproduced with permission from Kammona, O. and Kiparissides, C., *J. Control. Release*, 161, 781, 2012.)

Section II

Drug Delivery Systems

6 Parenteral Controlled Delivery and Pharmacokinetics of Therapeutic Peptides and Proteins

6.1 INTRODUCTION

Presently, peptide and protein drugs are almost exclusively marketed for parenteral administration. This is because these drugs face formidable enzymatic and penetration barriers, when administered orally. A limitation of the parenteral route for delivery of peptides and proteins is the extremely short half-lives of these drugs, being in the order of a few minutes. This demands repeated administration that is inconvenient to the patient. Also, parenteral administration is mostly done under the supervision of a health-care professional, thereby adding to the cost of the health-care system. For these reasons, nonparenteral routes of administration are being actively investigated. Another alternative is to develop controlled release parenteral formulations, where a single injection may release the drug over a month or even longer. Twice-a-year depot injections for leuprolide encapsulated in polymer are available as an alternative to once-a-day injections of the solution formulation (Schwendeman, Shah, Bailey, & Schwendeman, 2014).

The development and use of these controlled-release parenteral formulations of peptides and proteins are discussed in this chapter. General formulation principles for a simple parenteral formulation using conventional excipients have already been discussed in Chapter 4 and will not be discussed here. The concepts of sterility, clean rooms, removal of pyrogens, and related details are available in several other texts and will also not be discussed here. A parenteral formulation may be administered as a continuous intravenous (IV) infusion or as an IV, intramuscular (IM), or subcutaneous (SC) injection. Variables that can affect drug release after IM or SC injection have been reviewed (Zuidema, Kadir, Titulaer, & Oussoren, 1994). If the peptide or protein is degraded to any significant extent at the SC or IM injection site, then IV injection may be the best option. In this case, an IV injection may allow the administration of a smaller dosage as the drug is not lost by proteolytic degradation. However, IM and SC routes, if viable, will permit a longer duration of action, an important advantage considering the short half-lives of these drugs. Furthermore, a fraction of the drug may take the lymphatic route, and uptake into the blood may be delayed by several hours. Particulates and macromolecules (molecular weight [MW] >2000) favor drainage into the more open capillaries of the lymph from the interstitial spaces following SC or IM injection, and thus, lymphatic transport plays an important role in absorption of peptides and proteins following parenteral administration. It seems that more than 50% of the dose of molecules with an MW greater than 16 kDa would be absorbed by the regional lymphatics after SC administration. Particulate systems will also be taken up by the lymphatics, provided that their size is less than about 100 nm to allow drainage from the injection site. Thus, many liposomes would be taken up by the lymphatics. Microspheres would be too large for lymphatic uptake, but being biodegradable, their fragments may be taken up. Lymphatic transport of interferon and monoclonal antibodies has been extensively studied (Porter, 1997). Lymph has been reported to represent the predominant pathway for absorption of r-metHu-Leptin (MW 16.2 kDa) after SC administration (McLennan et al., 2003).

Controlled-release parenteral formulations may be made as implants, as oily injections, or as particulate systems. For example, a suspension of α-interferon in oil has been reported to prolong the activity of interferon when administered subcutaneously to beagles (Takenaka, Fujioka, & Takada, 1989). The particulate systems include microspheres, microparticles, microcapsules, nanocapsules, nanospheres, and nanoparticles (Kreuter, 1994). Recombinant polymers have been used as well (Price, Poursaid, & Ghandehari, 2014). Typically, a microcapsule contains the drug as a central core, and the drug release is zero order. In contrast, a microsphere has the drug dispersed throughout the particle, and drug release is first order. Particles, microspheres, and microcapsules smaller than about 1 μm are usually referred to as nanoparticles, nanospheres, or nanocapsules, respectively. The blood capillaries in the body have a diameter of about 5 μm so that only nanoparticles are generally administered by the IV route. Microparticles, in contrast, are typically around 100 μm in diameter and are administered by SC or IM injection. The various polymer materials used and the formulation of these controlled-release formulations will be briefly discussed. Controlled-release parenteral formulations often have mass limitations because of which they can only be developed for potent drugs. This may not be a constraint for peptide and protein drugs as these are generally potent. Tice and Tabibi (1992) have discussed that for drugs that are potent and need less amount of rate controlling excipient, a small mass is needed per day so that a longer duration of release is possible. For instance, consider a 1.0 mL volume of an IM injection with 20% solids. This injection will contain 200 mg of solids. Assuming that a 1:10 ratio of drug/excipient was required for adequate controlled release over a 2-month period, this will allow for a total of 20 mg of drug with a release rate of 0.33 mg/day for a 2-month period following a single administration.

6.2 PHARMACOKINETICS OF PEPTIDE AND PROTEIN DRUGS

An understanding of basic pharmacokinetics, especially as relevant to peptide/protein drugs, is essential to an understanding of their parenteral administration and disposition, as well as safety-related issues and various other delivery methods (Vugmeyster, Xu, Theil, Khawli, & Leach, 2012). Therapeutic proteins may also be used in combination with other proteins or small molecules, and protein–drug interactions resulting from overlap in mechanism of action, alteration in target, and/or drug–disease interaction are important considerations receiving increasing attention from regulatory agencies (Girish et al., 2011). However, a detailed discussion of these is outside the scope of this book. Pharmacokinetics is the study of the time course of drug absorption, distribution, metabolism, and excretion. In this section, the pharmacokinetic principles are briefly reviewed, and the unique differences for application of these principles to peptide and protein drugs are pointed out. Some specific examples for pharmacokinetics of individual proteins are provided in Section 6.6. For the most part, the discussion is confined to pharmacokinetic considerations following parenteral (IV, IM, or SC) delivery. Any unique pharmacokinetic consideration following oral, nasal, or other mucosal delivery routes has been discussed in Chapters 7 and 9 that discuss these routes. One important aspect to the pharmacokinetics of peptides and proteins that is not being discussed here is the importance of having a sensitive, stability-indicating, validated bioanalytical assay. Since proteins are very potent drugs, only small quantities are used, and once distributed in the volume of distribution, the plasma concentrations may be below the assay limits for some analytical methods. Also, the use of radiotracers can lead to erroneous results as the assay may be measuring proteolytic fragments rather than the intact protein. Furthermore, the presence of endogenous proteins must be taken into account, and these may exist in several forms that may affect immunoassay results. The administration of pharmacological doses of exogenous proteins may also affect the production of the endogenous protein (Howanitz, 1993; Wills & Ferraiolo, 1992). Analytical techniques such as radioimmunoassay and enzyme-linked immunosorbent assay should be used to characterize the pharmacokinetics of proteins, and these assay techniques are discussed in Chapter 2. Approaches based on mass spectroscopy can generally provide the assay sensitivity required (Mesmin, Fenaille, Ezan, & Becher, 2011).

The rate and extent of drug absorption, often referred to as bioavailability, can be estimated by comparing the total area under the drug concentration in the plasma versus time curve. A drug is generally available 100% after IV administration so that a comparison of AUC after oral administration with that after IV administration gives us the fraction (F) of oral dose available to systemic circulation:

$$F = (AUC)_{\text{oral}} / (AUC)_{\text{IV}} \tag{6.1}$$

Though the term *bioavailability* is generally used with oral absorption, it includes absorption from all routes of administration. For peptides and proteins, the availability from an SC injection may be reduced due to proteolytic degradation at the injection site. Furthermore, even the IV route cannot be assumed to be 100% available for proteins because degradation by proteolytic enzymes may start in the blood itself. Similarly, bioavailability of peptides and proteins from mucosal route is reduced due to poor membrane permeability of these drugs. The transfer of drug from blood to the extravascular fluids and tissues is referred to as drug distribution. The enzymatic or biochemical transformation of drug is referred to as drug metabolism, and the transfer of drug and its metabolites to urine or other excretory components is called drug elimination. Though distribution and elimination occur simultaneously, some drugs attain distribution equilibrium before a measurable fraction of dose is eliminated. In such cases, the body behaves as a single compartment. However, for many drugs, the drug distributes slowly so that some fraction of a dose is eliminated before distribution equilibrium is achieved. For such drugs, the body behaves like a multicompartment system. In such systems, the semilogarithmic plot of drug concentration in plasma versus time cannot be described by a single exponential expression. Following IV bolus dosing, most peptides and proteins exhibit a biexponential decline (Colburn, 1991). In some cases, a triexponential decline may also be observed. For example, the plasma data of ^{131}I-labeled α1-antitrypsin in humans following IV administration were fitted to a sum of three exponential terms in accordance with a three-compartment model (Constans et al., 1992).

6.2.1 Pharmacokinetic Parameters

Pharmacokinetic evaluation of drugs involves the determination of four critical parameters: volume of distribution, clearance, half-life, and bioavailability. Bioavailability considerations have already been discussed, and the other three important pharmacokinetic parameters (volume of distribution, clearance, and half-life) will now be discussed.

6.2.1.1 Volume of Distribution

Once a drug has reached its distribution equilibrium, its concentration in plasma reflects these distribution factors. The amount of drug in the body (A) may then be related to the drug concentration in the plasma (C) as

$$A = V \cdot C \tag{6.2}$$

where V is the apparent volume of distribution. Since the amount of drug in the body immediately after IV administration is equivalent to dose, the following relationship exists:

$$V = \frac{i.v.\ dose}{(AUC)k} \tag{6.3}$$

where k is the first-order elimination rate constant. The plasma volume of a normal 70 kg man is 3 L, blood volume is about 5.5 L, extracellular fluid outside the plasma is 12 L, and the total body water is about 42 L. However, the magnitude of V rarely corresponds to plasma volume, extracellular volume, or volume of total body water but may vary from a few liters to several hundred liters in a 70 kg man. Thus, volume of distribution does not represent a real volume and is strongly influenced by factors such as protein binding (Gloff & Benet, 1990). A drug with high degree of binding to plasma proteins will exhibit a small volume of distribution. For drugs that bind more extensively to extravascular sites than to plasma proteins, the apparent volume of distribution can be large, several fold higher than the total body water. Comparison of the volumes of distribution is made using the term volume of distribution at steady state (V_{ss}), which reflects the sum of the volumes of all the pools into which the drug may distribute. For a drug exhibiting multicompartment pharmacokinetics, the apparent volume of distribution, usually called V_β, is given by

$$V_\beta = \frac{i.v.\ dose}{(AUC)_\beta} \tag{6.4}$$

where β is the terminal first-order elimination rate constant. Thus, V_β relates the amount of drug in the body to the drug concentration in the plasma during the terminal (log-linear) phase of drug elimination (achieved at distribution equilibrium). V_β can overestimate volumes of distribution of a drug as it reflects not only distribution but elimination as well.

6.2.1.2 Clearance

Clearance measures the body's ability to eliminate a drug. It does not indicate how much drug is being removed but rather the volume of blood or plasma that would be completely cleared of drug. Thus, clearance is expressed as a volume per unit time, as a flow parameter. Clearance can occur at many sites in the body and is generally additive. Drug elimination by a single organ may be defined by the extraction ratio (*ER*), which is the difference between the drug concentration entering (C_A) via arterial blood and leaving (C_V) the organ via venous blood flow to that organ and may be calculated as

$$ER = \frac{C_A - C_V}{C_A} \tag{6.5}$$

Clearance by an organ such as the kidney (renal clearance), CL_R, is then equal to the product of blood flow to that organ (Q_R) and the *ER* as follows:

$$CL_R = Q_R \cdot ER \tag{6.6}$$

The total body or systemic clearance can be obtained by summing the individual organ clearances. When extraction is complete ($ER = 1$), the maximum clearance of an organ of elimination is blood flow to that organ, that is, $CL_{organ} = Q_{organ}$. However, in the case of peptides and proteins, there is often nonspecific or blood clearance so that clearance can exceed cardiac output (Colburn, 1991). For conventional or classical drugs, drug metabolism produces polar metabolites that concentrate in measurable biological fluids. However, for peptides and proteins, breakdown may produce less polar products, which because of their lipophilicity and/or small size are more difficult to isolate and quantitate. Also, while liver and kidney are important sites of drug elimination for both classical and protein drugs, an additional elimination pathway for protein drugs is the proteases, which are present throughout the body (Gloff & Benet, 1990). The clearance concept has also been utilized to study systemic disposition characteristics and targeting efficacy of various macromolecular drug carriers. The clearance values of the drug–carrier conjugate in various organs can be controlled by modulating the physicochemical properties of carriers in *passive* targeting or by the introduction of a homing device in *active* targeting (Hashida & Takakura, 1994). Drug targeting is also very important for the use of monoclonal antibodies against tumor cells. Antibody fragments may be used such as the use of antibody without the F_c portion. This will reduce nonspecific uptake, while at the same time, the reduced size may allow better penetration into the tumors. When tumor cells are circulating, as in leukemia and some lymphomas, binding to antibody is very rapid and associated with the clearance of tumor cells from the blood (Henry, Begent, & Pedley, 1992).

6.2.1.3 Half-Life

Since organs of clearance can only clear drug from blood or plasma in direct contact with the organ, the time course of drug in the body depends on both the volume of distribution and clearance:

$$t_{1/2} \approx 0.693\ V/CL \tag{6.7}$$

The half-life does not indicate the processes involved in drug elimination or distribution but is a very useful parameter to define dosing intervals. For drugs that exhibit multiple distribution pools, it is possible to calculate more than one half-life (Gloff & Benet, 1990). Proteins typically have very short half-lives with the exception of antibodies, with the latter having half-lives of days.

6.2.2 Other Pharmacokinetic/Pharmacodynamic Considerations

6.2.2.1 Pharmacodynamics

Pharmacodynamics relates the measured response to the pharmacokinetics of the drug. While pharmacokinetics describes what the body does to the drug, pharmacodynamics describes what the drug does to the body. Combined pharmacokinetic/pharmacodynamic models to evaluate the concentration–response relationships for insulin include the gamma-linear model and the sigmoidal E_{max} model (Brown, Nelson, & Bottoms, 1987; Hooper, Bowsher, & Howey, 1991). These studies related the pharmacodynamics of insulin to the pharmacokinetic model by using a hypothetical effect compartment linked to the central compartment or linked to the peripheral compartment. The parameters describing these models may differ depending on the type of diabetes mellitus, and these models may offer an alternative method for distinguishing rapid insulin metabolism from insulin resistance (Brown et al., 1987). Pharmacodynamics of growth hormone (GH) is mediated in most tissues by the local production of insulin-like growth factor I, also called somatomedin C. Using hypophysectomized rats as the animal model, recombinant human GH stimulates body weight gain with daily SC administrations in a log-linear manner (Cronin, Ferraiolo, & Moore, 1991).

6.2.2.2 Antibody Induction

For a general discussion of protein immunogenicity, see Chapter 1. Immunogenic responses following administration of proteins can be routinely expected.

While antibody formation may not directly affect the pharmacokinetics of the protein drug, the formation of an antigen–antibody complex may increase or decrease the clearance rates of the protein drug. If the antibody neutralizes the protein, it can directly nullify the pharmacological action of the drug. As would be evident from the remarks made so far, even a nonneutralizing antibody can reduce the activity of the protein if the antigen–antibody complex is cleared rapidly from the body (Wills & Ferraiolo, 1992; Working, 1992). Antibody induction depends both on the characteristics of the protein and the study design. Proteins administered by SC or IM routes may be more immunogenic as compared to IV administration, as aggregation may occur at the SC or IM site of injection, and aggregated proteins are generally more immunogenic. The underlying disease states can also affect the immunologic outcome of treatment with recombinant human proteins (Working, 1992). Administration of recombinant human interferon-gamma to rhesus monkeys has been shown to induce antibody formation. However, these antibodies were nonneutralizing and had no discernible effect on the calculated pharmacokinetic parameters (Ferraiolo, Fuller, Burnett, & Chan, 1988). Based on risk assessment, recommendations have been developed for antidrug antibodies relating to concentration/titer, affinity, immunoglobulin type, and neutralization capacity. This is required by regulatory agencies to understand the impact of immunogenicity on pharmacokinetics/pharmacodynamics and efficacy (Chirmule, Jawa, & Meibohm, 2012).

6.2.2.3 Interspecies Scaling

Animals commonly used in preclinical drug studies such as mice, rats, rabbits, monkeys, and dogs do not eliminate drugs at the same rate that humans eliminate drugs. Drug elimination is usually faster in small mammals as compared to large mammals. Scaling techniques are often required to extrapolate pharmacokinetic data for the drug from laboratory animals to humans. The use of scaling techniques in the early stages of drug development provides a rational basis for dose selection in the clinical environment. Interspecies pharmacokinetic scaling can be done by two approaches: an allometric approach and a physiologic approach. In the allometric approach, no attempt is made to give physiologic meaning to the pharmacokinetic parameters as the underlying anatomy, physiology, and biochemistry contribute to the shape of the profile. In the physiologic approach, pharmacokinetics of the drug in an animal species is reduced to physiologically, anatomically, and biochemically meaningful parameters, such as blood flow to eliminating organs, tissue, and fluid volumes, and drug protein binding (Mordenti, 1986).

Allometric scaling may be done using a power function written as follows:

$$Y = aW^b \tag{6.8}$$

where

Y is the parameter of interest
W is the body weight
a is the allometric coefficient
b is the allometric exponent

This technique has been used to analyze the clearance and volume of distribution of five human proteins in human and laboratory animals as a function of body weight. These proteins, recombinant CD4, CD4 immunoglobulin G, GH, tissue plasminogen activator, and relaxin, cover a 16-fold range of MW (6–98 kDa). It was found that the clearance and volume data for each protein were satisfactorily described by Equation 6.8. The allometric exponent for the various pharmacokinetic parameters ranged from 0.65 to 0.84 for CL, 0.83 to 1.05 for V_{initial}, and 0.84 to 1.02 for V_{ss} (Mordenti, Chen, Moore, Ferraiolo, & Green, 1991). In another study, the clearance of tissue plasminogen activator was found to be predicted by the equation $CL = 18.9\ W^{0.824}$, across five mammalian species (rats, rabbits, dogs, marmosets, and humans). The clearance occurred via hepatic elimination and was found to be a saturable process. However, enzyme saturation occurs only at high doses, which are much higher than the clinical doses that will normally be used (Tanswell, Heinzel, Greischel, & Krause, 1990). Later studies have used correction factors to improve prediction of clearance rather than using simple allometry as the only method of scaling. A study analyzed the data for 15 therapeutic proteins and concluded that simple allometry or brain weight can be used to predict clearance depending on the exponents of simple allometry and that more than two species are needed for a reliable prediction (Mahmood, 2004). Similarly, for pediatric dosing of antibody-based therapeutics, quantitative pharmacokinetic/pharmacodynamic modeling approaches have been used (Xu, Davis, & Zhou, 2013).

6.2.2.4 Chemical Modification to Control Protein Disposition

Chemical modification can be used to improve the general pharmacokinetic features of a protein, as well as for its specific delivery to a target site. Proteins or other macromolecules of differing MWs such as $(Asu^{1,7})$-eel calcitonin (ECT, 3.36 kDa), inulin (5 kDa), neocarzinostatin (NCS, 20 kDa), soybean trypsin inhibitor (STI, 20 kDa), superoxide dismutase (SOD, 32 kDa), bovine serum albumin (BSA, 67 kDa), and dextran (70 kDa) can be conjugated with polymers such as polyethylene glycol (PEG) and dextran to increase molecular size. They can also be conjugated with anionic carboxymethyl-dextran (CM-dex) or cationic diethylaminoethyl-dextran (DEAE-dex) to introduce electrical charges. Protein with modified charge such as cationized BSA (cBSA) can also be synthesized directly. Furthermore, proteins can be modified with galactose (Gal) and mannose (Man) moieties to bestow an affinity for receptor-mediated endocytosis in cells. By these modifications, the in vivo disposition features of proteins can be extensively changed (Hashida, Nishikawa, Yamashita, & Takakura, 1994; Takakura, Mihara, & Hashida, 1994).

Proteins with an MW of less than 30 kDa are susceptible to glomerular filtration and then undergo metabolic degradation in kidney. The renal disposition characteristics of model protein drugs and macromolecules using isolated perfused rat kidney have been studied. The kidney was perfused under filtering and nonfiltering conditions. In the nonfiltering kidney, glomerular filtration was inhibited so that uptake of test compounds only occurs from the capillary side. This way, the effects of glomerular filtration, tubular reabsorption, and uptake from the capillary side can be quantitatively evaluated. Larger molecules had smaller or zero urine recoveries, while inulin, a marker for glomerular filtration, had 2.5% of dose recovery in urine. Besides size, charge also plays a role, as seen by the renal disposition profiles for BSA versus cBSA. Three protein drugs (NCS, STI, and SOD) accumulated in the filtering kidney only, suggesting that they are reabsorbed in the renal tubules after glomerular filtration. Of these proteins, SOD was selected for chemical modification to create conjugates. The conjugates SOD-PEG and SOD-CM-dex had higher molecular size and showed significantly smaller urinary excretion than native SOD. The other conjugates, Gal-SOD and Man-SOD, had decreased tubular reabsorption and thus enhanced the exposure of the luminal surface to SOD (Takakura et al., 1994). The possibility of targeting of protein drugs to the liver nonparenchymal cells by direct succinylation has also been investigated since the liver sinusoidal cells have scavenger receptors for polyanionic macromolecules. The proteins SOD, BSA, and uricase (UC) were succinylated to get Suc-BSA (70 kDa), Suc-UC (130 kDa), and Suc-SOD (34 kDa). All succinylated proteins had pI values of less than 4.0. It seems that proteins with an MW greater than 70 kDa can be targeted to the liver via direct succinylation, while some sulfated polysaccharides such as DS (8 kDa) have a large CL_{liver} value regardless of MW. Targeting of the antioxidant enzyme, SOD, to the liver nonparenchymal cells would be useful clinically for hepatic diseases mediated by reactive oxygen species (Takakura et al., 1994).

Protein engineering techniques can also be used to modify the pharmacokinetic parameters of a protein. Using a series of deletion mutants, regions of tissue plasminogen activator responsible for its liver clearance, its fibrin affinity, and its fibrin specificity have been identified. Thus, it becomes possible to alter specific fractions of the protein related to its poor pharmacokinetics without affecting the biological efficacy of the protein. Several recombinant hybrid and mutant plasminogen activators have been created by these techniques, with some molecules being cleared up to about 100-fold more slowly than the native protein (Browne et al., 1993; Tomlinson & Livingstone, 1989). The glycosylation status of a protein will also affect its residence time in the body, and it is sometimes believed that selective glycosylation has evolved as a means of controlling the clearance of proteins from the circulation (Jenkins & Curling, 1994). Protein PEGylation is another important means to change its disposition, and several products are commercially available. This topic is discussed later in Section 6.5.1.

6.2.2.5 Transport of Peptides across the Blood–Brain Barrier

In recent years, it has been increasingly recognized that peptides can cross the blood–brain barrier (BBB) as intact molecules. However, for the development of peptides as neuropharmaceuticals, BBB is still a formidable barrier, and strategies must be developed to overcome this barrier (Yi, Manickam, Brynskikh, & Kabanov, 2014). Proteins such as nerve growth factor could be useful for the treatment of degenerative brain diseases like Alzheimer's (Marx, 1990), while other agents may have relevance to therapy for infections of the central nervous system (CNS) (Scheld, 1989). The BBB represents a complex system of mechanisms that act together to regulate the exchange of fluids and substances between the CNS and blood. The barrier consists of endothelial or capillary barrier and the ependymal barrier found at the circumventricular organs and choroid plexus. Circumferential *tight junctions* exist between the cerebral capillary endothelial cells that do not allow paracellular transport and are responsible for the production of a high electrical resistance across the endothelium. However, it is now known that small amounts of peptides may cross the BBB in the direction of either blood to brain or brain to blood by nonsaturable diffusion through the membranes or by saturable carrier-mediated transport (Banks & Kastin, 1991; Banks, Kastin, & Barrera, 1991). A few precautions need to be taken to interpret the literature in the area of blood–brain transport (Zlokovic, 1995). First, the appearance of CNS effects does not necessarily imply that the peptide has crossed the BBB since some CNS effects may be mediated through peripheral pathways. Also, peptides found in the blood have ready access to circumventricular organs, which are areas of the brain with a capillary bed that does not participate in the BBB, but have neural connections to other parts of the brain. Many early studies that concluded a lack of transport across BBB had used insensitive analytical techniques that led to erroneous conclusions. Several in vitro and in vivo models, coupled with sensitive analytical techniques, are available to study drug transport to the brain (Banks et al., 1991; Deboer et al., 1994).

Pharmacokinetic aspects of peptides in the plasma and CNS will be influenced by the factors that influence the transport of a peptide across the BBB. These include MW, charge, degree of protein binding, peptide aggregation status, and, perhaps most importantly, lipophilicity. A large volume of distribution, rapid clearance, extensive degradation, or protein binding can all reduce blood concentration of the peptide and thus result in low availability to the CNS. In an early study by Hoffman, Walter, and Bulat (1977), the infusion of a radiolabeled diketopiperazine cyclo (Leu-Gly) peptide resulted in a relatively constant level of radioactivity in plasma and colony-stimulating factor (CSF), with a plasma/CSF gradient of about 2.7. Chromatographic evaluation of ethanolic extracts of these fluids revealed single spots, suggesting that the peptide was not metabolized during transport. Several strategies can be used to enhance the transport of peptides to the CNS. For example, peptide analogs with increased plasma half-life and increased stability in brain extracellular fluid can be used. Similarly, the affinity of the peptide for plasma membrane can be enhanced by manipulating molecular structure to enhance lipid solubility, provided that the central activity of the analog is retained (Banks et al., 1991; Begley, 1994). Bodor has described another approach to target peptides to the CNS by using a chemical delivery system. In this system, an enkephalin analog was *packaged* with a lipophilic function (cholesterol) at one end and a redox targetor linked via an amino acid spacer to the end terminal of the peptide. The large, lipophilic cholesterol group increases the lipid solubility of the drug, while the targetor group achieves specific targeting to the brain tissue. Once the packaged drug passes the BBB, the targetor is oxidized enzymatically to a positively charged salt that traps the peptide on the brain side of the barrier as the BBB allows lipid-soluble neutral species to flow in but does not allow positively charged species to flow out (Bodor, 1994). A study has suggested that permeability of peptides across the BBB may be related to their hydrogen-bonding potential rather than their lipophilicity. Using model peptides and in vitro as well as in situ BBB model, a good correlation between the permeabilities of these peptides and the number of potential hydrogen bonds they can make with water was observed (Chikhale, Ng, Burton, & Borchardt, 1994).

This result is similar to the conclusions drawn by several studies on the role of the hydrogen-bonding potential in the oral absorption of peptides (Chapter 7). The blood–CSF barrier and peptides that can transport across it have been reviewed (Hirsch, Upasani, & Banga, 2005).

6.3 POLYMERS USED FOR CONTROLLED DELIVERY OF PEPTIDES AND PROTEINS

For use in a controlled-release system, a polymer must generally be biocompatible and have reasonable mechanical strength. Other desirable properties depend on the specific usage and include release of the drug by permeation through a polymeric membrane or by degradation of a polymeric matrix (Langer, 1993). These polymers may be broadly classified as nondegradable and degradable polymers. Degradable polymers used in pharmaceutical systems often imply that these are biodegradable. For IM or SC injection, a biodegradable polymer as a particulate formulation is likely to be used. Both biodegradable and nondegradable polymers may potentially be used as implants. For its use as an implant, the polymer must, however, be biocompatible. After implantation, the biocompatibility of the implant may be assessed in terms of the acute and chronic local inflammatory responses and the subsequent fibrous capsule formation (Banga & Chien, 1988). Commercially available block copolymer, poloxamer 407 (Pluronic F-127), a nonionic surfactant, has been investigated for its potential dual role to preserve activity and allow controlled release of proteins from the semisolid matrix. Pluronic F-127 shows the property of reverse thermal gelation, that is, the polymer exists as a mobile viscous liquid at reduced temperatures but forms a rigid semisolid gel network when the temperature is increased. The formulation may be designed such that it is liquid at room temperature but gels at body temperature once injected. Its lack of irritancy to muscle tissue and its rapid sol-to-gel transition make it an attractive vehicle for sustained release of proteins. It has been shown to be a promising vehicle for formulation and sustained delivery of recombinant proteins such as interleukin-2 and urease, a model enzyme (Fults & Johnston, 1990; Johnston, Punjabi, & Froelich, 1992; Pec, Wout, & Johnston, 1992). Blends of polymers including pluronics have also been used for controlled release of proteins (DesNoyer & McHugh, 2003). It should be noted that proteins themselves are also used as carriers for other conventional or protein-based drugs. Synthetically designed protein-based materials consisting of repeating amino acid sequences, that is, engineered protein polymers, are now being used for drug delivery and biomedical applications.

6.3.1 Nondegradable Polymers

These include hydrophilic polymers such as polyacrylamide or hydrophobic polymers such as ethylene vinyl acetate copolymer (EVAc) and silicone elastomers. Various commercial grades of silicone elastomers have been fabricated into matrices, which can be mixed with silicone fluids as required. These matrices can release macromolecules for more than 100 days, with release rates often showing a deviation from the square-root-of-time relationship expected for matrix release. The addition of a hydrophilic pore-building excipient to polydimethylsiloxane matrices, control of total matrix loading, and other conditions can be used to adjust release rates. Peptides and proteins investigated for their release behavior from silicone elastomers include a dipeptide (Gly-Tyr), insulin, BSA, chymotrypsin, and pepsin (Hoth & Merkle, 1991; Hsieh, Chiang, & Desai, 1985). Microparticles prepared from polyvinyl alcohol (PVA), a linear polymer, have also been used. Since traditional cross-linking agents can leave behind toxic residues, cross-linking was achieved by a novel freeze–thaw process in the absence of a cross-linking agent. When PVA is exposed to repeated freeze–thaw cycles, crystallites form and the system behaves as a chemically cross-linked system. These microparticles were found to release albumin for up to 7 days (Ficek & Peppas, 1993). Hydroxyapatite is another biocompatible material that has been made as microcarriers for controlled release of proteins (Fu, Rahaman, Day, & Brown, 2011; Ijntema, Heuvelsland, Dirix, & Sam, 1994).

EVAc polymer is one of the more commonly used nondegradable polymers as it is biocompatible and offers control on the release rates of embedded polypeptides. The incorporation of a powdered polypeptide into EVAc polymer matrix creates a series of interconnecting channels. Water can diffuse into the matrix from these channels to dissolve the polypeptide, which is then released through the porous matrix. Because release occurs only when pores are interconnected, there is a minimum loading dose below which no release takes place (Banga & Chien, 1988; Heller, 1993b). EVAc matrix incorporating peptides or proteins can be fabricated by several techniques as described in the literature (Cohen, Siegel, & Langer, 1984; Rhine, Hsieh, & Langer, 1980). The incorporation of insulin into EVAc matrix has been evaluated, and its release kinetics was enhanced by increasing the solubility, particle size, or the level of insulin loading within the matrix (Brown, Siemer, Munoz, & Langer, 1986). When insulin-loaded EVAc devices were implanted in diabetic rats, controlled release of insulin took place for up to 105 days from a single implant (Brown, Munoz, Siemer, Edelman, & Langer, 1986). The in vivo release rates of macromolecules from EVAc matrices may be predicted by in vitro release kinetics studies as it appears that a good in vitro–in vivo correlation exists (Brown, Wei, & Langer, 1983). Diffusion-controlled matrix devices such as EVAc normally do not deliver the drug at zero-order rates. However, it may be possible to vary matrix geometry to get a zero-order release. A hemisphere matrix has been made based on theoretical and experimental analysis. This matrix was laminated with an impermeable coating, except for an exposed cavity in the center, to get an inwardly releasing hemisphere. BSA was released from this matrix at a zero-order rate of 0.5 mg/day for over 60 days (Hsieh, Rhine, & Langer, 1983).

6.3.2 Biodegradable Polymers

For the controlled delivery of peptides and proteins, the most promising systems are lactide/glycolide copolymers, poly(ortho esters), and polyanhydrides (Heller, 1993b). In addition, cyanoacrylates and natural polymers can also be used. While most promising degradable polymers such as PLGA are hydrophobic systems, hydrophilic polymers such as some biodegradable hydrogels and dextran may offer advantages in protein loading (Gehrke, Uhden, & McBride, 1998).

Poly(D,L-lactide-co-glycolide) (PLGA) is a linear polyester that hydrolyzes by an acid- or base-catalyzed reaction to form the natural metabolites, glycolic and lactic acids. L-lactic and glycolic acids will be metabolized by the Krebs cycle, while D-lactic acid will be excreted intact. Thus, these polymers are biocompatible, biodegradable, and considered safe and have been used as bioerodible sutures. As discussed in Section 6.4.4, several marketed microsphere peptide formulations contain this polymer. The mole ratio of lactide to glycolide in the PLGA polymer can be changed, and it is commercially available in several such ratios. Increasing the amount of glycolide in the polymer increases its rate of biodegradation, for example, a 50:50 ratio of lactide/glycolide will degrade more rapidly than a 75:25 ratio. For polymers with the same lactide/glycolide ratio, the polymers with the lowest molecular mass degrade most rapidly, though some variation may be seen from one commercial source to another. However, the lactide/glycolide ratio has a greater influence on the rate of polymer degradation than the polymer molecular mass (O'Hagan, Jeffery, & Davis, 1994). The components of the PLGA polymer, poly(L-lactic acid) (L-PLA), and polyglycolic acid (PGA) can also be used separately as biodegradable polymers. However, their degradation time is rather long, being as long as several years for L-PLA and several months for PGA. The lactide/glycolide copolymers have shorter biodegradation times, of the order of several weeks (Holland, Tighe, & Gould, 1986; Mehta et al., 1994). The lactide/glycolide copolymers that have been discussed so far undergo a hydrolysis process that occurs throughout the bulk of the material. In contrast, poly(ortho esters) and polyanhydrides undergo a hydrolysis process that is largely confined to the surface of the polymer. In the case of PLGA polymers, the permeability was changing with time as the polymer underwent bulk erosion, thus rendering the drug release rate somewhat unpredictable. However, the poly(ortho esters) and polyanhydrides can provide precise release rates as they undergo only surface erosion (Banga & Chien, 1988; Heller, 1993b). Other biodegradable polymers such as polyhydroxyalkanonate and pseudo-poly(amino acids) are also being developed. Phosphorus-containing biodegradable polymers such as polyphosphazenes, polyphosphonates, and poly(phosphate esters) have also been developed. Polyphosphazenes provide an excellent platform for controlled drug delivery due to their chemical flexibility and other desirable attributes (Chaubal, Gupta, Lopina, & Bruley, 2003).

Natural polymers such as gelatin, serum albumin, and starch can also be used to prepare microspheres and nanospheres (Artursson, Edman, Laakso, & Sjoholm, 1984; Thies & Bissery, 1984). Recombinant gelatin and collagen are now available for drug delivery and tissue engineering applications and will provide a better control on quality, purity, and predictability of performance (Olsen et al., 2003; Yang et al., 2004). IV injection of albumin microspheres can be used to target either the lung (15–30 μm) or the liver (1–3 μm) depending on the size of the microspheres. Albumin microspheres can be labeled with radioisotopes to produce diagnostic agents. Albumin is also being used to produce fusion proteins by fusing the gene for the protein with the albumin gene (Chuang, Kragh-Hansen, & Otagiri, 2002). Gelatin microspheres have been prepared by a desolvation process in which warm aqueous gelatin solutions are added to a cold solvent. The drug can then be loaded into the completed microparticles (Oner & Groves, 1993). Gelatin microspheres with a diameter of less than 2 μm have also been synthesized by cross-linking with glutaraldehyde (Tabata & Ikada, 1987, 1989). The release of interferon from these microspheres was regulated by the extent of cross-linking with glutaraldehyde. However, the microspheres were phagocytosed by macrophages, regardless of the extent of cross-linking, and this resulted in the slow release of interferon in the cells (Tabata & Ikada, 1989). Several complex coacervate systems can also be made of natural polymers. Albumin/acacia coacervate microgels were found to be stable to aging, temperature, pH, and glutaraldehyde cross-linking. The microgel coacervate particles could be recovered as a free-flowing powder with a mean particle size of about 6 μm (Burgess, 1990; Burgess & Singh, 1993). Starch microparticles can also be linked to recombinant mouse interferon-γ by covalent linkage for targeting to the reticuloendothelial system (RES) of mice for antileishmanial effect (Degling, Stjarnkvist, & Sjoholm, 1993). A fibrin matrix has also been shown to be biodegradable and biocompatible matrix that can be used as a surgically implantable or injectable delivery device (Senderoff, Sheu, & Sokoloski, 1991). Liposomes sequestered in a collagen gel matrix have also been investigated as a sustained-release drug delivery system. Following IM or SC injection, insulin and GH were slowly released into circulation from this matrix over 5 or 14 days, respectively (Weiner, Carpenter-Green, Soehngen, Lenk, & Popescu, 1985).

Biodegradable hydrogels may be made using synthetic or natural polymers. These hydrogels may exist as covalently cross-linked chemical gel network or as physical gel without covalent cross-linking. Physical gels form by linking polymer chains through entanglements, ionic forces, or hydrophobic interactions. Hydrogels may biodegrade due to the presence of degradable polymer backbones, degradable cross-links, or pendant chains that can be cleaved from the polymer backbone. Water-soluble polymers may biodegrade simply due to their ability to absorb water and dissolve. Polymers composed of proteins or polysaccharides can undergo enzymatic hydrolysis. The presence of a specific enzyme at some site may be exploited to develop hydrogels for site-specific delivery (Kamath & Park, 1993). A biodegradable hydrogel offers advantages for incorporation of the water-soluble peptides and proteins as it is composed largely of water. Also, the synthesis of the hydrogel may avoid exposure to organic solvents and/or high temperatures, unlike the synthesis of some hydrophobic polymers. Cross-linked poly(hydroxypropyl L-glutamine) hydrogel microspheres have been used as a carrier for drug targeting. These microspheres were synthesized by treating the hydrophobic synthetic polyamino acid, poly(benzyl L-glutamate) microspheres with aminopropyl alcohol containing 5% (w/v) hexanediamine as a cross-linker. As more phenyl groups were substituted by hydroxypropyl groups, the microspheres became increasingly hydrophilic, thereby resulted in a substantial reduction in liver uptake (Li, Yang, Kuang, & Wallace, 1993). A polyester-based bioerodible hydrogel has been made as dried microspheres that can be resuspended in water. The microspheres could be passed through a 22-gauge hypodermic needle and exhibited a reasonably linear release of albumin by matrix erosion (Heller, Helwing, Baker, & Tuttle, 1983). Bioerodible non-cross-linked poly(methyl methacrylate-co-methacrylic acid) beads have also been used to achieve a synchronization of swelling and dissolution fronts of the spherical bead for the release of a GH–releasing peptide (Lee, 1993). Hydrogel made of recombinant human serum albumin cross-linked by dithiothreitol has been reported to act as a sustained-release reservoir for drugs that have binding affinity for albumin (Hirose & Tachibana, 2010).

6.4 PARENTERAL CONTROLLED-RELEASE SYSTEMS

Several controlled-release parenteral delivery systems have been investigated for the delivery of therapeutic peptides and proteins. These include microspheres, implants, liposomal delivery systems, nanoparticles, vaccines, and pulsatile drug delivery systems (Jorgensen, Moeller, van de Weert, Nielsen, & Frokjaer, 2006). Besides these approaches, cellular carriers such as erythrocyte vesicles can also be used for controlled release of proteins. The binding of human insulin to erythrocyte membrane by encapsulation or adsorption has been described (Al-Achi & Greenwood, 1993). Similarly, exosome mimetics have been explored as drug delivery systems (Kooijmans, Vader, van Dommelen, van Solinge, & Schiffelers, 2012). Exosomes (30–100 nm) are extracellular vesicles that are known to carry proteins and nucleic acids in biological fluids and have specialized functions. Also, antibodies are being investigated to provide target-site specificity, especially for anticancer drugs to minimize their side effects.

6.4.1 MICROSPHERES

Microspheres are free-flowing powders, ideally less than about 125 μm in diameter, which can be suspended in suitable aqueous vehicles for injection with a conventional syringe using an 18- or 20-gauge needle. Biodegradable microspheres have been used for delivery of peptides and proteins (Sinha & Trehan, 2003). The first injectable peptide microsphere formulation was marketed by IPSEN Biotech of France in 1986. It delivered the luteinizing hormone–releasing hormone (LHRH) agonist, [D-Trp^{6}]-LHRH, over a 1-month period for treatment of prostate cancer, following a single injection. It used 50:50 poly(DL-lactide-co-glycolide) and was made by a phase-separation microencapsulation technique and terminally sterilized by gamma radiation (Tice & Tabibi, 1992). Some of the initial formulations contain a synthetic analog of LHRH, leuprolide acetate (Lupron Depot®). Another LHRH analog, nafarelin, has also been incorporated into microspheres (Sanders et al., 1989). Human GH was also available in a PLGA microsphere formulation (Nutropin Depot), but the formulation was discontinued due to problems and costs associated with manufacture. It was packaged in vials as a sterile, preservative-free powder that was suspended in a sterile diluent containing a suspending agent when needed for SC injection. Several other sustained-release formulations for recombinant human GH have been investigated, and a hyaluronate microparticle formulation is available in Korea from LG Life Science (Kwak et al., 2009).

Also, microspheres of octreotide acetate (Sandostatin LAR® Depot, Novartis) are commercially available. Other proteins that have been investigated for formulation into microspheres include thyrotropin-releasing hormone (Heya et al., 1994; Heya, Okada, Ogawa, & Toguchi, 1994), recombinant human epidermal growth factor (Han, Lee, Gao, & Park, 2001), recombinant human granulocyte colony-stimulating factor (GCSF) (Gibaud et al., 1998), recombinant human interferon-γ (Yang & Cleland, 1997), as well as neuropeptides that were formulated into microspheres and then stereotactically implanted in rat brain for slow delivery in the brain (Couvreur, BlancoPrieto, Puisieux, Roques, & Fattal, 1997).

Nondegradable polymers, such as EVAc, have also been used to prepare microspheres. EVAc microspheres containing cyclosporine were found to retain a spherical shape and give optimum release when they contained 25% cyclosporine. A lower drug loading (5%) provided deformed microspheres, while a higher loading (50%) produced tear drop–shaped microspheres (D'Souza, 1988). Research on cyclosporin A (CyA) microspheres or other alternative dosage forms is continuing. CyA-loaded microspheres and nanospheres based on the biodegradable PLGA polymer have been designed and were found to allow for controlled release of CyA over a prolonged period of time (Sanchez, Vilajato, & Alonso, 1993).

6.4.2 Release of Drugs from Microspheres

Since most proteins are insoluble in PLGA, their release from the microspheres by classical partition-dependent diffusion is negligible. Instead, their release is dependent on the degradation of PLGA (Cohen, Yoshioka, Lucarelli, Hwang, & Langer, 1991). The release of a peptide or protein from a PLGA microsphere may occur in three stages. First, an initial *burst* release may be expected from the protein on the surface. This is followed by a *lag phase* of least release during which time the polymer degradation occurs. Finally, a third phase of continuous release results when sufficient polymer degradation has occurred. Control of the release rates can be achieved by selecting the appropriate lactide/glycolide ratio, the polymer molecular mass, and drug loading levels. As the MW of the polymer is reduced by bulk hydrolysis, the polymer matrix becomes more hydrophilic, allowing more water to penetrate and allowing for protein release (O'Hagan et al., 1994). A similar release profile was obtained for the release of lanreotide, a somatostatin analog, from PLGA microspheres (Kuhn et al., 1994). It has been suggested that *burst* release and damaging effects of protein interface during manufacturing can be avoided by entrapping PEGylated proteins in microspheres (Pai, Tilton, & Przybycien, 2009).

It should be noted that the stability of the protein at the temperature used (typically 37°C to simulate body temperature) as well as the composition of the release media such as the buffer used, ionic strength, and its pH and the presence of any surfactants can have a significant effect on in vitro release rates (Yang & Cleland, 1997). The size and properties of the protein may also affect release though a reported study that used linear hexapeptides did not find any effect of size, charge, and conformational flexibility of the peptide on its release kinetics from PLGA microspheres (Okumu, Cleland, & Borchardt, 1997).

Park, Cohen, and Langer (1992) have reported that the initial *burst* release may be decreased by coating the PLGA matrix with a water-soluble adsorptive polymer, polyethyleneimine (PEI). It appears that PEI can diffuse into the polymer matrices and cross-link protein molecules by ionic interactions. The resulting PEI–protein network at the surface of the matrix acts as a diffusional barrier to further release. Degradation of PLGA microparticles may also be affected by plasma proteins as the degradation in aqueous buffers has been reported to be accelerated by the addition of albumin, γ-globulins, and fibrinogen (Makino, Ohshima, & Kondo, 1987). For some small peptides, release from the matrix by diffusion may also occur. For glycine and its homopeptides, di- and triglycine were released from PLA microspheres predominantly by a matrix-controlled diffusion process. For tetra- and pentaglycine, dissolution played a rate-limiting role due to their low aqueous solubility (Pradhan & Vasavada, 1994).

Polyanhydride microspheres have also been used to achieve a near-constant sustained release of water-soluble molecules (Tabata & Langer, 1993). In a study by Tabata, Gutta, and Langer (1993), polyanhydride microspheres, when prepared by the double-emulsion solvent evaporation method, encapsulated more than 80% of protein. Proteins of different molecular sizes, lysozyme, trypsin, heparinase, ovalbumin, and immunoglobulin, were released at a near-constant rate from these microspheres without any large initial burst. The microspheres protected the encapsulated protein from activity loss. For instance, free trypsin lost 80% of its activity in solution at 37°C at pH 7.4 in 12 h, while trypsin inside microspheres lost less than 10% activity under these conditions (Tabata et al., 1993).

6.4.3 Preparation of Microspheres

The processes commonly used to make parenteral microspheres include spray drying, spray-freeze-drying, solvent evaporation, and phase separation techniques. During spray drying, microcapsules or microspheres may be produced. If the drug is suspended in a solution of the polymer, microcapsules will result. If the drug is dissolved in a solution of the polymer, microspheres will result as the drug gets uniformly dispersed in the polymer (Conte, Conti, Giunchedi, & Maggi, 1994). Though spray drying and several other processes exist to make microspheres, they may not be suitable for parenteral products. Solvent evaporation method has been used for preparation of microspheres, but drug loadings are usually less than 10%. For microspheres containing peptides or proteins, a solvent evaporation method using a double emulsion is often used. This method results in loading as high as 90%. The protein is dissolved in water and emulsified in a solution of PLGA in dichloromethane. The resulting w/o emulsion is then emulsified with an aqueous solution of PVA to produce a w/o/w emulsion. This multiple emulsion is then stirred under ambient conditions to allow solvent evaporation. The resulting microspheres are isolated by filtration or centrifugation, washed, and freeze dried (Couvreur et al., 1997; Guo, 1994; Iwata & Mcginity, 1993; O'Hagan et al., 1994; Ogawa, Yamamoto, Okada, Yashiki, & Shimamoto, 1988; Prieto et al., 1994; Sah & Chien, 1993). For a lipophilic peptide such as CyA, the peptide can be directly dissolved in methylene chloride along with PLGA, and a single w/o emulsion can be made for solvent evaporation (Sanchez et al., 1993). During the preparation of microspheres, the agitation speed is very important as higher agitation will produce smaller droplets, thus reducing the particle size of the final product.

Microencapsulation of proteins into microspheres may have more limitation as compared to peptides. This is because proteins have a more complex structure, characterized by a well-defined tertiary structure that may be affected by the harsh manufacturing conditions of microsphere preparation. Contact with organic solvents or exposure to air–water interfaces during mixing may also induce self-aggregation of proteins. Preparation of microspheres by the commonly used solvent evaporation method has the potential to cause degradation of the protein by the harsh processing conditions such as high shear forces and organic solvents. For albumin, however, no changes in MWs or conformation could be observed following its processing under these conditions (Hora et al., 1990; Jeffery, Davis, & O'Hagan, 1993; Sah & Chien, 1993; Sah, Toddywala, & Chien, 1994). Processing conditions such as viscosity of the polymer solution and concentration of methanol and peptide in the dispersed phase have been reported to affect the physicochemical properties of PLGA microspheres (Jeyanthi, Mehta, Thanoo, & DeLuca, 1997). The multiple-emulsion technique results in high encapsulation efficiency but involves complex manufacturing conditions.

Investigators are also trying to improve the efficiency of the relatively simple o/w solvent evaporation method. One approach used the addition of a fatty acid salt in the oily phase for successful entrapment of a hexapeptide, which is a neurotensin analog. The addition of fatty acid salt to oily phase and the use of alkaline water phase slowed down the release of the peptide into the outer water phase during manufacture. Also, the interaction between the cationic peptide and the anionic PLA was considered to contribute to increasing the entrapment of the peptide into the microspheres (Yamakawa, Tsushima, Machida, & Watanabe, 1992). A peptide such as the salmon calcitonin, which has a strong hydrophobic region, can also bind to the polymer by hydrophobic and ionic forces, resulting in high incorporation. Such adsorption also allowed the incorporation of calcitonin into preformed microspheres (Mehta et al., 1994; Tsai, Mehta, & DeLuca, 1996). Hayashi, Yoshioka, Aso, Po, and Terao (1994) have proposed a reverse micelle solvent evaporation technique as an alternative to agitation to prepare the w/o emulsions. In this method, an aqueous solution of the protein is solubilized in an organic solvent as reverse micelles by using surfactants. The polymer was dissolved in this micellar solution, which was then added to a 1% PVA aqueous solution, and the organic solvent was then removed under reduced pressure. Using this micelle solvent evaporation technique, model proteins, SOD, β-galactosidase, and recombinant human tumor necrosis factor were successfully encapsulated in L-PLA microspheres in their active form. Another way to incorporate a peptide or protein into microspheres without using any harsh processing conditions is to first produce porous microspheres. These porous microspheres can then be loaded with the peptide or protein by a simple and mild process in which the drug remains in a stable aqueous environment during loading. The microspheres can then be freeze dried so that the drug can remain in solid state within microspheres until use. These types of microspheres will also allow the loading of a sterile filtered protein solution into presterilized porous microspheres so that terminal sterilization is not required. Alternatively, a spray-freeze-drying process can be used (Cleland, Johnson, Putney, & Jones, 1997; Johnson & Herbert, 2003; Supersaxo, Kou, Teitelbaum, & Maskiewicz, 1993). A report has compared processing of darbepoetin alfa by spray drying and spray freeze drying (Nguyen, Herberger, & Burke, 2004), and factors affecting spray freeze drying have been investigated (Costantino et al., 2002; Sonner, Maa, & Lee, 2002).

For preparing microspheres from natural polymers, a typical procedure involves preparation of a w/o emulsion, with the drug and the polymer being dissolved in the water phase. Once the emulsion is formed, the water-soluble polymer is precipitated by one of several possible approaches. In the case of gelatin, simply chilling to 4°C will result in a gel network. For albumin, thermal denaturation can be used, but the peptide or protein is unlikely to be stable at the temperatures required. Chemical cross-linking agents can also be used, for example, glutaraldehyde for albumin or epichlorohydrin for starch. Biodegradability of glutaraldehyde-treated serum albumin microspheres depends on the amount of glutaraldehyde added to the aqueous phase (Thies & Bissery, 1984).

6.4.4 Microsphere Formulations of LHRH and Analogs

LHRH, also known as gonadotropin-releasing hormone (GnRH), is a naturally occurring decapeptide that stimulates the release of the pituitary gonadotropins, LH and FSH. LHRH is secreted in a pulsatile manner, and its replacement for treatment of primary infertility of hypothalamic origin requires pulsatile administration every 60–90 min through programmable pumps (Vickery, 1991). LHRH is readily degraded in the body by proteolytic enzymes and has a half-life of only 8 min. Several agonists and antagonists of LHRH have been synthesized and are promising agents for a range of clinical applications. Compared to LHRH, these agonists have relatively long half-lives, and they can mimic the effects of continuous LHRH infusion if administered twice daily. Such administration stimulates the release of pituitary gonadotropins initially, but repeated doses abolish the stimulatory effects. Due to receptor desensitization and downregulation, a stage of chemical castration results in about 4 weeks. Chronic application of LHRH agonists in microsphere formulations has proved effective in treating hormone-dependent cancers and prostate and breast cancers and in the treatment of sex-hormone-dependent disorders, endometriosis, and uterine fibroids (Okada et al., 1994). Microsphere formulations of a potent analog of LHRH, leuprolide acetate, that release drug over 1–4 months have been investigated (Okada, 1997; Okada, Doken, Ogawa, & Toguchi, 1994a, 1994b; Okada, Heya, Ogawa, Toguchi, & Shimamoto, 1991; Woo et al., 2001). An in-water drying method using a (w/o)/w emulsion was first developed for efficient encapsulation of leuprolide acetate into PLGA microspheres (Ogawa et al., 1988). A once-a-month injectable microsphere formulation was produced, and its effective human dose was estimated to be about 3.2–8.1 mg analog/month (Okada et al., 1991). In clinical studies, sustained serum levels of the peptide for more than 1 month were obtained after a single injection. In these studies, serum testosterone suppression with the 7.5 mg microsphere depot formulation was the same as that resulting from a 1 mg daily injection for 1 month. Thus, the 7.5 mg microsphere formulation can effectively reduce the dose of the peptide to one-fourth of that required with daily injection (PDR Staff, 2014; Okada et al., 1994). Currently, Lupron Depot (AbbVie) is available on the market. The product is available in 3.75 mg (for 1-month release) or 11.25 mg (for 3-month release) strength to improve anemia due to vaginal bleeding from fibroids. The product consists of lyophilized microspheres that are reconstituted and administered monthly as a single IM injection. Reconstitution is done with a diluent that contains the suspending agent, carboxymethylcellulose sodium, along with D-mannitol, polysorbate 80, water for injection USP, and acetic acid, to control pH. The surfactant polysorbate 80 helps to wet the microsphere mass rapidly. Also, Lupron Depot-PED™ (depot suspension) is available for the treatment of children with central precocious puberty.

Sanders and others (1984) have investigated microsphere formulations of another LHRH analog, nafarelin acetate. Substitution of glycine in natural LHRH by the hydrophobic D-substitute, 3-(2-naphthyl) alanine, results in a biological potency about 200 times higher and a half-life 5–7 times higher than that of the natural LHRH. Injection of nafarelin acetate in PLGA microspheres to rats resulted in a continuous suppression of estrous cyclicity for 24 days. Also, a single SC injection of 1 mg of nafarelin acetate in PLGA microspheres in rhesus monkeys resulted in detectable plasma levels for over 40 days, during which time ovulation was completely inhibited. In subsequent clinical studies, normal men over the age of 50 were given a single IM injection of a suspension of PLGA–nafarelin microspheres, containing 4 mg of nafarelin. Drug loading levels of 2%, 4%, and 7% nafarelin in PLGA were evaluated. At 2% drug loading, most of the release took place in the tertiary erosion-controlled phase, with circulating levels of >60 days. A more desirable release pattern was seen with 4% drug loading. At 7% loading, release shifted from the erosional to the diffusional phase (Sanders et al., 1989).

6.4.5 Implants

Implants are often placed subcutaneously by a large bore needle, pellet injectors, or minor surgery. An implant of goserelin acetate is commercially available (Zoladex®; AstraZeneca) as a sterile, biodegradable, 1.5 mm diameter cylinder, which is preloaded in a special single-use syringe with a 14-gauge needle. It contains goserelin acetate, a potent synthetic decapeptide (MW 1269 Da) analog of LHRH, dispersed in a matrix of D,L-lactic acid and glycolic acid copolymer for administration every 12 weeks for prostatic carcinoma. It is administered into the anterior abdominal wall below the navel line using an aseptic technique under the supervision of a physician (PDR Staff, 2014). Bioneedles made from biodegradable polymers have also been investigated as small hollow mini-implants, which can be injected subcutaneously using compressed air. Once implanted, they will dissolve to release the protein/antigen (Hirschberg, van de Wijdeven, Kraan, Amorij, & Kersten, 2010).

If a nonbiodegradable polymer is used, the implant will also need surgical removal at the end of the release period. If feasible based on dose and other considerations, biodegradable microspheres, injected subcutaneously, can also be used and will overcome many disadvantages associated with implants. In situ forming depots can provide an attractive alternate such as the Saber™ (sucrose acetate isobutyrate [SAIB] extended release) technology platform. SAIB is a fully esterified sucrose derivative used as a food additive, which is used by DURECT Corporation (Cupertino, CA) for sustained drug delivery due to its high hydrophobicity and high viscosity. The drug is dissolved or dispersed in an SAIB/solvent solution, and upon SC or IM administration, the solvent diffuses out to leave a viscous depot of SAIB and drug (Agarwal & Rupenthal, 2013; Okumu et al., 2002; Tipton, 2003).

A peptide or protein drug can be formulated into an implant by standard tableting or compression techniques (see Section 5.4.3). Certain lipid materials can be directly mixed with the peptide or protein drug to form these implants (Weiner, 1990). Another approach used for L-asparaginase was its immobilization in spherical polyacrylamide microparticles, which were then inserted in polyacrylamide gel for implantation in rats (Edman & Sjoholm, 1981). Collagen obtained from the connective tissue of animals is a biocompatible and biodegradable protein that has been used to develop minipellets containing α-interferon. These minipellets, when administered subcutaneously in mice, were found to release interferon for extended periods (Takenaka et al., 1989). A biodegradable pellet system has been made for $GRF29NH_2$ by compression molding of powdered $GRF29NH_2$ mixed with a lactic/glycolic acid copolymer. $GRF29NH_2$ is a fragment or smaller analog of human growth hormone–releasing hormone (GHRH or $GRF44NH_2$), composed of the first 29 amino acids of the 44 amino acid $hGRF44NH_2$. Entrapment of $GRF29NH_2$ within the polymer matrix protects it from degradation and delivers it slowly to result in long-term stimulation of the secretion of GH. Pellets, 1 mm thick × 7 mm diameter, were made of a lactic/glycolic acid copolymer that can keep its geometrical integrity for 2 weeks but is resorbed after 6weeks. Thus, release was designed to be by dissolution–diffusion from a percolating network of $GRF29NH_2$ particles rather than a matrix degradation release (Mariette, Coudane, Vert, Gautier, & Moneton, 1993). An implant for the release of vasopressin has also been described. The Accurel polypropylene/collodion device caused a decrease in urine production for at least 50 days, when implanted SC in vasopressin-deficient Brattleboro rats (Kruisbrink & Boer, 1984, 1986). PLGA implants containing the LHRH agonist, nafarelin, have also been studied both in vivo and in vitro. A triphasic release profile was observed as in the case of microspheres. The release profile could be controlled by modifying the physical properties of the polymer such as MW or the ratio of the more hydrophobic lactic acid monomer to the less hydrophobic glycolic acid monomer (Sanders, Kell, McRae, & Whitehead, 1986). A novel approach for an injectable implant has been reported by Shah et al. (1993). In this approach, PLGA was formulated as solution or suspension such that it could be injected through a 22-/23-gauge needle. Once injected, the formulation formed a gel matrix upon contact with aqueous environment. The drug is then released slowly from this matrix. Drug release for small molecules and proteins was influenced by the concentration of the polymer, physicochemical properties of the drug, method of drug incorporation in formulation, and excipients. A polymeric implant for leuprolide based on a similar in situ precipitation of PLGA has also been reported (Mashayekhi, Mobedi, Najafi, & Enayati, 2013).

Biologically oriented microelectromechanical systems (BioMEMS) can be used to make low-volume pumps and reservoirs that need low power and can be implanted in places where conventional pumps cannot be implanted (Nuxoll, 2013). Mammalian cells encapsulated within polymeric membranes are also being investigated as a means to achieve controlled release of peptides and proteins. The polymeric membrane would allow peptides or small proteins produced by encapsulated cells to diffuse out, while acting as a permselective barrier to the entry of large proteins such as antibodies. This will protect the encapsulated cells from the cellular and humoral components of the immune system, without the need to use immunosuppressants. For long-term success of transplantation of such microcapsules, the polymeric membrane must be biocompatible. Also, the fragile nature of mammalian cells imposes some limitations on the process conditions of encapsulation and on the properties of the resulting microcapsules. A promising approach using an interfacial precipitation technique for encapsulation of mammalian cells in a polyacrylate membrane has been reviewed (Uludag, Kharlip, & Sefton, 1993). The use of self-regulating implantable delivery systems for insulin will be discussed in Section 6.4.9.3. Degarelix (FIRMAGON, Ferring Pharmaceuticals) is a synthetic decapeptide with seven unnatural amino acids, five of which are D-amino acids. Due to its tendency to self-aggregate, it forms a gel upon injection, providing a sustained-release effect over a month or more.

6.4.6 Liposomes

The formulation of peptides and proteins into liposomal dosage forms has been discussed in Chapter 4. Liposomes are biodegradable and can be used for controlled release or drug targeting of entrapped drug (Bergers, Otter, & Crommelin, 1994). If administered by IV route, liposomes will be taken up by the RES due to their particulate nature. Recently, the use of specialized phospholipids, esterified with hydrophilic groups such as polyethylene derivatives, is being investigated as a means to avoid RES elimination and thus prolong the in vivo circulation time. When injected by the SC route, liposomes up to 200 nm may be preferentially drained by the lymphatic system (Zuidema et al., 1994). Liposomes have been used to prolong the duration of effect of antidiuretic hormone, Arg^8-vasopressin in vasopressin-deficient Brattleboro rats. While free vasopressin gave reduced urine production for less than 24 h, vasopressin formulated in phosphatidylserine (PS) liposomes was effective up to 4 days. When the PS component was replaced by either phosphatidylglycerol or a phospholipid derivatized with PEG, long-circulating liposomes were produced that further increased the duration of action of vasopressin (Woodle et al., 1992). Liposomes have also been conjugated with specific ligands to enable targeting to cancer cells (Kawahara et al., 2013; Wang et al., 2010).

6.4.7 Nanoparticles

Nanoparticles are colloidal polymer particles with a size of less than about 1 μm. They can be used as carriers for drugs and vaccines for controlled release or for drug targeting (Bruno, Miller, & Lim, 2013; Han et al., 2014; Martins, Sarmento, Ferreira, & Souto, 2007; Pisal, Kosloski, & Balu-Iyer, 2010). IV injection of ^{14}C-labeled PLGA nanoparticles to mice has been reported to result in the highest uptake by liver, the main organ of the RES, and significant uptake by the spleen, the second major organ of the RES system (Le Ray, Vert, Gautier, & Benoit, 1994). The interaction of a particulate system with the RES may be altered by modification of surface characteristics of the particle. Coating with the nonionic surfactant, Pluronic F108, has been shown to result in a significant reduction in liver uptake (Illum & Davis, 1982). Nanoparticles may be prepared by interfacial polymerization, emulsion polymerization, and denaturation or desolvation of natural proteins or carbohydrates (Couvreur & Puisieux, 1993; Kreuter, 1991). Nanoparticles based on cyanoacrylates have been used because alkyl α-cyanoacrylate monomers can polymerize spontaneously at oil/water interfaces. The procedure involves preparation of a w/o emulsion with the protein drug dissolved in the water. As the cyanoacrylate monomer is added to the oil phase, spontaneous polymerization occurs at the water/oil interface, resulting in a polymer film at the interface (Chouinard, Buczkowski, & Lenaerts, 1994; Gibaud et al., 1998; Thies & Bissery, 1984). Polycyanoacrylate nanocapsules are biodegradable, with a typical diameter being about 200 nm, with an inner structure that is highly porous. Because of their porous structure, the nanocapsules have a high-sorption capacity (Couvreur et al., 1979). Though the cyanoacrylate polymers are biodegradable, they may cause inflammatory responses in tissues as they produce formaldehyde as a by-product. It has been reported that as the homologous series is ascended, the inflammatory response is decreased with the butyl derivatives and higher homologs being well tolerated by the tissues (Leonard, Kulkarni, Brandes, Nelson, & Cameron, 1966). Niwa, Takeuchi, Hino, Kunou, and Kawashima (1994) have described a novel spontaneous emulsification solvent diffusion method for preparation of PLGA nanospheres using nafarelin acetate as a model peptide. The peptide and PLGA were dissolved in an acetone–dichloromethane–water mixture, which was then poured under moderate stirring into an aqueous solution of PVA. Rapid diffusion of acetone from organic to aqueous phase resulted in spontaneous emulsification. Subsequent evaporation of dichloromethane produced nanospheres of about 200–300 nm in size. However, the encapsulation efficiency of this process is low. The entrapment efficiency was increased when oil was employed as the outer phase to prevent leakage of the drug out of the nanospheres (Niwa, Takeuchi, Hino, Nohara, & Kawashima, 1995). Lipid–polymer hybrid nanoparticles (LPNs) have a polymeric core with a lipid/lipid–PEG shell, which gives them some characteristics of both polymeric nanoparticles and liposomes (Hadinoto, Sundaresan, & Cheow, 2013).

6.4.8 Vaccines

The development of controlled-release vaccines is receiving increasing attention in recent years. Immunization will be discussed in detail in Chapter 10, while some brief comments on parenteral vaccines will be made here. Normally, immunization requires multiple injections, and dropout rates from the first to the second dose can be as high as 70% in developing countries. A controlled-release vaccine may deliver the antigen to achieve a long-lasting effect following a single injection. Development of a pulsatile release pattern may also allow a single injection to provide both the primary and booster immunization (Heller, 1993b). Many antigens are often only weakly immunogenic, and the use of adjuvants may be required to enhance the antibody response. Currently, aluminum salts are used as adjuvants. Particulate carriers based on synthetic polymers can also act as adjuvants in addition to providing controlled release. Adsorption and incorporation of influenza virus antigen into poly(methyl methacrylate) nanoparticles have been shown to result in very efficient vaccines with a good adjuvant effect and antibody response in mice. The antibody response in mice protected the mice against infection with mice-adapted influenza virus. Also, the polymer vaccine was more stable against thermal inactivation as compared to vaccines with aluminum hydroxide or those without adjuvant (Kreuter & Liehl, 1981). Similarly, biodegradable polyglycolic acid microspheres containing hepatitis B surface antigen were successfully used for immunization against hepatitis B. When injected intraperitoneally into guinea pigs, antibody response to microspheres was higher than an alum-adsorbed antigen used as a positive control (Nellore, Pande, Young, & Bhagat, 1992). Alginate–polylysine microcapsules containing Bacillus Calmette–Guerin (BCG) vaccine have also been prepared. The microcapsules had a size range of 5–15 μm, and the BCG organisms retained their viability following processing by aseptic production and freeze drying (Kwok & Burgess, 1992; Kwok, Groves, & Burgess, 1991).

6.4.9 Pulsatile Drug Delivery Systems

Many physiological peptides and proteins are released in the body in a pulsatile fashion as opposed to a continuous release. Examples of peptides and proteins that are released in a pulsatile pattern include vasopressin (Koch & Lutz-Bucher, 1985), LHRH (Vickery, 1991), insulin (Matthews et al., 1987), and GH (Hartman, Iranmanesh, Thorner, & Veldhuis, 1993), among others. GH is secreted in discrete pulses by the anterior pituitary gland in response to a complex collection of external and internal stimuli (Hartman et al., 1993). As previously discussed, a different pharmacological response may result depending on the mode of administration in the case of LHRH and its analogs. Thus, it is obvious that many therapeutic peptides and proteins need to be administered in a pulsatile pattern rather than in a continuous manner. Pulsatile delivery systems should ideally function in an on–off manner in response to internal or external stimuli (Yoshida, Sakai, Okano, & Sakurai, 1993). Externally regulated pulsatile delivery systems may be considered as open-loop systems, while the internal or self-regulated systems may be considered closed-loop systems. The closed-loop systems are based on the fact that many peptides and proteins in the human body are released under the control of a feedback system called *homeostasis*. This feedback system is essential to maintain health and a normal metabolic balance. Such systems are under development for the delivery of insulin tied to the feedback of changing blood glucose levels.

6.4.9.1 Externally Regulated or Open-Loop Systems

In these systems, the release is controlled by a user-generated external signal, such as electric current, magnetic field, or ultrasound (Langer, 1989). Electric current and ultrasound have been used to modulate transdermal drug delivery. These techniques, called iontophoresis and phonophoresis, respectively, will be discussed in Chapter 8. Magnetically modulated delivery systems incorporate small magnetic beads within a polymer matrix that also contains the drug. Release of BSA from EVAc matrices was enhanced about 30-fold when exposed to magnetic field. It is believed that the beads, when exposed to magnetic field, compress and expand the pores in the polymer matrix, thereby squeezing drug out of the pores. In the absence of magnetic field, the drug is released slowly by diffusion through this network of interconnecting pores. Such systems can allow the development of demand-responsive drug delivery systems (Heller, 1993a). The use of hydrogels sensitive to pH, temperature, or electric field to achieve pulsatile drug delivery has been reviewed (Yoshida et al., 1993). Slightly cross-linked poly(N-acrylolyl pyrrolidine) polymer has thermosensitive properties, and the permeation of insulin across this polymer has been investigated (Bae, Okanko, & Kim, 1989). Poly(N-isopropyl acrylamide-co-alkyl methacrylate) (poly(IPAAm-co-RMA)), a thermoresponsive hydrogel, can cause *on–off* regulation of drug release. As the temperature is increased, a dense skin layer forms on the surface and stops drug release from the hydrogel (Yoshida et al., 1993). Pulsed release of albumin and horseradish peroxidase has been achieved by using a system of enzymatically activated microencapsulated liposomes (Kibat, Igari, Wheatley, Eisen, & Langer, 1990). Control of delivery by programmable pumps may also be considered as externally regulated systems. Though some implantable pumps may not be under the direct control of the patients, pumps in general will be considered in the following discussion.

6.4.9.2 Pumps

The driving force for drug delivery from a pump is a pressure difference. In this respect, a pump differs from other drug delivery systems in which drug is released due to diffusion, driven by a concentration gradient. The driving force for pump delivery in a pump may be generated by osmotic effects or by direct mechanical action. Also, these pumps may be portable external pumps or implantable systems (Banerjee, Hosny, & Robinson, 1991; Heller, 1993a). An example of an external pump is a syringe pump that has been used for continuous, SC insulin infusion (CSII) for several years (Leichter, Schreiner, Reynolds, & Bolick, 1985). Pumps that can intermittently deliver very small volumes are also available and can be used for pulsatile delivery of peptides such as LHRH. Patch pumps have also been marketed and can be worn directly on the skin to infuse drugs into the body and can be controlled wirelessly by a device, for example, OmniPod for insulin administration (Narasimhan, Mach, & Shameem, 2012).

Several implantable pumps are on the market. An example of an early configuration of an implantable pump is the Infusaid® pump that consists of a disk made of titanium, divided into two chambers by a bellow. One compartment contains a propellant that, under the influence of body heat, pushes against the other compartment to push the drug into a catheter, which is inserted into a vein or artery. The pump is placed in a pocket surgically fashioned under the skin and can be refilled by a direct percutaneous injection into a self-sealing septum at the top of the pump (Heller, 1993a). The Alza (now marketed by DURECT Corporation, CA) osmotic minipump (Alzet®) can also be implanted subcutaneously and has been widely used in research investigations in animal models for the delivery of peptides and proteins. This pump delivers the drug under a gradient of osmotic pressure (Banerjee et al., 1991). An osmotic pump from Alza for human use (Viadur™) is also now on the market that delivers leuprolide for a once-yearly implant for the palliative treatment of advanced prostate cancer. This is the first product to use Alza's DUROS® implant technology and provides continuous, 12-month suppression of testosterone with a single treatment. DURECT Corporation is developing further uses of the DUROS implant technology, including the development of catheterized DUROS systems for intratumoral delivery or targeted delivery to the heart for continuous intrapericardial infusion, for example, for delivery of growth factors to induce new blood vessel formation (Verity, 2003). Pulsatile release can also be achieved by a programmed microinfusion apparatus that can be implanted subcutaneously (Lynch, Rivest, & Wurtman, 1980). A portable device consisting of a peristaltic pump and a computerized timing device has been used to deliver LHRH in a pulsatile fashion. The chronic, intermittent administration of LHRH by this device was used as a new mode for treatment for infertility, and successful induction of pregnancy in patients with severe hypothalamic amenorrhea has been reported (Leyendecker, Wildt, & Hansmann, 1980).

6.4.9.3 Self-Regulated or Closed-Loop Systems

These systems have mostly been developed for the delivery of insulin regulated by the blood glucose concentrations (Chien & Lin, 2002). In body, glucose levels are maintained in the range of 70–110 mg/dL by an autofeedback mechanism of the pancreas to release insulin. For diabetic patients who cannot control blood glucose levels, external biofeedback approaches are being investigated. One biochemical approach is based on the principles of competitive and complementary binding behavior of concanavalin A (Con A) with glucose and glycosylated insulin (G-insulin). A device has been developed in which Con A is covalently bound to Sepharose beads and enclosed within a polymer membrane that is permeable to glucose and insulin but not to Con A. In studies with pancreatectomized dogs implanted with this delivery system, increased blood glucose levels displaced G-insulin from Con A and G-insulin then diffuses out from the device to control the blood glucose levels. The device, which functions like an artificial pancreas, has been reported to successfully control blood glucose levels in response to feeding challenges for up to 48 h but failed after this time. Postmortem, the devices were found to be infiltrated with proteins and covered with fibrous tissue, and efforts are underway to resolve these issues (Kim & Jacobs, 1994).

Another approach for self-regulated delivery of insulin is based on the enzyme–substrate reaction. In this approach, the pH change resulting from a reaction between glucose and glucose oxidase is used to induce a change in an acid-sensitive polymer. This principle may be used to develop membrane-controlled, erosion-controlled, or solubility-controlled devices. In the membrane-controlled devices, the membrane is fabricated from a hydrogel polymer containing pendant amine groups and entrapped glucose oxidase. As glucose diffuses into the membrane, it is oxidized by glucose oxidase to gluconic acid that protonates the amine groups. The porosity of the membrane increases due to electrostatic repulsion between protonated amine groups, which allow insulin to diffuse through the membrane. This type of glucose-sensitive membrane has been extensively investigated to develop bioresponsive insulin delivery systems (Albin, Horbett, Miller, & Ricker, 1987; Albin, Horbett, & Ratner, 1985; Horbett, Ratner, Kost, & Singh, 1984; Ishihara, Kobayashi, Ishimaru, & Shinohara, 1984; Kost, Horbett, Ratner, & Singh, 1985). In an erosion-controlled device, the lowered pH accelerates the rate of erosion of a hydrolytically labile, pH-sensitive, erodible polymer containing the drug. The solubility approach uses trilysyl insulin that exhibits a large change in solubility between pH 7 and 5, thus changing the rate of dissolution to allow self-regulated release (Fischel-Ghodsian, Brown, Mathiowitz, Brandenburg, & Langer, 1988; Heller, 1993b). Recently, the development of a pH-sensitive liposome system has been reported. In this system, glucose oxidase and insulin are coencapsulated in pH-sensitive liposomes. As glucose enters the liposome, gluconic acid is produced, which lowers the pH so that the pH-sensitive liposome can release insulin (Kim, Im, Lim, Oh, & Han, 1994). While these self-regulated insulin delivery systems have not yet been commercialized, several insulin formulations that can provide a sustained release of insulin are already on the market.

An ideal approach to self-regulated insulin delivery would be to transplant the whole pancreas. Since organ transplants have to contend with rejection by host or side effects of immunosuppression, islets of Langerhans have been used. Again, supply of human islets is limited, and use of islets from different species may lead to immune responses. A way around this problem is to microencapsulate these islets using a polymer membrane. The membrane is such that it allows glucose to diffuse in and insulin to diffuse out but does not allow antibodies or other agents that trigger rejection. Such microencapsulation of islets of Langerhans to create an artificial pancreas for the treatment of type I diabetic patients is being investigated. Excellent in vitro results have been achieved using sodium alginate and poly-L-lysine to form the capsules. However, this success has not yet been matched in transplantation experiments due to pericapsular fibrosis. The presence of fibrosis around encapsulated islets causes a failure of the graft as the supply of nutrients to the islets is cut off, along with a buildup of waste products and the inability of insulin to diffuse out of the capsule (Clayton, James, & London, 1993; Heller, 1993a). Self-regulated or closed-loop delivery systems are very promising, but several technical and toxicological problems need to be resolved before these can be commercialized (Heller, 1993a). These are similar approaches to develop an artificial pancreas have been underway for several years now.

6.5 INNOVATIONS IN PARENTERAL ADMINISTRATION OF PROTEINS

The use of nonaqueous solvents and related product has been discussed in Chapter 5. Other novel technologies are also investigated, for example, producing pure-protein microspheres by adding water-soluble polymers to a solution of the protein. The polymers do not encapsulate the protein but instead induce precipitation of the protein from the solution as they compete for water that was keeping the protein in solution (Rios, 2004). Chemical approaches other than PEGylation are also being investigated, for example, developing polymer-modified proteins by chemical synthesis. Kochendoerfer et al. (2003) have reported the chemical synthesis of a synthetic erythropoiesis protein, a 51 kDa protein–polymer conjugate consisting of a 166-amino acid polypeptide chain (similar to the sequence of human erythropoietin) and two covalently attached, branched, negatively charged, and monodisperse polymer moieties. Dendrimers are being prepared and investigated for drug delivery applications. These are made of synthetic, highly branched, monodisperse polymers that assemble into nanometer size range (Sakthivel & Florence, 2003).

6.5.1 PEGylated Proteins

Conjugation of proteins to PEG is perhaps the most common approach to chemical modification of a protein for several potential beneficial effects. These can include increased bioavailability, increased plasma half-lives, decreased immunogenicity, reduced proteolysis, and enhanced solubility and stability, though potential toxicity of side products also needs to be considered (Knop, Hoogenboom, Fischer, & Schubert, 2010). PEGylation reaction engineering and subsequent purification steps must be optimized not only for quality but also for costs (Peng et al., 2014; Pfister & Morbidelli, 2014).

PEG is a linear polymer based on the $-CH_2CH_2O-$ repeat unit and is commercially available in varying MWs. The covalent conjugation of PEG to proteins is referred to as *PEGylation* and involves the lysine amino groups of proteins. Site-specific conjugation chemistries are important since undesirable isomers can result from nonspecific chemistries. The organic and polymer chemistry of PEG activation has now matured, and protein PEGylation is becoming viable commercially. However, the technique needs significant know-how, and the modified protein is considered to be a new chemical entity from a regulatory point of view. The MW of the PEG used as well as the extent of PEGylation gives control over the biophysical properties of PEG protein produced such as size, hydrophobicity, and charge. Typically, at least a 20 kDa PEG is needed to reduce renal clearance. The hydrodynamic radius of PEGylated proteins is higher than one would expect based on MW due to the hydration of the PEG moiety. The PEG protein may have a plasma half-life as much as 3–486-fold higher than the native protein. This is believed to result because the heavily hydrated PEG molecule excludes the protein from immediate uptake by organs. Similarly, the PEG molecules shield the antigenic sites, thereby reducing the immunogenicity of the protein (Katre, 1993). In the past, problems resulted due to PEG polymers being a heterogeneous population of variable chain lengths, but now monodisperse PEG oligomers are available.

PEGylated proteins that have been on the market for some time now include PEG-adenosine deaminase (PEG-ADA, Adagen®) and PEG-asparaginase (PEG-ALL, Oncaspar®), both from Sigma-Tau Pharmaceuticals (Gaithersburg, MD). PEG-ADA is used to trigger immune defense mechanisms in severe combined immunodeficiency syndrome (SCIDS), while PEG-ALL is available for the treatment of acute lymphoblastic leukemia (Groves, Alkan, & Hickey, 1992). Interferon products, PEG-interferon as Peginterferon alfa-2b (PEG-INTRON®; Merck) and Peginterferon alfa-2a (PEGASYS®, Genentech), are also available for the treatment of chronic hepatitis C by parenteral administration. Due to an increase in the size of the interferon molecule, PEGylated interferon has slower absorption, prolonged half-life, and reduced clearance. It thus needs less frequent administration. Tolerability of the PEGylated interferons is comparable to the non-PEGylated formulations. PEG-INTRON was developed by attaching a 12 kDa monomethoxy PEG to the protein, and it has demonstrated enhanced pharmacokinetic profile in both animal and human studies. Due to its improved clinical benefits, PEG-INTRON plus ribavirin combination therapy can replace INTRON® and ribavirin for the treatment of chronic hepatitis C (Baker, 2001; Bukowski et al., 2002; Wang et al., 2002). Other PEGylated proteins on the market include pegfilgrastim (Neulasta®; Amgen) and pegvisomant (Somavert®; Pfizer). Pegvisomant is a PEGylated GH antagonist used for the treatment of acromegaly, while pegfilgrastim is PEGylated GCSF. Pegfilgrastim had a U.S. market of $4.1 billion in 2012 (Bruno et al., 2013).

6.5.2 Protein Crystals

Proteins are amorphous, but they can be crystallized into fragile crystals held together by weak intermolecular forces. Protein stability could possibly be improved by crystallization since crystalline form is generally considered to be an energetically favorable state. The protein is well hydrated within the crystal and can stay folded in its native state and may be less likely to undergo aggregation. While small-scale crystallization for structural analysis has been done for years (Chapter 1), large-scale crystallization efforts for improved formulation and/or delivery are more recent (with the exception of insulin crystal formulations). These crystals could also possibly provide a controlled-release dosage form due to slow dissolution of the protein. Fragile forms of protein crystals can be made more stable by cross-linking or coprecipitation with metal ions or other suitable ligands like stabilizing agents (Jen & Merkle, 2001). The former Altus Pharmaceuticals, Inc. had crystallized several proteins, including therapeutic monoclonal antibodies. The crystals are typically of 0.1–100 μm, and the process can be controlled and is scalable and optimized to get high yield at low process cost. Protein crystallization can be very beneficial as monoclonal antibodies typically need high doses to be administered, and these crystals could possibly be injected subcutaneously in high concentrations without increased viscosity and can still maintain excellent syringeability. Crystalline monoclonal antibodies have been shown to have similar pharmacokinetic profiles as soluble monoclonal antibodies following IV injection, and crystallization did not affect the physicochemical properties, secondary structure, or the bioactivity of the antibody (Shenoy, 2002). However, when cross-linked enzyme crystals are developed, it is possible that the protein may be inactive in the crystalline state. Also, a tedious approach is needed to find the right cross-linking method and reagent (Park, Kwon, & Park, 2011). Microcrystals of bovine insulin have been recently reported to be encapsulated into PLGA microspheres to preserve stability and as a model to investigate encapsulation of protein crystals into microspheres (Choi, Kwon, & Kim, 2004). Protein crystals can also be potentially useful for oral or other delivery means.

6.5.3 Prefilled Syringes and Needle-Free Injections

Most biotech molecules are now available in prefilled syringes as these offer several advantages, including rapid and accurate dosing, reduced chance of dosing errors, reduced overfill requirements, and reduced risk of microbial contamination (Majumdar et al., 2011). Reduced costs have been reported when hemodialysis patients on epoetin alfa were switched from multidose vials to prefilled syringes (Wazny, Raymond, Do, & Skwarchuk, 2009). However, there can be some drawbacks as well. The possible impact of siliconized syringes on protein aggregation has been discussed in Section 4.3.6. Prefilled syringes also need to meet guidelines in elemental impurity limits (Van, Catry, & Vanhaecke, 2013), and potential oxidation of the protein by the tip cap material needs to be evaluated (Badkar, Wolf, Bohack, & Kolhe, 2011). Injector-pen devices have been available for several years for insulin administration and can be filled with disposable insulin cartridges. The device has a built-in dosing meter and a disposable needle tip and allows for convenient pen-like transport by the patient for multiple administrations. Therefore, compared to conventional administration of insulin, these devices may be exposed to higher temperatures and more agitation, and thus, suitable stress testing has been developed to ensure stability under these conditions (Shnek, Hostettler, Bell, Olinger, & Frank, 1998). A precision one-time-use dosing pen is available for peginterferon therapy for chronic hepatitis C (PEG-INTRON REDIPEN™) with a dial-up dosing button for individualized weight-based dosing and a small needle size (30-gauge) to minimize patient discomfort. Mixing occurs by pushing down on the pen to combine the PEG-INTRON powder with sterile water, both of which are stored in the body of the pen. Technological advances recently been made in the needle-free drug delivery have also make this mode of drug delivery more promising. Earlier devices were large, complex, and expensive and were made by a handful of small companies. However, the new generation of devices is simpler and more compact, with some being the size and shape of a small flashlight. Upon actuation, these devices force the liquid through the syringe orifice to form a *liquid needle* at pressures that can penetrate the skin. Such insulin *pens* are now commonly available in the European and U.S. market (Pass & Hayes, 2003).

6.6 EXAMPLES OF PROTEIN PHARMACOKINETICS

Having discussed the critical pharmacokinetic parameters and how they may differ for protein drugs in Section 6.2, specific examples of interferon, interleukins, insulin, and epoetin will now be discussed. It should be realized that more than one protein may be used in combination for some specific disease states. Preclinical and clinical studies with some proteins such as cytokines and growth factors have focused on their utility not only as single agents but also as the use of multiple cytokines or a cytokine in combination with chemotherapeutic agents (Talmadge, 1993).

6.6.1 Interferon

The pharmacokinetics of interferons has been studied by several investigators and has been reviewed. Following IV administration, there is a rapid decline in serum concentrations, and the pharmacokinetic parameters vary widely across the family of interferons (Wills, 1990). Following IV administration, serum concentration of recombinant human interferon-β_{ser} (rIFN-β_{ser}) in healthy male volunteers was found to decline biexponentially, with a mean serum clearance of 0.76 ± 0.28 L/h-kg, a mean steady-state volume of distribution of 2.88 ± 1.81 L/kg, and a mean terminal half-life of 4.22 ± 2.29 h, as determined by noncompartmental analysis. Calculations were done by fitting concentration-time data to a two-compartment model function. The single IV injection was followed by single or eight consecutive daily SC doses. Following SC administration, absorption of rIFN-β_{ser} was prolonged, but no accumulation in serum was noted after eight daily SC injections (Chiang, Gloff, Yoshizawa, & Williams, 1993). The pharmacokinetics of recombinant human interferon-gamma (rIFN-gamma) in rhesus monkeys has also been investigated. Clearance after IV administration (0.1 mg/kg) was 18.7 mL/min/kg, and V_{ss} value was 510 mL/kg. The peak serum concentrations after IM and SC administration of 0.25 mg/kg rIFN-gamma were 50.7 and 52.3 ng/mL, respectively, and time to reach peak serum concentration was 480 min for both routes. The mean bioavailabilities after IM and SC administrations were 109% and 90%, respectively (Ferraiolo et al., 1988). The in vivo half-life of human interferon alpha-2b (hIFN-α-2b) has been increased by the development of an extended-release formulation (Depo/IFN) by encapsulation of hIFN-α-2b into a lipid-based drug delivery system. The release of free hIFN-α-2b from Depo/IFN into the peritoneal cavity of mice was slow, with a 10-fold lower peak concentration and a 13-fold longer apparent half-life (20 vs. 1.5 h) (Bonetti & Kim, 1993).

6.6.2 Interleukins

Interleukins have had only a modest success in the treatment of cancer and infectious diseases, largely because of a narrow therapeutic index and toxicity. In the body, interleukins are normally produced and released in microenvironments, in which they exert their effects and are metabolized. Thus, pharmacokinetics is irrelevant in a physiological environment. Furthermore, any trace amounts that may spill into circulation are diluted and eliminated so rapidly that they are undetectable. With the availability of interleukins as drugs, pharmacokinetic characterization becomes important though rapid removal from plasma is still a barrier to investigations. The disappearance curve appears to be multiexponential, but a two-compartment model is sufficient in most cases (Bocci, 1991). Interleukin-2 is known to be less rapidly cleared when administered by less traditional routes such as intraperitoneal or inhalation administration. Thus, these routes may provide an improved therapeutic index compared with that resulting after administration of the drug by IV administration (Anderson & Sorenson, 1994). Alternatively, sustained-release products can be developed for parenteral administration. Sustained-release products provide therapeutic levels of interleukin for long periods, which are more positively correlated with increased antitumor efficacy as compared to transient high blood levels. A sustained-release formulation of interleukin-2 was evaluated following IM injection in rats. The formulation contained poloxamer 407, a block copolymer that shows the property of reverse thermal gelation. The formulation could be injected as a viscous mobile solution but gels in vivo once injected. This formulation resulted in a decreased blood concentration of interleukin-2, but the effect was prolonged so that the bioavailability of the protein was not affected (Wang & Johnston, 1995).

6.6.3 Insulin

When insulin was discovered in 1921 by Banting and Best, it was thought that a simple and final treatment of diabetes mellitus was achieved. While insulin has alleviated the sufferings of millions of diabetics since then, its optimum delivery has still not been achieved (Chien & Banga, 1989; Thomas, 1986). Coupled with the drug delivery efforts, pharmacokinetics of insulin has also been extensively investigated over the years. The duration of action of regular insulin typically lasts for 4–8 h, and longer-acting preparations are often desired, for example, to avoid having to wake up children at night for injections (see also Section 1.2.1.1.2). Products based on neutral protamine Hagedorn (NPH) insulin have been used to prepare long-acting preparations (Chien, 1996). *Humulin* N (NPH human insulin isophane suspension) is a crystalline suspension of human insulin with the highly basic peptide protamine and zinc providing an intermediate-acting insulin with a duration of activity up to 24 h. An extended zinc suspension for human insulin is also available and consists of an amorphous and crystalline (*Humulin* L *Lente*) or crystalline suspension of human insulin with zinc (*Humulin* U *Ultralente*). These preparations have duration of activity up to 24 or 28 h, respectively. Unlike the NPH crystallization process, ultralente suspensions do not use protamine, and precipitation is initiated by adjusting the pH to the isoelectric point of insulin in the presence of zinc ions to take advantage of insulin's unique self-association and ligand-binding abilities. Other than nanocarrier- or microcarrier-based systems, insulin is the only protein commercialized as a suspension formulation.

Earlier studies used radiolabeled products, and some proposed a three-compartment insulin model (Berger et al., 1979; Lauritzen, Faber, & Binder, 1979; Sherwin et al., 1974). Studies in animal models have generally indicated that proteolytic degradation occurs at the site of injection following SC administration. Such degradation could, however, be minimized by the use of protease inhibitors (Hori, Komada, & Okumura, 1983; Takeyama et al., 1991). Studies in dogs have shown that following SC injection of regular insulin, the rate and extent of absorption were variable, with half-life values of 2.3 ± 1.3 h for the absorption. The extent of absorption, which was found to be first order, was $78\% \pm 15\%$ (Ravis, Comerci, & Ganjam, 1986). A recent study administered regular human insulin to healthy men, and the resulting data on serum insulin concentration versus time curve were fitted to a two-compartment open pharmacokinetic model and came up with mean half-lives of 5.09 and 57.7 min for distribution and elimination phases, respectively. The mean systemic clearance was 29.2 L/h, while the mean values for the apparent volumes of distribution were as follows: central compartment 4.75 L, steady state 14.2 L, and β-phase 45.2 L (Hooper et al., 1991). Since early 1980s, techniques have been made available to administer insulin to ambulatory diabetics by programmable SC infusions. This technique, known as SC insulin infusion (CSII), supplies insulin by SC route using an external electromechanical device containing soluble or neutral insulin. Most insulins infused subcutaneously in this way reach the systemic circulation, and SC insulin degradation is low in most diabetics. However, the absorption is slow, and it may take 6–8 h to reach a steady state following a change in the SC infusion. Thus, meal insulin requirements are best met by bolus delivery, while basal insulin delivery is feasible by CSII (Kraegen & Chisholm, 1985). Absorption of insulin after SC injection or infusion depends on both the volume and concentration of the injected or infused insulin. A model has been described to explain the absorption process by taking into account the aggregation status of insulin and its binding in tissue. Insulin solutions of normal therapeutic strength exist as hexamers, while absorption takes place in a molecular form no greater than dimer. Thus, insulin forms a depot that is cleared only slowly, and this accounts for the experimentally observed variations in the absorption rate over a wide range of volumes and of concentrations (Mosekilde, Jensen, Binder, Pramming, & Thorsteinsson, 1989). Following absorption, the major enzyme involved in the cellular degradation of insulin is insulin protease, also referred to as insulin-degrading enzyme. While insulin-degrading enzyme causes proteolytic cleavage in the A and B chains of insulin, insulin can also be acted upon by the enzyme glutathione insulin transhydrogenase, which causes reduction of the disulfide bonds of insulin, catalyzing disulfide interchange (Lee, 1988). Even with regular SC injections of neutral regular human insulin (0.2 U/kg), the peak effects do not occur until 3–4 h after injection and are present for as long as 8 h. Thus, it is clear that with presently available insulins or insulin regimens, simulation of basal or meal-stimulated components of normal insulin secretion is not feasible. For this reason, insulin manufacturers are engaged in developing new analogs that can more closely replicate endogenous insulin secretion or its effects. An insulin analog [Lys(B28), Pro(B29)]-human insulin (LYSPRO) was developed by inverting the natural amino acid sequence of the B chain at positions 28 and 29. This molecule was developed using computational chemistry (Chapter 1) and has reduced capacity for self-association in solution (Galloway, 1993). In clinical studies, LYSPRO produced significantly different pharmacokinetic and pharmacodynamic profiles from that of human regular insulin after SC injection. Serum concentrations of LYSPRO peaked about twice as high and in less than half the time than regular human insulin. LYSPRO and human insulin appear to be equipotent, and thus, the more rapid absorption and shorter duration of LYSPRO may offer an advantage over human regular insulin in the control of blood glucose after meals (Howey, Bowsher, Brunelle, & Woodworth, 1994). The rapid absorption of LYSPRO is attributed to its reduced capability for self-association. In contrast, the long-acting nature of neutral regular insulin is due to its tendency to self-associate into dimers, tetramers, hexamers, and polymers, with absorption of SC insulin occurring only after it has dissociated into a less aggregated form (Galloway, 1993). Other insulin analogs with similar properties have been discussed in Section 1.2.1.1.2. In addition to the development of insulin analogs with improved pharmacokinetics, other approaches to improve the quality of life for a diabetic include the availability of self-monitoring blood glucose devices, pen injectors, and premixes of NPH and regular insulin. The use of the insulin pen has allowed widespread acceptance of multiple insulin injection regimens, resulting in less dietary restrictions and a more flexible lifestyle (Galloway, 1993; Gordon, 1992; Lindmayer et al., 1986).

6.6.4 Epoetin

Epoetin (recombinant human erythropoietin) is an effective therapeutic agent for correcting anemia of chronic renal failure in patients on maintenance hemodialysis. After IV administration, epoetin is distributed in volume comparable to plasma volume, and plasma concentrations decay monoexponentially with a half-life of 4–12 h.

For patients maintained on continuous ambulatory peritoneal dialysis, who have no ready vascular access, the IV route becomes impractical. Thus, intraperitoneal and SC routes of administration have also been used for delivery of epoetin. The bioavailability of epoetin administered intraperitoneally in dialysis fluids is very low (3%–8%), but bioavailability following SC administration can reach 20%–30%. The pharmacokinetics of epoetin following SC administration is different from that obtained with IV dosing. The C_{max} is lower following SC administration, and plasma clearance is slower, allowing a buildup of concentrations following repeated administration. These changes represent a therapeutic advantage from the clinical perspective (Abraham, St Peter, Redic-kill, & Halstenson, 1992; Macdougall, Roberts, & Coles, 1991). Long-term treatment of patients with epoetin for hemodialysis and hemofiltration does not appear to change the pharmacokinetics and metabolism of epoetin. Thus, dosage adjustments or substitution is not necessary (Gladziwa et al., 1993).

6.6.5 Miscellaneous

The pharmacokinetics of several other peptides and proteins such as thyrotropin-releasing hormone (Miyamoto et al., 1993), LHRH (Handelsman, Jansen, Boylan, Spaliviero, & Turtle, 1984; Handelsman & Swerdloff, 1986), vasopressin (Davison, Sheills, Barron, Robinson, & Lindheimer, 1989), atrial natriuretic peptide (Tan, Russel, Thien, & Benraad, 1993), recombinant factor IX (McCarthy et al., 2002), somatostatin analogs lanreotide and octreotide (Gancel et al., 1994; Kuhn et al., 1994), aprotinin (Levy, Bailey, & Salmenpera, 1994), tumor necrosis factor (Ferraiolo et al., 1988), GH (Chalew, Phillip, & Kowarski, 1993), α_1-antitrypsin (Constans et al., 1992), and CSF (Yoon et al., 1993) has been described in the literature. Dose titration of recombinant factor VIII or other coagulation factors requires estimation of pharmacokinetic parameters in the individual patient (Bjorkman et al., 2010). The pharmacokinetics of LHRH was influenced by the route of administration. The paired steady-state IV and SC infusions indicated an irreversible loss of about one-third of the drug infused via SC route. In addition, the prolonged delayed response of LHRH after SC injection resulted in a different time course of plasma profiles as compared to IV administration (Handelsman et al., 1984). In contrast, the absorption of recombinant human GH following SC injection was very efficient as suggested by the similarity of clearance rate following SC or IV administration. SC injection of 0.06 mg/kg of GH was effective in accelerating growth rate in GH-deficient patients and produced plasma-integrated concentrations higher than those typically observed in normally growing children (Chalew et al., 1993). The pharmacokinetics of human granulocyte/macrophage CSF following IV administration to rabbits at doses of 0.05–2.5 mg/kg was found to be dose dependent. This suggests a saturable metabolism of the factor at the dose ranges studied (Yoon et al., 1993). The pharmacokinetic disposition of recombinant human tumor necrosis factor-α in mice has been studied and was found to be biexponential. However, the terminal phase accounted for less than 1% of total AUC. After IV or IM administration of the ^{125}I-labeled protein, its rank order of accumulation in the major organs was liver > kidney > lung > heart > spleen (Ferraiolo et al., 1988).

6.7 CONCLUSIONS

Peptide and protein drugs are currently administered almost exclusively by the parenteral route. However, repeated administrations are required due to the short half-lives of these drugs. Thus, the development of parenteral controlled-release systems offers much promise. Polymers are an integral part of many of these systems. For use in body, these polymers must be biocompatible and preferably biodegradable. Some of the more promising polymers include the poly(D,L-lactide-co-glycolide), poly(ortho esters), polyanhydrides, and ethylene vinyl acetate copolymer. Various controlled-release systems that have been developed include oil-based injections, implants, liposomes, nanoparticles, microspheres, and pulsatile delivery systems. Also, PEGylated proteins and protein crystals can provide controlled release. Some microsphere peptide/protein formulations are already on the market. Pulsatile delivery systems may be externally regulated or self-regulated. A self-regulated or closed-loop system for the delivery of insulin in response to changing blood glucose levels will be desirable, and research is continuing in this direction. Prefilled syringes and needle-free injections provide added convenience to parenteral delivery of proteins. Pharmacokinetics of peptides and proteins involves several unique considerations. Important pharmacokinetic parameters may be affected by proteolytic degradation at the site of injection, binding to plasma proteins, and the generation of degradation products that are smaller and more lipophilic than the parent peptide or protein. The induction of immunogenic responses following administration of proteins is a major problem and may be minimized by conjugation with polymers such as PEG. Such chemical modification can also alter protein disposition.

REFERENCES

Abraham, P. A., St Peter, W. L., Redic-kill, K. A., & Halstenson, C. E. (1992). Controversies in determination of epoetin (recombinant human erythropoietin) dosages. *Clin. Pharmacokinet., 22,* 409–415.

Agarwal, P. & Rupenthal, I. D. (2013). Injectable implants for the sustained release of protein and peptide drugs. *Drug Discov. Today, 18,* 337–349.

Al-Achi, A. & Greenwood, R. (1993). Human insulin binding to erythrocyte-membrane. *Drug Dev. Ind. Pharm., 19,* 673–684.

Albin, G., Horbett, T. A., & Ratner, B. D. (1985). Glucose sensitive membranes for controlled delivery of insulin: Insulin transport studies. *J. Contr. Rel., 2,* 153–164.

Albin, G. W., Horbett, T. A., Miller, S. R., & Ricker, N. L. (1987). Theoretical and experimental studies of glucose sensitive membranes. *J. Contr. Rel., 6,* 267–291.

Anderson, P. M. & Sorenson, M. A. (1994). Effects of route and formulation on clinical pharmacokinetics of interleukin-2. *Clin. Pharmacokinet., 27,* 19–31.

Artursson, P., Edman, P., Laakso, T., & Sjoholm, I. (1984). Characterization of polyacryl starch microparticles as carriers for proteins and drugs. *J. Pharmaceut. Sci., 73,* 1507–1513.

Badkar, A., Wolf, A., Bohack, L., & Kolhe, P. (2011). Development of biotechnology products in pre-filled syringes: Technical considerations and approaches. *AAPS Pharm. Sci. Tech., 12,* 564–572.

Bae, Y. H., Okanko, T., & Kim, S. W. (1989). Insulin permeation through thermo-sensitive hydrogels. *J. Contr. Rel., 9,* 271–279.

Baker, D. E. (2001). Pegylated interferons. *Rev. Gastroenterol. Disord., 1,* 87–99.

Banerjee, P. S., Hosny, E. A., & Robinson, J. R. (1991). Parenteral delivery of peptide and protein drugs. In V. H. L. Lee (Ed.), *Peptide and protein drug delivery* (pp. 487–543). New York: Marcel Dekker, Inc.

Banga, A. K. & Chien, Y. W. (1988). Systemic delivery of therapeutic peptides and proteins. *Int. J. Pharmaceut., 48,* 15–50.

Banks, W. A. & Kastin, A. J. (1991). Regulation of the passage of peptides across the blood-brain barrier. In P. D. Gorzone, W. A. Colburn, & M. Mokotott (Eds.), *Pharmacokinetics and pharmacodynamics. Vol.3: Peptides, peptoids and proteins* (pp. 148–153). Cincinnati, OH: Harvey Whitney Books Co.

Banks, W. A., Kastin, A. J., & Barrera, C. M. (1991). Delivering peptides to the central nervous system: Dilemmas and strategies. *Pharmaceut. Res., 8,* 1345–1350.

Begley, D. J. (1994). Strategies for delivery of peptide drugs to the central nervous system: Exploiting molecular structure. *J. Contr. Rel., 29,* 293–306.

Berger, M., Halban, P. A., Girardier, L., Seydoux, J., Offord, R. E., & Renold, A. E. (1979). Absorption kinetics of subcutaneously injected insulin. *Diabetologia, 17,* 97–99.

Bergers, J. J., Otter, W. D., & Crommelin, D. J. A. (1994). Vesicles for tumour-associated antigen presentation to induce protective immunity: Preparation, characterization and enhancement of the immune response by immunomodulators. *J. Contr. Rel., 29,* 317–327.

Bjorkman, S., Blanchette, V. S., Fischer, K., Oh, M., Spotts, G., Schroth, P. et al. (2010). Comparative pharmacokinetics of plasma- and albumin-free recombinant factor VIII in children and adults: The influence of blood sampling schedule on observed age-related differences and implications for dose tailoring. *J. Thromb. Haemost., 8,* 730–736.

Bocci, V. (1991). Interleukins. Clinical pharmacokinetics and practical implications. *Clin. Pharmacokinet., 21,* 274–284.

Bodor, N. (1994). Drug targeting and retrometabolic drug design approaches: Introduction. *Adv. Drug Deliv. Rev., 14,* 157–166.

Bonetti, A. & Kim, S. (1993). Pharmacokinetics of an extended-release human interferon alpha-2b formulation. *Canc. Chemother. Pharmacol., 33,* 258–261.

Brown, L., Munoz, C., Siemer, L., Edelman, E., & Langer, R. (1986). Controlled release of insulin from polymer matrices: Control of diabetes in rats. *Diabetes, 35,* 692–697.

Brown, L., Siemer, L., Munoz, C., & Langer, R. (1986). Controlled release of insulin from polymer matrices: In-vitro kinetics. *Diabetes, 35,* 684–691.

Brown, L. R., Wei, C. L., & Langer, R. (1983). In vivo and in vitro release of macromolecules from polymeric drug delivery systems. *J. Pharmaceut. Sci., 72,* 1181–1185.

Brown, S. A., Nelson, R. W., & Bottoms, G. D. (1987). Models for the pharmacokinetics and pharmacodynamics of insulin in alloxan-induced diabetic dogs. *J. Pharmaceut. Sci., 76,* 295–299.

Browne, M. J., Carey, J. E., Chapman, C. G., Dodd, I., Esmail, A. F., Lawrence, G. M. P. et al. (1993). Protein engineering and comparative pharmacokinetic analysis of a family of novel recombinant hybrid and mutant plasminogen activators. *Fibrinolysis, 7,* 357–364.

Bruno, B. J., Miller, G. D., & Lim, C. S. (2013). Basics and recent advances in peptide and protein drug delivery. *Ther. Deliv., 4,* 1443–1467.

Bukowski, R. M., Tendler, C., Cutler, D., Rose, E., Laughlin, M. M., & Statkevich, P. (2002). Treating cancer with PEG Intron: Pharmacokinetic profile and dosing guidelines for an improved interferon-alpha-2b formulation. *Cancer, 95,* 389–396.

Burgess, D. J. (1990). Practical analysis of complex coacervate systems. *J. Colloid Interf. Sci., 140,* 227–238.

Burgess, D. J. & Singh, O. N. (1993). Spontaneous formation of small sized albumin/acacia coacervate particles. *J. Pharm. Pharmacol., 45,* 586–591.

Chalew, S. A., Phillip, M., & Kowarski, A. (1993). Plasma integrated concentration of growth hormone after recombinant human growth hormone injection—Implications for determining an optimal dose. *Am. J. Dis. Child, 147,* 274–278.

Chaubal, M. V., Gupta, A. S., Lopina, S. T., & Bruley, D. F. (2003). Polyphosphates and other phosphorus-containing polymers for drug delivery applications. *Crit. Rev. Ther. Drug Carrier Syst., 20,* 295–315.

Chiang, J., Gloff, C. A., Yoshizawa, C. N., & Williams, G. J. (1993). Pharmacokinetics of recombinant human interferon-beta(ser) in healthy volunteers and its effects on serum neopterin. *Pharmaceut. Res., 10,* 567–572.

Chien, Y. W. (1996). Human insulin: Basic sciences to therapeutic uses. *Drug Dev. Ind. Pharm., 22,* 753–789.

Chien, Y. W. & Banga, A. K. (1989). Potential developments in systemic delivery of insulin. *Drug Develop. Ind. Pharm., 15,* 1601–1634.

Chien, Y. W. & Lin, S. (2002). Optimisation of treatment by applying programmable rate-controlled drug delivery technology. *Clin. Pharmacokinet., 41,* 1267–1299.

Chikhale, E. G., Ng, K. Y., Burton, P. S., & Borchardt, R. T. (1994). Hydrogen bonding potential as a determinant of the in vitro and in situ blood-brain barrier permeability of peptides. *Pharmaceut. Res., 11,* 412–419.

Chirmule, N., Jawa, V., & Meibohm, B. (2012). Immunogenicity to therapeutic proteins: Impact on PK/PD and efficacy. *AAPS J., 14,* 296–302.

Choi, S. H., Kwon, J. H., & Kim, C. W. (2004). Microencapsulation of insulin microcrystals. *Biosci. Biotechnol. Biochem., 68,* 749–752.

Chouinard, F., Buczkowski, S., & Lenaerts, V. (1994). Poly(alkylcyanoacrylate) nanocapsules: Physicochemical characterization and mechanism of formation. *Pharmaceut. Res., 11,* 869–874.

Chuang, V. T., Kragh-Hansen, U., & Otagiri, M. (2002). Pharmaceutical strategies utilizing recombinant human serum albumin. *Pharmaceut. Res., 19,* 569–577.

Clayton, H. A., James, R. F. L., & London, N. J. M. (1993). Islet microencapsulation—A review. *Acta Diabetol., 30,* 181–189.

Cleland, J. L., Johnson, O. L., Putney, S., & Jones, A. J. S. (1997). Recombinant human growth hormone poly(lactic-co-glycolic acid) microsphere formulation development. *Adv. Drug Deliv. Rev., 28,* 71–84.

Cohen, J., Siegel, R. A., & Langer, R. (1984). Sintering technique for the preparation of polymer matrices for the controlled release of macromolecules. *J. Pharmaceut. Sci., 73,* 1034–1037.

Cohen, S., Yoshioka, T., Lucarelli, M., Hwang, L. H., & Langer, R. (1991). Controlled delivery systems for proteins based on poly(lactic/glycolic acid) microspheres. *Pharmaceut. Res., 8,* 713–720.

Colburn, W. A. (1991). Peptide, peptoid, and protein pharmacokinetics/pharmacodynamics. In P. D. Garzone, W. A. Colburn, & M. Mokotott (Eds.), *Pharmacokinetics and pharmacodynamics. Vol. 3: Peptides, peptoids and proteins* (pp. 94–115). Cincinnati, OH: Harvey Whitney Books Co.

Constans, J., Carles, P., Boneu, A., Arnaud, J., Tufenkji, A. E., Pujazon, M. et al. (1992). Clinical pharmacokinetics of a1-antitrypsin in homozygous PiZ deficient patients. *Clin. Pharmacokinet., 23,* 161–168.

Conte, U., Conti, B., Giunchedi, P., & Maggi, L. (1994). Spray dried polylactide microsphere preparation—Influence of the technological parameters. *Drug Dev. Ind. Pharm., 20,* 235–258.

Costantino, H. R., Firouzabadian, L., Wu, C., Carrasquillo, K. G., Griebenow, K., Zale, S. E. et al. (2002). Protein spray freeze drying. 2. Effect of formulation variables on particle size and stability. *J. Pharmaceut. Sci., 91,* 388–395.

Couvreur, P., BlancoPrieto, M. J., Puisieux, F., Roques, B., & Fattal, E. (1997). Multiple emulsion technology for the design of microspheres containing peptides and oligopeptides. *Adv. Drug Deliv. Rev., 28,* 85–96.

Couvreur, P., Kante, B., Roland, M., Guiot, P., Bauduin, P., & Speiser, P. (1979). Polycyanoacrylate nanocapsules as potential lysosomotropic carriers: Preparation, morphological and sorptive properties. *J. Pharm. Pharmacol., 31,* 331–332.

Couvreur, P. & Puisieux, F. (1993). Nanoparticles and microparticles for the delivery of polypeptides and proteins. *Adv. Drug Deliv. Rev., 10,* 141–162.

Cronin, M. J., Ferraiolo, B. L., & Moore, J. A. (1991). Contemporary issues involving the activities of recombinant human hormones. In P. D. Garzone, W. A. Colburn, & M. Mokotoff (Eds.), *Pharmacokinetics and pharmacodynamics* (pp. 139–146). Cincinnati, OH: Harvey Whitney Books Co.

D'Souza, M. J. (1988). Controlled release of cyclosporine from microspheres. *Drug Dev. Ind. Pharm., 14,* 1351–1357.

Davison, J. M., Sheills, E. A., Barron, W. M., Robinson, A. G., & Lindheimer, M. D. (1989). Changes in the metabolic clearance of vasopressin and in plasma vasopressinase throughout human pregnancy. *J. Clin. Invest., 83,* 1313–1318.

Deboer, A. G., Devries, H. E., Delange, E. C. M., Danhof, M., Kuiper, J., & Breimer, D. D. (1994). Drug transport to the brain—In vitro versus in vivo approaches. *J. Contr. Rel., 28,* 259–263.

Degling, L., Stjarnkvist, P., & Sjoholm, I. (1993). Interferon-Y in starch microparticles: Nitric oxide-generating activity in vitro and antileishmanial effect in mice. *Pharmaceut. Res., 10,* 783–790.

DesNoyer, J. R. & McHugh, A. J. (2003). The effect of Pluronic on the protein release kinetics of an injectable drug delivery system. *J.Contr. Rel., 86,* 15–24.

Edman, P. & Sjoholm, I. (1981). Prolongation of effects of asparaginase by implantation in polyacrylamide in rats. *J. Pharmaceut. Sci., 70,* 684–685.

Ferraiolo, B. L., Fuller, G. B., Burnett, B., & Chan, E. (1988). Pharmacokinetics of recombinant human interferon-gamma in the rhesus monkey after intravenous, intramuscular, and subcutaneous administration. *J. Biol. Response Modif., 7,* 115–122.

Ferraiolo, B. L., Moore, J. A., Crase, D., Gribling, P., Wilking, H., & Baughman, R. A. (1988). Pharmacokinetics and tissue distribution of recombinant human tumor necrosis factor-alpha in mice. *Drug Metabol. Dispos., 16,* 270–275.

Ficek, B. J. & Peppas, N. A. (1993). Novel preparation of poly(vinyl alcohol) microparticles without crosslinking agent for controlled drug delivery of proteins. *J. Contr. Rel., 27,* 259–264.

Fischel-Ghodsian, F., Brown, L., Mathiowitz, E., Brandenburg, D., & Langer, R. (1988). Enzymatically controlled drug delivery. *Proc. Natl. Acad. Sci., 85,* 2403–2406.

Fu, H., Rahaman, M. N., Day, D. E., & Brown, R. F. (2011). Hollow hydroxyapatite microspheres as a device for controlled delivery of proteins. *J. Mater. Sci. Mater. Med., 22,* 579–591.

Fults, K. A. & Johnston, T. P. (1990). Sustained-release of urease from a poloxamer gel matrix. *J. Parent. Sci. Tech., 44(2),* 58–65.

Galloway, J. A. (1993). New directions in drug development—Mixtures, analogues, and modeling. *Diabetes Care, 16,* 16–23.

Gancel, A., Vuillermet, P., Legrand, A., Catus, F., Thomas, F., & Kuhn, J. M. (1994). Effects of a slow-release formulation of the new somatostatin analogue lanreotide in tsh-secreting pituitary adenomas. *Clin. Endocrinol., 40,* 421–428.

Gehrke, S. H., Uhden, L. H., & McBride, J. F. (1998). Enhanced loading and activity retention of bioactive proteins in hydrogel delivery systems. *J. Contr. Rel., 55,* 21–33.

Gibaud, S., Rousseau, C., Weingarten, C., Favier, R., Douay, L., Andreux, J. P. et al. (1998). Polyalkylcyanoacrylate nanoparticles as carriers for granulocyte-colony stimulating factor (G-CSF). *J. Contr. Rel., 52,* 131–139.

Girish, S., Martin, S. W., Peterson, M. C., Zhang, L. K., Zhao, H., Balthasar, J. et al. (2011). AAPS workshop report: Strategies to address therapeutic protein-drug interactions during clinical development. *AAPS J., 13,* 405–416.

Gladziwa, U., Klotz, U., Baumer, K., Zollinger, R., Mann, H., & Sieberth, H. (1993). Pharmacokinetics of epoetin (recombinant human erythropoietin) after long term therapy in patients undergoing haemodialysis and haemofiltration. *Clin. Pharmacokinet., 25,* 145–153.

Gloff, C. A. & Benet, L. Z. (1990). Pharmacokinetics and protein therapeutics. *Adv. Drug Deliv. Rev., 4,* 359–386.

Gordon, D. (1992). Insulin pens. Is delivery sacrificed to improve patient compliance? *Clin. Pharmacokinet., 23,* 249–252.

Groves, M. J., Alkan, M. H., & Hickey, A. J. (1992). The formulation of proteins and peptides. In M. E. Klegerman & M. J. Groves (Eds.), *Pharmaceutical biotechnology: Fundamentals and essentials* (pp. 218–248). Buffalo Grove, IL: Interpharm Press, Inc.

Guo, J. H. (1994). Preparation methods of biodegradable microspheres on bovine serum albumin loading efficiency and release profiles. *Drug Dev. Ind. Pharm., 20,* 2535–2545.

Hadinoto, K., Sundaresan, A., & Cheow, W. S. (2013). Lipid-polymer hybrid nanoparticles as a new generation therapeutic delivery platform: A review. *Eur. J. Pharm. Biopharm., 85,* 427–443.

Han, J. A., Kang, Y. J., Shin, C., Ra, J. S., Shin, H. H., Hong, S. Y. et al. (2014). Ferritin protein cage nanoparticles as versatile antigen delivery nanoplatforms for dendritic cell (DC)-based vaccine development. *Nanomedicine, 10,* 561–569.

Han, K., Lee, K. D., Gao, Z. G., & Park, J. S. (2001). Preparation and evaluation of poly(L-lactic acid) microspheres containing rhEGF for chronic gastric ulcer healing. *J. Contr. Rel., 75,* 259–269.

Handelsman, D. J., Jansen, R. P. S., Boylan, L. M., Spaliviero, J. A., & Turtle, J. R. (1984). Pharmacokinetics of gonadotropin-releasing hormone: Comparison of subcutaneous and intravenous routes. *J. Clin. Endocrinol. Metabol., 59,* 739–746.

Handelsman, D. J. & Swerdloff, R. S. (1986). Pharmacokinetics of gonadotropin-releasing hormone and its analogs. *Endocrine Rev., 7,* 95–105.

Hartman, M. L., Iranmanesh, A., Thorner, M. O., & Veldhuis, J. D. (1993). Evaluation of pulsatile patterns of growth hormone release in humans—A brief review. *Am. J. Hum. Biol., 5,* 603–614.

Hashida, M., Nishikawa, M., Yamashita, F., & Takakura, Y. (1994). Targeting delivery of protein drugs by chemical modification. *Drug Dev. Ind. Pharm., 20(4),* 581–590.

Hashida, M. & Takakura, Y. (1994). Pharmacokinetics in design of polymeric drug delivery systems. *J. Contr. Rel., 31,* 163–171.

Hayashi, Y., Yoshioka, S., Aso, Y., Po, A. L. W., & Terao, T. (1994). Entrapment of proteins in poly(L-lactide) microspheres using reversed micelle solvent evaporation. *Pharmaceut. Res., 11,* 337–340.

Heller, J. (1993a). Modulated release from drug delivery devices. *Crit. Rev. Ther. Drug Carrier Syst., 10,* 253–305.

Heller, J. (1993b). Polymers for controlled parenteral delivery of peptides and proteins. *Adv. Drug Deliv. Rev., 10,* 163–204.

Heller, J., Helwing, R. F., Baker, R. W., & Tuttle, M. E. (1983). Controlled release of water-soluble macromolecules from bioerodible hydrogels. *Biomaterials, 4,* 262–266.

Henry, R., Begent, J., & Pedley, B. (1992). Monoclonal antibody administration. Current clinical pharmacokinetic status and future trends. *Clin. Pharmacokinet., 23,* 85–89.

Heya, T., Mikura, Y., Nagai, A., Miura, Y., Futo, T., Tomida, Y. et al. (1994). Controlled release of thyrotropin releasing hormone from microspheres: Evaluation of release profiles and pharmacokinetics after subcutaneous administration. *J. Pharmaceut. Sci., 83,* 798–801.

Heya, T., Okada, H., Ogawa, Y., & Toguchi, H. (1994). In vitro and in vivo evaluation of thyrotrophin releasing hormone release from copoly(dl-lactic/glycolic acid) microspheres. *J. Pharmaceut. Sci., 83,* 636–640.

Hirose, M. & Tachibana, A. T. T. (2010). Recombinant human serum albumin hydrogel as a novel drug delivery vehicle. *Mater. Sci. Eng. C, 30,* 664–669.

Hirsch, A. C., Upasani, R. S., & Banga, A. K. (2005). Factorial design approach to evaluate interactions between electrically assisted enhancement and skin stripping for delivery of tacrine. *J. Contr. Rel., 103,* 113–121.

Hirschberg, H. J., van de Wijdeven, G. G., Kraan, H., Amorij, J. P., & Kersten, G. F. (2010). Bioneedles as alternative delivery system for hepatitis B vaccine. *J. Contr. Rel., 147,* 211–217.

Hoffman, P. L., Walter, R., & Bulat, M. (1977). An enzymatically stable peptide with activity in the central nervous system: Its penetration through the blood-CSF barrier. *Brain Res., 122,* 87–94.

Holland, S. J., Tighe, B. J., & Gould, P. L. (1986). Polymers for biodegradable medical devices. 1. The potential of polyesters as controlled macromolecular release systems. *J. Contr. Rel., 4,* 155–180.

Hooper, S. A., Bowsher, R. R., & Howey, D. C. (1991). Pharmacokinetics and pharmacodynamics of intravenous regular human insulin. In P. D. Garzone, W. A. Colburn, & M. Mokotott (Eds.), *Pharmacokinetics and pharmacodynamics. Vol.3: Peptides, peptoids and proteins.* Cincinnati, OH: Harvey Whitney Books Co.

Hora, M. S., Rana, R. K., Nunberg, J. H., Tice, T. R., Gilley, R. M., & Hudson, M. E. (1990). Release of human serum albumin from poly(lactide-co-glycolide) microspheres. *Pharmaceut. Res., 7,* 1190–1194.

Horbett, T. A., Ratner, B. D., Kost, J., & Singh, M. (1984). A bioresponsive membrane for insulin delivery. In J. M. Anderson & S. W. Kim (Eds.), *Recent advances in drug delivery systems* (pp. 209–220). New York: Plenum Publishing Corp.

Hori, R., Komada, F., & Okumura, K. (1983). Pharmaceutical approach to subcutaneous dosage forms of insulin. *J. Pharmaceut. Sci., 72,* 435–439.

Hoth, M. & Merkle, H. P. (1991). Formulation of silicone matrix systems for long term constant release of peptides. *Drug Dev. Ind. Pharm., 17,* 985–999.

Howanitz, J. H. (1993). Review of the influence of polypeptide hormone forms on immunoassay results. *Arch. Pathol. Lab. Med., 117,* 369–372.

Howey, D. C., Bowsher, R. R., Brunelle, R. L., & Woodworth, J. R. (1994). [Lys(B28), Pro(B29)]—Human insulin: A rapidly absorbed analogue of human insulin. *Diabetes, 43,* 396–402.

Hsieh, D. S. T., Chiang, C. C., & Desai, D. S. (1985). Controlled release of macromolecules from silicone elastomer. *Pharmacuet. Tech., 9(6),* 39–49.

Hsieh, D. S. T., Rhine, W. D., & Langer, R. (1983). Zero-order controlled-release polymer matrices for micro- and macromolecules. *J. Pharmaceut. Sci., 72,* 17–22.

Ijntema, K., Heuvelsland, W. J. M., Dirix, C. A. M. C., & Sam, A. P. (1994). Hydroxyapatite microcarriers for biocontrolled release of protein drugs. *Int. J. Pharmaceut., 112,* 215–224.

Illum, L. & Davis, S. S. (1982). The targeting of drugs parenterally by use of microspheres. *J. Parenter. Sci. Technol., 36,* 242–248.

Ishihara, K., Kobayashi, M., Ishimaru, N., & Shinohara, I. (1984). Glucose induced permeation control of insulin through a complex membrane consisting of immobilized glucose oxidase and a poly(amine). *Polymer J., 16,* 625–631.

Iwata, M. & Mcginity, J. W. (1993). Dissolution, stability, and morphological properties of conventional and multiphase poly(DL-lactic-co-glycolic acid) microspheres containing water-soluble compounds. *Pharmaceut. Res., 10,* 1219–1227.

Jeffery, H., Davis, S. S., & O'Hagan, D. T. (1993). The preparation and characterization of poly(lactide-co-glycolide) microparticles. II. The entrapment of a model protein using a (water-in-oil)-in-water emulsion solvent evaporation technique. *Pharmaceut. Res., 10,* 362–368.

Jen, A. & Merkle, H. P. (2001). Diamonds in the rough: Protein crystals from a formulation perspective. *Pharmaceut. Res., 18,* 1483–1488.

Jenkins, N. & Curling, E. M. A. (1994). Glycosylation of recombinant proteins—Problems and prospects. *Enzym. Microb. Tech., 16,* 354–364.

Jeyanthi, R., Mehta, R. C., Thanoo, B. C., & DeLuca, P. P. (1997). Effect of processing parameters on the properties of peptide-containing PLGA microspheres. *J. Microencapsul., 14,* 163–174.

Johnson, O. L. & Herbert, P. (2003). Long acting protein formulations—ProLease technology. In M. J. Rathbone, J. Hadgraft, & M. S. Roberts (Eds.), *Modified-release drug delivery technology* (pp. 671–677). New York: Marcel Dekker, Inc.

Johnston, T. P., Punjabi, M. A., & Froelich, C. J. (1992). Sustained delivery of interleukin-2 from a poloxamer 407 gel matrix following intraperitoneal injection in mice. *Pharm. Res., 9,* 425–434.

Jorgensen, L., Moeller, E. H., van de Weert, M., Nielsen, H. M., & Frokjaer, S. (2006). Preparing and evaluating delivery systems for proteins. *Eur. J. Pharm. Sci., 29,* 174–182.

Kamath, K. R. & Park, K. (1993). Biodegradable hydrogels in drug delivery. *Adv. Drug Deliv. Rev., 11,* 59–84.

Katre, N. V. (1993). The conjugation of proteins with polyethylene glycol and other polymers—Altering properties of proteins to enhance their therapeutic potential. *Adv. Drug Deliv. Rev., 10,* 91–114.

Kawahara, H., Naito, H., Takara, K., Wakabayashi, T., Kidoya, H., & Takakura, N. (2013). Tumor endothelial cell-specific drug delivery system using apelin-conjugated liposomes. *PLoS One, 8,* e65499.

Kibat, P. G., Igari, Y., Wheatley, M. A., Eisen, H. N., & Langer, R. (1990). Enzymatically activated microencapsulated liposomes can provide pulsatile drug release. *FASEB, 4,* 2533–2539.

Kim, C. K., Im, E. B., Lim, S. J., Oh, Y. K., & Han, S. K. (1994). Development of glucose-triggered pH-sensitive liposomes for a potential insulin delivery. *Int. J. Pharmaceut., 101,* 191–197.

Kim, S. W. & Jacobs, H. A. (1994). Self-regulated insulin delivery—Artificial pancreas. *Drug Dev. Ind. Pharm., 20(4),* 575–580.

Knop, K., Hoogenboom, R., Fischer, D., & Schubert, U. S. (2010). Poly(ethylene glycol) in drug delivery: Pros and cons as well as potential alternatives. *Angew. Chem. Int. Ed. Engl., 49,* 6288–6308.

Koch, B. & Lutz-Bucher, B. (1985). Specific receptors for vasopressin in the pituitary gland: Evidence for down-regulation and desensitisation to adrenocorticotropin-releasing factors. *Endocrinology, 116,* 671–676.

Kochendoerfer, G. G., Chen, S. Y., Mao, F., Cressman, S., Traviglia, S., Shao, H. et al. (2003). Design and chemical synthesis of a homogeneous polymer-modified erythropoiesis protein. *Science, 299,* 884–887.

Kooijmans, S. A., Vader, P., van Dommelen, S. M., van Solinge, W. W., & Schiffelers, R. M. (2012). Exosome mimetics: A novel class of drug delivery systems. *Int. J. Nanomedi, 7,* 1525–1541.

Kost, J., Horbett, T. A., Ratner, B. D., & Singh, M. (1985). Glucose-sensitive membranes containing glucose oxidase: Activity, swelling, and permeability studies. *J. Biomed. Mat. Res., 19,* 1117–1133.

Kraegen, E. W. & Chisholm, D. J. (1985). Pharmacokinetics of insulin: Implications for continuous subcutaneous insulin infusion therapy. *Clin. Pharmacokin., 10,* 303–314.

Kreuter, J. (1991). Nanoparticle-based drug delivery systems. *J. Contr. Rel., 16,* 169–176.

Kreuter, J. (1994). Nanoparticles. In J. Kreuter (Ed.), *Colloidal drug delivery systems* (pp. 219–342). New York: Marcel Dekker, Inc.

Kreuter, J. & Liehl, E. (1981). Long-term studies of microencapsulated and adsorbed influenza vaccine nanoparticles. *J. Pharmaceut. Sci., 70,* 367–371.

Kruisbrink, J. & Boer, G. J. (1984). Controlled long-term release of small peptide hormones using a new microporous polypropylene polymer: Its application for vasopressin in the brattleboro rat and potential perinatal use. *J. Pharmaceut. Sci., 73,* 1713–1718.

Kruisbrink, J. & Boer, G. J. (1986). The use of [3H] vasopressin for in-vivo studies of controlled delivery from an accurel/collodion device in the brattleboro rat. *J. Pharm. Pharmacol., 38,* 893–897.

Kuhn, J. M., Legrand, A., Ruiz, J. M., Obach, R., Deronzan, J., & Thomas, F. (1994). Pharmacokinetic and pharmacodynamic properties of a long-acting formulation of the new somatostatin analogue, lanreotide, in normal healthy volunteers. *Br. J. Clin. Pharmacol., 38,* 213–219.

Kwak, H. H., Shim, W. S., Choi, M. K., Son, M. K., Kim, Y. J., Yang, H. C. et al. (2009). Development of a sustained-release recombinant human growth hormone formulation. *J. Contr. Rel., 137,* 160–165.

Kwok, K. K. & Burgess, D. J. (1992). A novel method for determination of sterility of microcapsules and measurement of viability of encapsulated organisms. *Pharmaceut. Res., 9,* 410–413.

Kwok, K. K., Groves, M. J., & Burgess, D. J. (1991). Production of 5–15 mm diameter alginate-polylysine microcapsules by an air-atomization technique. *Pharmaceut. Res., 8,* 341–344.

Langer, R. (1989). Biomaterials in controlled drug delivery: New perspectives from biotechnological advances. *Pharm. Technol., 13(8),* 18–30.

Langer, R. (1993). Polymer-controlled drug delivery systems. *Acc. Chem. Res., 26,* 537–542.

Lauritzen, T., Faber, O. K., & Binder, C. (1979). Variation in 125I-insulin absorption and blood glucose concentration. *Diabetologia, 17,* 291–295.

Le Ray, A. M., Vert, M., Gautier, J. C., & Benoit, J. P. (1994). Fate of [C-14]poly(DL-lactide-co-glycolide) nanoparticles after intravenous and oral administration to mice. *Int. J. Pharmaceut., 106,* 201–211.

Lee, P. I. (1993). Swelling and dissolution kinetics during peptide release from erodible anionic gel beads. *Pharmaceut. Res., 10,* 980–985.

Lee, V. H. L. (1988). Enzymatic barriers to peptide and protein absorption. *CRC Crit. Rev. Ther. Drug Carrier Syst., 5,* 68–97.

Leichter, S. B., Schreiner, M. E., Reynolds, L. R., & Bolick, T. (1985). Long term follow up of diabetic patients using insulin infusion pumps: Considerations for future clinical application. *Arch. Intern. Med., 145,* 1409–1412.

Leonard, F., Kulkarni, R. K., Brandes, G., Nelson, J., & Cameron, J. J. (1966). Synthesis and degradation of poly(alkyl alpha-cyanoacrylates). *J. Appl. Pol. Sci., 10,* 259–272.

Levy, J. H., Bailey, J. M., & Salmenpera, M. (1994). Pharmacokinetics of aprotinin in preoperative cardiac surgical patients. *Anesthesiology, 80,* 1013–1018.

Leyendecker, G., Wildt, L., & Hansmann, M. (1980). Pregnancies following chronic intermittent (pulsatile) administration of Gn-RH by means of a portable Pump ("ZYKLOMAT")—A new approach to the treatment of infertility in hypothalamic amenorrhea. *J. Clin. Endocrinol. Metab., 51,* 1214–1216.

Li, C., Yang, D. J., Kuang, L. R., & Wallace, S. (1993). Polyamino acid microspheres—Preparation, characterization and distribution after intravenous injection in rats. *Int. J. Pharmaceut., 94,* 143–152.

Lindmayer, I., Menassa, K., Lambert, J., Legendre, A., Legault, C., Letendre, M. et al. (1986). Development of a new jet injection for insulin therapy. *Diabetes Care, 9,* 294–297.

Lynch, H. J., Rivest, R. W., & Wurtman, R. J. (1980). Artificial induction of melatonin rhythms by programmed microinfusion. *Neuroendocrinology, 31,* 106–111.

Macdougall, I. C., Roberts, D. E., & Coles, G. A. (1991). Clinical pharmacokinetics of epoetin (recombinant human erythropoietin). *Clin. Pharmacokinet., 20,* 99–113.

Mahmood, I. (2004). Interspecies scaling of protein drugs: Prediction of clearance from animals to humans. *J. Pharmaceut. Sci., 93,* 177–185.

Majumdar, S., Ford, B. M., Mar, K. D., Sullivan, V. J., Ulrich, R. G., & D'Souza, A. J. (2011). Evaluation of the effect of syringe surfaces on protein formulations. *J. Pharmaceut. Sci., 100,* 2563–2573.

Makino, K., Ohshima, H., & Kondo, T. (1987). Effects of plasma proteins on degradation properties of poly(L-lactide) microcapsules. *Pharmaceut. Res., 4,* 62–65.

Mariette, B., Coudane, J., Vert, M., Gautier, J. C., & Moneton, P. (1993). Release of the GRF29NH(2) analog of human GRF44NH(2) from a PLA/GA matrix. *J. Contr. Rel., 24,* 237–246.

Martins, S., Sarmento, B., Ferreira, D. C., & Souto, E. B. (2007). Lipid-based colloidal carriers for peptide and protein delivery—Liposomes versus lipid nanoparticles. *Int. J. Nanomed., 2,* 595–607.

Marx, J. (1990). NGF and alzheimer's: Hopes and fears. *Science, 247,* 408–410.

Mashayekhi, R., Mobedi, H., Najafi, J., & Enayati, M. (2013). In-vitro/in-vivo comparison of leuprolide acetate release from an in-situ forming plga system. *Daru, 21,* 57.

Matthews, D. R., Hermansen, K., Connolly, A. A., Gray, D., Schmitz, O., Clark, A. et al. (1987). Greater in vivo than in vitro pulsatility of insulin secretion with synchronized insulin and somatostatin secretory pulses. *Endocrinology, 120,* 2272–2278.

McCarthy, K., Stewart, P., Sigman, J., Read, M., Keith, J. C., Jr., Brinkhous, K. M. et al. (2002). Pharmacokinetics of recombinant factor IX after intravenous and subcutaneous administration in dogs and cynomolgus monkeys. *Thromb. Haemost., 87,* 824–830.

McLennan, D. N., Porter, C. J., Edwards, G. A., Brumm, M., Martin, S. W., & Charman, S. A. (2003). Pharmacokinetic model to describe the lymphatic absorption of r-metHu-leptin after subcutaneous injection to sheep. *Pharmaceut. Res., 20,* 1156–1162.

Mehta, R. C., Jeyanthi, R., Calis, S., Thanoo, B. C., Burton, K. W., & DeLuca, P. P. (1994). Biodegradable microspheres as depot system for parenteral delivery of peptide drugs. *J. Contr. Rel., 29,* 375–384.

Mesmin, C., Fenaille, F., Ezan, E., & Becher, F. (2011). MS-based approaches for studying the pharmacokinetics of protein drugs. *Bioanalysis, 3,* 477–480.

Miyamoto, M., Hirai, K., Takahashi, H., Kato, K., Nishiyama, M., Okada, H. et al. (1993). Effects of sustained release formulation of thyrotropin-releasing hormone on learning impairments caused by scopolamine and AF64A in rodents. *Eur. J. Pharmacol., 238,* 181–189.

Mordenti, J. (1986). Man versus beast: Pharmacokinetic scaling in mammals. *J. Pharmaceut. Sci., 75,* 1028–1040.

Mordenti, J., Chen, S. A., Moore, J. A., Ferraiolo, B. L., & Green, J. D. (1991). Interspecies scaling of clearance and volume of distribution data for five therapeutic proteins. *Pharmaceut. Res., 8,* 1351–1359.

Mosekilde, E., Jensen, K. S., Binder, C., Pramming, S., & Thorsteinsson, B. (1989). Modeling absorption kinetics of subcutaneous injected soluble insulin. *J. Pharmacokinet. Biopharm., 17,* 67–87.

Narasimhan, C., Mach, H., & Shameem, M. (2012). High-dose monoclonal antibodies via the subcutaneous route: Challenges and technical solutions, an industry perspective. *Ther. Deliv., 3,* 889–900.

Nellore, R. V., Pande, P. G., Young, D., & Bhagat, H. R. (1992). Evaluation of biodegradable microspheres as vaccine adjuvant for hepatitis B surface antigen. *J. Parenter. Sci. Technol., 46(5),* 176–180.

Nguyen, X. C., Herberger, J. D., & Burke, P. A. (2004). Protein powders for encapsulation: A comparison of spray-freeze drying and spray drying of darbepoetin alfa. *Pharmaceut. Res., 21,* 507–514.

Niwa, T., Takeuchi, H., Hino, T., Kunou, N., & Kawashima, Y. (1994). In vitro drug release behavior of D,L-lactide/glycolide copolymer (PLGA) nanospheres with nafarelin acetate prepared by a novel spontaneous emulsification solvent diffusion method. *J. Pharmaceut. Sci., 83,* 727–732.

Niwa, T., Takeuchi, H., Hino, T., Nohara, M., & Kawashima, Y. (1995). Biodegradable submicron carriers for peptide drugs: Preparation of DL-lactide/glycolide copolymer (PLGA) nanospheres with nafarelin acetate by a novel emulsion-phase separation method in an oil system. *Int. J. Pharmaceut., 121,* 45–54.

Nuxoll, E. (2013). BioMEMS in drug delivery. *Adv. Drug Deliv. Rev., 65,* 1611–1625.

O'Hagan, D. T., Jeffery, H., & Davis, S. S. (1994). The preparation and characterization of poly(lactide-co-glycolide) microparticles. III. Microparticle/polymer degradation rates and the in vitro release of a model protein. *Int. J Pharmaceut., 103,* 37–45.

Ogawa, Y., Yamamoto, M., Okada, H., Yashiki, T., & Shimamoto, T. (1988). A new technique to efficiently entrap leuprolide acetate into microcapsules of polylactic acid or copoly(lactic/glycolic) acid. *Chem. Pharm. Bull., 36,* 1095–1103.

Okada, H. (1997). One- and three-month release injectable microspheres of the LH-RH superagonist leuprorelin acetate. *Adv. Drug Deliv. Rev., 28,* 43–70.

Okada, H., Doken, Y., Ogawa, Y., & Toguchi, H. (1994a). Preparation of three-month depot injectable microspheres of leuprorelin acetate using biodegradable polymers. *Pharmaceut. Res., 11,* 1143–1147.

Okada, H., Doken, Y., Ogawa, Y., & Toguchi, H. (1994b). Sustained suppression of the pituitary-gonadal axis by leuprorelin three-month depot microspheres in rats and dogs. *Pharmaceut. Res., 11,* 1199–1203.

Okada, H., Heya, T., Ogawa, Y., Toguchi, H., & Shimamoto, T. (1991). Sustained pharmacological activities in rats following single and repeated administration of once-a-month injectable microspheres of leuprolide acetate. *Pharmaceut. Res., 8,* 584–587.

Okada, H., Inoue, Y., Heya, T., Ueno, H., Ogawa, Y., & Toguchi, H. (1991). Pharmacokinetics of once-a-month injectable microspheres of leuprolide acetate. *Pharmaceut. Res., 8,* 787–791.

Okada, H., Yamamoto, M., Heya, T., Inoue, Y., Kamei, S., Ogawa, Y. et al. (1994). Drug delivery using biodegradable microspheres. *J. Contr. Rel., 28,* 121–129.

Okumu, F. W., Cleland, J. L., & Borchardt, R. T. (1997). The effect of size, charge and cyclization of model peptides on their in vitro release from DL-PLGA microspheres. *J. Contr. Rel., 49,* 133–140.

Okumu, F. W., Dao, L. N., Fielder, P. J., Dybdal, N., Brooks, D., Sane, S. et al. (2002). Sustained delivery of human growth hormone from a novel gel system: SABER. *Biomaterials, 23,* 4353–4358.

Olsen, D., Yang, C., Bodo, M., Chang, R., Leigh, S., Baez, J. et al. (2003). Recombinant collagen and gelatin for drug delivery. *Adv. Drug Deliv. Rev., 55,* 1547–1567.

Oner, L. & Groves, M. J. (1993). Optimization of conditions for preparing 2-micron-range to 5- micron-range gelatin microparticles by using chilled dehydration agents. *Pharmaceut. Res., 10,* 621–626.

Pai, S. S., Tilton, R. D., & Przybycien, T. M. (2009). Poly(ethylene glycol)-modified proteins: Implications for poly(lactide-co-glycolide)-based microsphere delivery. *AAPS J., 11,* 88–98.

Park, K., Kwon, I., & Park, K. (2011). Oral protein delivery: Current status and future prospects. *React. Funct. Polym., 71,* 280–287.

Park, T. G., Cohen, S., & Langer, R. (1992). Controlled protein release from polyethyleneimine-coated poly(L lactic acid)/pluronic blend matrices. *Pharmaceut. Res., 9,* 37–39.

Pass, F. & Hayes, J. (2003). Needle-free drug delivery. In M. J. Rathbone, J. Hadgraft, & M. S. Roberts (Eds.), *Modified-release drug delivery technology* (pp. 599–606). New York: Marcel Dekker, Inc.

PDR Staff. *Physicians' Desk Reference* (2014). (68th ed.) Montvale, NJ: PDR Network, LLC.

Pec, E. A., Wout, Z. G., & Johnston, T. P. (1992). Biological activity of urease formulated in poloxamer 407 after intraperitoneal injection in the rat. *J. Pharmaceut. Sci., 81,* 626–630.

Peng, F., Liu, Y., Li, X., Sun, L., Zhao, D., Wang, Q. et al. (2014). PEGylation of G-CSF in organic solvent markedly increase the efficacy and reactivity through protein unfolding, hydrolysis inhibition and solvent effect. *J. Biotechnol., 170,* 42–49.

Pfister, D. & Morbidelli, M. (2014). Process for protein PEGylation. *J. Contr. Rel., 180C,* 134–149.

Pisal, D. S., Kosloski, M. P., & Balu-Iyer, S. V. (2010). Delivery of therapeutic proteins. *J. Pharmaceut. Sci., 99,* 2557–2575.

Porter, C. J. (1997). Drug delivery to the lymphatic system. *Crit. Rev. Ther. Drug Carrier Syst., 14,* 333–393.

Pradhan, R. S. & Vasavada, R. C. (1994). Formulation and in vitro release study on poly(DL-lactide) microspheres containing hydrophilic compounds—Glycine homopeptides. *J. Contr. Rel., 30,* 143–154.

Price, R., Poursaid, A., & Ghandehari, H. (2014). Controlled release from recombinant polymers. *J. Contr. Rel., 190,* 304–313.

Prieto, M. J. B., Delie, F., Fattal, E., Tartar, A., Puisieux, F., Gulik, A. et al. (1994). Characterization of V3 BRU peptide-loaded small PLGA microspheres prepared by a (w1/o)w2 emulsion solvent evaporation method. *Int. J. Pharmaceut., 111,* 137–145.

Ravis, W. R., Comerci, C., & Ganjam, V. K. (1986). Pharmacokinetics of insulin following intravenous and subcutaneous administration in canines. *Biopharm. Drug Dispos., 7,* 407–420.

Rhine, W. D., Hsieh, D. S. T., & Langer, R. (1980). Polymers for sustained macromolecule release: Procedures to fabricate reproducible delivery systems and control release kinetics. *J. Pharmaceut. Sci., 69,* 265–270.

Rios, M. (2004). Strategies for micro-and nanoparticle development. *Pharm. Technol., 28,* 40–53.

Sah, H. K. & Chien, Y. W. (1993). Evaluation of a microreservoir-type biodegradable microcapsule for controlled release of proteins. *Drug Dev. Ind. Pharm., 19,* 1243–1263.

Sah, H. K., Toddywala, R., & Chien, Y. W. (1994). The influence of biodegradable microcapsule formulations on the controlled release of a protein. *J. Contr. Rel., 30,* 201–211.

Sakthivel, T. & Florence, A. T. (2003). Dendrimer applications. *Drug Del. Technol., 3,* 50–55.

Sanchez, A., Vilajato, J. L., & Alonso, M. J. (1993). Development of biodegradable microspheres and nanospheres for the controlled release of cyclosporin-A. *Int. J. Pharmaceut., 99,* 263–273.

Sanders, L. M., Kell, B. A., McRae, G. I., & Whitehead, G. W. (1986). Prolonged controlled-release of nafarelin, a luteinizing hormone-releasing hormone analogue, from biodegradable polymeric implants: Influence of composition and molecular weight of polymer. *J. Pharmaceut. Sci., 75,* 356–360.

Sanders, L. M., Kent, J. S., McRae, G. I., Vickery, B. H., Tice, T. R., & Lewis, D. H. (1984). Controlled release of a luteinizing hormone releasing hormone analogue from poly(D,L-lactide-co-glycolide) microspheres. *J. Pharmaceut. Sci., 73,* 1294–1297.

Sanders, L. M., McRae, G., Vitale, K., Burns, R., Hoffman, P., & Shek, E. (1989). Design and performance of controlled release biodegradable delivery system of nafarelin: An overview. *J. Pharmaceut. Sci., 78,* 888–890.

Scheld, W. M. (1989). Drug delivery to the central nervous system: General principles and relevance to therapy for infections of the central nervous system. *Rev. Infect. Dis., 11,* S1669–S1690.

Schwendeman, S. P., Shah, R. B., Bailey, B. A., & Schwendeman, A. S. (2014). Injectable controlled release depots for large molecules. *J. Contr. Rel., 190,* 240–253.

Senderoff, R. I., Sheu, M.-T., & Sokoloski, T. D. (1991). Fibrin based drug delivery systems. *J. Parent. Sci. Tech., 45(1),* 2–6.

Shah, N. H., Railkar, A. S., Chen, F. C., Tarantino, R., Kumar, S., Murjani, M. et al. (1993). A biodegradable injectable implant for delivering micromolecules and macromolecules using poly(lactic-co-glycolic) acid (PLGA) copolymers. *J. Contr. Rel., 27,* 139–147.

Shenoy, B. C. (2002). Crystallization-based protein drug delivery, protein & peptide drug delivery. *IBCs Second International Conference*, Boston MA.

Sherwin, R. S., Kramer, K. J., Tobin, J. D., Insel, P. A., Liljenuist, J. E., Berman, M. et al. (1974). A model of the kinetics of insulin in man. *J. Clin. Invest., 53,* 1481–1492.

Shnek, D. R., Hostettler, D. L., Bell, M. A., Olinger, J. M., & Frank, B. H. (1998). Physical stress testing of insulin suspensions and solutions. *J. Pharmaceut. Sci., 87,* 1459–1465.

Sinha, V. R. & Trehan, A. (2003). Biodegradable microspheres for protein delivery. *J. Contr. Rel., 90,* 261–280.

Sonner, C., Maa, Y. F., & Lee, G. (2002). Spray-freeze-drying for protein powder preparation: Particle characterization and a case study with trypsinogen stability. *J. Pharmaceut. Sci., 91,* 2122–2139.

Supersaxo, A., Kou, J. H., Teitelbaum, P., & Maskiewicz, R. (1993). Preformed porous microspheres for controlled and pulsed release of macromolecules. *J. Contr. Rel., 23,* 157–164.

Tabata, Y., Gutta, S., & Langer, R. (1993). Controlled delivery systems for proteins using polyanhydride microspheres. *Pharmaceut. Res., 10,* 487–496.

Tabata, Y. & Ikada, Y. (1987). Macrophage activation through phagocytosis of muramyl dipeptide encapsulated in gelatin microspheres. *J. Pharm. Pharmacol., 39,* 698–704.

Tabata, Y. & Ikada, Y. (1989). Synthesis of gelatin microspheres containing interferon. *Pharmaceut. Res., 6,* 422–427.

Tabata, Y. & Langer, R. (1993). Polyanhydride microspheres that display near-constant release of water-soluble model drug compounds. *Pharmaceut. Res., 10,* 391–399.

Takakura, Y., Fujita, T., Furitsu, H., Nishikawa, M., Sezaki, H., & Hashida, M. (1994). Pharmacokinetics of succinylated proteins and dextran sulfate in mice—Implications for hepatic targeting of protein drugs by direct succinylation via scavenger receptors. *Int. J. Pharmaceut., 105,* 19–29.

Takakura, Y., Mihara, K., & Hashida, M. (1994). Control of the disposition profiles of proteins in the kidney via chemical modification. *J. Contr. Rel., 28,* 111–119.

Takenaka, H., Fujioka, K., & Takada, Y. (1989). New formulations of interferon. In D. Marshak & D. Liu (Eds.), *Therapeutic peptides and proteins: Formulation, delivery and targeting* (pp. 37–40). Cold Spring Harbor, NY: Cold Spring Harbor Laboratory.

Takeyama, M., Ishida, T., Kokubu, N., Komada, F., Iwakawa, S., Okumura, K. et al. (1991). Enhanced bioavailability of subcutaneously injected insulin by pretreatment with ointment containing protease inhibitors. *Pharmaceut. Res., 8,* 60–64.

Talmadge, J. E. (1993). The pharmaceutics and delivery of therapeutic polypeptides and proteins. *Adv. Drug Del. Rev., 10,* 247–299.

Tan, A. C. I. T. L., Russel, F. G. M., Thien, T., & Benraad, T. J. (1993). Atrial natriuretic peptide. An overview of clinical pharmacology and pharmacokinetics. *Clin. Pharmacokinet., 24,* 28–45.

Tanswell, P., Heinzel, G., Greischel, A., & Krause, J. (1990). Nonlinear pharmacokinetics of tissue-type plasminogen activator in three animal species and isolated perfused rat liver. *J. Pharmacol. Exp. Ther., 255,* 318–324.

Thies, C. & Bissery, M. (1984). Biodegradable microspheres for parenteral administration. In F. Lim (Ed.), *Biomedical applications of microencapsulation* (pp. 53–74). Boca Raton, FL: CRC Press, Inc.

Thomas, J. (1986). Optimum administration of insulin: A complex story. *Aust. J. Pharm., 67,* 745–746.

Tice, T. R. & Tabibi, S. E. (1992). Parenteral drug delivery: Injectables. In A. Kydonieus (Ed.), *Treatise on controlled drug delivery* (pp. 315–339). New York: Marcel Dekker, Inc.

Tipton, A. J. (2003). Sucrose acetate isobutyrate (SAIB) for parenteral delivery. In M. J. Rathbone, J. Hadgraft, & M. S. Roberts (Eds.), *Modified-release drug delivery technology* (pp. 679–687). New York: Marcel Dekker, Inc.

Tomlinson, E. & Livingstone, C. (1989). Therapeutic peptides and proteins. *Pharm. J., 243,* 646–648.

Tsai, T. M., Mehta, R. C., & DeLuca, P. P. (1996). Adsorption of peptides to poly(D,L-lactide-co-glycolide). 1. Effect of physical factors on the adsorption. *Int. J. Pharmaceut., 127,* 31–42.

Uludag, H., Kharlip, L., & Sefton, M. V. (1993). Protein delivery by microencapsulated cells. *Adv. Drug Deliv. Rev., 10,* 115–130.

Van, H. K., Catry, C., & Vanhaecke, F. (2013). Determination of elemental impurities in leachate solutions from syringes using sector field ICP-mass spectrometry. *J. Pharm. Biomed. Anal., 77,* 139–144.

Verity, N. (2003). Continuous targeted delivery of therapeutic proteins. Protein and peptide drug delivery: Scientific advances in enabling novel approaches & improved products. *IBC Meeting*, Philadelphia, PA, September 22–23.

Vickery, B. H. (1991). Biological actions of synthetic analogs of luteinizing hormone- releasing hormone. In P. D. Garzone, W. A. Colburn, & M. Mokotoff (Eds.), *Pharmacokinetics and pharmacodynamics. Vol. 3: Peptides, peptoids and proteins* (pp. 41–49). Cincinnati, OH: Harvey Whitney Books Co.

Vugmeyster, Y., Xu, X., Theil, F. P., Khawli, L. A., & Leach, M. W. (2012). Pharmacokinetics and toxicology of therapeutic proteins: Advances and challenges. *World J. Biol. Chem., 3,* 73–92.

Wang, P. L. & Johnston, T. P. (1995). Sustained-release interleukin-2 following intramuscular injection in rats. *Int. J. Pharmaceut., 113,* 73–81.

Wang, T., D'Souza, G. G., Bedi, D., Fagbohun, O. A., Potturi, L. P., Papahadjopoulos-Sternberg, B. et al. (2010). Enhanced binding and killing of target tumor cells by drug-loaded liposomes modified with tumor-specific phage fusion coat protein. *Nanomedicine (Lond), 5,* 563–574.

Wang, Y. S., Youngster, S., Grace, M., Bausch, J., Bordens, R., & Wyss, D. F. (2002). Structural and biological characterization of pegylated recombinant interferon alpha-2b and its therapeutic implications. *Adv. Drug Deliv. Rev., 54,* 547–570.

Wazny, L. D., Raymond, C. B., Do, M. K., & Skwarchuk, D. E. (2009). Reduced drug costs from switching hemodialysis patients from epoetin alfa in multidose vials to pre-filled syringes. *CANNT J., 19,* 39–41.

Weiner, A. L. (1990). Lipid-based vehicles for peptide and protein drugs: Part II: Manufacturing variables. *Biopharm, 3(4),* 16–21.

Weiner, A. L., Carpenter-Green, S. S., Soehngen, E. C., Lenk, R. P., & Popescu, M. C. (1985). Liposome-collagen gel matrix: A novel sustained drug delivery system. *J. Pharmaceut. Sci., 74,* 922–925.

Wills, R. J. (1990). Clinical pharmacokinetics of interferons. *Clin. Pharmacokinet., 19,* 390–399.

Wills, R. J. & Ferraiolo, B. L. (1992). The role of pharmacokinetics in the development of biotechnologically derived agents. *Clin. Pharmacokinet., 23,* 406–414.

Woo, B. H., Kostanski, J. W., Gebrekidan, S., Dani, B. A., Thanoo, B. C., & DeLuca, P. P. (2001). Preparation, characterization and in vivo evaluation of 120-day poly(D,L-lactide) leuprolide microspheres. *J. Contr. Rel., 75,* 307–315.

Woodle, M. C., Storm, G., Newman, M. S., Jekot, J. J., Collins, L. R., Martin, F. J. et al. (1992). Prolonged systemic delivery of peptide drugs by long-circulating liposomes: Illustration with vasopressin in the brattleboro rat. *Pharmaceut. Res., 9,* 260–265.

Working, P. K. (1992). Potential effects of antibody induction by protein drugs. In B. L. Ferraiolo, M. A. Mohler, & C. A. Gloff (Eds.), *Pharmaceutical biotechnology. Vol. 1: Protein pharmacokinetics and metabolism* (pp. 73–92). New York: Plenum Press.

Xu, Z., Davis, H. M., & Zhou, H. (2013). Rational development and utilization of antibody-based therapeutic proteins in pediatrics. *Pharmacol. Ther., 137,* 225–247.

Yamakawa, I., Tsushima, Y., Machida, R., & Watanabe, S. (1992). Preparation of neurotensin analogue-containing poly(DL-lactic acid) microspheres formed by oil-in-water solvent evaporation. *J. Pharmaceut. Sci., 81,* 899–903.

Yang, C., Hillas, P. J., Baez, J. A., Nokelainen, M., Balan, J., Tang, J. et al. (2004). The application of recombinant human collagen in tissue engineering. *BioDrugs, 18,* 103–119.

Yang, J. & Cleland, J. L. (1997). Factors affecting the in vitro release of recombinant human interferon-gamma (rhIFN-gamma) from PLGA microspheres. *J. Pharmaceut. Sci., 86,* 908–914.

Yi, X., Manickam, D. S., Brynskikh, A., & Kabanov, A. V. (2014). Agile delivery of protein therapeutics to CNS. *J. Contr. Rel., 190,* 637–663.

Yoon, E. J., Bae, E. J., Jeong, Y. N., Kim, M. M., Kim, B. S., Lee, S. H. et al. (1993). Dose-dependent pharmacokinetics of human granulocyte/macrophage colony-stimulating factor in rabbits. *Int. J. Pharmaceut., 97,* 213–218.

Yoshida, R., Sakai, K., Okano, T., & Sakurai, Y. (1993). Pulsatile drug delivery systems using hydrogels. *Adv. Drug Deliv. Rev., 11,* 85–108.

Zlokovic, B. V. (1995). Cerebrovascular permeability to peptides: Manipulations of transport systems at the blood-brain barrier. *Pharmaceut. Res., 12,* 1395–1406.

Zuidema, J., Kadir, F., Titulaer, H. A. C., & Oussoren, C. (1994). Release and absorption rates of intramuscularly and subcutaneously injected pharmaceuticals (II). *Int. J. Pharmaceut., 105,* 189–207.

7 Oral Delivery of Peptide and Protein Drugs

7.1 INTRODUCTION

The gastrointestinal (GI) tract, being rich in proteolytic enzymes, digests therapeutic proteins with the same efficiency as it digests dietary proteins. The odds against peptide and protein drug delivery by the oral route are formidable. This, however, has not discouraged researchers from attempting oral delivery of peptides and proteins (Choonara et al., 2014). The oral route is the most popular and convenient route of delivery for traditional drugs (Lipka, Crison, & Amidon, 1996; Wang, 1996). Tremendous effort has thus been directed toward oral delivery of peptide and protein drugs, and some promising results have begun to appear. Broadly speaking, oral delivery refers to absorption from the buccal through the rectal mucosa. However, buccal and rectal delivery will be discussed separately in Chapter 9. In this chapter, the focus is on absorption from the small and large intestine and on strategies to avoid drug release in stomach following oral intake. The enzymatic and penetration barriers to oral delivery will be briefly reviewed followed by a discussion of the approaches to overcome or minimize these barriers.

The intestine is a tubelike structure that progresses from the duodenum to the rectum via the jejunum, ileum, ascending colon, transverse colon, and descending colon. The walls of the small intestine have several folds and are lined with millions of tiny fingerlike projections called villi. There are about 20–40 villi/mm^2 of mucosa, and each villus is about 0.5–1 mm long. The villus is covered by a single layer of columnar epithelium. The free edges of the cells of the epithelium of the villi are divided into minute microvilli, which form a brush border. The villi, microvilli, and the folds of the small intestine collectively increase the absorption surface by about 600-fold. The fully differentiated, polarized cells located at the upper two-thirds of the villi are called villus cells. Each villus core contains capillary blood vessels and a terminal branch of the lymphatic plexus, the central lacteal. The intestinal absorptive cell or *enterocyte* is not a uniform cell type; its phenotype changes during intestinal development of the fetus and neonate and from one site to another. Peyer's patches exist in the form of white spots in the duodenum, jejunum, and ileum and contain special cells known as M-cells. The M-cells constitute the major cell type in Peyer's patch. These cells can take up macromolecules (e.g., antigens) through pinocytosis and deliver them to lymphocytes. The GI tract is lined with a layer of mucus that provides a barrier to acid in the stomach. The mucus contains glycoproteins and bicarbonate ions and forms a viscoelastic gel that resists the autodigestion of the GI tract. A slow baseline secretion of mucus is maintained by exocytosis from goblet cells in the GI tract (Macadam, 1993). A space (unstirred water layer) exists between the mucus itself and the apical surface of the epithelial cell monolayer. This monolayer rests upon a noncellular matrix termed the basement membrane. Underneath the basement membrane lies the submucosal space, which has blood vessels, nerves, and lymphatic ducts (Mrsny, 1991). It should be realized that a drug must first diffuse across the mucus layer before absorption across the epithelia is possible. The aqueous boundary or unstirred water layer can be rate limiting for highly lipophilic compounds. It is less likely to be a barrier to the hydrophilic peptides and proteins.

As discussed, the center of each villus contains a lymphatic vessel. Once a protein crosses the monolayer of intestinal epithelial cells, it can enter either the capillaries of the portal venous system or the lymphatic lacteal (Porter, 1997). The relative proportion of drug uptake by these two routes depends on the drug and its formulation. A more lipophilic drug or a formulation containing lipid components will be more likely to be absorbed by lymphatic route. Uptake of cytokines and lymphokines by lymphatic route will allow them to target white cell populations in the lymph system (Mrsny, Cromwell, Villagran, & Rubas, 1993). Also, the lymphatic circulation bypasses the liver and thus offers an attractive approach to delivery of peptides and proteins. Absorption into the lymphatic lacteals provides a very slow systemic delivery over several hours as the lymph moves at a slow rate of only about 1–2 mL/min. In contrast, absorption into the portal venous system results in rapid delivery within minutes to systemic circulation after an initial hepatic pass. As can be expected, the liver will result in a significant metabolic breakdown. For example, the oral bioavailability of the hexapeptide, His-D-Trp-Ala-Trp-D-Phe-Lys-NH_2, is less than 1% in both animals and man due to significant first-pass elimination (Smith et al., 1994). Similarly, the pentapeptide, Tyr-D-Ala-Gly-Phe-*N*-Me-Met-NH_2, has been reported to be extracted by a hepatic first-pass effect, thus limiting its concentration in the systemic circulation following colonic administration. In the concentration range investigated, there was no indication of a saturability of the hepatic extraction, indicating that this elimination pathway is very significant (Langguth, Merkle, Stockli, & Wolffram, 1994).

7.2 BARRIERS TO PROTEIN ABSORPTION AND PATHWAYS OF PENETRATION

Barriers to protein absorption include the extracellular and cellular barriers. Although formidable barriers exist to the oral absorption of therapeutic peptides and proteins, the development of oral replacements for injectable peptides is a high-priority research area for the pharmaceutical industry and will lead to a wider acceptance of some of the biotechnology products by the public (Kleinert, Baker, & Stein, 1993; Mahato, Narang, Thoma, & Miller, 2003; Shah, Ahsan, & Khan, 2002). The oral bioavailability of most peptides and proteins is less than 1%. However, exceptions exist and these exceptions provide encouragement for further research. For example, cyclosporine, a lipophilic cyclic peptide with 11 amino acids used as an immunosuppressant, can have bioavailability of 50% or more when administered in suitable vehicles. Its commercial product, Sandimmune® (Novartis), is available as an oral solution or soft gelatin capsules for oral administration and also as a solution for parenteral administration. However, the absorption of cyclosporine during chronic administration of Sandimmune soft gelatin capsules and oral solution can be erratic, and patients need to be monitored for blood levels. Another product, also from Novartis, Neoral®, has an increased bioavailability as compared to Sandimmune. Neoral, which contains corn oil-mono-di-triglycerides and polyoxyl 40 hydrogenated castor oil, immediately forms a microemulsion in an aqueous environment (PDR Staff, 2014). The original product, Sandimmune, was designed as an emulsion preconcentrate, which results in an emulsion with dispersed droplets about 1 μm size. In contrast, the new product, Neoral, is a microemulsion preconcentrate, which results in dispersed droplets that are less than 100 nm in size. Other cyclosporine formulations are also available on the worldwide market, and bioavailability comparisons have been made (Andrysek, 2001). Arginine and lysine vasopressins (AVP and LVP) and the synthetic analogue 1-deamino-8-D-arginine vasopressin (DDAVP) are also active by oral administration (Saffran, Bedra, Kumar, & Neckers, 1988; Vilhardt & Bie, 1983). Inhibition of proteolytic enzymes and opening of tight junctions to allow for paracellular transport have been reported to increase the intestinal absorption of peptides (Guo et al., 2004).

7.2.1 Extracellular Barriers

Peptide and protein absorption through the oral route is faced with several enzymatic and penetration barriers. Hydrolysis of peptides and proteins in the GI tract can occur in the lumen, at the brush border, or intracellularly at the cytosol of the enterocyte (Langguth et al., 1997). As the protein is ingested, it reaches the stomach where it is acted upon by gastric juice in a very acidic environment. Gastric juice contains a family of aspartic proteinases called pepsins, which are most active at pH 2–3. Pepsins break the protein into a mixture of polypeptides, which then move down to the duodenum. As the protein enters the duodenum, the pH rises to about 6.0–8.0 (Mrsny, 1991). This sharp pH change from about 2 to about 8 can also cause precipitation of the protein as it may pass through its isoelectric point and may not then easily redissolve (Mrsny, 1991). In the duodenum, the polypeptides are acted upon by pancreatic proteases. These proteases may be endopeptidases like trypsin, chymotrypsin, and elastase or exopeptidases like carboxypeptidase A. These enzymes degrade the polypeptides into smaller peptides. These peptides are then further degraded by the proteases in the brush border and the cytosol of the enterocyte. For peptides having four or more residues, more than 90% of the proteolytic activity is in the brush-border membrane. For tripeptides, the activity is 10%–60%, whereas for dipeptides, it is only about 10%. Lysosomes and other cell organelles can also act as potential sites of peptide and protein degradation (Bai & Amidon, 1992; Lee, 1988; Lee, Dodda-Kashi, Grass, & Rubas, 1991; Sasaki et al., 1994; Wearley, 1991). The brush border contains exopeptidases that act at the N-terminal end of the protein (aminopeptidases). Aminopeptidases present in the intestine include leucine aminopeptidase, aminopeptidase N, aminopeptidase A, and aminopeptidase B. The aminopeptidase activity in Peyer's patches of the jejunum and ileum is only about 20%–30% of that in the neighboring patch-free areas (Hayakawa & Lee, 1992). Knowledge of the distribution of brush-border membrane peptidases along the intestine helps to predict the preferential uptake of peptides and proteins from various intestinal regions (Bai, 1993, 1994a). Since this distribution may vary for different enzymes, site-specific oral delivery may be dependent on the amino acid sequence of the peptide. This is due to the substrate specificity of the brush-border membrane peptidases. Also, the activities of brush-border membrane peptidases may be controlled by the surface pH of mucosal cells, rather than the luminal pH (Bai, 1993). Finally, it should be realized that several enzymes that act on carbohydrates may also affect the drug if it is a glycoprotein.

7.2.2 Cellular Barriers

Peptides and proteins can move across the epithelia by two routes: transcellular or paracellular (Figure 7.1). The transcellular route involves intracellular transfer from the apical to the basolateral surface of an individual epithelial cell. This transport can take place either through specific uptake mechanisms of the cell or through sequential partitioning events from aqueous to lipid to aqueous environments. The transcellular route is important for the uptake of lipophilic drugs by sequential partitioning events. Sugars and amino acids are known to be transported by carrier-mediated processes. While L-amino acids are absorbed by an active transport mechanism, D-amino acids are absorbed by passive diffusion alone. There are four separate transport systems for amino acids, while a separate system transports dipeptides and tripeptides into the mucosal cells. These carrier systems can even transport amino acid–type or peptide-type drugs, such as β-lactam antibiotics, angiotensin-converting enzyme (ACE) inhibitors, and α-methyldopa (Bai & Amidon, 1992; Yuasa, Amidon, & Fleisher, 1993).

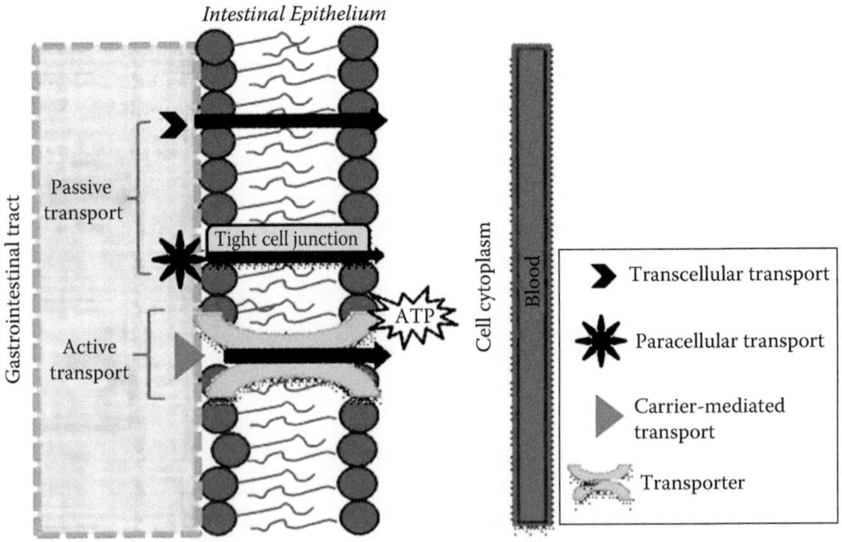

FIGURE 7.1 **(See color insert.)** Intestinal drug transport mechanisms. (Reproduced from Renukuntla, J. et al., *Int. J. Pharm.*, 447, 75, 2013. With permission.)

P-glycoprotein (P-gp) transporter associated with multidrug resistance in tumor cells may also be implicated in oral delivery of peptides that are P-gp substrates, and overlapping substrates for different enzymes may result in some drugs affecting the oral bioavailability of certain proteins (Bruno, Miller, & Lim, 2013; Mahato et al., 2003). Active transport mechanisms for transcellular route include Na^+-coupled glucose transport and H^+-coupled dipeptide transport. Studies on the kinetics of glycyl-proline transport in intestinal brush-border vesicles have shown that dipeptide transport is a saturable process and follows Michaelis–Menten kinetics (Ganapathy & Leibach, 1982). Thwaites, Hirst, and Simmons (1993), Thwaites, Simmons, and Hirst (1993) have investigated whether thyrotropin-releasing hormone (TRH; pGlu-His-Pro-NH_2) is absorbed by a carrier-mediated process. Using brush-border membrane vesicles (BBMVs) prepared from rabbit duodenum and jejunum and rat upper small intestine, no evidence was found for the oral absorption of TRH by an Na^+- or H^+-dependent carrier system in the brush-border membrane. Instead, it appears that the TRH absorption in vivo may be accounted for by passive absorption of the peptide across a paracellular route and its resistance to luminal hydrolysis.

The paracellular route involves transfer between adjacent cells. The villus cells have tight intracellular junctions, which prevent paracellular transport of solutes. Movement through this route is limited by the junctional complex to molecules with a radius less than about 8 Å (Mrsny et al., 1993). If a peptide can pass through the paracellular route, it will not be subject to degradation by the intracellular proteases. While the paracellular route may be preferable for this reason, structural features of peptides that may encourage their paracellular transport are not well understood. The partition coefficient between n-octanol and water may be one predictor, and drugs with a log P of less than zero (i.e., hydrophilic molecules such as most peptides and proteins) are more likely to follow a paracellular pathway. Peptides following a paracellular pathway may also be more affected by penetration enhancers such as zonula occludens toxin (ZOT), a protein from *Vibrio cholerae* that can reversibly open tight junctions between intestinal cells (Fasano, 1998; Mahato et al., 2003). Paracellular transport is constrained by the physical properties of the permeant. Larger peptides (greater than three amino acids) are usually absorbed in small amounts by passive diffusion via the paracellular route. While larger proteins are not typically absorbed in the GI tract, protein antigens can be taken up by M-cells, which are specialized intestinal epithelial cells that overlie aggregates of lymphoid tissue (Peyer's patches). In a study by Yen and Lee (1994), it has been shown that a pentapeptide, 4-phenylazobenzyloxy-carbonyl-L-Pro-L-Leu-Gly-L-Pro-D-Arg (Pz-peptide), penetrated the colonic and intestinal segments of the albino rabbit via the paracellular route. The Pz-peptide, when incubated with the homogenates and subcellular fractions of the various intestinal segments, underwent substantial degradation. However, it penetrated all intestinal segments with more than 80% of the peptide in the intact form. This suggests that the peptide permeated through a paracellular route and thus avoided intracellular proteolysis. It is suggested that the Pz-peptide can enhance the tight junctional permeability by a yet unknown mechanism (Yen & Lee, 1994). It has been suggested that nonpolar compounds with a molecular weight (MW) of 3–4 kDa can be absorbed orally to a significant extent, provided that they are stable and have the desired partition coefficient (Lipka et al., 1996). The tight junctions for paracellular diffusion have been reported to be generally impermeable to molecules with a radii of more than 11–15 Å, which may represent the limit for hydrodynamic radius for oral delivery of spherical rigid molecules. However, peptides have some conformational flexibility, and even larger molecules can perhaps permeate the tight junctions (Pauletti et al., 1996).

7.3 APPROACHES TO IMPROVE ORAL DELIVERY

Various approaches are being investigated for oral delivery of proteins (Lee, 2002; Morishita & Peppas, 2006; Renukuntla, Vadlapudi, Patel, Boddu, & Mitra, 2013). Broadly speaking, these approaches may modify the drug or its delivery system or the oral mucosa (Al-Hilal, Alam, & Byun, 2013). Some of the approaches include site-specific drug delivery, chemical modification, bioadhesive polymers, penetration enhancers, protease inhibitors, carrier systems, and the use of novel formulation approaches, and these approaches are all discussed in the following. Multifunctional matrices that may offer protection against almost all the barriers at the same time are also being investigated. These matrices are based on polymers that exhibit mucoadhesive properties, a penetration-enhancing effect, and protease inhibitor properties. In addition, they may provide sustained delivery. Examples include noncovalent binding polymers such as the anionic (e.g., alginate, polyacrylic acid, chitosan-EDTA, or sodium carboxymethyl cellulose), cationic (e.g., chitosans), or nonionic mucoadhesive polymers. Also, covalent-binding polymers such as the thiolated polymers or thiomers can be used. Thiomers are formed by the immobilization of thiol-bearing ligands to mucoadhesive polymeric excipients, and they improve mucoadhesive properties by over 130-fold due to the formation of disulfide bonds with mucus glycoproteins (Bernkop-Schnurch & Walker, 2001). Cell-penetrating peptides (see Section 8.2.4) have also been investigated to enhance the oral delivery of peptides and proteins (Khafagy & Morishita, 2012). Significant work has been reported to investigate oral delivery of insulin (Rekha & Sharma, 2013). However, despite considerable efforts, no major breakthrough has been achieved for oral delivery of proteins. For proteins that may have some oral absorption, bioavailability is typically less than 1%–2% (Park, Kwon, & Park, 2011; Renukuntla et al., 2013).

7.3.1 Site-Specific Drug Delivery

Absorption of peptides and proteins from different regions of the intestine is not uniform. For example, the preferred absorption site is the duodenum for cyclosporine, the upper GI tract and rectum for tetragastrin, the duodenum and the ileocecal junction for desmopressin, and the upper GI tract for octreotide (Lee, 1992; Pantzar, Lundin, & Westrom, 1995). Oxytocin and vasopressin analogues have a higher transport rate across distal intestinal segments than across proximal small intestine of rat. This is believed to be due to a decreased proteolytic activity in the distal region. Also, the distal region has a higher paracellular permeability despite a decreased absorption area (Lundin, Pantzar, Broeders, Ohlin, & Westrom, 1991). The protease activity in the cytosol does not show regional variation, but the same is not true for the brush border or for the luminal fluid. The stomach, with its low pH and enzymatic activity, presents very harsh conditions for a protein drug. A typical approach to prevent dissolution of a dosage form in the stomach is to use enteric coating.

The apparent permeability of insulin from rat intestine shows a site-dependent variation as measured by everted rat gut sac technique. The permeability was significantly greater in the jejunum and ileum than in the duodenum. In these in vitro experiments, insulin was remarkably stable. This suggests that insulin metabolism at the brush border is not significant. However, insulin was metabolized almost completely in intestinal homogenates. Thus, it appears that degradation of insulin under an in vivo situation would be due to luminal and cytosolic enzymes (Schilling & Mitra, 1990). In situ experiments have shown that the absolute bioavailability for insulin was higher when administered to the more distal region of the rat intestine (0.133%) than that absorbed from a more proximal region (0.059%) of the intestine (Schilling & Mitra, 1992). Absorption of insulin from isobutyl cyanoacrylate nanocapsules administered to diabetic rats was dependent on the site of absorption. The hypoglycemic effect following absorption from various sites was ileum (65%), stomach (59%), duodenum and jejunum (52%), and colon (34%). The insulin was protected by the nanocapsules in this study, and the hypoglycemic effect, which started on the second day, lasted for 3–18 days (Michel, Aprahamian, Defontaine, Couvreur, & Damge, 1991).

The intestinal segments have progressively fewer and smaller villi in the more distal sections. This leads to a progressively reducing surface area, with the colon having the lowest surface area for a particular length. The colon also has a variable pH and the presence of solid fecal matter that may interfere with drug absorption. However, the colon has a relatively low enzymatic activity and is promising in this regard. Using isolated luminal enzymes and studies in intact mucosa, calcitonin was found to degrade much more in small intestine as compared to the colon (Lu, Kopeckova, & Kopecek, 1999). The colon has a high population of bacteria, largely anaerobic species. This fact has been exploited for an ingenious approach to target peptides and proteins to the colon (Ashford & Fell, 1994). In a study by Saffran et al. (1986), polypeptides such as insulin or vasopressin were coated with polymers cross-linked with azoaromatic groups to protect orally administered polypeptides from digestion in the stomach and small intestine of rats. Once the polypeptide reached the colon, the indigenous microflora reduced the azo bonds, thus breaking the cross-links and releasing the polypeptide for absorption. More recently, freeze-dried plant cells are being investigated as a platform for oral delivery of proteins and vaccines. They release the protein/vaccine when plant cell walls are digested by microbes that colonize the gut (Kwon, Verma, Singh, Herzog, & Daniell, 2013).

The upper half of the large intestine is drained by hepatic portal veins, while the lower half is drained by lymphatics. If a polypeptide is destroyed in the liver, it may be possible to adjust the thickness or composition of coating so that the drug is released in the lower colon, where it will bypass the hepatic veins. Delivery of insulin and an absorption promoter to the colon has also been attempted using a soft gelatin capsule coated with a polyacrylic polymer (Eudragit) having pH-dependent properties (Touitou & Rubinstein, 1986). Delivery of an insulin-like growth factor I (IGF-I) to rat and minipig colonic mucosa under in vitro conditions has been investigated. IGF-I is a 7649 Da protein of 70 amino acids that exerts its biological actions through specific IGF-I receptors. It has been found useful to lower blood glucose levels in insulin-resistant diabetic patients in clinical studies. IGF-I was absorbed intact across rat colonic mucosa as determined by RP-HPLC, SDS-PAGE, and Western blotting (Quadros, Landzert, LeRoy, Gasparani, & Worosila, 1994). A time-based drug release system for colon-specific delivery has also been developed. This system exploits the relatively constant small intestine transit time of dosage forms (Gazzaniga, Iamartino, Maffione, & Sangalli, 1994). Time-based systems can be designed to release their drug after a predetermined lag time, with the lag time being independent of normal physiological conditions such as pH, digestive state of the subject, and anatomical position at the time of release (Pozzi, Furlani, Gazzaniga, Davis, & Wilding, 1994; Wilding, Davis, Pozzi, Furlani, & Gazzaniga, 1994).

7.3.2 Chemical Modification

Biocon (Bangalore, India), which has acquired the intellectual property rights of the former Nobex Corporation (North Carolina), has used an approach to attach low-MW polymers at specific sites on proteins to create a conjugated molecule with improved oral absorption. The carrier molecules used reversibly destabilize the native protein, favoring a partially unfolded conformation. This noncovalent interaction between the carrier and partially unfolded protein conformation increases solvent exposure of hydrophobic side chains, thereby increasing lipid solubility and oral absorption of the protein by a passive and transcellular route (Milstein et al., 1998). Nobex and Biocon have developed a hexyl insulin monoconjugate 2 (HIM2) in which a single amphiphilic oligomer is covalently linked to the free amino group on the Lys-B29 residue of insulin via an amide bond (Figure 7.2). Lysine residue was an important site for enzymatic attack, so the conjugate has improved stability, and also the conjugate is more soluble in water, and about 30%–35% of the insulin surface is covered by the polymer (Mahato et al., 2003; Park et al., 2011; Price, 2003). Conjugated calcitonin is resistant to chymotrypsin degradation, and the stability of calcitonin tablets made from the conjugate is good (Price, 2003). A recent clinical study has suggested that oral modified insulin, when added to a basal insulin regimen, was safe and may prove to be effective in controlling postprandial hyperglycemia in patients with type I diabetes mellitus (Clement, Dandona, Still, & Kosutic, 2004).

Generally, a chemical modification approach is more applicable to peptides than to proteins because of the structural complexity of proteins. A peptide can be chemically modified to improve its enzymatic stability and/or membrane permeation. For example, substitution of D-amino acids for L-amino acids in the primary structure may improve the enzymatic stability of the peptide. An example of an instance where chemical modification of a peptide results in increased enzymatic stability without affecting its membrane permeability is the various analogues of the naturally occurring pentapeptide, met-enkephalin. Metabolism of these analogues by BBMV shows large differences in degradation rates, while they all have similar effective permeabilities across Caco-2 cells (Bohner, Langguth, Biber, & Merkle, 1994).

Another clue to chemical modification comes from the fact that a lipophilic peptide, cyclosporin A, is readily absorbed from the GI tract. Thus, efforts have been directed toward imparting lipid solubility to peptides by bonding an acyl group of a fatty acid to an amino terminus of the peptide. Using a range from tripeptides to proteins, lipid solubility was achieved for TRH, tetragastrin, insulin, and lysozyme. These new derivatives maintained their biological activity and had increased absorption from the intestine (Muranishi, 1991). Another approach to increase lipophilicity can be cyclization, which will remove charged N- and C-terminal groups, reducing overall solvent accessible surface area of the molecule. Also, more lipophilic synthetic amino acids such as *t*-butylglycine, β-naphthylalanine, and *p*-phenylphenylalanine can be used to synthesize peptide analogues, provided that biological activity is not lost. The use of a conjugate system, which combines structural features of

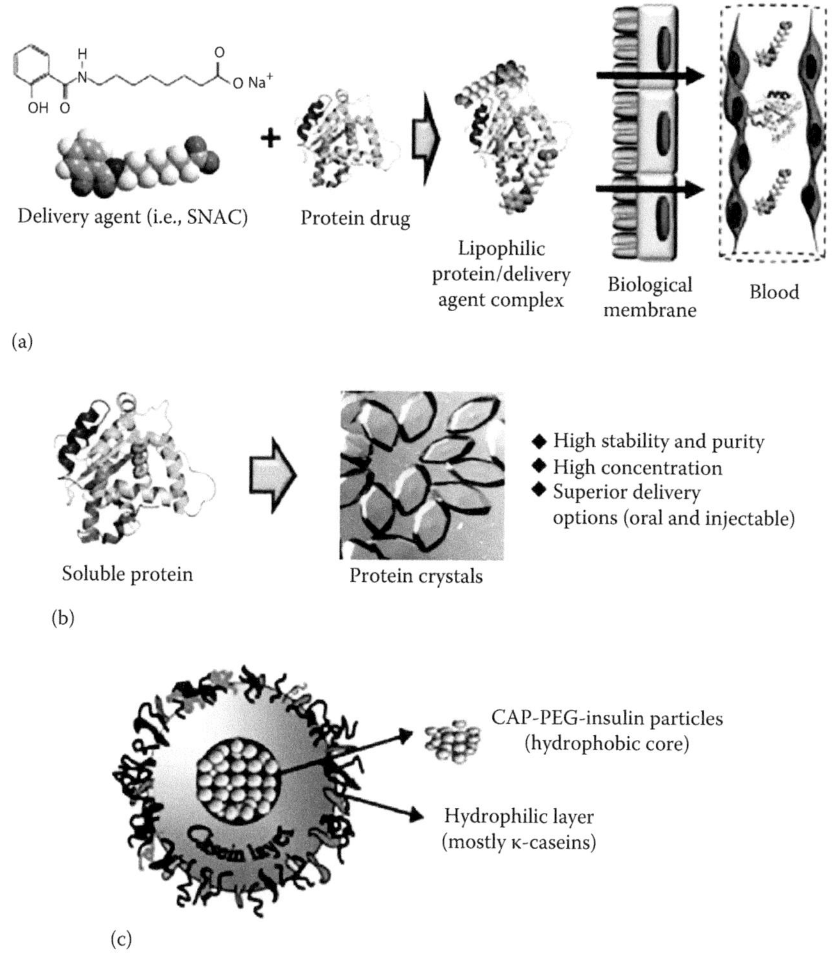

FIGURE 7.2 (See color insert.) Examples of oral protein delivery technologies developed for clinical applications. (a) Emisphere: Eligen™ technology. (b) Altus: Cross-linked enzyme crystal (CLEC®) technology. (c) BioSante: BioOral system (calcium phosphate nanoparticles).

(Continued)

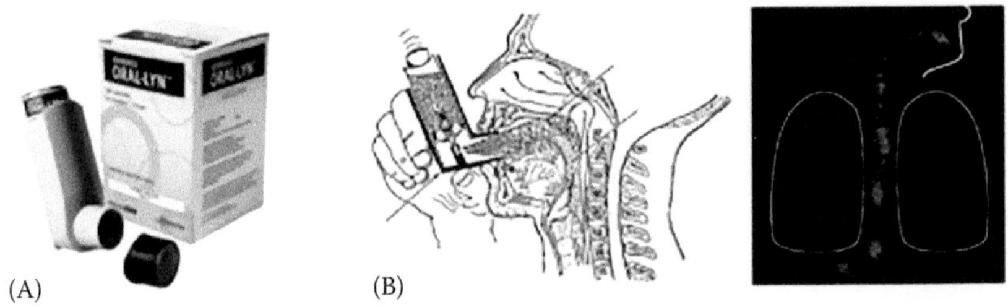

(A) (B)

(d)

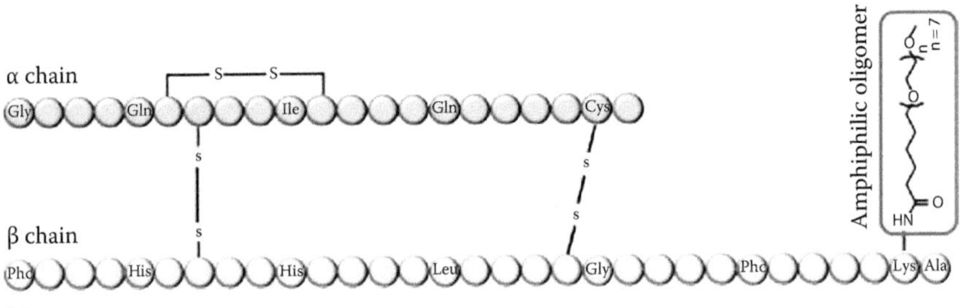

(e)

FIGURE 7.2 (Continued) **(See color insert.)** Examples of oral protein delivery technologies developed for clinical applications. (d) Generex: Oral-lyn™ (A) the RapidMist™ device and (B) the formulation is delivered to the oral cavity for absorption via the oral mucosa. (e) NOBEX and Biocon: Amphiphilic oligomers. (Reproduced from Park, K. et al., *React. Funct. Polym.*, 71, 280, 2011. With permission.)

lipids with those of amino acids and peptides, is likely to provide a high degree of membrane-like character for the conjugate, which may allow its passage across membranes (Toth, 1994). In one study, the enzymatically labile peptides, luteinizing hormone–releasing hormone (LHRH) and TRH, were conjugated to various lipidic peptides. The lipidic peptides were oligomers of lipidic amino acids and thus combined the structural features of lipids with those of amino acids. Because of their bifunctional nature, these lipidic peptides could be conjugated to LHRH or TRH. Conjugation to these lipidic peptides increased the half-life of LHRH and TRH during incubation with Caco-2 cell homogenates. The half-life of unconjugated LHRH was about 5 min, and it increased to about 45 min when conjugated to a lipidic amino acid. When conjugated to two lipidic amino acids, the half-life increased further to 360 min. Enzymatic activity of the Caco-2 cell homogenates released the parent peptide from the conjugate. The released LHRH or TRH demonstrated a longer half-life than when present alone in the homogenate, possibly because the cleaved lipidic peptide acted as a protease inhibitor due to its long alkyl side chains (Toth, Flinn, Hillery, Gibbons, & Artursson, 1994). Chemical modification of salmon calcitonin has also been done to make its oral absorption feasible. In a study by Devogelaer et al. (1994), six patients suffering from active Paget's disease of bone were administered a highly modified calcitonin analog, ASC 710. An oral capsule of 20 mg of

ASC 710 provoked similar biological modifications as those achieved after a single s.c. injection of 50 IU of salmon calcitonin or a s.c. injection of 0.2 mg of ASC 710. Another study modified salmon calcitonin by a new method to prepare fatty acid–polypeptide conjugates that can be carried out in aqueous solutions and can regenerate the original active polypeptide in tissues or blood. Using this reversible aqueous lipidization approach, the AUC of modified calcitonin delivered orally was about 20 times higher than that of unmodified calcitonin (Wang, Chow, Heiati, & Shen, 2003). PEGylated proteins (see Section 6.5.1) may also have a potential for oral delivery. PEGylation of recombinant human granulocyte colony-stimulating factor (rhG-CSF) has been reported to increase its stability and in vivo bioactivity when administered by the intraduodenal route. Its bioavailability by the enteral route was 1.8%–3.5%, while the unmodified protein did not produce any quantifiable response (Jensen-Pippo, Whitcomb, DePrince, Ralph, & Habberfield, 1996).

Chemical modification does not always lead to improved oral absorption. Diacyl derivatives of insulin exhibit a higher proteolysis than native insulin in the small intestine of rat under in vitro conditions. This was because insulin association was inhibited by diacylation, making more monomers available for proteolysis (Asada et al., 1994). The structural requirements for intestinal absorption of peptide drugs have been reviewed (Pauletti et al., 1996). Barlow and Satoh (1994) have conducted a series of elementary analysis to define the basic design features for a potent, specific, and absorbable peptide drug. Recognition of the peptide by its target receptor seems to need about 4–6 amino acid residues, and the rest of the structure may be *redundant* for bioactivity. The resulting structure is still too big for transport by paracellular transport. Also, peptide transporters for molecules larger than the three residues are unlikely to exist so that active transport is also not feasible. For transport by simple diffusion, the lipophilicity of the peptide needs to be increased. Based on molecular modeling, it was predicted that an active absorbable peptide should have a total surface area of around 350 Å^2, of which the polar surface area should be 50 Å^2. This could be attempted by methylating the peptide NH groups, eliminating charged termini, and/or cyclization of the molecule so that peptide CO and NH groups are made inaccessible to solvent due to intramolecular hydrogen bonding. As the authors discussed, all these changes will result in a peptide drug that is *not a peptide at all* (Barlow & Satoh, 1994). Computer simulations to design peptides or to predict their oral absorption may also be possible. A theoretical analysis to estimate the extent of peptide absorption has been developed on the basis of a mass balance approach. Using this analysis, simulations showed that the intestinal absorption of insulin is approximately 1% of the administered dose (Sinko, Leesman, & Amidon, 1993). Since some peptide transporters are known to exist in the intestinal mucosa, knowledge of its structure will lead to rational design of peptide mimetics having affinity for this receptor. However, passive diffusion will still be of primary interest since the utilization of the transport will be limited because of substrate specificity. Structural features of peptides to achieve drug delivery are not necessarily the same as those required for bioactivity. Therefore, a collaborative effort by a multidisciplinary team is required for the rational design of peptide mimetics with adequate oral absorption (Borchardt, 1994).

7.3.3 Bioadhesive Drug Delivery Systems

Bioadhesive drug delivery systems have been widely investigated to prepare oral sustained release preparations for conventional dosage forms (Duchene, Touchard, & Peppas, 1988; Park & Robinson, 1984; Plate, Valuev, Sytov, & Valuev, 2002). Such systems may anchor the drug-containing delivery system to the GI tract. This increases the overall time for drug absorption as the delivery system will not be dependent on the GI transit time for removal. Also, the drug will not have to diffuse through luminal contents or mucus layer in order to reach the mucosal epithelium. Because of intimate contact with the mucosa, a high drug concentration is presented for absorption. Also, site-specific delivery may be possible if bioadhesion can occur at a particular site in the GI tract. Bioadhesive delivery systems may be affected by the mucus turnover time in the GI tract, which varies based on the site. In the GI tract of rats, the colon and cecum were found to be the best location for mucoadhesion of polycarbophil disks. This was largely attributed to the fact that mucus turnover was low in these regions (Rubinstein & Tirosh, 1994). Mucoadhesive intestinal patches have also been investigated for oral delivery of conventional drug molecules (Shen & Mitragotri, 2002). Bioadhesive polymers can be used to improve the oral absorption of peptide drugs. The bioadhesive polymers, polycarbophil, and chitosan derivatives have been used to enhance the absorption of the peptide drug 9-desglycinamide, 8-arginine vasopressin (DGAVP), in the vertically perfused intestinal loop model of rat (LueBen et al., 1994). Chitosan–EDTA conjugates have also been synthesized and have been reported to have excellent mucoadhesive properties as well as protease inhibitor properties and may also function as penetration enhancers (Bernkop-Schnurch & Krajicek, 1998). Minitablets of salmon calcitonin, chitosan, and chitosan conjugates have been made, and plasma calcium levels were monitored after oral administration to rats. The strongest and most sustained reduction in plasma calcium levels was observed with a formulation containing chitosan–pepstatin A conjugate, chitosan–4-thiobutylamidine conjugate, and reduced glutathione (Guggi, Krauland, & Bernkop-Schnurch, 2003). A conjugate of chitosan with exendin-4, a glucagon-like peptide 1 mimetic, has been reported to have an oral bioavailability of 6.4% relative to subcutaneous administration (Ahn et al., 2013). Lectin-conjugated silicon microdevices that have bioadhesive properties and can allow highly localized and unidirectional release of drugs have also been investigated, and these micromachined drug delivery vehicles can be made using microfabrication technology (Ahmed, Bonner, & Desai, 2002).

7.3.4 Penetration Enhancers

Penetration enhancers can enhance oral absorption by their action on the transcellular and/or paracellular pathway. For effects on the transcellular pathway, surfactants and fatty acids may alter membrane lipid organization and may thus increase oral transport. Surfactants can get incorporated into lipid bilayers, thus changing the physical properties of the cell membranes. For effects on the paracellular pathway, chelating agents can disrupt the integrity of occluding junctional complexes by chelating calcium or magnesium around tight junctions. Exposure of Caco-2 cell monolayers to sodium dodecyl sulfate (SDS) has been reported to increase the absorption of mannitol, vasopressin, and polyethylene glycol. Transepithelial electrical resistance (TEER) measurements, together with fluorescence and transmission electron microscopy, suggest that exposure of Caco-2 cell culture to SDS resulted in apical membrane wounds, a shortening of the microvilli of the cells, and structural separation of the tight junctions. It was suggested that SDS enhances peptide absorption across intestinal epithelium by the paracellular pathway (Anderberg & Artursson, 1993). Bile salts such as sodium deoxycholate and sodium cholate can also be used in the formulation to promote the absorption of insulin from the colon (Ziv, Kidron, Berry, & Bar-On, 1981). The penetration enhancement of insulin by surfactants and bile salts may also be related to their ability to dissociate insulin aggregates. SDS, at 5 mM concentration, has been reported to completely dissociate porcine-zinc insulin hexamers into monomers. When insulin was injected into the distal jejunum/proximal ileum segment of rat, 5 mM SDS increased its absorption from a negligible value to 2.8%. Similarly, 30 mM concentration of sodium glycocholate also increased insulin absorption to 2.3% (Shao, Li, Krishnamoorthy, Chermak, & Mitra, 1993).

Mixed micellar systems have also been used to enhance the oral absorption of polypeptides (Hochman & Artursson, 1994). Such systems are also known to form naturally in the GI tract to aid the absorption of lipids. The dietary fats are first emulsified by bile salts in the intestine and then acted upon by pancreatic lipase to produce monoglycerides and free fatty acids. The monoglycerides, free fatty acids, and bile salts then all combine to form mixed micelles from which the digested lipids are then absorbed. Mixed micelles have been used to enhance the oral absorption of human calcitonin (hCT) across the rat colon. In both rat and man, the bioavailability of intracolonically administered hCT is normally less than 1%. By using 40 mM monoolein/40 mM sodium taurocholate, the absorption of hCT was increased by ninefold. This absorption enhancer was shown to increase the intestinal absorption of molecules over the MW range of 4,000–40,000 without causing acute tissue damage (Hastewell et al., 1994). Lipoidal dispersions of insulin in fatty acids using sodium glycocholate as an emulsifier and absorption promoter have been investigated. The hypoglycemic effects after oral administration to rabbits were found to be dependent on the fatty acid used. A palmitic acid system reduced blood glucose from 105 to 75 mg/dL in 30 min. This effect was significantly higher than that obtained with unsaturated fatty acids or with non–fatty acid systems (Mesiha, Plakogiannis, & Vejosoth, 1994). Penetration enhancers may enhance the absorption of drugs preferentially in some specific region of the GI tract (Tomita et al., 1992). Cyclodextrins have also been used to enhance the absorption of insulin from the lower jejunal/upper ileal segments of the rat by an in situ closed-loop method. By using 10% w/v dimethyl β-cyclodextrin (DMβCD) with 0.5 mg/mL of porcine-zinc insulin, the bioavailability was increased from a negligible value (≈0.06%) to 5.63%. The mechanism probably involves dissociation of insulin oligomers by DMβCD, since DMβCD increases the rate constant for degradation of insulin hexamers by α-chymotrypsin (Shao, Li, Chermak, & Mitra, 1994). In reported studies, a polyoxyethylated castor oil derivative (HCO-60®) and organic acids (e.g., citric acid) have been reported to markedly increase the pharmacological activity of rhG-CSF in rats. The solution or tablets were administered into the rat's intestine, and the blood total leukocyte counts were measured as a pharmacological index of rhG-CSF (Takada et al., 1994; Takaya, Niwa, & Takada, 1994). When penetration enhancers are used to enhance oral absorption, it should be realized that they have limitations that may prevent their general acceptance for usage. Also, the potential lack of specificity of penetration enhancers may have long-term toxicity implications that can only be evaluated in chronic studies. The potential lytic nature of surfactants raises safety concerns, since the intestinal epithelium provides a barrier to the entry of toxins, bacteria, and viruses. Similarly, chelators that cause Ca^{2+} depletion do not act specifically on tight junctions but rather may induce global changes in cells such as disruption of actin filaments or adherent junctions. Thus, it will be difficult to induce opening of tight junctions in a rapid, reversible, and reproducible manner (Hochman & Artursson, 1994).

7.3.5 Protease Inhibitors

Protease inhibitors may also promote oral absorption of therapeutic peptides and proteins by reducing their proteolytic breakdown in the GI tract (Friedman & Amidon, 1991; Su et al., 2012; Taki et al., 1994). Generally, inhibitory agents may be classified as (1) polypeptide protease inhibitors (e.g., aprotinin), (2) peptide and modified peptides (e.g., bacitracin, chymostatin, and amastatin), (3) amino acids and modified amino acids (e.g., α-aminoboronic acid derivatives), and (4) others (e.g., *p*-aminobenzamidine and camostat mesilate) (Bernkop-Schnurch, 1998). An aminopeptidase inhibitor, amastatin, has been reported to reduce the hydrolysis of the pentapeptide, Leu-enkephalin (YGGFL), at a high pH. At lower pH (below 5.0), the endopeptidase inhibitors, tripeptides YGG and GGF, were found to be effective. Coperfusion of YGGFL with a combination of amino- and endopeptidase inhibitors was most effective in inhibiting hydrolysis in the rat jejunum. In the absence of these inhibitors, extensive hydrolysis of YGGFL was observed in the rat jejunum primarily by brush-border enzymes and secondarily by luminal peptidases (Friedman & Amidon, 1991). In another study, an aminopeptidase inhibitor (puromycin) was able to increase the absorption of metkephamid (MKA), a stable analogue of met-enkephalin, across the rat intestine. However, in this study, an endopeptidase inhibitor (thiorphan) was found to be ineffective. This is because the dominant enzyme participating in MKA metabolism during absorption is aminopeptidase (Taki et al., 1994). Bile salts, in addition to being penetration enhancers, can also act as protease inhibitors to enhance oral absorption. Bile salts have been shown to inhibit brush-border membrane and cytosolic proteolytic hydrolysis and would thus be useful to reduce intestinal degradation of peptide drugs (Bai, 1994b). In an in situ study with closed small and large intestinal loops in rats, no marked hypoglycemic response was observed when insulin alone was administered. However, a significant hypoglycemic effect was obtained following large intestinal administration of insulin with 20 mM of Na glycocholate, camostat mesilate, and bacitracin (Yamamoto et al., 1994).

It has been suggested that if a protease inhibitor such as soybean trypsin inhibitor is used to prevent the proteolysis of insulin in the rat intestine, then its absorption is promoted by the endogenous bile acid present in the intestine (Kidron, Bar-On, Berry, & Ziv, 1982). Very small doses (about 1 μg) of vasopressin in solution produced antidiuresis in rats following oral administration. The biological response was enhanced for AVP and LVP by the simultaneous administration of 1000 units of aprotinin, a protease inhibitor. The synthetic analogue DDAVP was more active than the natural hormones, but the effect of aprotinin with DDAVP was inconsistent. The relatively greater oral activity of DDAVP is due to the unnatural D-arginine, which makes it resistant to attack by trypsin (Saffran et al., 1988). Starch-*g*-poly(acrylic acid) copolymers and starch/poly(acrylic acid) mixtures have been synthesized and may have potential for enabling oral peptide delivery due to their proteolytic enzyme inhibition activity and ion binding capacity (Ameye et al., 2001).

7.3.6 Carrier Systems

Carrier systems, such as nanoparticles, microspheres, liposomes, or erythrocytes, or pH-sensitive polymers can also be used to improve the oral absorption of peptides and proteins and release entrapped drug via a variety of mechanisms (Figure 7.3) (Wang & Zhang, 2012). Other particle designs using biopolymers have also been reported (Lim, Tey, & Chan, 2014). Oral administration of a particulate formulation of recombinant human growth hormone resulted in observed efficacy, which corresponds to a relative pharmacological bioavailability of 3.4% relative to subcutaneous administration (Salmaso, Bersani, Elvassore, Bertucco, & Caliceti, 2009). Some of the carrier-based approaches are shown in Figure 7.2. Emisphere Technologies, Inc. (Tarrytown, NY) has conducted clinical trials for oral delivery of insulin using their carrier Eligen® technology. The company has also used their technology to investigate oral delivery of calcitonin, GLP-1, human growth hormone, and parathyroid hormone. These carrier molecules, in high concentrations, cause the protein to undergo a conformational change to a partially unfolded or molten globule state that has a higher oral permeability. The carrier molecules are small organic molecules with an MW of about 200–400 Da. The protein is used in its native state rather than a chemical modification approach, and the interaction between carrier and protein is noncovalent (Goldberg, 2003; Goldberg & Gomez-Orellana, 2003; Park et al., 2011; Stoll, Leipold, Milstein, & Edwards, 2000). The company is in the process of launching an Eligen®B12 product. Generex Biotechnology Corporation (Ontario, Canada) has developed an Oral-lyn™ formulation, in which insulin is formulated into micelles larger than 7 μm that are delivered by a RapidMist™ device (Figure 7.2). The drug is absorbed from the buccal mucosa in the mouth as the micelles are too large to enter the lung (Park et al., 2011). The product is in phase III clinical trials around the world.

BioSante Pharmaceuticals, later acquired by ANI Pharmaceuticals (Baudette, microneedles), has developed an insulin delivery system based on calcium phosphate (CAPOral) for oral insulin (CAPIC™). The formulation is created through

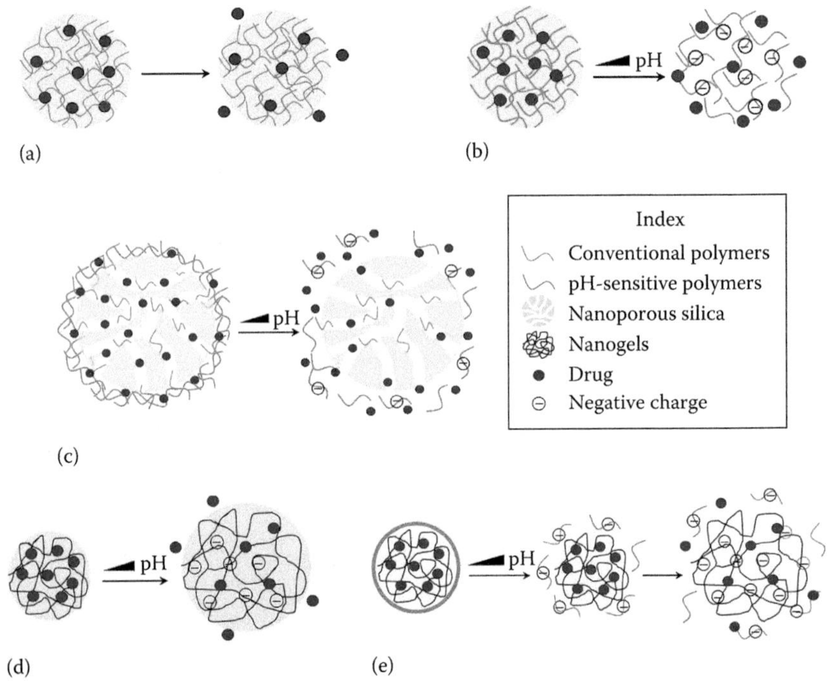

FIGURE 7.3 **(See color insert.)** Drug release mechanism from carrier systems. (a) Release from conventional nanoparticles by diffusion, (b) release from pH-sensitive nanoparticles after the material dissolution at a specific pH, (c) release from pH-sensitive nanomatrix, (d) release from pH-sensitive nanoparticles after the material swelling at a certain pH, and (e) release from pH-sensitive nanoparticles as the coeffect of dissolution and swelling of the materials at specific pH. (Reproduced from Wang, X.Q. and Zhang, Q., *Eur. J. Pharm. Biopharm.*, 82, 219, 2012. With permission.)

nanoparticulate technology with particles composed of calcium phosphate, PEG, and insulin and coated with caseins (Park et al., 2011). The use of different carrier systems to improve the absorption of insulin in anesthetized diabetic rats following intraduodenal administration from a midline incision has been evaluated (Alachi & Greenwood, 1993). Several erythrocyte membrane carrier systems were tested. These included erythrocyte ghosts (EGs) prepared by hemolysis of human red blood cells, erythrocyte vesicles (EVs) prepared by sonication of EG suspension, and liposome-incorporating ghosts or vesicles (LEGs and LEVs). Compared to a control group, these carriers enhanced oral absorption of insulin, with LEV being the best carrier for most efficient delivery. Similarly, hydrogel-based devices have been investigated for oral delivery of insulin to protect it from enzymatic degradation in the acidic stomach and deliver effectively into the intestine (Chaturvedi, Ganguly, Nadagouda, & Aminabhavi, 2013; Kamei et al., 2009). Formulation of multilayered mucoadhesive hydrogel films has also been reported (Ding, He, Zhou, Tang, & Yin, 2012). Chitosan-based nano- or microparticles can be used for mucosal delivery of proteins and vaccines and can be made in sizes ranging from 300 nm to 3 μm, depending on formulation factors such as concentration of chitosan, its MW, and addition rate of precipitant salt (Koppolu et al., 2014). Nanoparticles with peptidic ligands are particularly promising to achieve specific targeting in the GI tract (Yun, Cho, & Park, 2013). The influence of size and surface properties of nanoparticles on their nonspecific or targeted uptake by enterocytes and/or M-cells has been reviewed (des, Fievez, Garinot, Schneider, & Preat, 2006).

Uptake of liposomes by Peyer's patch can increase the uptake of any entrapped drug. Negatively charged liposomes with at least 25 mol% phosphatidylserine have been reported to be taken up readily by the rat Peyer's patches following intraluminal administration (Tomizawa et al., 1993). Proteins such as albumin have also been used to prepare microparticles to improve the stability of drugs in the GI tract (Mora & Pato, 1993). Thermally condensed amino acids (proteinoids) can spontaneously form microspheres when exposed to an acidic medium. Proteinoid microspheres have been used to deliver encapsulated calcitonin to rats and monkeys, with positive results. In rats, the serum calcium levels decreased by 23 μg/mL 1 h after dosing encapsulated calcitonin. In contrast, rats receiving control calcitonin (no microspheres) had a decrease of only 6.5 μg/mL (Sarubbi et al., 1994).

The absorption and tissue distribution of ^{14}C-labeled poly(DL-lactide-*co*-glycolide) nanoparticles after oral administration to the mice have been determined in comparison to the intravenous route. The GI transit of the nanoparticles was very fast, with most of the radioactivity appearing rapidly in the colon 4 h after administration and in the feces 24 h after administration. Of the amount absorbed through the intestinal barrier (about 2%), most was found in the carcass and liver (Le Ray, Vert, Gautier, & Benoit, 1994). Nanocapsules may be preferentially absorbed through Peyer's patches and may be visible in M-cells and intercellular spaces around lymph cells. It seems that this uptake by Peyer's patches is especially important in the ileum. Absorption of nanocapsules in the jejunum may be by a paracellular pathway, possibly through the intercellular spaces formed by the desquamation of well-differentiated absorptive cells at the tip of the villi (Couvreur & Puisieux, 1993). Polyalkylcyanoacrylate nanocapsules loaded with insulin have been administered to diabetic rats (Damge, Michel, Aprahamian, & Couvreur, 1988; Michel et al., 1991). A single intragastric dose of 50 U/kg reduced glycemia by 50%–60%, measured after an overnight fast. The effect appeared 2 days after administration and was maintained as long as 20 days. Free insulin did not affect glycemia when administered orally under the same experimental conditions (Damge et al., 1988). The intestinal absorption of insulin and calcitonin encapsulated in polyisobutylcyanoacrylate nanoparticles has been investigated in rats, and the resulting pharmacokinetic profiles were characteristic of sustained delivery. A relatively higher plasma concentration was seen at the later time points but was balanced by lower initial concentrations, and thus, there was no significant net enhancement of absorption. This suggests that the nanocapsules slowly released the peptide into the intestinal lumen, with small amounts being absorbed (Lowe & Temple, 1994). Hydrogel nanospheres composed of poly(methacrylic acid-grafted-poly(ethylene glycol)) have also been investigated for oral protein delivery and have been reported to be capable of opening the tight junctions between epithelial cells in Caco-2 cell monolayers (Torres-Lugo, Garcia, Record, & Peppas, 2002).

Lactococcus lactis, a food-grade GRAS status bacterium, can also potentially be used as a delivery vector for oral delivery of proteins/vaccines. It has dimensions similar to microparticulate vaccine delivery systems that are taken up by M-cells of the intestinal mucosa (Bahey-El-Din, Gahan, & Griffin, 2010; Bermudez-Humaran, Kharrat, Chatel, & Langella, 2011).

7.3.7 Other Formulation Approaches

Several rather unique formulation approaches have been reported for the oral absorption of peptides and proteins. In one approach, an emulsion is made in which the peptide is located in the aqueous phase in the presence of aprotinin. The oil phase contains a lipid composition similar to that of chylomicrons. Aprotinin, being a protease inhibitor, will prevent peptide degradation, while chylomicrons will improve absorption into the enterocyte. The emulsion is coated on carrier powders that are then filled in hard gelatin capsules. The capsules are then enteric coated to prevent dissolution in stomach. Another approach involves noncovalent linking of the peptide to phospholipids so that the complex can be absorbed into the enterocyte by endocytosis. These approaches are being used to develop oral formulations for insulin, calcitonin, porcine somatotropin, erythropoietin, and α-interferon. Oral administration of insulin in solid form to nondiabetic and diabetic dogs has been attempted by mixing insulin with cholate and soybean trypsin inhibitor and delivered orally as enterocoated microtablets. Following administration of the drug, plasma insulin levels increased and plasma glucose levels decreased after a gap of about 60–140 min. Since delivery of insulin by the oral route leads to targeting of the enterohepatic pathway, the authors of this study felt that this or a similar formulation may serve as an adjuvant treatment for patients with type II diabetes mellitus (Ziv et al., 1994). Protein crystals may also help to stabilize protein for oral delivery (Figure 7.2).

7.4 METHODS TO STUDY ORAL ABSORPTION

7.4.1 Intestinal Segments

Intestinal segments can be isolated for in vitro studies, or in situ perfusion studies can be performed. A commonly used in vitro technique uses an everted gut sac apparatus. Such a technique has been used to study site-specific absorption of insulin from various intestinal regions of the rat. The cut intestinal segments were everted on a thin stainless steel rod and ligated on the other end with a silk thread. The gut sac was filled with a known volume of modified Krebs-Ringer phosphate bicarbonate buffer and placed inside a test tube containing the test solution at 37°C. A weight was also tied to the everted sac to prevent peristaltic muscular contractions. The test solution was continuously bubbled with 95% O_2/5% CO_2, and samples could be drawn from both serosal and mucosal compartments. The gut sac used in this study remained viable for at least 20 min (Schilling & Mitra, 1990). Even though the viability may be as high as 3 h in some cases, this is still a relatively short time period and constitutes one of the main drawbacks of the in vitro technique. The in situ method has certain advantages over in vitro and in vivo methods. Advantages over in vitro methods include the presence of intact lymphatic and blood vessels for solute uptake and extended tissue viability. Advantages over in vivo methods include bypassing drug dissolution and stomach emptying steps, avoiding degradation of peptides and proteins in stomach, and allowing control over drug input and choice of the intestinal region to be perfused. It has been suggested that active and paracellular solute transport is not compromised during in situ experiments (Lu, Thomas, Tukker, & Fleisher, 1992).

7.4.2 In Vivo Studies

Preclinical absorption screening of orally delivered drugs is commonly performed in dogs. Metabolic and physiological differences between dogs and humans must be well characterized for such studies to be useful. A study characterizing the upper GI pH, volumetric flow rate, and activity of chymotrypsin in mongrel fistulated dogs as a function of fasted GI motility phase has been reported (Sinko, 1992). The anatomical site where a peptide is released from its formulation prior to its absorption can be determined by radioiodination of peptide in conjunction with gamma scintigraphy. It should be realized that the muscular activity in the gut can affect the drug absorption. Thus, the drug uptake in an anesthetized animal may be different from that in a conscious animal. An in vivo approach in humans, using a regional perfusion technique, has also been reported. This technique uses a multichannel perfusion tube, which can be introduced orally. The tube has two inflatable balloons, 10 cm apart, enabling perfusion of a 10 cm long segment. The occlusion of an intestinal segment between these two intraluminal balloons prevents the contamination of luminal contents both proximally and distally into the isolated segment (Lennernas et al., 1992).

7.4.3 Diffusion Cells

In these studies, an intestinal segment is used as a semipermeable membrane between a donor and a receiver chamber of a side-by-side diffusion cell system. A widely used cell of this type is the Ussing chamber. Ussing chambers have been used to evaluate the transepithelial transport and metabolism of several compounds (Smith, Mirabelli, Fondacaro, Ryan, & Dent, 1988). Modifications to Ussing chambers have also been reported. A modified cell with reduced leakage, better mixing, increased working tissue surface area, and easier cleaning than Ussing cells has been reported. This cell can also be equipped to obtain electrophysiologic measurements such as the TEER, which can be a useful index of tissue integrity and viability (Sutton, Forbes, Cargill, Hochman, & LeCluyse, 1992).

7.4.4 Cell Cultures

Monolayers of intestinal cells grown in culture have been widely used as a method to predict intestinal drug absorption (Artursson, 1990; Buur & Mork, 1992; Cogburn, Donovan, & Schasteen, 1991; Hidalgo, Hillgren, Grass, & Borchardt, 1991; Laboisse,

Jarry, Bouhanna, Merlin, & Vallette, 1994). Unlike isolated tissues, cell cultures can have good viability when supplied with nutrients and oxygen. While it is difficult to culture intestinal cells, the availability of a well-differentiated human intestinal cell line, Caco-2, makes such studies possible. The Caco-2 cell line is obtained from human colon carcinoma cells that are known to undergo enterocytic differentiation in culture. By day 3 in culture, these cells develop occluding junctional complexes between adjacent cells. After about 15 days in culture, the cells develop morphological characteristics of the small intestine such as microvilli, desmosomes, occluding junctions, and cell polarity. The cells can be grown on microporous membranes and can be used for drug diffusion experiments for at least 20 days.

Careful selection of the microporous membrane used to support cell attachment and growth is important. It should be readily permeable to hydrophilic or hydrophobic solutes of low or high MWs. Also, it should be sufficiently translucent to allow microscopic examination of the cell monolayer. These Caco-2 cells also have major drug-metabolizing enzymes as well as an array of transporters found in intestinal cells. Though the Caco-2 cells are now widely used, literature has some conflicting reports on the permeability of marker compounds across monolayers. Interlaboratory comparisons will be more meaningful if standardized protocols are developed which specify the passage number, time of usage postseeding, and cell source. Also, electrophysiology should be used to assess cell viability as well as monolayer integrity (Delie & Rubas, 1997). The integrity of these monolayers can be evaluated by measuring TEER. TEER values have been reported to increase from 96.6 ± 22 at day 3 to 173.5 ± 10.9 Ω/cm^2 at day 6 and remain constant through 30 days (Borchardt, 1994). With the exception of the mucin layer, the Caco-2 model contains three major barriers to oral absorption: the *unstirred* water layer, the tight junctions, and the cell membranes (Artursson, 1990). Within certain limits, the permeability of Caco-2 monolayers may be used to estimate human drug absorption. Permeability through Caco-2 monolayers can also be related to permeabilities through several other animal models (Rubas, Jezyk, & Grass, 1993).

Monolayers of Caco-2 cells have been used to study the transport of several peptides and proteins. The permeability values determined in Caco-2 cell culture model have been found to be a good predictor of the intestinal permeability of peptides (Conradi et al., 1993; Kim, Burton, & Borchardt, 1993). Two types of diffusion apparatus have been used in these studies: the unstirred cell-insert system and the side-by-side diffusion system stirred by gas lift. The gas lift provides stirring by O_2/CO_2, thereby avoiding potential damage to the cell monolayer and minimizing the thickness of the aqueous boundary layer (Borchardt, 1994). Studies done with the Caco-2 cell line have investigated the in vitro permeability of the met-human growth hormone, insulin-like growth factor, interferon gamma, and an HIV surface coat protein, gp 120. The permeability for these proteins was extremely low and was not dependent on the MW. In contrast, arginine–glycine–aspartic acid (RGD) tripeptides and their analogues had significant transport rates. Studies with these analogues indicated that chemical modifications to reduce the number of potential hydrogen binding sites enhanced the uptake rates. This modification increased the internal, rather than external, hydrogen bonds so that the peptide molecule has less water molecules associated with its structure and can thus partition more easily into lipid bilayers (Mrsny et al., 1993). In another study, a series of peptides prepared from D-phenylalanine and glycine were used to study their permeation across a confluent monolayer of Caco-2 cells. Again, the flux was found to correlate with the total number of hydrogen bonds the peptide could make with water. In contrast, no correlation was found between the flux and the apparent lipophilicity of the peptides (Conradi, Hilgers, Ho, & Burton, 1991). Thus, it seems that a reduction in the hydrogen bonding potential of a peptide will improve its oral absorption. This presents a challenge since by its very nature, the amide backbone of a peptide will have at least two hydrogen bonds for each amino acid in the primary sequence. One possible approach to reduce the hydrogen bonding is by alkylation of the nitrogen in the amide. Using a series of tetrapeptide analogues prepared from D-phenylalanine, a substantial increase in transport across the Caco-2 cell culture has been reported with each methyl group added to the amide nitrogens (Conradi, Hilgers, Ho, & Burton, 1992). While this approach may be promising, the modified structure may not be biologically active. This might be overcome if a reversible modification of the amide bond is made feasible by a prodrug approach. Some natural peptides may have a unique secondary structure that results in a significant number of intramolecular hydrogen bonds. Delta-sleep-inducing peptide (DSIP) is an endogenous neuropeptide consisting of nine amino acids. Based on its physicochemical properties, DSIP is not expected to permeate through biomembranes. However, it is known to be readily permeable to biomembranes by passive diffusion. A study of its solution conformation has revealed that it has two β-turns, which involve eight of its nine residues, thus resulting in a high degree of intramolecular hydrogen bonding. Partitioning of solutes between heptane and ethylene glycol has been recently used to predict the oral absorption potential of peptides. The log partition coefficient in this system was found to correlate ($r^2 = 0.86$, $n = 11$) with the oral permeability coefficients for a group of model peptides. The partitioning is related to hydrogen bond breaking, and thus, oral absorption was again related to hydrogen bond breaking in this study (Paterson, Conradi, Hilgers, Vidmar, & Burton, 1994).

7.4.5 Brush-Border Membrane Vesicles

Brush border, that is, microvillus membranes, can be obtained from human intestinal epithelial cells. Isolated purified BBMVs have been widely used as a model for intestinal transport studies (Thwaites, Hirst et al., 1993; Thwaites, Simmons et al., 1993). The procedure was initially described by Schmitz et al. (1973). A modification of this procedure for simplification has been described by Kessler et al. (1978). Calcium ions (Ca^{2+}) play an important role in the preparation of brush-border fragments. Brush-border fragments are not readily separable from microsomal fragments (endoplasmic reticulum) unless Ca^{2+} ions are used. Ca^{2+} acts to aggregate the sedimentable endoplasmic reticulum and the mitochondria into larger particles that can then settle at lower gravitational forces. Thus, after addition of $CaCl_2$, centrifugation at 2,000× g can pelletize the microsomal fragments, and then centrifugation at 20,000× g can collect brush-border fragments (Schmitz et al., 1973).

7.5 ORAL IMMUNIZATION

Use of proteins as vaccines is discussed in Chapter 10, but a few brief comments on oral immunization will be made here. Even though most vaccines are given by parenteral administration, most bacteria and viruses gain entry into the host through mucosal tissues. It has been suggested that to confer the highest level of protection, the site of immunization should parallel the site of infection. Oral immunization is clinically underexploited. A wider use of oral immunization will improve access to vaccination, particularly in regions or countries where health-care professionals are in short supply (Manganaro, Ogra, & Ernst, 1994; O'Hagan, 1992). Oral immunization may be feasible because Peyer's patches in the intestinal tract can sample antigens. Antigens are believed to be captured through pinocytosis by M-cells localized in Peyer's patches of the intestine. Following their entry into Peyer's patch, antigens may enter the lymphatics or portal circulation and redistribute to different sites. Thus, widespread induction of immune responses may occur following the oral administration of antigens. In order to prevent frequent adverse reactions, the intestine provides a low immune response to food and other environmental antigens. Therefore, one challenge to oral vaccine development is to augment an immune response in a relatively nonresponsive environment. Secondly, although intestine will respond to a sudden increase in antigen load, chronic administration of antigens will induce oral tolerance. This oral tolerance presents another challenge to the development of oral peptide vaccines. Cholera toxin has been used as a potent oral adjuvant for the induction of mucosal IgA antibody responses to protein antigens (Snider, Marshall, Perdue, & Liang, 1994).

Targeting of oral vaccines to inductive sites has been investigated with replicating agents such as relatively safe bacteria or viruses or nonreplicating killed pathogens or carriers such as erythrocytes, liposomes, microparticles, or nanoparticles (Manganaro et al., 1994). A size range of 5–10 μm seems to be optimum for micro- or nanoparticles to stimulate a mucosal response. The efficacy of influenza virus antigen entrapped in proteinoid microspheres as oral vaccines has been evaluated in rats. A single oral dose of the antigen entrapped in proteinoid microspheres was able to induce a significant IgG response as early as 2 weeks after the administration. Rats dosed orally with the free antigen (no microspheres) showed no detectable antibody response (Santiago et al., 1993). Normally, such carrier systems still need repeated exposures over a period of weeks to induce a significant immune response. Polymers with the greatest hydrophobicity seem to be absorbed more readily. Polylactic-*co*-glycolic acid (PLGA) microspheres of 5–10 μm size seem to be an attractive carrier for oral delivery of antigens as this size allows adequate absorption of the antigen. The incorporation of antigens into microspheres also protects the antigen from proteolysis and allows for possible coincorporation of immunological adjuvants and cytokines that may further enhance the immune response. Also, a single dose of microspheres containing a mixture of microspheres may result in a primary immune response from some microspheres, followed by a long-lasting secondary immune response from the remaining microspheres, which were designed to have a longer biodegradation time. This will eliminate the need for a booster immunization. Microspheres containing antigens are stable for many months when stored as a dry powder.

The PLGA polymer has the additional advantage of biocompatibility and biodegradability (Couvreur & Puisieux, 1993; Mestecky et al., 1994).

7.6 CONCLUSIONS

Successful peptide or protein delivery by oral route needs a succession of events to bypass the various penetration or enzymatic barriers at each stage. Often, a combination of two or more approaches may be required. The goal is to increase the typical less than 1% bioavailability to at least 10%–20%. A site-specific delivery system and approaches to minimize proteolytic degradation are required. The use of penetration enhancers, protease inhibitors, carrier systems, or chemical modification of the peptide or protein offers promising approaches to enhance their oral delivery. Several methods are being used to study oral absorption using these approaches. Design of peptides that penetrated the intestinal mucosa by the paracellular pathway seems to be a promising approach. Promising results have begun to appear, and oral delivery of insulin is currently in clinical trials.

REFERENCES

Ahmed, A., Bonner, C., & Desai, T. A. (2002). Bioadhesive microdevices with multiple reservoirs: A new platform for oral drug delivery. *J. Control. Release, 81,* 291–306.

Ahn, S., Lee, I. H., Lee, E., Kim, H., Kim, Y. C., & Jon, S. (2013). Oral delivery of an antidiabetic peptide drug via conjugation and complexation with low molecular weight chitosan. *J. Control. Release, 170,* 226–232.

Al-Hilal, T. A., Alam, F., & Byun, Y. (2013). Oral drug delivery systems using chemical conjugates or physical complexes. *Adv. Drug Deliv. Rev., 65,* 845–864.

Alachi, A. & Greenwood, R. (1993). Intraduodenal administration of biocarrier-insulin systems. *Drug Dev. Ind. Pharm., 19,* 1303–1315.

Ameye, D., Voorspoels, J., Foreman, P., Tsai, J., Richardson, P., Geresh, S. et al. (2001). Trypsin inhibition, calcium and zinc ion binding of starch-*g*-poly(acrylic acid) copolymers and starch/poly(acrylic acid) mixtures for peroral peptide drug delivery. *J. Control. Release, 75,* 357–364.

Anderberg, E. K. & Artursson, P. (1993). Epithelial transport of drugs in cell culture. VIII: Effects of sodium dodecyl sulfate on cell membrane and tight junction permeability in human intestinal epithelial (Caco-2) cells. *J. Pharm. Sci., 82,* 392–398.

Andrysek, T. (2001). The role of particle size distribution on bioavailability of cyclosporine: Novel drug delivery system. *Biomed. Pap., 145,* 3–8.

Artursson, P. (1990). Epithelial transport of drugs in cell culture. I: A model for studying the passive diffusion of drugs over intestinal absorbtive (Caco-2) cells. *J. Pharm. Sci., 79,* 476–482.

Asada, H., Douen, T., Mizokoshi, Y., Fujita, T., Murakami, M., Yamamoto, A. et al. (1994). Stability of acyl derivatives of insulin in the small intestine: Relative importance of insulin association characteristics in aqueous solution. *Pharm. Res., 11,* 1115–1120.

Ashford, M. & Fell, J. T. (1994). Targeting drugs to the colon: Delivery systems for oral administration. *J. Drug Target., 2,* 241–257.

Bahey-El-Din, M., Gahan, C. G., & Griffin, B. T. (2010). *Lactococcus lactis* as a cell factory for delivery of therapeutic proteins. *Curr. Gene Ther., 10,* 34–45.

Bai, J. P. F. (1993). Distribution of brush-border membrane peptidases along the rabbit intestine—Implication for oral delivery of peptide drugs. *Life Sci., 52,* 941–947.

Bai, J. P. F. (1994a). Distribution of brush-border membrane peptidases along the rat intestine. *Pharm. Res., 11,* 897–900.

Bai, J. P. F. (1994b). Effects of bile salts on brush-border and cytosolic proteolytic activities of intestinal enterocytes. *Int. J. Pharm., 111,* 147–152.

Bai, J. P. F. & Amidon, G. L. (1992). Structural specificity of mucosal-cell transport and metabolism of peptide drugs: Implication for oral peptide drug delivery. *Pharm. Res., 8,* 969–978.

Barlow, D. & Satoh, T. (1994). The design of peptide analogues for improved absorption. *J. Control. Release, 29,* 283–291.

Bermudez-Humaran, L. G., Kharrat, P., Chatel, J. M., & Langella, P. (2011). Lactococci and lactobacilli as mucosal delivery vectors for therapeutic proteins and DNA vaccines. *Microb. Cell Fact., 10(Suppl. 1),* S4.

Bernkop-Schnurch, A. (1998). The use of inhibitory agents to overcome the enzymatic barrier to perorally administered therapeutic peptides and proteins. *J. Control. Release, 52,* 1–16.

Bernkop-Schnurch, A. & Krajicek, M. E. (1998). Mucoadhesive polymers as platforms for peroral peptide delivery and absorption: Synthesis and evaluation of different chitosan-EDTA conjugates. *J. Control. Release, 50,* 215–223.

Bernkop-Schnurch, A. & Walker, G. (2001). Multifunctional matrices for oral peptide delivery. *Crit. Rev. Ther. Drug Carrier Syst., 18,* 459–501.

Bohner, V., Langguth, P., Biber, J., & Merkle, H. P. (1994). Intestinal mucosal transport and metabolism of pentapeptides. In *Proceedings of the International Symposium on Controlled Release on Bioactive Materials* (21st ed., pp. 895–896). Controlled Release Society, Inc., St. Paul, MN.

Borchardt, R. T. (1994). Rational delivery strategies for the design of peptides with enhanced oral delivery. *Drug Dev. Ind. Pharm., 20(4),* 469–483.

Bruno, B. J., Miller, G. D., & Lim, C. S. (2013). Basics and recent advances in peptide and protein drug delivery. *Ther. Deliv., 4,* 1443–1467.

Buur, A. & Mork, N. (1992). Metabolism of testosterone during in vitro transport across CACO-2 cell monolayers: Evidence for beta-hydroxysteroid dehydrogenase activity in differentiated CACO-2 cells. *Pharm. Res., 9,* 1290–1294.

Chaturvedi, K., Ganguly, K., Nadagouda, M. N., & Aminabhavi, T. M. (2013). Polymeric hydrogels for oral insulin delivery. *J. Control. Release, 165,* 129–138.

Choonara, B. F., Choonara, Y. E., Kumar, P., Bijukumar, D., Du Toit, L. C., & Pillay, V. (2014). A review of advanced oral drug delivery technologies facilitating the protection and absorption of protein and peptide molecules. *Biotechnol. Adv., 32(7),* 1269–1282.

Clement, S., Dandona, P., Still, J. G., & Kosutic, G. (2004). Oral modified insulin (HIM2) in patients with type 1 diabetes mellitus: Results from a phase I/II clinical trial. *Metabolism, 53,* 54–58.

Cogburn, J. N., Donovan, M. G., & Schasteen, C. S. (1991). A model of human small intestinal absorptive cells. 1. Transport barrier. *Pharm. Res., 8,* 210–216.

Conradi, R. A., Hilgers, A. R., Ho, N. F. H., & Burton, P. S. (1991). The influence of peptide structure on transport across Caco-2 cells. *Pharm. Res., 8,* 1453–1460.

Conradi, R. A., Hilgers, A. R., Ho, N. F. H., & Burton, P. S. (1992). The influence of peptide structure on transport across Caco-2 cells. II. Peptide bond modification which results in improved permeability. *Pharm. Res., 9,* 435–439.

Conradi, R. A., Wilkinson, K. F., Rush, B. D., Hilgers, A. R., Ruwart, M. J., & Burton, P. S. (1993). In vitro/in vivo models for peptide oral absorption: Comparison of Caco-2 cell permeability with rat intestinal absorption of renin inhibitory peptides. *Pharm. Res., 10,* 1790–1792.

Couvreur, P. & Puisieux, F. (1993). Nanoparticles and microparticles for the delivery of polypeptides and proteins. *Adv. Drug Deliv. Rev., 10,* 141–162.

Damge, C., Michel, C., Aprahamian, M., & Couvreur, P. (1988). New approach for oral administration of insulin with polyalkylcyanoacrylate nanocapsules as drug carrier. *Diabetes, 37,* 246–251.

Delie, F. & Rubas, W. (1997). A human colonic cell line sharing similarities with enterocytes as a model to examine oral absorption: Advantages and limitations of the Caco-2 model. *Crit. Rev. Ther. Drug Carrier Syst., 14,* 221–286.

Des, R. A., Fievez, V., Garinot, M., Schneider, Y. J., & Preat, V. (2006). Nanoparticles as potential oral delivery systems of proteins and vaccines: A mechanistic approach. *J. Control Release, 116,* 1–27.

Devogelaer, J. P., Azria, M., Attinger, M., Abbiati, G., Castiglioni, C., & Dedeuxchaisnes, C. N. (1994). Comparison of the acute biological action of injectable salmon calcitonin and an injectable and oral calcitonin analogue. *Calcif. Tissue Int., 55,* 71–73.

Ding, J., He, R., Zhou, G., Tang, C., & Yin, C. (2012). Multilayered mucoadhesive hydrogel films based on thiolated hyaluronic acid and polyvinylalcohol for insulin delivery. *Acta Biomater., 8,* 3643–3651.

Duchene, D., Touchard, F., & Peppas, N. A. (1988). Pharmaceutical and medical aspects of bioadhesive systems for drug administration. *Drug Dev. Ind. Pharm., 14,* 283–318.

Fasano, A. (1998). Novel approaches for oral delivery of macromolecules. *J. Pharm. Sci., 87,* 1351–1356.

Friedman, D. I. & Amidon, G. L. (1991). Oral absorption of peptides: Influence of pH and inhibitors on the intestinal hydrolysis of leu-enkephalin and analogues. *Pharm. Res., 8,* 93–96.

Ganapathy, V. & Leibach, F. H. (1982). Peptide transport in intestinal and renal brush border membrane vesicles. *Life Sci., 30,* 2137–2146.

Gazzaniga, A., Iamartino, P., Maffione, G., & Sangalli, M. E. (1994). Oral delayed-release system for colonic specific delivery. *Int. J. Pharm., 108,* 77–83.

Goldberg, M. & Gomez-Orellana, I. (2003). Challenges for the oral delivery of macromolecules. *Nat. Rev. Drug Discov., 2,* 289–295. (http://www.emisphere.com). (Accessed August 30, 2014.)

Goldberg, M. M. (September 22–23, 2003). Review of the large and growing database of oral delivery of macromolecules using eligen (TM). *Protein and Peptide Drug Delivery: Scientific Advances in Enabling Novel Approaches & Improved Products IBC Meeting*, Philadelphia, PA.

Guggi, D., Krauland, A. H., & Bernkop-Schnurch, A. (2003). Systemic peptide delivery via the stomach: In vivo evaluation of an oral dosage form for salmon calcitonin. *J. Control. Release, 92,* 125–135.

Guo, J., Ping, Q., Jiang, G., Dong, J., Qi, S., Feng, L. et al. (2004). Transport of leuprolide across rat intestine, rabbit intestine and Caco-2 cell monolayer. *Int. J. Pharm., 278,* 415–422.

Hastewell, J., Lynch, S., Fox, R., Williamson, I., Skelton-Stroud, P., & Mackay, M. (1994). Enhancement of human calcitonin absorption across the rat colon in vivo. *Int. J. Pharm., 101,* 115–120.

Hayakawa, E. & Lee, V. H. L. (1992). Aminopeptidase activity in the jejunal and ileal Peyer's patches of the albino rabbit. *Pharm. Res., 9,* 535–540.

Hidalgo, I. J., Hillgren, K. M., Grass, G. M., & Borchardt, R. T. (1991). Characterization of the unstirred water layer in Caco-2 cell monolayers using a novel diffusion apparatus. *Pharm. Res., 8,* 222–227.

Hochman, J. & Artursson, P. (1994). Mechanisms of absorption enhancement and tight junction regulation. *J. Control. Release, 29,* 253–267.

Jensen-Pippo, K. E., Whitcomb, K. L., DePrince, R. B., Ralph, L., & Habberfield, A. D. (1996). Enteral bioavailability of human granulocyte colony stimulating factor conjugated with poly(ethylene glycol). *Pharm. Res., 13,* 102–107.

Kamei, N., Morishita, M., Chiba, H., Kavimandan, N. J., Peppas, N. A., & Takayama, K. (2009). Complexation hydrogels for intestinal delivery of interferon beta and calcitonin. *J. Control. Release, 134,* 98–102.

Kessler, M., Acuto, O., Storelli, C., Murer, H., Muller, M., & Semenza, G. (1978). A modified procedure for the rapid preparation of efficiently transporting vesicles from small intestinal brush border membranes and their use in investigating some properties of D-glucose and choline transport systems. *Biochica et Biophysica Acta, 506,* 136–154.

Khafagy, E. & Morishita, M. (2012). Oral biodrug delivery using cell-penetrating peptide. *Adv. Drug Deliv. Rev., 64,* 531–539.

Kidron, M., Bar-On, H., Berry, E. M., & Ziv, E. (1982). The absorption of insulin from various regions of the rat intestine. *Life Sci., 31,* 2837–2841.

Kim, D., Burton, P. S., & Borchardt, R. T. (1993). A correlation between the permeability characteristics of a series of peptides using an in vitro cell culture model (Caco-2) and those using an in situ perfused rat ileum model of the intestinal mucosa. *Pharm. Res., 10,* 1710–1714.

Kleinert, H. D., Baker, W. R., & Stein, H. H. (1993). Orally bioavailable peptide like molecules: A case history. *Biopharm, 6(1),* 36–41.

Koppolu, B. P., Smith, S. G., Ravindranathan, S., Jayanthi, S., Suresh Kumar, T. K., & Zaharoff, D. A. (2014). Controlling chitosan-based encapsulation for protein and vaccine delivery. *Biomaterials, 35,* 4382–4389.

Kwon, K. C., Verma, D., Singh, N. D., Herzog, R., & Daniell, H. (2013). Oral delivery of human biopharmaceuticals, autoantigens and vaccine antigens bioencapsulated in plant cells. *Adv. Drug Deliv. Rev., 65,* 782–799.

Laboisse, C. L., Jarry, A., Bouhanna, C., Merlin, D., & Vallette, G. (1994). Intestinal cell culture models. *Eur. J. Pharm. Sci., 2,* 36–38.

Langguth, P., Bohner, V., Heizmann, J., Merkle, H. P., Wolffram, S., Amidon, G. L. et al. (1997). The challenge of proteolytic enzymes in intestinal peptide delivery. *J. Control. Release, 46,* 39–57.

Langguth, P., Merkle, H. P., Stockli, A., & Wolffram, S. (1994). Absorption from ascending colon and first-pass extraction of a pentapeptide drug. In *Proceedings of the International Symposium on Controlled Release Bioactive Materials* (21 ed., pp. 802–803). Controlled Release Society, Inc., St. Paul, MN.

Le Ray, A. M., Vert, M., Gautier, J. C., & Benoit, J. P. (1994). Fate of [C-14]poly(DL-lactide-*co*-glycolide) nanoparticles after intravenous and oral administration to mice. *Int. J. Pharm., 106,* 201–211.

Lee, H. J. (2002). Protein drug oral delivery: The recent progress. *Arch. Pharm. Res., 25,* 572–584.

Lee, V. H. L. (1988). Enzymatic barriers to peptide and protein absorption. *CRC Crit. Rev. Ther. Drug Carrier Syst., 5,* 68–97.

Lee, V. H. L. (1992). Oral route of peptide and protein drug delivery. *Biopharm, 5(6),* 39–50.

Lee, V. H. L., Dodda-Kashi, S., Grass, G. M., & Rubas, W. (1991). Oral route of peptide and protein drug delivery. In V. H. L. Lee (Ed.), *Peptide and protein drug delivery* (pp. 691–738). New York: Marcel Dekker, Inc.

Lennernas, H., Ahrenstedt, O., Hallgren, R., Knutson, L., Ryde, M., & Paalzow, L. K. (1992). Regional jejunal perfusion, a new in vivo approach to study oral drug absorption in man. *Pharm. Res., 9,* 1243–1251.

Lim, H. P., Tey, B. T., & Chan, E. S. (2014). Particle designs for the stabilization and controlled-delivery of protein drugs by biopolymers: A case study on insulin. *J. Control. Release, 186,* 11–21.

Lipka, E., Crison, J., & Amidon, G. L. (1996). Transmembrane transport of peptide type compounds: Prospects for oral delivery. *J. Control. Release, 39,* 121–129.

Lowe, P. J. & Temple, C. S. (1994). Calcitonin and insulin in isobutylcyanoacrylate nanocapsules: Protection against proteases and effect on intestinal absorption in rats. *J. Pharm. Pharmacol., 46,* 547–552.

Lu, H., Thomas, J. D., Tukker, J. J., & Fleisher, D. (1992). Intestinal water and solute absorption studies: Comparison of in situ perfusion with chronic isolated loops in rats. *Pharm. Res., 9,* 894–900.

Lu, R. H., Kopeckova, P., & Kopecek, J. (1999). Degradation and aggregation of human calcitonin in vitro. *Pharm. Res., 16,* 359–367.

LueBen, H. L., Lehr, C. M., Rentel, C. O., Noach, A. B. J., Deboer, A. G., Verhoef, J. C. et al. (1994). Bioadhesive polymers for the peroral delivery of peptide drugs. *J. Control. Release, 29,* 329–338.

Lundin, S., Pantzar, N., Broeders, A., Ohlin, M., & Westrom, B. R. (1991). Differences in transport rate of oxytocin and vasopressin analogues across proximal and distal isolated segments of the small intestine of the rat. *Pharm. Res., 8,* 1274–1280.

Macadam, A. (1993). The effect of gastro-intestinal mucus on drug absorption. *Adv. Drug Deliv. Rev., 11,* 201–220.

Mahato, R. I., Narang, A. S., Thoma, L., & Miller, D. D. (2003). Emerging trends in oral delivery of peptide and protein drugs. *Crit. Rev. Ther. Drug Carrier Syst., 20,* 153–214.

Manganaro, M., Ogra, P. L., & Ernst, P. B. (1994). Oral immunization—Turning fantasy into reality. *Int. Arch. Allergy Immunol., 103,* 223–233.

Mesiha, M., Plakogiannis, F., & Vejosoth, S. (1994). Enhanced oral absorption of insulin from desolvated fatty acid sodium glycocholate emulsions. *Int. J. Pharm., 111,* 213–216.

Mestecky, J., Moldoveanu, Z., Novak, M., Huang, W. Q., Gilley, R. M., Staas, J. K. et al. (1994). Biodegradable microspheres for the delivery of oral vaccines. *J. Control. Release, 28,* 131–141.

Michel, C., Aprahamian, M., Defontaine, L., Couvreur, P., & Damge, C. (1991). The effect of site of administration in the gastrointestinal tract on the absorption of insulin from nanocapsules in diabetic rats. *J. Pharm. Pharmacol., 43,* 1–5.

Milstein, S. J., Leipold, H., Sarubbi, D., Leone-Bay, A., Mlynek, G. M., Robinson, J. R. et al. (1998). Partially unfolded proteins efficiently penetrate cell membranes—Implications for oral drug delivery. *J. Control. Release, 53,* 259–267.

Mora, M. & Pato, J. (1993). Preparation of egg albumin microparticles for oral application. *J. Control. Release, 25,* 107–113.

Morishita, M. & Peppas, N. A. (2006). Is the oral route possible for peptide and protein drug delivery? *Drug Discov. Today, 11,* 905–910.

Mrsny, R. J. (1991). Challenges for the oral delivery of proteins and peptides. In *Peptides: Theoretical and practical approaches to their delivery* (pp. 45–52). Greenwood, SC: Capsugel Symposia Series.

Mrsny, R. J., Cromwell, M., Villagran, J., & Rubas, W. (1993). Oral delivery of peptides to the small intestine. In Capsugel library (Eds.), *Current status on targeted drug delivery to the gastrointestinal tract* (pp. 45–56). Greenwood, SC: Capsugel Symposia Series.

Muranishi, S. (1991). Improvement of intestinal absorption of peptides by chemical modification with fatty acids. In *Peptides: Theoretical and practical approaches to their delivery* (pp. 73–77). Greenwood, SC: Capsugel Symposia Series.

O'Hagan, D. T. (1992). Oral delivery of vaccines formulation and clinical pharmacokinetic considerations. *Clin. Pharmacokinet., 22,* 1–10.

Pantzar, N., Lundin, S., & Westrom, B. R. (1995). Different properties of the paracellular pathway account for the regional small intestinal permeability to the peptide desmopressin. *J. Pharm. Sci., 84,* 1245–1248.

Park, K., Kwon, I., & Park, K. (2011). Oral protein delivery: Current status and future prospects. *React. Funct. Polym., 71,* 280–287. (http://www.generex.com). (Accessed August 30, 2014.)

Park, K. & Robinson, J. R. (1984). Bioadhesive polymers as platforms for oral-controlled drug delivery: Method to study bioadhesion. *Int. J. Pharm., 19,* 107–127.

Paterson, D. A., Conradi, R. A., Hilgers, A. R., Vidmar, T. J., & Burton, P. S. (1994). A non-aqueous partitioning system for predicting the oral absorption potential of peptides. *Quant. Struct.-Act. Relat., 13,* 4–10.

Pauletti, G. M., Gangwar, S., Knipp, G. T., Nerurkar, M. M., Okumu, F. W., Tamura, K. et al. (1996). Structural requirements for intestinal absorption of peptide drugs. *J. Control. Release, 41,* 3–17.

PDR Staff. *Physicians' Desk Reference* (2014). (68th ed.) Montvale, NJ: PDR Network, LLC.

Plate, N. A., Valuev, I. L., Sytov, G. A., & Valuev, L. I. (2002). Mucoadhesive polymers with immobilized proteinase inhibitors for oral administration of protein drugs. *Biomaterials, 23,* 1673–1677.

Porter, C. J. (1997). Drug delivery to the lymphatic system. *Crit. Rev. Ther. Drug Carrier Syst., 14,* 333–393.

Pozzi, F., Furlani, P., Gazzaniga, A., Davis, S. S., & Wilding, I. R. (1994). The TIME CLOCK system: A new oral dosage form for fast and complete release of drug after a predetermined lag time. *J. Control. Release, 31,* 99–108.

Price, C. H. (September 22–23, 2003). Oral delivery of macromolecules to eliminate invasive delivery. In *Protein and Peptide Drug Delivery: Scientific Advances in Enabling Novel Approaches & Improved Products IBC Meeting*, Philadelphia, PA.

Quadros, E., Landzert, N. M., LeRoy, S., Gasparani, F., & Worosila, G. (1994). Colonic absorption of insulin-like growth factor I in vitro. *Pharm. Res., 11,* 226–230.

Rekha, M. R. & Sharma, C. P. (2013). Oral delivery of therapeutic protein/peptide for diabetes—Future perspectives. *Int. J. Pharm., 440,* 48–62.

Renukuntla, J., Vadlapudi, A. D., Patel, A., Boddu, S. H., & Mitra, A. K. (2013). Approaches for enhancing oral bioavailability of peptides and proteins. *Int. J. Pharm., 447,* 75–93.

Rubas, W., Jezyk, N., & Grass, G. M. (1993). Comparison of the permeability characteristics of a human colonic epithelial (Caco-2) cell line to colon of rabbit, monkey, and dog intestine and human drug absorption. *Pharm. Res., 10,* 113–118.

Rubinstein, A. & Tirosh, B. (1994). Mucus gel thickness and turnover in the gastrointestinal tract of the rat: Response to cholinergic stimulus and implication for mucoadhesion. *Pharm. Res., 11,* 794–799.

Saffran, M., Bedra, C., Kumar, G. S., & Neckers, D. C. (1988). Vasopressin: A model for the study of effects of additives on the oral and rectal administration of peptide drugs. *J. Pharm. Sci., 77,* 33–38.

Saffran, M., Kumar, G. S., Savariar, C., Burnham, J. C., Williams, F., & Neckers, D. C. (1986). A new approach to the oral administration of insulin and other peptide drugs. *Science, 233,* 1081–1084.

Salmaso, S., Bersani, S., Elvassore, N., Bertucco, A., & Caliceti, P. (2009). Biopharmaceutical characterisation of insulin and recombinant human growth hormone loaded lipid submicron particles produced by supercritical gas micro-atomisation. *Int. J. Pharm., 379,* 51–58.

Santiago, N., Milstein, S., Rivera, T., Garcia, E., Zaidi, T., Hong, H. et al. (1993). Oral immunization of rats with proteinoid microspheres encapsulating influenza virus antigens. *Pharm. Res., 10,* 1243–1247.

Sarubbi, D., Variano, B., Haas, S., Maher, J., Falzarano, L., Burnett, B. et al. (1994). Oral delivery of calcitonin in rats and primates using proteinoid microspheres. In *Proceedings of the International Symposium on Controlled Release Bioactive Materials* (21st ed., pp. 288–289). Controlled Release Society, Inc., St. Paul, MN.

Sasaki, I., Fujita, T., Murakami, M., Yamamoto, A., Nakamura, E., Imasaki, H. et al. (1994). Intestinal absorption of azetirelin, a new thyrotropin-releasing hormone (TRH) analogue. I. Possible factors for the low oral bioavailability in rats. *Biol. Pharm. Bull., 17,* 1256–1261.

Schilling, R. J. & Mitra, A. K. (1990). Intestinal mucosal transport of insulin. *Int. J. Pharm., 62,* 53–64.

Schilling, R. J. & Mitra, A. K. (1992). Pharmacodynamics of insulin following intravenous and enteral administrations of porcine-zinc insulin to rats. *Pharm. Res., 9,* 1003–1009.

Schmitz, J., Preiser, H., Maestracci, D., Ghosh, B. K., Cerda, J. J., & Crane, R. K. (1973). Purification of the human intestinal brush border membrane. *Biochim. Biophys. Acta, 323,* 98–112.

Shah, R. B., Ahsan, F., & Khan, M. A. (2002). Oral delivery of proteins: Progress and prognostication. *Crit. Rev. Ther. Drug Carrier Syst., 19,* 135–169.

Shao, Z., Li, Y., Krishnamoorthy, R., Chermak, T., & Mitra, A. K. (1993). Differential effects of anionic, cationic, nonionic, and physiologic surfactants on the dissociation, alpha-chymotryptic degradation, and enteral absorption of insulin hexamers. *Pharm. Res., 10,* 243–251.

Shao, Z. H., Li, Y. P., Chermak, T., & Mitra, A. K. (1994). Cyclodextrins as mucosal absorption promoters of insulin. II. Effects of beta-cyclodextrin derivatives on alpha-chymotryptic degradation and enteral absorption of insulin in rats. *Pharm. Res., 11,* 1174–1179.

Shen, Z. & Mitragotri, S. (2002). Intestinal patches for oral drug delivery. *Pharm. Res., 19,* 391–395.

Sinko, P. J. (1992). Intestinal absorption of peptides and peptide analogues: Implications of fasting pancreatic serine protease levels and pH on the extent of oral absorption in dogs and humans. *Pharm. Res., 9,* 320–325.

Sinko, P. J., Leesman, G. D., & Amidon, G. L. (1993). Mass balance approaches for estimating the intestinal absorption and metabolism of peptides and analogues: Theoretical development and applications. *Pharm. Res., 10,* 271–275.

Smith, P., Mirabelli, C., Fondacaro, J., Ryan, F., & Dent, J. (1988). Intestinal 5-fluorouracil absorption: Use of ussing chambers to assess transport and metabolism. *Pharm. Res., 5,* 598–603.

Smith, P. L., Yeulet, S. E., Citerone, D. R., Drake, F., Cook, M., Wall, D. A. et al. (1994). SK&F 110679—Comparison of absorption following oral or respiratory administration. *J. Control. Release, 28,* 67–77.

Snider, D. P., Marshall, J. S., Perdue, M. H., & Liang, H. (1994). Production of IgE antibody and allergic sensitization of intestinal and peripheral tissues after oral immunization with protein Ag and cholera toxin. *J. Immunol., 153,* 647–657.

Stoll, B. R., Leipold, H. R., Milstein, S., & Edwards, D. A. (2000). A mechanistic analysis of carrier-mediated oral delivery of protein therapeutics. *J. Control. Release, 64,* 217–228.

Su, F. Y., Lin, K. J., Sonaje, K., Wey, S. P., Yen, T. C., Ho, Y. C. et al. (2012). Protease inhibition and absorption enhancement by functional nanoparticles for effective oral insulin delivery. *Biomaterials, 33,* 2801–2811.

Sutton, S. C., Forbes, A. E., Cargill, R., Hochman, J. H., & LeCluyse, E. L. (1992). Simultaneous in vitro measurement of intestinal tissue permeability and transepithelial electrical resistance (TEER) using sweetana-grass diffusion cells. *Pharm. Res., 9,* 316–319.

Takada, K., Nakahigashi, Y., Tanaka, T., Kishimoto, S., Ushirogawa, Y., & Takaya, T. (1994). Pharmacological activity of tablets containing recombinant human granulocyte colony-stimulating factor (rhG-CSF) in rats. *Int. J. Pharm., 101,* 89–96.

Takaya, T., Niwa, K., & Takada, K. (1994). Pharmacological effect of recombinant human colony-stimulating factor (rhG-CSF) after administration into rat large intestine. *Int. J. Pharm., 110,* 47–53.

Taki, Y., Yamashita, S., Sakane, T., Nadai, T., Sezaki, H., Langguth, P. et al. (1994). Gastrointestinal absorption of metkephamid—Quantitative evaluation of degradation and permeation. In *Proceedings of the International Symposium on Controlled Release Bioactive Materials* (21st ed., pp. 814–815). Controlled Release Society, Inc., St. Paul, MN.

Thwaites, D. T., Hirst, B. H., & Simmons, N. L. (1993). Passive transepithelial absorption of thyrotropin-releasing hormone (TRH) via a paracellular route in cultured intestinal and renal epithelial cell lines. *Pharm. Res., 10,* 674–681.

Thwaites, D. T., Simmons, N. L., & Hirst, B. H. (1993). Thyrotropin-releasing hormone (TRH) uptake in intestinal brush-border membrane vesicles: Comparison with proton-coupled dipeptide and Na^+-coupled glucose transport. *Pharm. Res., 10,* 667–673.

Tomita, M., Sawada, T., Ogawa, T., Ouchi, H., Hayashi, M., & Awazu, S. (1992). Differences in the enhancing effects of sodium caprate on colonic and jejunal drug absorption. *Pharm. Res., 9,* 648–653.

Tomizawa, H., Aramaki, Y., Fujii, Y., Hara, T., Suzuki, N., Yachi, K. et al. (1993). Uptake of phosphatidylserine liposomes by rat Peyer's patches following intraluminal administration. *Pharm. Res., 10,* 549–552.

Torres-Lugo, M., Garcia, M., Record, R., & Peppas, N. A. (2002). Physicochemical behavior and cytotoxic effects of p(methacrylic acid-*g*-ethylene glycol) nanospheres for oral delivery of proteins. *J. Control. Release, 80,* 197–205.

Toth, I. (1994). A novel chemical approach to drug delivery: Lipidic amino acid conjugates. *J. Drug Target., 2,* 217–239.

Toth, I., Flinn, N., Hillery, A., Gibbons, W. A., & Artursson, P. (1994). Lipidic conjugates of luteinizing hormone releasing hormone $(LHRH)^+$ and thyrotropin releasing hormone $(TRH)^+$ that release and protect the native hormones in homogenates of human intestinal epithelial (Caco-2) cells. *Int. J. Pharm., 105,* 241–247.

Touitou, E. & Rubinstein, A. (1986). Targeted enteral delivery of insulin to rats. *Int. J. Pharm., 30,* 95–99.

Vilhardt, H. & Bie, P. (1983). Antidiuretic response in conscious dogs following peroral administration of arginine vasopressin and its analogues. *Eur. J. Pharmacol., 93,* 201–204.

Wang, J., Chow, D., Heiati, H., & Shen, W. C. (2003). Reversible lipidization for the oral delivery of salmon calcitonin. *J. Control. Release, 88,* 369–380.

Wang, W. (1996). Oral protein drug delivery. *J. Drug Target., 4,* 195–232.

Wang, X. Q. & Zhang, Q. (2012). pH-sensitive polymeric nanoparticles to improve oral bioavailability of peptide/protein drugs and poorly water-soluble drugs. *Eur. J. Pharm. Biopharm., 82,* 219–229.

Wearley, L. L. (1991). Recent progress in protein and peptide delivery by noninvasive routes. *Crit. Rev. Ther. Drug Carrier Syst., 8,* 331–394.

Wilding, I. R., Davis, S. S., Pozzi, F., Furlani, P., & Gazzaniga, A. (1994). Enteric coated timed release systems for colonic targeting. *Int. J. Pharm., 111,* 99–102.

Yamamoto, A., Taniguchi, T., Rikyuu, K., Tsuji, T., Fujita, T., Murakami, M. et al. (1994). Effects of various protease inhibitors on the intestinal absorption and degradation of insulin in rats. *Pharm. Res., 11,* 1496–1500.

Yen, W. C. & Lee, V. H. L. (1994). Paracellular transport of a proteolytically labile pentapeptide across the colonic and other intestinal segments of the albino rabbit—Implications for peptide drug design. *J. Control. Release, 28,* 97–109.

Yuasa, H., Amidon, G. L., & Fleisher, D. (1993). Peptide carrier-mediated transport in intestinal brush border membrane vesicles of rats and rabbits: Cephradine uptake and inhibition. *Pharm. Res., 10,* 400–404.

Yun, Y., Cho, Y. W., & Park, K. (2013). Nanoparticles for oral delivery: Targeted nanoparticles with peptidic ligands for oral protein delivery. *Adv. Drug Deliv. Rev., 65,* 822–832.

Ziv, E., Kidron, M., Berry, E. M., & Bar-On, H. (1981). Bile salts promote the absorption of insulin from the rat colon. *Life Sci., 29,* 803–809.

Ziv, E., Kidron, M., Raz, I., Krausz, M., Blatt, Y., Rotman, A. et al. (1994). Oral administration of insulin in solid form to nondiabetic and diabetic dogs. *J. Pharm. Sci., 83,* 792–794.

8 Transdermal and Topical Delivery of Therapeutic Peptides and Proteins

8.1 INTRODUCTION

The skin provides an appealing site for the noninvasive entry of drugs into systemic circulation (Brown & Langer, 1988; Guy & Hadgraft, 1992; Michaels, Chandrasekaran, & Shaw, 1975; Paudel et al., 2010; Walters, 1986; Wester & Maibach, 1992). Delivery of drugs through the skin for systemic effect, called transdermal delivery, came into therapeutics in 1978 when Transderm V® (now called Transderm Scop®) was marketed for motion sickness. Transdermal delivery has become very popular since the introduction of several nitroglycerin products in 1981, and since then, several more products have been introduced to the market. Currently, several drugs such as buprenorphine, capsaicin, clonidine, estradiol, fentanyl, granisetron, methylphenidate, rivastigmine, rotigotine, selegiline, nicotine, nitroglycerin, oxybutynin, scopolamine, and testosterone have been successfully marketed for transdermal delivery, with several brands available for many of these drugs. Patches with estrogen/progestin combination for birth control and lidocaine patches for topical use are also available. Nitroglycerin patches for prophylaxis of angina pectoris are available as several different brands of nitroglycerin TDS. These brands differ in their system design, formulation composition, and technology basis. Transdermal patches are considered by the regulatory agencies to be drug–device combination products (Couto, Perez-Breva, Saraiva, & Cooney, 2012).

Although several drugs are now available on the market as transdermal patches, none of these is a peptide or protein drug. The reasons are not difficult to guess. The skin is ordinarily permeable only to small lipophilic molecules, a criterion readily fulfilled by drugs such as nitroglycerin, scopolamine, clonidine, nicotine, and other drugs on the market. Peptide and protein drugs, being hydrophilic and macromolecular in nature, do not readily permeate the skin. However, recently some enhancement techniques such as iontophoresis are becoming available, which are likely to make transdermal delivery a promising approach for delivery of peptide/protein and other biotechnology-derived drugs (Kasha & Banga, 2008; Nolen, Catz, & Friend, 1994; Panchagnula, Pillai, Nair, & Ramarao, 2000; Pillai, Nair, Poduri, & Panchagnula, 1999; Wallace & Lasker, 1993). Other promising enhancement techniques include use of ultrasound energy (phonophoresis) and creation of microscopic holes (microporation) by microneedles or other thermal means (Banga, 2011; Schoellhammer, Blankschtein, & Langer, 2014).

Transdermal route offers some distinct advantages for the delivery of peptide drugs. Since these drugs have short half-lives, the greatest benefit would be the fact that the transdermal route provides a continuous mode of administration, somewhat similar to that provided by an intravenous (IV) infusion. However, unlike an IV infusion, delivery is noninvasive and no hospitalization is required. Also, the skin is very low in proteolytic activity, as compared to other mucosal routes, thereby reducing degradation at the site of administration. Once absorbed, the hepatic circulation is bypassed, thus avoiding another major site of potential degradation (Berner & John, 1994). However, it should be realized that the skin is an immunocompetent organ and irritancy and allergy can be a stringent constraint on transdermal delivery. The skin has Langerhans and other kinds of cells that could be involved in the immunological process. This is because topical or transdermal delivery leads to drug accumulation in the skin, which is several times higher than that ever achieved in the skin through systemic administration of that drug, thus raising tissue concentration above the threshold for a response (Flynn & Stewart, 1988). The peptide molecules may serve as antigens or haptens to the immunological apparatus of the skin (Merkle, 1989). This could result in delayed hypersensitivity, allergic contact dermatitis, or even loss of biological activity of the drug.

Although systemic transdermal delivery using enhancement techniques such as iontophoresis will probably be limited to peptide or small polypeptide drugs only, topical application or delivery for localized effects on the skin would extend to protein drugs as well. An example for topical delivery would be the use of growth factors for wound healing. However, systemic delivery of larger protein molecules or even vaccines may be feasible with other enhancement techniques such as microporation or ultrasound enhancement. Both systemic and topical deliveries of peptides and proteins through the skin have been investigated (see Table 8.1) and are discussed in this chapter.

8.1.1 Skin Structure

The skin may be divided into two layers: the outer epidermis and the inner dermis. Epidermis can be subdivided into several layers of cells. These cells originate in the stratum germinativum and undergo differentiation as they move up through the layers of spinosum, granulosum, and lucidum. The uppermost layer, the stratum corneum, consists of dead, flattened keratin-filled cells (corneocytes) in a lipid matrix. Each corneocyte is bounded by a thick, proteinaceous envelope with the tough fibrous protein (keratin) as the main component. The intercellular spaces are filled with broad multiple lamellae. The lipids constituting these lamellae are unique in that they do not contain phospholipids. Instead, they are composed of ceramides, cholesterol, fatty acids, and cholesteryl esters. Epidermis also contains Langerhans cells, which can initiate skin immune responses. The entire epidermis is avascular, and the underlying dermis provides its physiological support by supplying it with nerve endings, blood vessels, and lymphatic vessels. Thus, dermis is often referred to as *viable dermis*.

TABLE 8.1
Transdermal/Topical Delivery of Peptide and Protein Drugs

Peptide/Protein	Animal Model/ Experimental Design	Transdermal/ Topical	Enhancement Mechanism	References
Calcitonin	Hairless rats	Transdermal	Iontophoresis Microneedles Cell-penetrating peptide	Banga and Chien (1993), Manosroi et al. (2013b), Morimoto et al. (1992), Tas et al. (2012), Thysman et al. (1994), Tomohira, Machida, Onishi, and Nagai (1997), Vemulapalli et al. (2012)
Cholecystokinin-8 analog (CCK-8)	Human cadaver skin	Transdermal	Chemical enhancer/ iontophoresis	Srinivasan et al. (1990)
Erythropoietin	Rats	Transdermal	Microneedles	Peters, Ameri, Wang, Maa, and Daddona (2012)
EGF		Topical	Protease inhibitor Bioadhesive gel	Celebi et al. (1994), Kiyohara et al. (1993), Okumura et al. (1990)
Growth hormone–releasing factor analog (GRF/Ro 23-7861) and peptide (GHRP)	Hairless guinea pig/ in vitro Rats	Transdermal	Iontophoresis	Ellens et al. (1997) and Kumar et al. (1992)
Hisetal (α-MSH)	Hairless mouse/ in vitro Human skin/in vitro	Transdermal	Chemical enhancers	Ruland et al. (1994a, 1994b)
Human growth hormone	Yorkshire swine	Transdermal	Hollow or dissolving microneedles	Fukushima et al. (2010), Harvey et al. (2010)

(Continued)

TABLE 8.1 (*Continued*)
Transdermal/Topical Delivery of Peptide and Protein Drugs

Peptide/Protein	Animal Model/ Experimental Design	Transdermal/ Topical	Enhancement Mechanism	References
Insulin	Pigs Diabetic rabbits Hairless rats Human cadaver skin Hairless mice Yorkshire swine	Transdermal	Iontophoresis Chemical enhancer Microporation Phonophoresis Microneedles	Banga and Chien (1993), Harvey et al. (2010), Kanikkannan et al. (1999), Kari (1986), Langkjaer et al. (1998), Liu et al. (1988, 2012), Malakar et al. (2011), Martanto et al. (2004), Meyer et al. (1989), Pillai and Panchagnula (2003), Rastogi and Singh (2003), Sintov and Wormser (2007), Srinivasan et al. (1989), Stephen et al. (1984), Sun and Xue (1991), Tachibana (1992), Tachibana and Tachibana (1991), and Zakzewski et al. (1998)
Interferon	Guinea pig	Topical Transdermal	Liposomes Thermal microporation	Badkar et al. (2007), Egbaria et al. (1990), King et al. (2013), Weiner et al. (1989)
LHRH/analogs	Humans volunteers Hairless mice Pig/perfused flap Nude mouse/snake/ cadaver skin	Transdermal	Iontophoresis Microneedles Chemical enhancers	Delgadocharro et al. (1995), Heit et al. (1993, 1994), Hoogstraate et al. (1994), Knoblauch and Moll (1993), Lu et al. (1992), Meyer et al. (1988), Miller et al. (1990), Roy and Degroot (1994), Sachdeva, Zhou, and Banga (2013), Srinivasan et al. (1990)
Thyrotropin-releasing hormone (TRH)	Nude mice Human cadaver skin	Transdermal	Iontophoresis Prodrug	Burnette and Marrero (1986), Magnusson et al. (1997), Moss and Bundgaard (1990)
Vasopressin and analogs	Hairless or Wistar rat Human cadaver skin	Transdermal	Iontophoresis Chemical enhancer Microneedles	Banerjee and Ritschel (1989a, 1989b), Banga and Chien (1993), Fukushima et al. (2010), Hoogstraate et al. (1991), Lelawongs et al. (1989, 1990), Nakakura et al. (1998)

Human skin is about 2–3 mm thick, while the stratum corneum over most of the body is only about 15 μm thick. However, most of the epidermal mass is concentrated in the stratum corneum, which is considered to form the principal or rate-limiting barrier to the permeation of drugs. The stratum corneum shows significant chemical and structural differences from one site to another. The stratum corneum over the plantar and palmar skin is much thicker but has relatively high diffusivity. The skin also contains several appendages such as hair follicles and sweat glands. Associated with the hair follicles are sebaceous glands, which secrete sebum that covers the epidermis. These appendages extend through the epidermis and dermis to subcutaneous tissue but occupy only about 0.1% of the total human skin surface.

8.1.2 Pathways of Transdermal Delivery

Permeation of drugs through the stratum corneum can take place via a transepidermal or a transappendageal route. The transepidermal route may involve intercellular or transcellular pathways, while the transappendageal route involves transport across follicles and sweat glands. The intercellular route has also been implicated for the iontophoretic delivery of some ions (Monteiro-Riviere, Inman, & Riviere, 1994). The relative importance of the intercellular and transcellular paths depends on the physicochemical properties of the penetrant, especially the partition coefficients for contiguous phases and diffusion coefficients within the protein or lipid regions.

In normal skin, the lipid morphology plays an important role in the barrier function of the stratum corneum (Denda, Koyama, Namba, & Horii, 1994). For the absorption of macromolecules from the skin, the pore size of the skin becomes very important. The effective pore size of ethanol-pretreated human epidermal membrane has been estimated to be between 22 and 54 Å (Inamori, Ghanem, Higuchi, & Srinivasan, 1994). Once a drug has crossed the epidermis, it can enter the systemic circulation via the papillary plexus in the upper dermis. Many models assume that this is the end point for percutaneous absorption, but it should be realized that some molecules will bypass the circulation and diffuse deeper into the dermis (Flynn & Stewart, 1988).

Normally, the contribution of the appendageal permeation to total passive transport is very small and may be negligible. However, the appendageal route becomes very important during transdermal iontophoretic transport, as it provides the shortest shunt pathway to the flow of electric current and ions (Banga & Chien, 1988). Laboratory animals, such as rats, mice, and rabbits, lack sweat glands but have more hair follicles than human skin. Hairless animals are sometimes used in iontophoresis research for this reason, but it should be realized that even these animals may have rudimentary follicles even though the shafts are absent. Larger animals have more anatomical features in common with humans, and thus, weanling pigs and rhesus monkey may be good models for transdermal studies (Shah et al., 1991). The pig skin is considered to be morphologically and functionally similar to human skin and has a hair follicle density identical to that of human skin (Bartek, LaBudde, & Maibach, 1972; Monteiro-Riviere et al., 1994).

8.1.3 Enzymatic Barrier of Skin

The spectrum of reactions that can take place in the skin are similar to those that occur in the liver. However, the blood perfusion of the skin is poor, with the total skin blood flow representing only about 6.25% of the total liver blood flow. Therefore, the skin metabolism does not contribute much to systemic drug disposition. However, for topical and transdermal delivery, the drug has to pass into or through the skin, and metabolism in the skin will take place. Such metabolic activity includes oxidative, reductive, hydrolytic, and conjugative reactions. Enzymatic hydrolysis of salicylate esters in the skin has been reported (Guzek, Kennedy, McNeill, Wakshull, & Potts, 1989). Major differences were observed in the metabolism of these esters in human skin under in vitro versus in vivo conditions. In vitro measurements were found to overestimate metabolic activity, possibly due to lack of perfusion and/or release of enzymes. Skin grafted to athymic mice and rats may be more reflective of the metabolic and transport properties of human skin in vivo. Such skin maintains vascular perfusion and is not subjected to surgical procedures that would release enzymatic activity.

As discussed by Hopsu-Havu, Fraki, and Jarvinen (1977), the metabolic activity of the skin includes considerable proteolytic activity. Of the exopeptidases in the skin, aminopeptidases are the best known, while carboxypeptidases are relatively in smaller quantity or absent. The endopeptidases include the proteinases such as the caseinolytic enzymes, chymotrypsin- and trypsin-substrate hydrolyzing enzymes, thiol proteinase, and carboxyl proteinases. In addition, proteinases attacking specific substrates such as collagenases, elastase, fibrinolysins, and enzymes of the kallikrein–kinin system are present. It must be realized that studies with skin homogenates may not be a good predictor of proteolytic degradation actually encountered during transport. This is because if proteolysis in the skin is discussed based on studies with skin homogenates, the structural complexity of the skin is ignored. The subcellular compartmentalization of the proteolytic enzymes may be such that these enzymes do not encounter the peptide or protein drug as the pathways of penetration may not pass through the cells that hold these enzymes. Once a skin homogenate has been made, the origins of the individual enzymes in the homogenate cannot be determined.

8.2 APPROACHES TO ENHANCE TRANSDERMAL PEPTIDE DELIVERY

Approaches to enhance transdermal delivery may either modify the skin to make it more permeable or may act on the drug to provide a driving force to enter the skin (Banga, 2011; Gratieri, Alberti, Lapteva, & Kalia, 2013). An example of the latter approach would be the use of electric current (iontophoresis) to push a like charged drug molecule into the skin. Approaches that act on the skin can use either chemical enhancers or minimally invasive technologies such as skin microporation by microneedles or by laser or thermal ablation. Electroporation and sonophoresis also primarily act on the skin. Iontophoresis will be discussed in more details in Section 8.3, while other approaches are discussed in this section.

8.2.1 Skin Microporation

In recent years, a very promising technology is being developed for transdermal delivery of macromolecules such as proteins and vaccines. It involves a minimally invasive technique in which transport pathways of microns dimension are created in the skin. These micron-sized holes are huge compared to drug dimensions but still much smaller than the holes made by hypodermic needles. These micron-sized holes can be created in the skin using mechanical microneedles (Prausnitz, 2001, 2004; van der Maaden, Jiskoot, & Bouwstra, 2012) or by using a thermal microporation process (Bramson et al., 2003). The thermal microporation process involves the application of rapid and controlled pulses of thermal energy, by means of tiny resistive elements, to a matrix of microscopic sites on the skin surface. A short-duration (milliseconds) electric current is passed through the array. At each site, due to resistive heating, a micropore (around 100 μm wide and 40 μm deep) is created by flash vaporization of stratum corneum cells in an area about the width of a human hair. Using thermal microporation, interferon alpha-2b was reported to be delivered in vivo to hairless rats (Badkar, Smith, Eppstein, & Banga, 2007), and reporter gene expression in mice has been reported to be increased 100-fold following application of an adenovirus vector to microporated skin when compared to intact skin (Bramson et al., 2003). Clinical data for delivery of insulin and interferon through thermally created micropores have been reported, and the technique was investigated for vaccine delivery (Eppstein, 2002). The thermal microporation process can also be used for diagnostic applications such as glucose monitoring based on sampling skin interstitial fluid as an alternative method to the fingerstick blood glucose monitoring (Smith et al., 1999). Laser energy or radio-frequency ablation has also been used to create micropores in the skin. Companies that were developing thermal microporation or radio-frequency ablation techniques have suffered financial setbacks though the technology has been licensed out to other companies (Banga, 2011; Kalia, Bachhav, Bragagna, & Bohler, 2008). The microchannels created in the skin by any of these techniques will be temporary since the stratum corneum will be replaced through the natural process of desquamation.

8.2.1.1 Microneedles

An array of microneedles, much smaller in length than a hypodermic needle, can be used to bypass the stratum corneum and, depending on the length of the microneedle, may deliver the drug to the epidermis or dermis. Microneedles are typically less than 500 μm in length (Kalluri & Banga, 2009; Sachdeva & Banga, 2011). The drug may be placed on the skin in a patch or formulation after creating the microchannels in the skin using microneedles, or the drug may be coated on the microneedles. Soluble microneedles can also be used in which drug is incorporated into the microneedles and is released once the microneedles dissolve in the skin. Alternatively, the drug solution can be delivered into the skin using hollow microneedles (Figure 8.1). The microchannels created by microneedles can stay open for several days if the area is occluded, for example, with a formulation or patch, but will seal off within a few hours once the patch is removed or if the skin is not occluded after microporation (Gupta, Gill, Andrews, & Prausnitz, 2011; Kalluri & Banga, 2010).

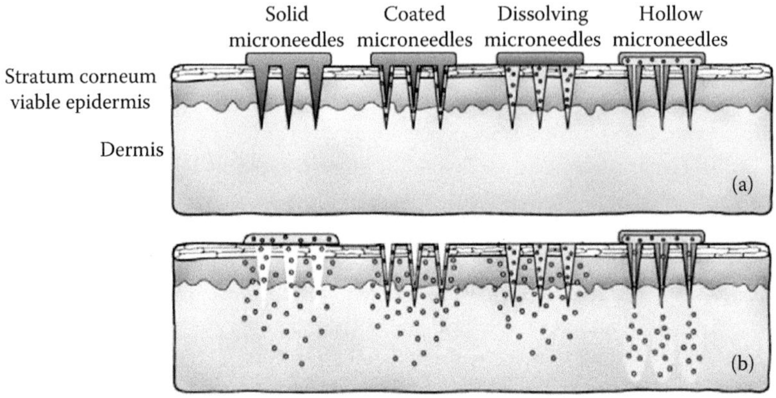

FIGURE 8.1 **(See color insert.)** Microneedles (MN) applied on the skin (a) to achieve drug delivery (b). The skin can be pretreated with solid microneedles followed by drug application, or the drug may be coated on microneedles or incorporated within dissolving microneedles or injected into the skin using hollow microneedles. (Reproduced from Kim, Y.C. et al., *Adv. Drug Deliv. Rev.*, 64, 1547, 2012. With permission.)

These microchannels can be visualized by staining the skin with methylene blue or using calcein dye followed by examination by fluorescent microscope, and creation of microchannels in the skin results in a significant increase in the transepidermal water loss. The microchannels can be further characterized by histology, confocal microscopy, and computation of a pore permeability index, and transport of proteins or antibodies through these microchannels can be visualized by immunohistochemical studies (Kolli & Banga, 2008; Li, Badkar, Kalluri, & Banga, 2009; Li, Badkar, Nema, Kolli, & Banga, 2009). Using immunohistochemical staining techniques and using IgG as a model drug for monoclonal antibodies, it was shown that proteins are transported across these hydrophilic microchannels created in the skin by microneedles (Figure 8.2) (Li et al., 2009).

Several companies including 3M, Apogee Technology, Becton Dickinson, Corium (acquired technology from Procter and Gamble), Debiotech, Elegaphy, Imtek, ISSYS, Kumetrix, Nanopass, Norwood Abbey, Silex Microsystems, Theraject, Valeritas (a subsidiary of BioValve), Zeopane, and Zosano (formerly Macroflux® and Alza/J&J spin-off) are exploring or have recently explored the use of microneedles for drug delivery. The microelectronics industry is making available the microfabrication tools (Tao & Desai, 2003) needed to make these small microneedles. Microfabrication uses tools used to make integrated circuits, that is, micromachining or microelectromechanical systems (MEMS), and consists of technologies supporting the core technology of microlithographic pattern transfer (McAllister, Allen, & Prausnitz, 2000). It has been reported that insertion of these microneedles is not painful in humans, and no erythema, edema, or other reaction to microneedles was observed (Bal, Caussin, Pavel, & Bouwstra, 2008; Henry, McAllister, Allen, & Prausnitz, 1998). These microneedles will typically enter through the stratum corneum and into the epidermis. The stratum corneum barrier has no nerves, which may partly

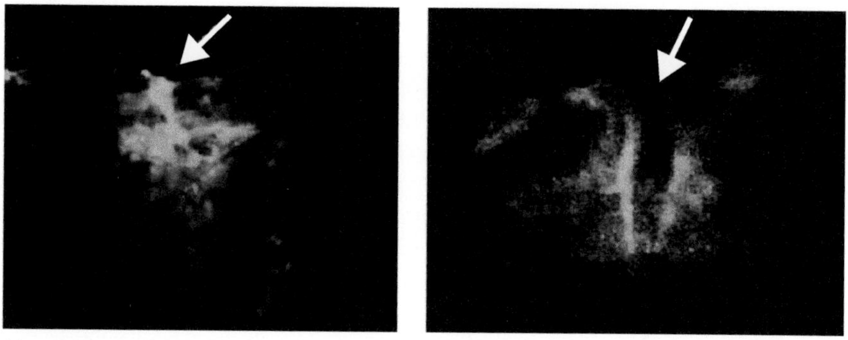

(a) (b)

FIGURE 8.2 (See color insert.) Fluorescent micrographs of IgG transport down through the microchannels (a) and radial diffusion (b) into the surrounding tissue in hairless rat skin following treatment with 500 μm long maltose microneedles. (Reproduced from Li, G. et al., *J. Pharm. Sci.*, 99, 1931, 2009. With permission.)

explain the lack of any pain sensation. Nerves are present in the epidermis, but the lack of pain may be attributed to their small size that possibly does not encounter or does not stimulate a nerve to produce a painful sensation (Prausnitz, 2004). In vitro studies using human epidermis have shown that skin permeability may be enhanced by as much as four orders of magnitude by using microneedles (Henry et al., 1998).

Typically, solid microneedles have been used in most reported studies. The first microneedles used were etched into a silicon wafer using lithography and reactive ion etching to form a 20-by-20 array where each needle measured 80 μm at the base and tapered to a height of 150 μm with a radius of curvature at the tip of about 1 μm. Solid metal microneedles have been reported to enable insulin delivery and lower blood glucose levels by as much as 80% in diabetic hairless rats in vivo (Martanto et al., 2004). More recently, the trend seems to be shifting toward the use of dissolving microneedles (Banga, 2011; Fukushima et al., 2010; Ito, Yoshimitsu, Shiroyama, Sugioka, & Takada, 2006; Lee, Park, & Prausnitz, 2008). Self-dissolving microneedles fabricated from hyaluronic acid (HA) have been demonstrated to enhance transdermal delivery of macromolecules (Liu et al., 2012, 2014). In one study, insulin-loaded HA microneedles (800 μm long) were dissolved in the skin within 1 h, and a dose-dependent hypoglycemic effect was observed (Liu et al., 2012). Self-dissolving microneedles made of thread-forming polymers such as dextrin, chondroitin sulfate, and albumin have also been used to show delivery of erythropoietin loaded within the microneedles (Ito et al., 2006). Hollow microneedles can also be used though they are more difficult to make and use (Hansen, Burton, & Tomai, 2009; Prausnitz, 2004). A longer hollow microneedle has been commercialized for influenza vaccination (Banga, 2011; Kim, Park, & Prausnitz, 2012). Intradermal delivery of proteins using hollow microneedles has been reported to change absorption kinetics by targeting a tissue bed better perfused with lymphatic and blood vessels as compared to subcutaneous space (Harvey et al., 2011).

Zosano Pharma has developed a Macroflux microprojection array system that incorporates a stainless steel or titanium microprojection array. For example, it may be fabricated from a 30 μm thick foil of stainless steel with a microprojection density of 240 cm^{-2} and a length of 430 μm. In vivo studies using hairless guinea pigs reported that this microprojection patch technology, by itself or in conjunction with iontophoresis, is capable of delivering therapeutically relevant quantities of oligonucleotides into and through the skin. Similarly, a titanium microprojection array (190 projections per square centimeter of 330 μm each) was used in hairless guinea pigs for vaccine delivery using a thin dry film coating of ovalbumin as a model protein antigen. The projections penetrated the skin to a depth of 100 μm with no projections deeper than 300 μm, and no responses visually assessed for skin tolerability varied from no detectable erythema to mild reactions that typically resolved within 24 h (Lin et al., 2001; Matriano et al., 2002; Widera et al., 2006). These microneedles were also coated with parathyroid hormone and tested in clinical trials for treatment of postmenopausal women with osteoporosis. The coating procedure, sterilization method, and clinical results have been described in literature (Ameri, Daddona, & Maa, 2009; Ameri, Fan, & Maa, 2010; Ameri, Wang, & Maa, 2010; Daddona, Matriano, Mandema, & Maa, 2011).

8.2.2 Phonophoresis

Phonophoresis, also referred to as sonophoresis, may be defined as the transport of drug molecules under the influence of ultrasound (Levy, Kost, Meshulam, & Langer, 1989; Machluf & Kost, 1993; Rao & Nanda, 2009; Sanghvi & Banakar, 1991; Tyle & Agrawala, 1989). Ultrasound is defined as sound with a frequency greater than 16 kHz, though a frequency of about 1 MHz is often used for diagnostic applications. The frequency used for phonophoresis has varied from 20 kHz to 10 MHz, though in recent years, it has been recognized that low-frequency sonophoresis (<100 kHz) may be most effective for drug delivery (Banga, 2011; Schoellhammer et al., 2014). The intensity used is also important and should normally not exceed 2.5 W/cm^2. Ultrasound has the potential to enhance delivery through the skin, though some of the reported results have been inconsistent. According to Bommannan, Menon, Okuyama, Elias, and Guy (1992), Bommannan, Okuyama, Stauffer, and Guy (1992), these inconsistent results may be attributed to investigators having used low ultrasound frequencies (1–3 MHz), applied at intensity levels of 1–3 W/cm^2. In order to expect biologic effects from ultrasound, the energy must be absorbed by the tissues. The depth of penetration in tissue is controlled by the attenuation of sound, which in turn is inversely related to frequency. Therefore, high-frequency ultrasound will increase energy disposition within the stratum corneum. Using frequencies of 2, 10, and 16 MHz at an intensity of 0.2 W/cm^2, the investigators successfully delivered salicylic acid or an electron dense, colloidal tracer, lanthanum hydroxide into hairless guinea pig skin. The tracer was found to penetrate epidermis and dermis with just 5 min of sonophoresis at 16 MHz. However, 20 min of sonophoresis at 16 MHz appeared to result in significant cytotoxic effects. Therefore, careful evaluation of the technique is required for acceptance of sonophoresis as a potential enhancement technique.

To transfer ultrasound energy to the body, a contact medium or coupling agent is required to transfer ultrasonic energy from the ultrasonic device to the skin. A gel, emulsion, or ointment could be used for such purpose (Kost, 1993). The coupling agent may also be used as a drug carrier, but the amount of coupling agent to be applied should be optimized. The mechanism of ultrasound-induced skin permeation enhancement is not clear. However, a combination of thermal, mechanical, or cavitational effects is involved. Also, an important question about the reversibility of skin permeation still needs to be answered. Phonophoresis, if found to be a viable technique for skin permeation enhancement, offers a potential for the delivery of peptides and polypeptides (Kost, 1993). Tachibana has reported successful delivery of insulin by ultrasound exposure to alloxan-diabetic rabbits (Tachibana, 1992) and to hairless mice (Tachibana & Tachibana, 1991). He used a device with a piezoelectric element and observed an elevation of plasma insulin and a decrease in the blood glucose level. Blood glucose levels continued to decrease after ultrasound and insulin exposure were stopped. This is likely due to a depot formation of insulin in the skin, similar to that observed with iontophoretic delivery, as will be discussed later. Any possible effect of ultrasound on the stability of peptide drugs must be evaluated. Even if the exposure itself does not cause any stability problems, cavitation or bubble formation may generate air/water interfaces. As we discussed in Chapter 3, air/water interfaces can lead to aggregation and other stability problems. Use of ultrasound to achieve quick skin analgesia by delivery of lidocaine is being developed by Echo Therapeutics (Philadelphia, PA). This may lay the foundation for future commercial development of phonophoresis for delivery of proteins.

8.2.3 Prodrug Approach

The use of prodrugs offers a promising biochemical approach for enhancing the skin permeation of drug molecules (Ghosh & Banga, 1993a). As discussed earlier, the low skin permeability of peptides is partly due to their hydrophilic nature. The synthesis of lipophilic prodrugs of such peptides is likely to improve their transdermal transport. After diffusing into or through the skin, the prodrug undergoes reconversion into the parent active drug molecule. Such an approach was found feasible for thyrotropin-releasing hormone (TRH). The prodrugs studied were the *N*-isobutyloxycarbonyl and *N*-octyloxycarbonyl derivatives of TRH (Moss & Bundgaard, 1990). Results of diffusion experiments using excised human skin indicate that the *N*-octyloxycarbonyl derivative showed an enhanced permeability. The prodrug penetrated into the receptor phase was found to exist primarily as TRH. The authors have shown that the quantities equivalent to those given by infusion or injection of TRH can be delivered in this manner.

8.2.4 Permeation Enhancers

Several chemical compounds can increase the transdermal permeation of drugs, primarily by altering the barrier properties of the skin (Barry, 1987, 1991; Bommannan, Potts, & Guy, 1991; Garrison, Doh, Potts, & Abraham, 1994). These accelerants usually function by altering the intercellular lipid or intracellular protein domains of the stratum corneum. Their interaction with intercellular lipids may disrupt the highly ordered lamellar structure, thus increasing the diffusivity of drugs through the skin. Alternatively, they may directly solubilize or plasticize skin tissue components. Examples of commonly investigated chemical permeation enhancers include dimethyl sulfoxide (DMSO), 1-dodecylazacycloheptan-2-one (Azone), surfactants, and phospholipids. An ideal enhancer must be nontoxic, nonirritating, nonallergenic, pharmacologically inert, and compatible with most drugs and excipients. However, no single agent meets all the desirable attributes of an enhancer. A combination of enhancers may thus be required (Ghosh & Banga, 1993b, 1993c). In addition to these chemical enhancers, many of the generally recognized as safe (GRAS) parenteral vehicles can also enhance percutaneous drug absorption. Many of the transdermal and dermal delivery products on the market are formulated with cosolvents (Pfister & Hsieh, 1990a). These cosolvents include propylene glycol, polyethylene glycol 400, isopropyl myristate, isopropyl palmitate, ethanol, water, and mineral oil. Since an enhancer is delivered to the skin, its pharmacokinetics must be determined so as to know its half-life in the skin, degree of absorption, mechanism of elimination, and metabolism. Also, the reversibility of skin barrier properties should be determined as any breach of the barrier properties of the stratum corneum could result in infection (Pfister & Hsieh, 1990b). Since use of a transdermal patch creates occlusive conditions on the skin, this leads to increased hydration and irritation of the skin underneath the patch. Increased skin hydration may increase the permeation of the enhancer itself, which may cause even more irritation or toxicity. Further, the enhancer may increase the permeation of the formulation components along with the drug. Therefore, a careful evaluation of the long-term local and systemic toxicity of the chemical enhancers in the final transdermal dosage form is required (Ghosh & Banga, 1993c).

A nonionic surfactant, *n*-decyl methyl sulfoxide, was found to increase the permeability of two amino acids, tyrosine and phenylalanine, through hairless mouse skin. This enhancer also increased the permeability of the dipeptide, Phe-Phe, and the pentapeptide, enkephalin (Choi, Flynn, & Amidon, 1990). The percutaneous absorption of a tetrapeptide, hisetal across hairless mouse skin, was found to be enhanced by penetration enhancers. Passive permeation of the tetrapeptide was found feasible with a flux of 0.22 $g/cm^2/h$. This was enhanced about 28-fold when oleic acid was used as an enhancer. While oleic acid was the most effective enhancer, Azone and dodecyl *N,N*-dimethylamino acetate (DDAA), a relatively new penetration enhancer, were also found to be effective (Ruland, Kreuter, & Rytting, 1994a). Compared to hairless mouse skin, human skin was much less permeable to hisetal. Also, the effectiveness of penetration enhancers was much greater with mouse skin as compared to human skin (Ruland, Kreuter, & Rytting, 1994b). While α-MSH has potential use in multiple sclerosis, an analog of α-MSH, (NLe^4, D-Phe^7)-α-MSH, has potential applications for hypopigmentary disorders and for the stimulation of skin tanning without ultraviolet light. This melanotropic peptide can be delivered transdermally through human skin, as suggested by some in vitro studies (Dawson et al., 1990). Effects of enhancers such as sodium lauryl sulfate (an anionic surfactant), DMSO, and Azone on the transdermal permeation of vasopressin across excised rat skin have been investigated (Banerjee & Ritschel, 1989a, 1989b). While sodium lauryl sulfate had no significant effect on permeation, Azone was found to be the most effective enhancer and increased flux about 15 times. These findings were confirmed by in vivo studies on Brattleboro rats. These rats are genetically deficient in vasopressin and secrete large volumes of urine with low osmolality. Significant reduction in urine volume and increase in osmolality over a 24 h period resulted upon enhancement of vasopressin absorption in the presence of Azone. Besides the use of enhancers, another factor that increased permeation was stripping of the skin with cellophane tape. In another study, Hoogstraate et al. (1991) investigated the potential of azacycloheptan-2-ones (Azones) as penetration enhancer for desglycinamide arginine vasopressin (DGAVP). Using human stratum corneum under in vitro conditions, it was found that the hydrocarbon chain length of Azone determined its effectiveness as a penetration enhancer. While pretreatment with hexyl- or octyl-Azone did not enhance the flux, the permeability increased 1.9-, 3.5-, or 2.5-fold after pretreatment with decyl-, dodecyl-, or tetradecyl-Azone, respectively. The morphological and kinetic data suggested that Azone acted by interference with the packing arrangement of the intercellular lamellar lipids in the stratum corneum. Chemical enhancers have also been reported to significantly enhance the permeability of the LHRH analogs, leuprolide acetate through nude mouse skin (Lu, Lee, & Rao, 1992), and nafarelin acetate through human cadaver skin (Roy & Degroot, 1994). Terpenes have also been used as enhancers though smaller terpenes with low lipophilicity should be used to preserve the conformational stability and biological activity of protein (Varman & Singh, 2012). Avicins, a family of naturally occurring glycosylated triterpenes with a molecular weight (MW) exceeding 2 kDa, as well as small lipophilic molecules, have been shown to permeate, and this was attributed to synergistic effects involving both hydrophobic and hydrophilic residues (Pino, Gutterman, Vonwil, Mitragotri, & Shastri, 2012).

In recent years, several publications have appeared on the use of cell-penetrating peptides (CPPs) to enhance topical and transdermal delivery (Candan et al., 2012; Desai, Patlolla, & Singh, 2010; Hou et al., 2007; Lopes et al., 2008; Nasrollahi, Taghibiglou, Azizi, & Farboud, 2012). CPPs or protein transduction domains (PTDs) are short peptides (up to 30 residues), typically with polycationic, amphipathic, or hydrophobic characteristics, that interact with the protein by covalent or noncovalent interactions to facilitate membrane transport at low micromolar concentrations. Enhancement can be observed both in vitro and in vivo, and receptors are not involved in the transport mechanism, which is not well understood. HIV1-*trans*-activating transcriptional (Tat) peptide has been used to enhance delivery of proteins and plasmids (Manosroi, Khositsuntiwong et al., 2013; Manosroi, Lohcharoenkal et al., 2013b). It has been suggested that positively charged arginine groups in Tat may bind peptide complexes to negatively charged surfaces and allow translocation of salmon calcitonin through excised skin (Manosroi et al., 2013b). A peptide identified by phage display and termed by authors as a skin penetrating and cell entering (SPACE) peptide has been reported to enable the delivery of siRNA (by direct conjugation) (Hsu & Mitragotri, 2011) and HA (Chen, Gupta, Anselmo, Muraski, & Mitragotri, 2014) into the skin. For delivery of HA (200–325 kDa),

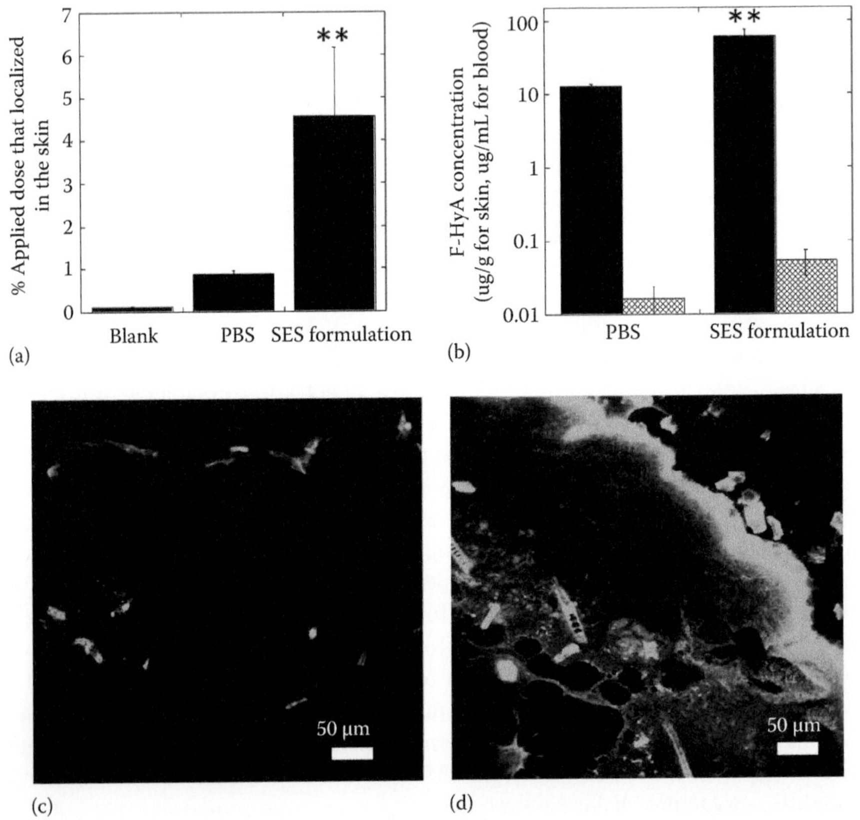

FIGURE 8.3 (See color insert.) SPACE-peptide ethosomal carrier (SES) enhances penetration of high-MW HA into hairless mice. The figure shows delivery into the skin as percentage of the applied dose (a) or as an amount in the skin and blood (b). Confocal microscopy shows that penetration was weaker from PBS buffer (c) and much stronger from SES formula (d). (Reproduced from Chen, M. et al., *J. Control. Release*, 173, 67, 2014. With permission.)

the SPACE peptide was conjugated to phospholipids and used to make an ethosomal carrier system (110 nm), which was found to penetrate deep into the skin, both in vitro and in vivo (Figure 8.3) (Chen et al., 2014).

8.2.5 Protease Inhibitors

Though the skin has less proteolytic activity compared to mucosal routes, it still is capable of hydrolyzing potential peptide drug candidates. Also, it should be noted that several animal species used in transdermal studies, such as hairless mouse skin, have significant proteolytic activity. Protease inhibitors can be used to avoid or reduce the enzymatic barrier of the skin. Protease inhibitors such as puromycin and amastatin were found to inhibit the degradation of leucine enkephalin in skin homogenates of hairless mouse skin. Interestingly, these inhibitors were not much effective in skin diffusion experiments (Choi et al., 1990). However, as discussed earlier, this could be because the proteolytic enzymes may be present in membranes outside the actual diffusion path. Other potentially useful protease inhibitors include aprotinin, bestatin, boroleucine, *p*-chloromercuribenzoate, leupeptin, pepstatin, and phenylmethylsulfonyl fluoride. A metalloprotease inhibitor, *o*-phenanthroline, has been shown to inhibit degradation of delta sleep-inducing peptide and increase its transdermal iontophoretic delivery eightfold at pH 7.4 (Chiang, Shao, & Chen, 1998). Sodium glycocholate, a penetration enhancer, may also act as a protease inhibitor. These agents could act by a variety of mechanisms, such as by tightly binding to or covalent modification of the active sites of proteases or by chelating metal ions essential for proteolytic activity. On a related mechanism, topical iodine was reported to allow transdermal delivery of insulin, presumably by inactivating endogenous sulfhydryls such as glutathione and gamma glutamylcysteine, which can reduce the disulfide bonds of the hormone, thereby retaining the potency of insulin (Sintov & Wormser, 2007).

8.2.6 Other Technologies

Electroporation, another electric enhancement technology, applies ultrashort pulses to the skin for a few milliseconds with an intensity of a few hundred volts. This opens pores in the skin that allows the drugs to pass. Normally, electroporation is used for reversibly increasing the permeability of lipid bilayers, such as those of living cells. The technique creates transient aqueous pores by the application of electrical pulses. Electroporation of multilamellar, nonphospholipid, intercellular lipid bilayers of the stratum corneum as an enhancement mechanism is a relatively new area of research. While iontophoresis acts primarily on the drug, electroporation acts primarily on the skin to enhance permeability (Prausnitz, Bose, Langer, & Weaver, 1993). Electroporation technique has promise for the delivery of both conventional and protein drugs (Banga & Prausnitz, 1998). However, the high voltages needed may not be suitable for routine drug delivery applications, and electroporation instead may find its niche in electrochemotherapy applications (Banga, 2011).

Norwood Abbey (Victoria, Australia) has received approval from U.S. FDA for laser-assisted delivery of a lidocaine formulation (Epiture Easytouch™). In a controlled, randomized, multicenter study of 320 subjects, the company reported that laser ablation of the stratum corneum facilitated the delivery of 4% lidocaine to achieve dermal anesthesia. The company was later acquired by another company. Proprietary vesicles from IDEA AG (Munich), Transfersomes®, contain a component that destabilizes the lipid bilayers and thus make the vesicle very deformable, and they seem to penetrate the skin barrier along the transcutaneous moisture gradient. Transfersomes have been reported to deliver insulin through intact skin with comparable pharmacokinetic and pharmacodynamic profile to ultralente insulin injected under the skin and are currently in clinical trials (Cevc, 2003a, 2003b).

Pressure waves generated by lasers and applied for short times (100 ns to 1 μs) have also been reported to reversibly permeabilize the skin without any accompanying pain or discomfort. These are different from ultrasound and are compression waves that exclude biological effects induced by cavitation. Delivery of insulin by pressure waves has been reported to drop the blood glucose of diabetic rats to about

80% of the initial levels (Doukas & Kollias, 2004). Nanovesicular and nanoparticle systems have been investigated for transdermal delivery (Choi et al., 2012; Marepally et al., 2013). Microemulsions for transdermal delivery of insulin have been investigated using isopropyl myristate or oleic acid as the oil phase, polysorbate 80 as surfactant, and isopropyl alcohol as the cosurfactant, as well as with other formulations (Malakar, Sen, Nayak, & Sen, 2011; Russell-Jones & Himes, 2011). Another very promising approach to transdermal peptide delivery is iontophoretic delivery. We will discuss this approach in greater details in Section 8.3.

8.3 IONTOPHORESIS

Iontophoresis is a technique that uses a constant, low-level electric current to push a charged drug through the skin (Banga, 2011; Banga & Chien, 1988; Kalia, Naik, Garrison, & Guy, 2004). For peptide drugs, the pH of the buffer used relative to the isoelectric point (pI) of the drug will determine whether the drug will have a positive or negative charge. Using the simple principle of charge repulsion, positively charged drugs can be delivered under the positive electrode (anode), while negatively charged drugs can be delivered under the negative electrode (cathode). Although clinicians have used iontophoresis for topical delivery for decades, its potential for systemic delivery has been recognized relatively recently (Chien & Banga, 1989). Several reviews have provided a comprehensive discussion of this area (Banga, 2011; Green, Flanagan, Shroot, & Guy, 1993; Kalia et al., 2004; Kasha & Banga, 2008; Singh & Maibach, 1994), and the reader is referred to these for a detailed treatment of the theoretical concepts involved in iontophoretic delivery. Iontophoresis will be a promising technique for the delivery of peptide drugs (Chien, Lelawongs, Siddiqui, Sun, & Shi, 1990; Chien, Siddiqui, Shi, Lelawongs, & Liu, 1989; Cullander & Guy, 1992; Parasrampuria & Parasrampuria, 1991). This section covers some of the more practical aspects, especially as they relate to peptide drugs. The advantages and limitations of iontophoretic delivery for peptide and protein drugs have been discussed (Sage, 1993). Besides the usual benefits of transdermal delivery, iontophoresis presents unique opportunity to provide programmed drug delivery. This is because the drug is delivered in proportion to the current, which can be readily adjusted. Such dependence on current is also likely to make drug absorption via iontophoresis to be almost independent of biological variables, unlike most other drug delivery systems. It may be noted that normally, the skin permeation data for ionic permeants through the skin are highly variable, as compared to that for neutral permeants (Liu, Nightingale, & Kuriharabergstrom, 1993). However, with iontophoresis, this variability is likely to be reduced. Also, patient compliance will improve, and since the dosage form includes electronics, means to remind patients to replace the dose can be built into the system. For peptide and protein drugs, their slow administration over several hours for an iontophoretic system provides an ideal situation. This is because even though an IV bolus injection may provide 100% bioavailability, clearance will be rapid because these drugs have very short half-lives. Iontophoretic enhancement provides further benefit for peptide delivery because, theoretically speaking, the rate of peptide delivery can be initiated, terminated, or accurately controlled/modulated merely by switching the current on and off or adjusting the current application parameters, respectively. This would be especially useful since a pulsatile delivery may be required for some peptides, as opposed to constant delivery. In the case of insulin, even though molecular size is high for iontophoretic delivery, pulsatile delivery may be required to deliver a higher dose of insulin after meals or when the blood glucose levels are elevated. It is also possible, at least theoretically, to link delivery to a biofeedback system that will detect the blood glucose levels and send an electronic signal to adjust the delivery rate of insulin. In fact, iontophoresis technique itself can be used to extract glucose from the blood in a noninvasive manner (Rao, Glikfeld, & Guy, 1993). This biofeedback system has the potential to overcome the disadvantages associated with a bolus subcutaneous injection of insulin and provides a means to have a variable rate of infusion that can be adjusted in a similar manner as that of endogenous insulin secretion by the pancreas. In addition, the delivery is in a noninvasive manner as opposed to an IV infusion. Iontophoresis also has potential applications in diagnostics. Reverse iontophoresis (Glikfeld, Hinz, & Guy, 1989; Rao et al., 1993) can be used for noninvasive sampling of biological fluids, which might be able to achieve iontophoretic extraction of small peptides from the subcutaneous tissue.

Iontophoresis studies may be done using several model systems. Commonly used models include the in vitro excised skin experimental setup, in vivo studies, or in vitro perfused systems (Glikfeld, Cullander, Hinz, & Guy, 1988; Sage & Riviere, 1992; Vollmer, Muller, Wilffert, & Peters, 1993). In vitro excised skin systems may be of the horizontal or vertical type, and these may be further classified into single-compartment, two-compartment, and three-compartment models (Chen & Chien, 1994). A typical horizontal-type in vitro system is shown schematically in Figure 8.4. The in vitro perfused systems, such as the Riviere porcine skin flap model, include all the anatomical structure of the skin, including vasculature, but avoid the large

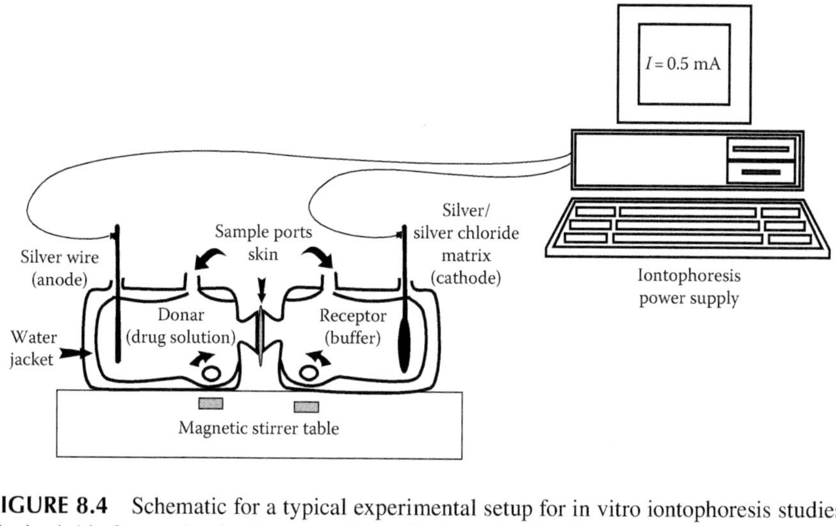

FIGURE 8.4 Schematic for a typical experimental setup for in vitro iontophoresis studies. Excised skin from animal or human cadaver skin is sandwiched between the donor and receptor cells, and current is supplied by the iontophoresis power source by reversible electrodes.

volume of distribution and rapid clearance that would result in an in vivo study. However, developing the skin flap models can be relatively expensive, technically demanding and labor intensive. Xenografts, that is, human skin transplanted onto animal models, can also be used for transdermal studies, but the animals must be treated systemically with immunosuppressive drugs (Shah et al., 1991). It has been reported that when human skin equivalents were transplanted onto athymic nude mice, the reconstructed epidermis showed barrier properties very similar to those of normal human skin (Higounenc et al., 1994). Cultured skin equivalents or living skin equivalents (LSEs) of human origin are also commercially available. These LSEs closely resemble human skin, consisting of a stratified epidermis with a well-differentiated stratum corneum and a dermis. Although these LSEs lack structural appendages such as hair follicles and sweat glands, they have been shown to be useful for iontophoresis studies. Results obtained using LSE correlated well with those obtained with guinea pig skin in vitro for salmon calcitonin and other compounds (Hager, Mancuso, Nazareno, Sharkey, & Siverly, 1994).

8.3.1 Factors Affecting Iontophoretic Delivery

8.3.1.1 Molecular Weight

As discussed earlier, the pathways of iontophoretic delivery are different from those for passive transdermal delivery. Since a pore pathway is most likely involved, the flux is expected to be size dependent with a cutoff at some point. Using eight tripeptides of the general structure alanine-X-alanine, Green, Hinz, Kim, Szoka, and Guy (1991) have investigated their delivery across hairless mouse skin in vitro. The normalized iontophoretic flux was found to be independent of lipophilicity but inversely related to MW. Although an upper size limit is not known, the largest polypeptide that has been delivered is around 13 kDa (Banga, 2011).

8.3.1.2 Drug Charge

The pH of the buffer used relative to the pI of the peptide controls the charge on the peptide. The peptide will be positively charged at a pH below its pI and should be delivered under the anode, whereas the peptide will have a negative charge if the pH is above pI, and in which case, it should be delivered under the cathode. The techniques of isoelectric focusing and capillary zone electrophoresis have also been used as tools to predict the ability of a peptide to be iontophoresed. Based on such studies, a native luteinizing hormone–releasing hormone (LHRH) was predicted to be better suited for iontophoretic delivery than its free-acid analog (Heit, Mcfarland, Bock, & Riviere, 1994). In order to stay charged in the skin environment, the pI of the protein should be away from the pI of the skin (pI 3–4). Typically, polypeptides with high pI values such as vasopressin or calcitonin are good candidates for transdermal iontophoretic delivery from a delivery efficiency point of view.

Another phenomenon that needs to be considered is electroosmotic flow. As discussed earlier, the skin has a pI that is typically between 3 and 4. At physiological pH, therefore, the skin is negatively charged. This makes the skin selectively permeable to cations or positively charged ions (Burnette & Ongpipattanakul, 1987). As these cations move into the skin, they drag along some solvent molecules, inducing a mass flow of water or other solvent. This phenomenon, called electroosmosis, enhances the transport of neutral species across the skin. Also, the anodal (anode-to-cathode direction) flux is typically higher, due to being aided by electroosmosis. At very low pH, the skin will be below its pI and will acquire a positive charge, causing a change in the direction of electroosmotic flow to favor cathodal flux. Also, adsorption of a positively charged drug on the negatively charged skin may lead to a change in the net charge of the skin, which may affect electroosmotic flow (Guy et al., 2000; Hirvonen & Guy, 1998). This has been shown for peptides like leuprolide and nafarelin. At the pH used, these cationic peptides strongly associate with the skin and neutralize its charge, which in turn leads to a gradual reduction and eventually a reversal of the convective (electroosmotic) flow (Delgadocharro, Rodriguezbayon, & Guy, 1995; Hoogstraate, Srinivasan, Sims, & Higuchi, 1994).

8.3.1.3 Current Density

As discussed earlier, an increase in current will increase drug delivery flux, thus allowing modulation of delivery in response to the current. However, in reality, the response may start to plateau at a certain current density, depending on the formulation and the physicochemical properties of the drug molecule being used. Once a limiting transport number is achieved, further increase in current does not increase delivery flux (Kalia et al., 2004).

8.3.1.4 Electrode Material

Appropriate choice of electrodes is a factor that is critical to successful iontophoretic delivery of a drug. The electrode material in an iontophoretic device is very important as it determines the type of electrochemical reaction taking place at the electrodes. The possibility of introducing metallic ions into the skin must be carefully considered. Historically, stainless steel, nickel, or other iron alloys have been used, but these are known dermal allergens and are no longer used. Platinum offers a more inert material for electrode construction. However, platinum causes electrolysis of water resulting in pH drifts. The oxidation reaction at the anode and reduction at the cathode may be depicted as

$$H_2O \rightarrow 2H^+ + \tfrac{1}{2}O_2 + 2\ e^- \text{ (at the anode)}$$

$$2H_2O + 2e^- \rightarrow H_2 + 2OH^- \text{ (at the cathode)}$$

As a function of iontophoresis time, the solution under the anode becomes increasingly acidic, while the solution under the cathode becomes increasingly basic. Since the type and magnitude of charge on a peptide are directly dependent on the solution pH, such pH shifts must be prevented. Also, the hydrogen and hydroxyl ions produced are small mobile ions and will thus carry a significant fraction of the current. They will thus compete with the peptide for electric current, thereby reducing the efficiency of iontophoretic transport. These pH changes may be avoided by using reversible electrodes such as silver–silver chloride electrodes. Reversible electrodes are consumed by electrochemistry, and thus, they do not force electrolysis of water to be a fuel for the electrochemistry. The use of a silver wire for the anode and chloridized silver for the cathode in a buffer containing chloride ion provides an ideal electrochemical system (Lattin, Padmanabhan, & Phipps, 1991). At the anode, the silver will react with chloride ions to form insoluble silver chloride. Simultaneously, the silver chloride cathode is reduced to silver metal, and these reactions prevent the electrolysis of water:

$$Ag + Cl^- \rightarrow AgCl + e^- \text{ (at the anode)}$$

$$AgCl + e^- \rightarrow Ag + Cl^- \text{ (at the cathode)}$$

Therefore, the use of silver wire as the anode and silver–silver chloride wire as the cathode is not accompanied by pH drifts. However, the system must provide some chloride ions to drive the electrochemistry as is evident from the aforementioned equations. There could be situations where silver–silver chloride electrodes may be unsuitable as they may react with some protein drugs or with hydrogels. In these situations, the use of platinum electrodes may be recommended, though other methods that can control pH would be preferred.

8.3.2 Examples of Iontophoretic Delivery of Therapeutic Peptides

8.3.2.1 Amino Acids and Small Peptides

Iontophoresis of the building blocks of peptides, that is, amino acids, may yield some information that may be useful to predict the delivery of small peptides. Amino acids may also have a direct benefit of moisturizing the skin (Coderch, Oliva, Pons, & Parra, 1994). Wearley, Tojo, and Chien (1990) have shown that the binding of a series of amino acids in excised abdominal skin of hairless rat decreased with an increase in the alkyl side chain. This suggests that binding is likely to be polar or electrostatic in nature. Green, Hinz, Cullander, Yamane, and Guy (1991) delivered a series of amino acids across excised hairless mouse skin to investigate the effects of permeant charge (neutral, +1, or −1), lipophilicity, and vehicle pH. As expected, the positively charged amino acids (i.e., positively charged at the vehicle pH used) had maximum flux under the anode, while the negatively charged amino acids had maximum flux under the cathode. For zwitterions (i.e., essentially neutral), the iontophoretic flux did not reach steady state under the experimental conditions used and was inversely proportional to the permeant octanol/pH 7.4 buffer distribution coefficient.

Iontophoretic transport of a tripeptide, TRH, across excised dorsal nude mouse skin, was found to be directly proportional to the applied current density. Even uncharged TRH was transported, presumably by the electroosmotic or convective flow that accompanies iontophoresis. However, in the absence of current, the flux of TRH across the skin was undetectable (Burnette & Marrero, 1986). Another study has shown that penetration enhancers can also enable transdermal delivery of TRH at physiologically relevant concentrations (Magnusson, Runn, Karlsson, & Koskinen, 1997). Another tripeptide, threonine-lysine-proline (Thr-Lys-Pro), was successfully delivered across nude rat skin by iontophoresis under both in vitro and in vivo conditions. The delivery of Thr-Lys-Pro was found to be directly proportional to the applied current density over the range of 0.18–0.36 mA/cm^2. Following 6 h of iontophoresis, 98.4% of the radioactivity in donor was still the intact peptide, while 94.0% of the radioactivity penetrated in the receptor phase was the parent Thr-Lys-Pro (Green, Shroot, Bernerd, Pilgrim, & Guy, 1992). Therefore, it appears that metabolism in the skin is not likely to be a significant problem for several peptides. Iontophoretic delivery of another tripeptide, enalaprilat, has also been investigated across hairless guinea pig skin (Gupta, Kumar, Bolton, Behl, & Malick, 1994). For the tetrapeptide, hisetal, iontophoresis increased its permeation rate across human skin by a factor of 30. The iontophoretic treatment was found to be much more effective than the use of penetration enhancers. Transport through skin under iontophoresis was shown to take place mainly through water-filled pores (Ruland, Rohr, & Kreuter, 1994).

8.3.2.2 Oligopeptides

Vasopressin, a nonapeptide antidiuretic hormone, and its analogs have been investigated for its transdermal iontophoretic delivery across the skin (Banga, Katakam, & Mitra, 1995; Craanevanhinsberg et al., 1994; Lelawongs, Liu, & Chien, 1990; Lelawongs, Liu, Siddiqui, & Chien, 1989; Nakakura, Kato, & Ito, 1998). Unlike some other drugs, such as insulin, the enhanced flux of vasopressin under iontophoresis was found to be reversible. This may be explained by the high pI (10.9) of vasopressin, which ensures that the molecule will stay highly charged at the pH environment of the skin. A decapeptide, LHRH, and its analogs have also been successfully delivered by iontophoresis (Heit, Monteiroriviere, Jayes, & Riviere, 1994). Meyer, Kreis, and Eschbach (1988) have delivered therapeutic doses of leuprolide, an LHRH analog, in 13 normal men using a double-blind, randomized, crossover study conducted under an IND granted by the FDA. Data analysis by ANOVA showed significant differences between the active and passive patches. The magnitude of elevation of LH produced by the active patches was in therapeutic range and comparable to that achieved by subcutaneous administration. The only adverse effect reported was a mild erythema at the site of the active patch in 6 of the 13 subjects. This erythema resolved rapidly without sequelae. Delivery of LHRH across excised hairless mouse skin (Miller, Kolaskie, Smith, & Rivier, 1990) and in the pig under in vivo conditions and using the skin flap (Heit, Williams, Jayes, Chang, & Riviere, 1993) has also been investigated. While LHRH cannot be delivered passively, 3 h of iontophoresis and a total of 5 h of sampling delivered 959 ($\pm$ 444) ng of LHRH using the IPPSF ($n=21$), under the conditions used. The IPPSF model was found to predict in vivo serum concentrations of the iontophoretically delivered peptide. Levels of LHRH in the IPPSF perfusate and in serum samples were quantitated by radioimmunoassay. In the in vivo studies, increases in FSH and LH concentrations occurred concurrently with increasing LHRH concentrations. Therefore, the authors concluded that the iontophoretically delivered LHRH is both immunologically and biologically active (Heit et al., 1993). A combination of ethanol pretreatment of the skin followed by iontophoresis has also been shown to be effective for leuprolide and a cholecystokinin-8 analog (MW 1150) (Srinivasan, Su, Higuchi, & Behl, 1990). Permeation of buserelin through isolated human stratum corneum by iontophoresis has also been reported. In this study, it was found that passive permeation of buserelin through human stratum corneum was not feasible (Knoblauch & Moll, 1993).

8.3.2.3 Polypeptides

Iontophoretic delivery of salmon calcitonin (Chang, Hofmann, Zhang, Deftos, & Banga, 2000a, 2000b; Morimoto, Iwakura, Nakatani, Miyazaki, & Tojima, 1992) and human calcitonin (Thysman, Hanchard, & Preat, 1994) has been investigated. Salmon calcitonin, when applied to the abdominal skin of rats, did not produce any hypocalcemic effect. However, when delivered under an anode as a cation, it produced a small hypocalcemic effect (Morimoto et al., 1992). The hypocalcemic effect was enhanced when aprotinin or camostat mesilate was used as proteolytic enzyme inhibitors but not when soybean trypsin inhibitor was used. This was apparently because aprotinin (MW 6500; pI = 10.5) existed as a cation at the pH used (pH 4.0) and was probably delivered into the skin during iontophoresis. In contrast, soybean trypsin inhibitor (MW 8000; pI = 4.0–4.2) was anionic or neutral at the pH used and also had a higher MW. Other polypeptides that have been delivered iontophoretically include an analog of the growth hormone–releasing factor with an MW of 3929 Da (Kumar et al., 1992).

8.3.2.4 Insulin

Iontophoretic delivery of insulin has generated considerable interest and has been extensively investigated and will thus be discussed in some detail in this section. Insulin (MW approx. 6 kDa), despite being a rather large molecule, has been widely investigated for iontophoretic delivery (Kari, 1986; Liu, Sun, Siddiqui, & Chien, 1988; Meyer et al., 1989; Siddiqui, Sun, Liu, & Chien, 1987; Srinivasan, Higuchi, Sims, Ghanem, & Behl, 1989; Stephen, Petelenz, & Jacobsen, 1984; Sun & Xue, 1991) (Table 8.1). The impetus for such research is obviously the importance of insulin for control of diabetes and the lack of any breakthrough drug delivery systems for such use as yet. In the first reported study on the transdermal iontophoretic delivery of insulin for systemic effect, Stephen et al. (1984) attempted to deliver regular soluble insulin to human volunteers but were unsuccessful. However, they were able to deliver a highly ionized and monomeric form of insulin to one pig and observed a decline in blood glucose levels and an increase in serum insulin levels. In 1986, Kari tried to deliver the regular, soluble insulin by cathodal iontophoresis to alloxan-induced diabetic New Zealand white rabbits. He found, however, that the stratum corneum has to be disrupted or removed in order to deliver enough insulin to reduce blood glucose levels (Kari, 1986). Increased penetration of insulin with application of a depilatory lotion or cream in combination with iontophoresis has been reported (Kanikkannan, Singh, & Ramarao, 1999; Rastogi & Singh, 2003; Zakzewski, Wasilewski, Cawley, & Ford, 1998). Conversely, other researchers (Chien, Siddiqui, Sun, Shi, & Liu, 1987; Meyer et al., 1989) reported that delivery of insulin through intact skin could be accomplished.

Meyer et al. (1989) have reported the successful transdermal delivery of a therapeutic dose of human insulin to diabetic albino rabbits across the skin with intact stratum corneum. In their study, they used 26 diabetic rabbits to which transdermal patches were applied continuously for 14 h. The active patches contained insulin, and an electrical circuit was established when the patch was applied on the skin, which delivered a direct current of 0.4 mA. Control patches, which had an equal amount of insulin but without electrical current delivery, were applied onto six animals. After placement of patches, animals with active patches were found to have significant elevation in serum insulin levels ($p<0.05$) with a corresponding reduction in blood glucose levels ($p<0.01$), while no changes were observed in controls. While iontophoresis overcomes the physical barrier to some extent, the question remains whether enough insulin can be delivered to be of therapeutic benefit to diabetic patients. A drug flux of 2–4 μg/cm^2-h is required to meet the basal insulin needs for a diabetic (Kalia et al., 2004). Although the proteolytic activity of the skin is low, it contains both exo- and endopeptidases and will cause degradation of insulin during delivery. Another problem is the formation of a depot in the skin. Kari (1986) observed that the blood glucose levels continued to decline even after the current was turned off, suggesting that insulin forms a reservoir and releases gradually from the skin. Similar observations on skin accumulation of insulin have been later made by other researchers (Liu et al., 1988; Siddiqui et al., 1987). Thus, it would seem that iontophoresis serves to load up the skin tissue with insulin to form a reservoir, from which insulin molecules continue to desorb slowly into the blood circulation, until several hours after the current application. This could be a potential disadvantage. Furthermore, most commercially available insulin products actually exist in hexameric form, so that we are really trying to deliver a protein with an MW of about 36 kDa, which is most likely too high to be within the scope of iontophoretic delivery. Iontophoretic delivery of monomeric insulin analogs has been investigated with better success than regular hexameric insulin (Kanikkannan et al., 1999; Langkjaer, Brange, Grodsky, & Guy, 1998). Furthermore, the pI of insulin (5.3) falls in the region of the skin pI (4.0–6.0). This poses major hurdles to its delivery since insulin would lose its charge in the skin environment, and the predominant driving force for delivery (electrical repulsion) would cease. As some insulin then diffuses toward the blood or physiological interior (pH 7.4), it will acquire a negative charge and may be drawn back toward the anode. Therefore, despite the considerable interest in iontophoretic delivery of insulin, it is not a model protein for delivery. Its pI and the tendency toward aggregation make successful iontophoretic delivery of regular insulin unlikely. However, other enhancement techniques discussed in this chapter, such as microporation, may be more promising for insulin delivery.

8.3.3 Commercialization of Iontophoretic Delivery

Several companies are trying to commercialize an iontophoretic skin patch. One company that had received approval for a prefilled iontophoresis drug/device system in 2004 was Vyteris (Fair Lawn, NJ). The LidoSite® product from Vyteris had 10% lidocaine hydrochloride with 0.1% epinephrine, and a 10 min application was used for superficial dermatological procedures such as venipuncture, IV cannulation, and laser ablation of superficial skin lesions. However, both the company and the product do not exist now. A prescription migraine patch, Zecuity, has been approved for marketing as an iontophoresis patch and contains sumatriptan for the treatment of acute migraine. It was developed by NuPathe (Malvern, PA) using their SmartRelief technology, but the company was acquired by Teva. An iontophoresis device that was successfully developed after years of research but then had to be withdrawn soon after marketing is the IONSYS™ product from the former Alza Corporation (part of Johnson & Johnson). Incline Therapeutics now owns the rights to the product and is trying to bring it back to the hospital market. There are no iontophoretic products for delivery of peptides/proteins on the market at this time.

Although an iontophoretic patch is likely to be more costly than a passive transdermal patch, this may not be a disadvantage for a peptide drug. This is because the drug itself is likely to be expensive, and passive transdermal or other alternative delivery systems may not be possible. However, the efficiency of delivery would be important. It should be realized that iontophoresis technology is not new. Iontophoresis has been used for several decades in therapeutic settings (Banga & Chien, 1988). Iontophoresis is approved for the topical delivery of local anesthetics and corticosteroids. Use of hydrogel formulations in such units will allow dosage replacement by removing and replacing a drug-loaded hydrogel patch while reusing the same iontophoretic device. Loading of protein drugs into hydrogels by the use of electrophoresis has also been described in the literature (Shalaby, Abdallah, Park, & Park, 1993). Polypeptide drugs can be reliably loaded into the aqueous environment of a hydrogel, and their release from the hydrogel can be controlled and modulated by controlling the iontophoresis parameters such as the current density. Release flux of insulin was found to increase under current application and then falls off when the current is turned off (Banga & Chien, 1993). In one study, insulin was formulated into a poloxamer gel for permeation studies with enhancers and/or iontophoresis (Pillai & Panchagnula, 2003). Integrated iontophoretic patches (IontoPatch®) have now become available from Travanti Medical, a business unit of the Tapemark company (St. Paul, microneedles). The IontoPatch is a single-use wearable electronic disposable drug delivery (WEDD®) system with a self-contained battery that is available in 4 or 14 h average patient wear time configurations for the physical medicine and rehabilitation market. The Companion 80™ patch is designed to deliver 80 mA-min dose over 24 h and has reserve battery capacity to compensate for patients with higher skin resistance.

For commercialization of iontophoretic delivery, any long-term effects of current on the skin will need to be carefully evaluated. The only short-term effects experienced following iontophoretic delivery seem to be a feeling of tingling or warmth and possibly general irritation and erythema. Iontophoresis will be contraindicated for patients sensitive to the current levels used and to patients carrying electrically sensitive devices such as cardiac pacemakers (Singh & Maibach, 1994). Based on DSC and ATR-FTIR studies, it appears that effects of electrical treatment are less disruptive to the stratum corneum as compared to the effects of penetration enhancers (Clancy, Corish, & Corrigan, 1994).

8.4 TOPICAL DELIVERY OF THERAPEUTIC PEPTIDES AND PROTEINS

Much of the literature for peptide drugs is currently focused on transdermal delivery, that is, percutaneous absorption for systemic effects. However, topical delivery of peptide and protein drugs is also of great importance, one example being the use of growth factor for wound healing. Proteins can also be used as antigens for topical delivery for vaccination, and such use is discussed in Chapter 10.

For conventional drugs, however, topical products to treat dermatological ailments have been in existence from the earliest times. Surprisingly, though transdermal delivery is relatively recent, the principals involved here are better understood than for topical delivery. This is largely because with transdermals, the blood concentration needed to achieve therapeutic effects is generally known (Shah, Behl, Flynn, Higuchi, & Schaefer, 1992). For topical delivery, the skin is the target organ. Unlike transdermal delivery, a non-steady-state transport generally characterizes a topical drug product. Also, unlike transdermal delivery, an optimal drug buildup in the skin with little or no flux through the skin is desirable. However, it is difficult to measure the drug concentration in the skin.

Traditionally, a skin stripping approach has been considered to be promising for characterizing dermatopharmacokinetics of drugs (Shah, 2001). However, with the skin stripping technique, only the stratum corneum is easily accessible, while deeper tissues, such as the dermis, are not easily accessible. Also, skin stripping does not always relate well with clinical studies on topical bioequivalence.

Microdialysis has also been evaluated for prediction of dermal pharmacokinetics. A microdialysis probe that functionally resembles a blood vessel is implanted under the skin (subcutaneous microdialysis; concentric probe) or into the dermis (cutaneous microdialysis; linear probe) and is perfused with a perfusion fluid. Following dialysis across the semipermeable membrane in the probe, the dialysate is collected and analyzed for the drug entering the skin and corrected by a probe recovery factor. Samples are protein free, and thus, analysis is easy (Kreilgaard, 2002). Microdialysis membranes with a high-MW cutoff are now available, and the technique works better with hydrophilic drugs and would thus be applicable for peptides and small proteins.

8.4.1 Growth Factors

The wound healing process in the body is accompanied by the release of a number of protein-based growth factors that aid the healing process (Lenz & Mansson, 1991). These growth factors can now be produced biotechnologically in a commercially feasible manner, thus raising the possibility of accelerated wound healing. The use of growth factors could accelerate healing following surgery, allowing patients to leave hospital earlier. Considering the health-care costs, this could result in immense savings. This is driving several companies and other academic and government organizations to develop growth factors (Ratafia, 1988).

Although growth factors also have application in ophthalmology, a major indication for their use is to accelerate wound healing in the skin. The most promising agent for such use is the epidermal growth factor (EGF), a polypeptide of 53 amino acid residues, which has shown promise to heal open and burn wounds (Celebi, Erden, Gonul, & Koz, 1994). It has been found that the presence of protease inhibitors is required in the formulation to stabilize EGF at the wound site (Kiyohara, Komada, Iwakawa, Fuwa, & Okumura, 1993; Okumura et al., 1990). Formulations of acidic fibroblast growth factor have also been reported to accelerate wound healing in a diabetic mouse model (Matuszewska et al., 1994). Similarly, transforming growth factor-β (TGF-β) can be used topically to accelerate wound healing. A patent describes a reconstitutable lyophilized protein formulation that contains a cellulose polymer as a gelling agent. When reconstituted, the formulation forms a gel, which can then be used for topical application (Hsu, Nguyen, & Wu, 1993). The use of hydrogel wound dressings has been described in the literature (Kickhofen, Wokalek, Scheel, & Ruh, 1986), and it should be possible to incorporate growth factors in such dressings to accelerate the wound healing process.

8.4.2 Liposomes

The literature on the use of liposomes for topical/transdermal delivery now seems to have reached a consensus that liposomes do not traverse the skin. Therefore, they will generally not enhance percutaneous absorption of entrapped drugs although they are likely to enhance topical delivery into epidermis and maybe to dermis as well (Ganesan, Weiner, Flynn, & Ho, 1984; Martin, 1993; Mezei & Gulasekharam, 1982). It has been suggested that liposomes may penetrate the lipid-rich outer skin layers and then get localized because of the aqueous environment of the dermis. However, it may be difficult to perceive how relatively large lipid vesicles can traverse densely packed epidermal tissue to arrive intact within the dermal layers.

A study has shown that smaller liposome particle sizes did not result in higher cyclosporin A levels in the deeper strata of hairless mice, hamster, and pig skin. This suggests that topical delivery of cyclosporin A was not a result of passage of intact liposomes into the skin. An intermediate particle size for liposomes (0.3 μm) was found to be optimum for drug delivery, suggesting that the size of the liposomes may regulate the location of the depot effect. However, the mechanism is not clear, and analytical techniques need to be developed, which can visualize and differentiate the skin and liposomes (Duplessis, Ramachandran, Weiner, & Muller, 1994). The reduced epidermal and dermal clearance of liposome-entrapped drugs has been speculated as being partly attributed to the unavailability of the metabolizing enzymes within the skin to the encapsulated drug. It has also been suggested that larger liposomes, being unable to penetrate the underlying blood vessels, act as localized sustained release vesicles (Martin, 1993). In addition, liposomes can serve to solubilize poorly soluble drugs, and their phospholipid constituents may serve as penetration enhancers to facilitate topical absorption (Schreier & Bouwstra, 1994).

Lipids as such and liposomes have been incorporated into many cosmetic formulations (Siciliano, 1985). Lecithin is commonly used in pharmaceutical, cosmetic, and food products and has GRAS status. Liposome-containing preparations, as opposed to lipid-containing preparations, have been marketed as cosmetics. Although any advantages of such products have not been proven, they are popularly used (Martin, 1993). There are over 100 liposome and niosome cosmetic products on the market. However, only a few liposome preparations containing a drug are on the market. These include the antifungal drugs, econazole (Schreier & Bouwstra, 1994) and amphotericin B.

Most of the work on the use of liposomes for topical delivery was with drugs that would be associated with the lipid bilayer. In contrast, the water-soluble peptide and protein drugs are likely to be in the aqueous entrapped environment within the liposome. Liposomal or other vesicular formulations of interferon have been used to examine topical delivery of encapsulated interferon to cutaneous herpes lesions or treatment of human papillomavirus (King, Kumar, Michel, Batta, & Foldvari, 2013; Weiner et al., 1989). While liposomal formulations caused reduction of lesion scores, solutions or emulsions containing interferon were not effective. The method of liposomal preparation was found to be the most important factor determining the effectiveness of the formulation. The dehydration–rehydration method was found to be most effective, presumably because dehydration and subsequent rehydration of the liposomes facilitate partitioning of the interferon into liposomal bilayers. In continuing studies using an in vitro diffusion setup, it was observed that liposomes made from *skin lipids* were twice as effective as those made from phospholipids, perhaps because they are better able to transfer the drug to the skin (Egbaria, Ramachandran, Kittayanond, & Weiner, 1990). Unlike other biological membranes, the stratum corneum does not contain phospholipids and is made primarily of ceramides, cholesterol, fatty acids, and cholesteryl sulfate. Although stratum corneum lipids do not contain phospholipids, liposomes can still be made from these skin lipids. Wertz, Abraham, Landmann, and Downing (1986) prepared stable, unilamellar liposomes using lipid mixtures similar to those found in the stratum corneum.

8.4.3 Iontophoresis and Phonophoresis

Investigations on the potential use of iontophoresis and phonophoresis techniques to achieve systemic delivery of drugs are relatively recent, having started over the last decade or two. However, these techniques have been used for topical delivery of drugs and in physical therapy for several decades. The technique of iontophoresis is used routinely in clinics by physical therapists for the delivery of corticosteroids and local anesthetics to treat inflammatory conditions of muscles and tendons, such as bursitis, tendonitis, arthritis, carpal tunnel syndrome, and temporomandibular joint dysfunction. Several devices and electrodes are commercially available for this purpose, and the most commonly used drug for iontophoretic delivery is dexamethasone sodium phosphate. Similarly, phonophoresis is also routinely used in clinics by physical therapists, where hydrocortisone is a commonly administered drug (Banga, 2011; Bogner & Banga, 1994). Use of iontophoresis and phonophoresis to achieve topical delivery may be extended to peptide and protein drugs. In fact, as discussed earlier, insulin forms a depot or reservoir in the skin upon iontophoretic or phonophoretic delivery. Although this may not be a good scenario for drugs delivered for systemic effects, depot formation may be advantageous for certain topical drugs.

8.5 CONCLUSIONS

Several conventional drugs are already available on the market for transdermal delivery. Transdermal and topical routes are also very promising for the delivery of therapeutic peptides and proteins. However, several penetration and enzymatic barriers need to be overcome, and approaches that have been investigated include the use of penetration enhancers, protease inhibitors, and phonophoresis. Other promising approaches are iontophoresis and skin microporation. Extensive research is underway in the area of transdermal iontophoretic delivery, and commercialization of an iontophoretic patch for peptide delivery is expected. Topical delivery of bioactive peptides is already feasible and has a great potential. The use of growth factors to accelerate wound healing could result in substantial savings to the cost of health care. Several companies are actively investigating the use of microporation for delivery of proteins and vaccines through the skin.

REFERENCES

Ameri, M., Daddona, P. E., & Maa, Y. F. (2009). Demonstrated solid-state stability of parathyroid hormone PTH(1–34) coated on a novel transdermal microprojection delivery system. *Pharm. Res., 26,* 2454–2463.

Ameri, M., Fan, S. C., & Maa, Y. F. (2010). Parathyroid hormone PTH(1–34) formulation that enables uniform coating on a novel transdermal microprojection delivery system. *Pharm. Res., 27,* 303–313.

Ameri, M., Wang, X., & Maa, Y. F. (2010). Effect of irradiation on parathyroid hormone PTH(1–34) coated on a novel transdermal microprojection delivery system to produce a sterile product-adhesive compatibility. *J. Pharm. Sci., 99(4),* 2123–2134.

Badkar, A. V., Smith, A. M., Eppstein, J. A., & Banga, A. K. (2007). Transdermal delivery of interferon alpha-2B using microporation and iontophoresis in hairless rats. *Pharm. Res., 24,* 1389–1395.

Bal, S. M., Caussin, J., Pavel, S., & Bouwstra, J. A. (2008). In vivo assessment of safety of microneedle arrays in human skin. *Eur. J. Pharm. Sci., 35,* 193–202.

Banerjee, P. S. & Ritschel, W. A. (1989a). Transdermal permeation of vasopressin. I. Influence of pH, concentration, shaving and surfactant on in vitro permeation. *Int. J. Pharm., 49,* 189–197.

Banerjee, P. S. & Ritschel, W. A. (1989b). Transdermal permeation of vasopressin. II. Influence of azone on in vitro and in vivo permeation. *Int. J. Pharm., 49,* 199–204.

Banga, A. K. (2011). *Transdermal and intradermal delivery of therapeutic agents: Application of physical technologies*. CRC Press/Taylor & Francis, Boca Raton, FL.

Banga, A. K. & Chien, Y. W. (1988). Iontophoretic delivery of drugs: Fundamentals, developments and biomedical applications. *J. Control. Release, 7,* 1–24.

Banga, A. K. & Chien, Y. W. (1993). Hydrogel-based iontotherapeutic delivery devices for transdermal delivery of peptide/protein drugs. *Pharm. Res., 10,* 697–702.

Banga, A. K., Katakam, M., & Mitra, R. (1995). Transdermal iontophoretic delivery and degradation of vasopressin across human cadaver skin. *Int. J. Pharm., 116,* 211–216.

Banga, A. K. & Prausnitz, M. R. (1998). Assessing the potential of skin electroporation for the delivery of protein- and gene-based drugs. *Trends Biotechnol., 16,* 408–412.

Barry, B. W. (1987). Mode of action of penetration enhancers in human skin. *J. Control. Release, 6,* 85–97.

Barry, B. W. (1991). Lipid-protein-partitioning theory of skin penetration enhancement. *J. Control. Release, 15,* 237–248.

Bartek, M. J., LaBudde, J. A., & Maibach, H. I. (1972). Skin permeability *in vivo*: Comparison in rat, rabbit, pig and man. *J. Invest. Dermatol., 58,* 114–123.

Berner, B. & John, V. A. (1994). Pharmacokinetic characterization of transdermal delivery systems. *Clin. Pharmacokinet., 26,* 121–134.

Bogner, R. H. & Banga, A. K. (1994). Iontophoresis and phonophoresis. *U.S. Pharm., 19,* H10–H26.

Bommannan, D., Menon, G. K., Okuyama, H., Elias, P. M., & Guy, R. H. (1992). Sonophoresis. II. Examination of the mechanism(s) of ultrasound-enhanced transdermal drug delivery. *Pharm. Res., 9,* 1043–1047.

Bommannan, D., Okuyama, H., Stauffer, P., & Guy, R. H. (1992). Sonophoresis. I. The use of high-frequency ultrasound to enhance transdermal drug delivery. *Pharm. Res., 9,* 559–564.

Bommannan, D., Potts, R. O., & Guy, R. H. (1991). Examination of the effect of ethanol on human stratum corneum in vivo using infrared spectroscopy. *J. Control. Release, 16,* 299–304.

Bramson, J., Dayball, K., Evelegh, C., Wan, Y. H., Page, D., & Smith, A. (2003). Enabling topical immunization via microporation: A novel method for pain-free and needle-free delivery of adenovirus-based vaccines. *Gene Ther., 10,* 251–260.

Brown, L. & Langer, R. (1988). Transdermal delivery of drugs. *Annu. Rev. Med., 39,* 221–229.

Burnette, R. R. & Marrero, D. (1986). Comparison between the iontophoretic and passive transport of thyrotropin releasing hormone across excised nude mouse skin. *J. Pharm. Sci., 75,* 738–743.

Burnette, R. R. & Ongpipattanakul, B. (1987). Characterization of the permselective properties of excised human skin during iontophoresis. *J. Pharm. Sci., 76,* 765–773.

Candan, G., Michiue, H., Ishikawa, S., Fujimura, A., Hayashi, K., Uneda, A. et al. (2012). Combining poly-arginine with the hydrophobic counter-anion 4-(1-pyrenyl)-butyric acid for protein transduction in transdermal delivery. *Biomaterials, 33,* 6468–6475.

Celebi, N., Erden, N., Gonul, B., & Koz, M. (1994). Effects of epidermal growth factor dosage forms on dermal wound strength in mice. *J. Pharm. Pharmacol., 46,* 386–387.

Cevc, G. (2003a). Transdermal drug delivery of insulin with ultradeformable carriers. *Clin. Pharmacokinet., 42,* 461–474.

Cevc, G. (2003b). Transfersomes: Innovative transdermal drug carriers. In: M. J. Rathbone, J. Hadgraft, & M. S. Roberts (Eds.), *Modified-release drug delivery technology* (pp. 533–546). New York: Marcel Dekker, Inc.

Chang, S., Hofmann, G. A., Zhang, L., Deftos, L. J., & Banga, A. K. (2000a). The effect of electroporation on iontophoretic transdermal delivery of calcium regulating hormones. *J. Control. Release, 66,* 127–133.

Chang, S., Hofmann, G. A., Zhang, L., Deftos, L. J., & Banga, A. K. (2000b). Transdermal iontophoretic delivery of salmon calcitonin. *Int. J. Pharm., 200,* 107–113.

Chen, L. H. & Chien, Y. W. (1994). Development of a skin permeation cell to simulate clinical study of iontophoretic transdermal delivery. *Drug Dev. Ind. Pharm., 20,* 935–945.

Chen, M., Gupta, V., Anselmo, A. C., Muraski, J. A., & Mitragotri, S. (2014). Topical delivery of hyaluronic acid into skin using SPACE-peptide carriers. *J. Control. Release, 173,* 67–74.

Chiang, C. H., Shao, C. H., & Chen, J. L. (1998). Effects of pH, electric current, and enzyme inhibitors on iontophoresis of delta sleep-inducing peptide. *Drug Dev. Ind. Pharm., 24,* 431–438.

Chien, Y. W. & Banga, A. K. (1989). Iontophoretic (transdermal) delivery of drugs: Overview of historical development. *J. Pharm. Sci., 78,* 353–354.

Chien, Y. W., Lelawongs, P., Siddiqui, O., Sun, Y., & Shi, W. M. (1990). Facilitated transdermal delivery of therapeutic peptides and proteins by iontophoretic delivery devices. *J. Control. Release, 13,* 263–278.

Chien, Y. W., Siddiqui, O., Shi, W. M., Lelawongs, P., & Liu, J. C. (1989). Direct current iontophoretic transdermal delivery of peptide and protein drugs. *J. Pharm. Sci., 78,* 376–383.

Chien, Y. W., Siddiqui, O., Sun, Y., Shi, W. M., & Liu, J. C. (1987). Transdermal iontophoretic delivery of therapeutic peptides/proteins: (I) Insulin. In: R. L. Juliano (Ed.), *Biological approaches to the controlled delivery of drugs* (pp. 32–51). *Annals of the New York Academy of Sciences*, vol. 507. New York: New York Academy of Sciences.

Choi, H., Flynn, G. L., & Amidon, G. L. (1990). Transdermal delivery of bioactive peptides: The effect of *n*-decylmethyl sulfoxide, pH, and inhibitors on enkephalin metabolism and transport. *Pharm. Res., 7,* 1099–1106.

Choi, W. I., Lee, J. H., Kim, J. Y., Kim, J. C., Kim, Y. H., & Tae, G. (2012). Efficient skin permeation of soluble proteins via flexible and functional nano-carrier. *J. Control. Release, 157,* 272–278.

Clancy, M. J., Corish, J., & Corrigan, O. I. (1994). A comparison of the effects of electrical current and penetration enhancers on the properties of human skin using spectroscopic (FTIR) and calorimetric (DSC) methods. *Int. J. Pharm., 105,* 47–56.

Coderch, L., Oliva, M., Pons, L., & Parra, J. L. (1994). Percutaneous penetration in vivo of amino acids. *Int. J. Pharm., 111,* 7–14.

Couto, D. S., Perez-Breva, L., Saraiva, P., & Cooney, C. L. (2012). Lessons from innovation in drug-device combination products. *Adv. Drug Deliv. Rev., 64,* 69–77.

Craanevanhinsberg, W. H. M., Bax, L., Flinterman, N. H. M., Verhoef, J., Junginger, H. E., & Bodde, H. E. (1994). Iontophoresis of a model peptide across human skin *in vitro*: Effects of iontophoresis protocol, pH, and ionic strength on peptide flux and skin impedance. *Pharm. Res., 11,* 1296–1300.

Cullander, C. & Guy, R. H. (1992). (D) Routes of delivery: Case studies: (6) Transdermal delivery of peptides and proteins. *Adv. Drug Deliv. Rev., 8,* 291–329.

Daddona, P. E., Matriano, J. A., Mandema, J., & Maa, Y. F. (2011). Parathyroid hormone (1–34)-coated microneedle patch system: Clinical pharmacokinetics and pharmacodynamics for treatment of osteoporosis. *Pharm. Res., 28,* 159–165.

Dawson, B. V., Hadley, M. E., Levine, N., Kreutzfeld, K. L., Don, S., Eytan, T. et al. (1990). In vitro transdermal delivery of a melanotropic peptide through human skin. *J. Invest. Dermatol., 94,* 432–435.

Delgadocharro, M. B., Rodriguezbayon, A. M., & Guy, R. H. (1995). Iontophoresis of nafarelin: Effects of current density and concentration on electrotransport in vitro. *J. Control. Release, 35,* 35–40.

Denda, M., Koyama, J., Namba, R., & Horii, I. (1994). Stratum-corneum lipid morphology and transepidermal water loss in normal skin and surfactant-induced scaly skin. *Arch. Dermatol. Res., 286,* 41–46.

Desai, P., Patlolla, R. R., & Singh, M. (2010). Interaction of nanoparticles and cell-penetrating peptides with skin for transdermal drug delivery. *Mol. Membr. Biol., 27,* 247–259.

Doukas, A. G. & Kollias, N. (2004). Transdermal drug delivery with a pressure wave. *Adv. Drug Deliv. Rev., 56,* 559–579.

Duplessis, J., Ramachandran, C., Weiner, N., & Muller, D. G. (1994). The influence of particle size of liposomes on the deposition of drug into skin. *Int. J. Pharm., 103,* 277–282.

Egbaria, K., Ramachandran, C., Kittayanond, D., & Weiner, N. (1990). Topical delivery of liposomally encapsulated interferon evaluated by in vitro diffusion studies. *Antimicrob. Agents Chemother., 34,* 107–110.

Ellens, H., Lai, Z. P., Marcello, J., Davis, C. B., Cheng, H. Y., Oh, C. K. et al. (1997). Transdermal iontophoretic delivery of [H-3]GHRP in rats. *Int. J. Pharm., 159,* 1–11.

Eppstein, D. A. (2002). MicroPor system for noninvasive delivery of protein and peptide pharmaceuticals through the skin. In *Protein and peptide drug delivery: Scientific advances enabling novel approaches and improved products, IBC meeting*, Boston, MA, September 26–27, 2002.

Flynn, G. L. & Stewart, B. (1988). Percutaneous drug penetration: Choosing candidates for transdermal development. *J. Drug Dev. Res., 13,* 169–185.

Fukushima, K., Ise, A., Morita, H., Hasegawa, R., Ito, Y., Sugioka, N. et al. (2010). Two-layered dissolving microneedles for percutaneous delivery of peptide/protein drugs in rats. *Pharm. Res., 28(1),* 7–21.

Ganesan, M. G., Weiner, N. D., Flynn, G. L., & Ho, N. F. H. (1984). Influence of liposomal drug entrapment on percutaneous absorption. *Int. J. Pharm., 20,* 139–154.

Garrison, M. D., Doh, L. M., Potts, R. O., & Abraham, W. (1994). Effect of oleic acid on human epidermis: Fluorescence spectroscopic investigation. *J. Control. Release, 31,* 263–269.

Ghosh, T. K. & Banga, A. K. (1993a). Methods of enhancement of transdermal drug delivery: Part I, physical and biochemical approaches. *Pharm. Technol., 17(3),* 72–98.

Ghosh, T. K. & Banga, A. K. (1993b). Methods of enhancement of transdermal drug delivery: Part IIA, chemical permeation enhancers. *Pharm. Technol., 17(4),* 62–90.

Ghosh, T. K. & Banga, A. K. (1993c). Methods of enhancement of transdermal drug delivery: Part IIB, chemical permeation enhancers. *Pharm. Technol., 17(5),* 68–76.

Glikfeld, P., Cullander, C., Hinz, R. S., & Guy, R. H. (1988). A new system for in vitro studies of iontophoresis. *Pharm. Res., 5,* 443–446.

Glikfeld, P., Hinz, R. S., & Guy, R. H. (1989). Noninvasive sampling of biological fluids by iontophoresis. *Pharm. Res., 6,* 988–990.

Gratieri, T., Alberti, I., Lapteva, M., & Kalia, Y. N. (2013). Next generation intra- and transdermal therapeutic systems: Using non- and minimally-invasive technologies to increase drug delivery into and across the skin. *Eur. J. Pharm. Sci., 50,* 609–622.

Green, P., Shroot, B., Bernerd, F., Pilgrim, W. R., & Guy, R. H. (1992). In vitro and in vivo iontophoresis of a tripeptide across nude rat skin. *J. Control. Release, 20,* 209–218.

Green, P. G., Flanagan, M., Shroot, B., & Guy, R. H. (1993). Iontophoretic drug delivery. In K. A. Walters & J. Hadgraft (Eds.), *Pharmaceutical skin penetration enhancement* (pp. 311–333). New York: Marcel Dekker, Inc.

Green, P. G., Hinz, R. S., Cullander, C., Yamane, G., & Guy, R. H. (1991). Iontophoretic delivery of amino acids and amino acid derivatives across the skin *in vitro*. *Pharm. Res., 8,* 1113–1120.

Green, P. G., Hinz, R. S., Kim, A., Szoka, F. C., & Guy, R. H. (1991). Iontophoretic delivery of a series of tripeptides across the skin *in vitro*. *Pharm. Res., 8,* 1121–1127.

Gupta, J., Gill, H. S., Andrews, S. N., & Prausnitz, M. R. (2011). Kinetics of skin resealing after insertion of microneedles in human subjects. *J. Control. Release, 154,* 148–155.

Gupta, S. K., Kumar, S., Bolton, S., Behl, C. R., & Malick, A. W. (1994). Optimization of iontophoretic transdermal delivery of a peptide and a non-peptide drug. *J. Control. Release, 30,* 253–261.

Guy, R. H. & Hadgraft, J. (1992). Rate control in transdermal drug delivery? *Int. J. Pharm., 82,* R1–R6.

Guy, R. H., Kalia, Y. N., Delgado-Charro, M. B., Merino, V., Lopez, A., & Marro, D. (2000). Iontophoresis: Electrorepulsion and electroosmosis. *J. Control. Release, 64,* 129–132.

Guzek, D. B., Kennedy, A. H., McNeill, S. C., Wakshull, E., & Potts, R. O. (1989). Transdermal drug transport and metabolism. I. Comparison of in vitro and in vivo results. *Pharm. Res., 6,* 33–39.

Hager, D. F., Mancuso, F. A., Nazareno, J. P., Sharkey, J. W., & Siverly, J. R. (1994). Evaluation of a cultured skin equivalent as a model membrane for iontophoretic transport. *J. Control. Release, 30,* 117–123.

Hansen, K., Burton, S., & Tomai, M. (2009). A hollow microstructured transdermal system (hMTS) for needle-free delivery of biopharmaceuticals. *Drug Deliv. Technol., 9,* 38–44.

Harvey, A. J., Kaestner, S. A., Sutter, D. E., Harvey, N. G., Mikszta, J. A., & Pettis, R. J. (2011). Microneedle-based intradermal delivery enables rapid lymphatic uptake and distribution of protein drugs. *Pharm. Res., 28(1),* 107–116.

Heit, M. C., Mcfarland, A., Bock, R., & Riviere, J. E. (1994). Isoelectric focusing and capillary zone electrophoretic studies using luteinizing hormone releasing hormone and its analog. *J. Pharm. Sci., 83,* 654–656.

Heit, M. C., Monteiroriviere, N. A., Jayes, F. L., & Riviere, J. E. (1994). Transdermal iontophoretic delivery of luteinizing hormone releasing hormone (LHRH): Effect of repeated administration. *Pharm. Res., 11,* 1000–1003.

Heit, M. C., Williams, P. L., Jayes, F. L., Chang, S. K., & Riviere, J. E. (1993). Transdermal iontophoretic peptide delivery—In vitro and in vivo studies with luteinizing hormone releasing hormone. *J. Pharm. Sci., 82,* 240–243.

Henry, S., McAllister, D. V., Allen, M. G., & Prausnitz, M. R. (1998). Microfabricated microneedles: A novel approach to transdermal drug delivery. *J. Pharm. Sci., 87,* 922–925.

Higounenc, I., Demarchez, M., Regnier, M., Schmidt, R., Ponec, M., & Shroot, B. (1994). Improvement of epidermal differentiation and barrier function in reconstructed human skin after grafting onto athymic nude mice. *Arch. Dermatol. Res., 286,* 107–114.

Hirvonen, J. & Guy, R. H. (1998). Transdermal iontophoresis: Modulation of electroosmosis by polypeptides. *J. Control. Release, 50,* 283–289.

Hoogstraate, A. J., Srinivasan, V., Sims, S. M., & Higuchi, W. I. (1994). Iontophoretic enhancement of peptides: Behaviour of leuprolide versus model permeants. *J. Control. Release, 31,* 41–47.

Hoogstraate, A. J., Verhoef, J., Brussee, J., IJzerman, A. P., Spies, F., & Bodde, H. E. (1991). Kinetics, ultrastructural aspects and molecular modelling of transdermal peptide flux enhancement by *N*-alkylazacycloheptanones. *Int. J. Pharm., 76,* 37–47.

Hopsu-Havu, V. K., Fraki, J. E., & Jarvinen, M. (1977). Proteolytic enzymes in the skin. In: A. J. Barrett (Ed.), *Proteinases in mammalian cells and tissues* (pp. 547–581). Amsterdam, the Netherlands: North-Holland Biomedical Press.

Hou, Y. W., Chan, M. H., Hsu, H. R., Liu, B. R., Chen, C. P., Chen, H. H. et al. (2007). Transdermal delivery of proteins mediated by non-covalently associated arginine-rich intracellular delivery peptides. *Exp. Dermatol., 16,* 999–1006.

Hsu, C. C., Nguyen, H. M., & Wu, S. S. (1993, March 9). Reconstitutable lyophilized protein formulation. U.S. Patent 5,192,743.

Hsu, T. & Mitragotri, S. (2011). Delivery of siRNA and other macromolecules into skin and cells using a peptide enhancer. *Proc. Natl. Acad. Sci. U.S.A., 108,* 15816–15821.

Inamori, T., Ghanem, A. H., Higuchi, W. I., & Srinivasan, V. (1994). Macromolecule transport in and effective pore size of ethanol pretreated human epidermal membrane. *Int. J. Pharm., 105,* 113–123.

Ito, Y., Yoshimitsu, J., Shiroyama, K., Sugioka, N., & Takada, K. (2006). Self-dissolving microneedles for the percutaneous absorption of EPO in mice. *J. Drug Target, 14,* 255–261.

Kalia, Y. N., Bachhav, Y. G., Bragagna, T., & Bohler, C. (2008). Intraepidermal delivery: P.L.E.A.S.E., a new laser microporation technology. *Drug Deliv. Technol., 8,* 26–31.

Kalia, Y. N., Naik, A., Garrison, J., & Guy, R. H. (2004). Iontophoretic drug delivery. *Adv. Drug Deliv. Rev., 56,* 619–658.

Kalluri, H. & Banga, A. (2009). Microneedles and transdermal drug delivery. *J. Drug Deliv. Sci. Technol., 19,* 303–310.

Kalluri, H. & Banga, A. K. (2010). Formation and closure of microchannels in skin following microporation. *Pharm. Res., 28,* 82–94.

Kanikkannan, N., Singh, J., & Ramarao, P. (1999). Transdermal iontophoretic delivery of bovine insulin and monomeric human insulin analogue. *J. Control. Release, 59,* 99–105.

Kari, B. (1986). Control of blood glucose levels in alloxan-diabetic rabbits by iontophoresis of insulin. *Diabetes, 35,* 217–221.

Kasha, P. C. & Banga, A. K. (2008). A review of patent literature for iontophoretic delivery and devices. *Recent Patents Drug Deliv. Formul., 2,* 41–50.

Kickhofen, B., Wokalek, H., Scheel, D., & Ruh, H. (1986). Chemical and physical properties of a hydrogel wound dressing. *Biomaterials, 7,* 67–72.

Kim, Y. C., Park, J. H., & Prausnitz, M. R. (2012). Microneedles for drug and vaccine delivery. *Adv. Drug Deliv. Rev., 64,* 1547–1568.

King, M., Kumar, P., Michel, D., Batta, R., & Foldvari, M. (2013). In vivo sustained dermal delivery and pharmacokinetics of interferon alpha in biphasic vesicles after topical application. *Eur. J. Pharm. Biopharm., 84,* 532–539.

Kiyohara, Y., Komada, F., Iwakawa, S., Fuwa, T., & Okumura, K. (1993). Systemic effects of epidermal growth factor (EGF) ointment containing protease inhibitor or gelatin in rats with burns or open wounds. *J. Biol. Pharm. Bull., 16,* 73–76.

Knoblauch, P. & Moll, F. (1993). In vitro pulsatile and continuous transdermal delivery of buserelin by iontophoresis. *J. Control. Release, 26,* 203–212.

Kolli, C. S. & Banga, A. K. (2008). Characterization of solid maltose microneedles and their use for transdermal delivery. *Pharm. Res., 25,* 104–113.

Kost, J. (1993). Ultrasound induced delivery of peptides. *J. Control. Release, 24,* 247–255.

Kreilgaard, M. (2002). Assessment of cutaneous drug delivery using microdialysis. *Adv. Drug Deliv. Rev., 54(Suppl. 1),* S99–S121.

Kumar, S., Char, H., Patel, S., Piemontese, D., Iqbal, K., Malick, A. W. et al. (1992). Effect of Iontophoresis on in vitro skin permeation of an analogue of growth hormone releasing factor in the hairless guinea pig model. *J. Pharm. Sci., 81,* 635–639.

Langkjaer, L., Brange, J., Grodsky, G. M., & Guy, R. H. (1998). Iontophoresis of monomeric insulin analogues in vitro: Effects of insulin charge and skin pretreatment. *J. Control. Release, 51,* 47–56.

Lattin, G. A., Padmanabhan, R. V., & Phipps, J. B. (1991). Electronic control of iontophoretic drug delivery. *Ann. N. Y. Acad. Sci., 618,* 450–464.

Lee, J. W., Park, J. H., & Prausnitz, M. R. (2008). Dissolving microneedles for transdermal drug delivery. *Biomaterials, 29,* 2113–2124.

Lelawongs, P., Liu, J., & Chien, Y. W. (1990). Transdermal iontophoretic delivery of arginine-vasopressin (II): Evaluation of electrical and operational factors. *Int. J. Pharm., 61,* 179–188.

Lelawongs, P., Liu, J., Siddiqui, O., & Chien, Y. W. (1989). Transdermal iontophoretic delivery of arginine-vasopressin (I): Physicochemical considerations. *Int. J. Pharm., 56,* 13–22.

Lenz, G. R. & Mansson, P. E. (1991). Growth factors as pharmaceuticals. *Pharm. Technol., 15(1),* 34–40.

Levy, D., Kost, J., Meshulam, Y., & Langer, R. (1989). Effect of ultrasound on transdermal drug delivery to rats and guinea pigs. *J. Clin. Invest., 83,* 2074–2078.

Li, G., Badkar, A., Kalluri, H., & Banga, A. K. (2009). Microchannels created by sugar and metal microneedles: Characterization by microscopy, macromolecular flux and other techniques. *J. Pharm. Sci., 99,* 1931–1941.

Li, G., Badkar, A., Nema, S., Kolli, C. S., & Banga, A. K. (2009). In vitro transdermal delivery of therapeutic antibodies using maltose microneedles. *Int. J. Pharm., 368,* 109–115.

Lin, W., Cormier, M., Samiee, A., Griffin, A., Johnson, B., Teng, C. L. et al. (2001). Transdermal delivery of antisense oligonucleotides with microprojection patch (Macroflux) technology. *Pharm. Res., 18,* 1789–1793.

Liu, J. C., Sun, Y., Siddiqui, O., & Chien, Y. W. (1988). Blood glucose control in diabetic rats by transdermal iontophoretic delivery of insulin. *Int. J. Pharm., 44,* 197–204.

Liu, P. C., Nightingale, J. A. S., & Kuriharabergstrom, T. (1993). Variation of human skin permeation in vitro—Ionic vs. neutral compounds. *Int. J. Pharm., 90,* 171–176.

Liu, S., Jin, M. N., Quan, Y. S., Kamiyama, F., Katsumi, H., Sakane, T. et al. (2012). The development and characteristics of novel microneedle arrays fabricated from hyaluronic acid, and their application in the transdermal delivery of insulin. *J. Control. Release, 161,* 933–941.

Liu, S., Jin, M. N., Quan, Y. S., Kamiyama, F., Kusamori, K., Katsumi, H. et al. (2014). Transdermal delivery of relatively high molecular weight drugs using novel self-dissolving microneedle arrays fabricated from hyaluronic acid and their characteristics and safety after application to the skin. *Eur. J. Pharm. Biopharm., 86,* 267–276.

Lopes, L. B., Furnish, E., Komalavilas, P., Seal, B. L., Panitch, A., Bentley, M. V. et al. (2008). Enhanced skin penetration of P20 phosphopeptide using protein transduction domains. *Eur. J. Pharm. Biopharm., 68,* 441–445.

Lu, M. F., Lee, D., & Rao, G. S. (1992). Percutaneous absorption enhancement of leuprolide. *Pharm. Res., 9,* 1575–1579.

Machluf, M. & Kost, J. (1993). Ultrasonically enhanced transdermal drug delivery—Experimental approaches to elucidate the mechanism. *J. Biomater. Sci. Polym. Edn., 5,* 147–156.

Magnusson, B. M., Runn, P., Karlsson, K., & Koskinen, L. O. D. (1997). Terpenes and ethanol enhance the transdermal permeation of the tripeptide thyrotropin releasing hormone in human epidermis. *Int. J. Pharm., 157,* 113–121.

Malakar, J., Sen, S. O., Nayak, A. K., & Sen, K. K. (2011). Development and evaluation of microemulsions for transdermal delivery of insulin. *ISRN Pharm., 2011,* 780150.

Manosroi, J., Khositsuntiwong, N., Manosroi, W., Gotz, F., Werner, R. G., & Manosroi, A. (2013). Potent enhancement of transdermal absorption and stability of human tyrosinase plasmid (pAH7/Tyr) by Tat peptide and an entrapment in elastic cationic niosomes. *Drug Deliv., 20,* 10–18.

Manosroi, J., Lohcharoenkal, W., Gotz, F., Werner, R. G., Manosroi, W., & Manosroi, A. (2013). Transdermal absorption and stability enhancement of salmon calcitonin by Tat peptide. *Drug Dev. Ind. Pharm., 39,* 520–525.

Marepally, S., Boakye, C. H., Shah, P. P., Etukala, J. R., Vemuri, A., & Singh, M. (2013). Design, synthesis of novel lipids as chemical permeation enhancers and development of nanoparticle system for transdermal drug delivery. *PLoS One, 8,* e82581.

Martanto, W., Davis, S. P., Holiday, N. R., Wang, J., Gill, H. S., & Prausnitz, M. R. (2004). Transdermal delivery of insulin using microneedles in vivo. *Pharm. Res., 21,* 947–952.

Martin, G. P. (1993). Phospholipids as skin penetration enhancers. In K. A. Walters & J. Hadgraft (Eds.), *Pharmaceutical skin penetration enhancement* (pp. 57–93). New York: Marcel Dekker, Inc.

Matriano, J. A., Cormier, M., Johnson, J., Young, W. A., Buttery, M., Nyam, K. et al. (2002). Macroflux microprojection array patch technology: A new and efficient approach for intracutaneous immunization. *Pharm. Res., 19,* 63–70.

Matuszewska, B., Keogan, M., Fisher, D. M., Soper, K. A., Hoe, C., Huber, A. C. et al. (1994). Acidic fibroblast growth factor: Evaluation of topical formulations in a diabetic mouse wound healing model. *Pharm. Res., 11(1),* 65–71.

McAllister, D. V., Allen, M. G., & Prausnitz, M. R. (2000). Microfabricated microneedles for gene and drug delivery. *Annu. Rev. Biomed. Eng., 2,* 289–313.

Merkle, H. P. (1989). Transdermal delivery systems. *Methods Find. Exp. Clin. Pharmacol., 11,* 135–153.

Meyer, B. R., Katzeff, H. L., Eschbach, J. C., Trimmer, J., Zacharias, S. B., Rosen, S. et al. (1989). Transdermal delivery of human insulin to albino rabbits using electrical current. *Am. J. Med. Sci., 297,* 321–325.

Meyer, B. R., Kreis, W., & Eschbach, J. (1988). Successful transdermal administration of therapeutic doses of a polypeptide to normal human volunteers. *Clin. Pharmacol. Ther., 44,* 607–612.

Mezei, M. & Gulasekharam, V. (1982). Liposomes—A selective drug delivery system for the topical route of administration: Gel dosage form. *J. Pharm. Pharmacol., 34,* 473–474.

Michaels, A. S., Chandrasekaran, S. K., & Shaw, J. E. (1975). Drug permeation through human skin: Theory and in vitro experimental measurement. *AIchE J., 21,* 985–996.

Miller, L. L., Kolaskie, C. J., Smith, G. A., & Rivier, J. (1990). Transdermal iontophoresis of gonadotropin releasing hormone (LHRH) and two analogues. *J. Pharm. Sci., 79,* 490–493.

Monteiro-Riviere, N. A., Inman, A. O., & Riviere, J. E. (1994). Identification of the pathway of iontophoretic drug delivery: Light and ultrastructural studies using mercuric chloride in pigs. *Pharm. Res., 11(2),* 251–256.

Morimoto, K., Iwakura, Y., Nakatani, E., Miyazaki, M., & Tojima, H. (1992). Effects of proteolytic enzyme inhibitors as absorption enhancers on the transdermal iontophoretic delivery of calcitonin in rats. *J. Pharm. Pharmacol., 44,* 216–218.

Moss, J. & Bundgaard, H. (1990). Prodrugs of peptides. 7. Transdermal delivery of thyrotropin-releasing hormone (TRH) via prodrugs. *Int. J. Pharm., 66,* 39–45.

Nakakura, M., Kato, Y., & Ito, K. (1998). Safe and efficient transdermal delivery of desmopressin acetate by iontophoresis in rats. *J. Biol. Pharm. Bull., 21,* 268–271.

Nasrollahi, S. A., Taghibiglou, C., Azizi, E., & Farboud, E. S. (2012). Cell-penetrating peptides as a novel transdermal drug delivery system. *Chem. Biol. Drug Des., 80,* 639–646.

Nolen, H. W., Catz, P., & Friend, D. R. (1994). Percutaneous penetration of methyl phosphonate antisense oligonucleotides. *Int. J. Pharm., 107,* 169–177.

Okumura, K., Kiyohara, Y., Komada, F., Iwakawa, S., Hirai, M., & Fuwa, T. (1990). Improvement in wound healing by epidermal growth factor (EGF) ointment. I. Effect of Nafamostat, Gabexate, or Gelatin on stabilization and efficacy of EGF. *Pharm. Res., 7,* 1289–1293.

Panchagnula, R., Pillai, O., Nair, V. B., & Ramarao, P. (2000). Transdermal iontophoresis revisited. *Curr. Opin. Chem. Biol., 4,* 468–473.

Parasrampuria, D. & Parasrampuria, J. (1991). Percutaneous delivery of proteins and peptides using iontophoretic techniques. *J. Clin. Pharmacol. Ther., 16,* 7–17.

Paudel, K. S., Milewski, M., Swadley, C. L., Brogden, N. K., Ghosh, P., & Stinchcomb, A. L. (2010). Challenges and opportunities in dermal/transdermal delivery. *Ther. Deliv., 1,* 109–131.

Peters, E. E., Ameri, M., Wang, X., Maa, Y. F., & Daddona, P. E. (2012). Erythropoietin-coated ZP-microneedle transdermal system: Preclinical formulation, stability, and delivery. *Pharm. Res., 29,* 1618–1626.

Pfister, W. R. & Hsieh, D. S. T. (1990a). Permeation enhancers compatible with transdermal drug delivery systems. Part I: Selection and formulation considerations. *Pharm. Technol., 14(9),* 132–140.

Pfister, W. R. & Hsieh, D. S. T. (1990b). Permeation enhancers compatible with transdermal drug delivery systems: Part II: System design considerations. *Pharm. Technol., 14(10),* 54–60.

Pillai, O., Nair, V., Poduri, R., & Panchagnula, R. (1999). Transdermal iontophoresis. Part II: Peptide and protein delivery. *Methods Find. Exp. Clin. Pharmacol., 21,* 229–240.

Pillai, O. & Panchagnula, R. (2003). Transdermal delivery of insulin from poloxamer gel: Ex vivo and in vivo skin permeation studies in rat using iontophoresis and chemical enhancers. *J. Control. Release, 89,* 127–140.

Pino, C. J., Gutterman, J. U., Vonwil, D., Mitragotri, S., & Shastri, V. P. (2012). Glycosylation facilitates transdermal transport of macromolecules. *Proc. Natl. Acad. Sci. U.S.A., 109,* 21283–21288.

Prausnitz, M. R. (2001). Overcoming skin's barrier: The search for effective and user-friendly drug delivery. *Diabetes Technol. Ther., 3,* 233–236.

Prausnitz, M. R. (2004). Microneedles for transdermal drug delivery. *Adv. Drug Deliv. Rev., 56,* 581–587.

Prausnitz, M. R., Bose, V. G., Langer, R., & Weaver, J. C. (1993). Electroporation of mammalian skin—A mechanism to enhance transdermal drug delivery. *Proc. Natl. Acad. Sci. U.S.A., 90,* 10504–10508.

Rao, G., Glikfeld, P., & Guy, R. H. (1993). Reverse iontophoresis: Development of a noninvasive approach for glucose monitoring. *Pharm. Res., 10,* 1751–1755.

Rao, R. & Nanda, S. (2009). Sonophoresis: Recent advancements and future trends. *J. Pharm. Pharmacol., 61,* 689–705.

Rastogi, S. K. & Singh, J. (2003). Passive and iontophoretic transport enhancement of insulin through porcine epidermis by depilatories: Permeability and Fourier transform infrared spectroscopy studies. *AAPS PharmSciTech, 4,* E29.

Ratafia, M. (1988, September). Growth markets: New drugs that heal tissues. *Pharm. Executive,* 74–80.

Roy, S. D. & Degroot, J. S. (1994). Percutaneous absorption of nafarelin acetate, an LHRH analog, through human cadaver skin and monkey skin. *Int. J. Pharm., 110,* 137–145.

Ruland, A., Kreuter, J., & Rytting, J. H. (1994a). Transdermal delivery of the tetrapeptide hisetal (melanotropin (6–9)). I. Effect of various penetration enhancers: In vitro study across hairless mouse skin. *Int. J. Pharm., 101,* 57–61.

Ruland, A., Kreuter, J., & Rytting, J. H. (1994b). Transdermal delivery of the tetrapeptide hisetal (melanotropin (6–9)): II. Effect of various penetration enhancers—In vitro study across human skin. *Int. J. Pharm., 103,* 77–80.

Ruland, A., Rohr, U., & Kreuter, J. (1994). Transdermal delivery of the tetrapeptide hisetal (melanotropin (6–9)) and amino acids: Their contribution to the elucidation of the existence of an 'Aqueous Pore' pathway. *Int. J. Pharm., 107,* 23–28.

Russell-Jones, G. & Himes, R. (2011). Water-in-oil microemulsions for effective transdermal delivery of proteins. *Expert Opin. Drug Deliv, 8,* 537–546.

Sachdeva, V. & Banga, A. K. (2011). Microneedles and their applications. *Recent Patents Drug Deliv. Formul., 5,* 95–132.

Sachdeva, V., Zhou, Y., & Banga, A. K. (2013). In vivo transdermal delivery of leuprolide using microneedles and iontophoresis. *Curr. Pharm. Biotechnol., 14,* 180–193.

Sage, B. H. (1993). Iontophoresis. In J. Swarbrick & J. C. Boylan (Eds.), *Encyclopedia of pharmaceutical technology.* (pp. 217–247). New York: Marcel Dekker Inc.

Sage, B. H. & Riviere, J. E. (1992). Model systems in iontophoresis—Transport efficacy. *Adv. Drug Deliv. Rev., 9,* 265–287.

Sanghvi, P. & Banakar, U. V. (1991). Ultrasonics: Principles and biomedical applications. *BioPharm, 4,* 32–40.

Schoellhammer, C. M., Blankschtein, D., & Langer, R. (2014). Skin permeabilization for transdermal drug delivery: Recent advances and future prospects. *Expert Opin. Drug Deliv., 11,* 393–407.

Schreier, H. & Bouwstra, J. (1994). Liposomes and niosomes as topical drug carriers—Dermal and transdermal drug delivery. *J. Control. Release, 30,* 1–15.

Shah, V. P. (2001). Progress in methodologies for evaluating bioequivalence of topical formulations. *Am. J. Clin. Dermatol., 2,* 275–280.

Shah, V. P., Behl, C. R., Flynn, G. L., Higuchi, W. I., & Schaefer, H. (1992). Principles and criteria in the development and optimization of topical therapeutic products. *Pharm. Res., 9,* 1107–1111.

Shah, V. P., Flynn, G. L., Guy, R. H., Maibach, H. I., Schaefer, H., Skelly, J. P. et al. (1991). In vivo percutaneous penetration/absorption. *Int. J. Pharm., 74,* 1–8.

Shalaby, W. S. W., Abdallah, A. A., Park, H., & Park, K. (1993). Loading of bovine serum albumin into hydrogels by an electrophoretic process and its potential application to protein drugs. *Pharm. Res., 10,* 457–460.

Siciliano, A. A. (1985). Topical liposomes—An update and review of uses and production methods. *Cosmet. Toiletries, 100,* 43–46.

Siddiqui, O., Sun, Y., Liu, J. C., & Chien, Y. W. (1987). Facilitated transdermal transport of insulin. *J. Pharm. Sci., 76,* 341–345.

Singh, P. & Maibach, H. I. (1994). Transdermal iontophoresis: Pharmacokinetic considerations. *Clin. Pharmacokinet., 26(5),* 327–334.

Sintov, A. C. & Wormser, U. (2007). Topical iodine facilitates transdermal delivery of insulin. *J. Control. Release, 118,* 185–188.

Smith, A., Yang, D., Delcher, H., Eppstein, J., Williams, D., & Wilkes, S. (1999). Fluorescein kinetics in interstitial fluid harvested from diabetic skin during fluorescein angiography: Implications for glucose monitoring. *Diabetes Technol. Ther., 1,* 21–27.

Srinivasan, V., Higuchi, W. I., Sims, S. M., Ghanem, A. H., & Behl, C. R. (1989). Transdermal iontophoretic drug delivery: Mechanistic analysis and application to polypeptide delivery. *J. Pharm. Sci., 78,* 370–375.

Srinivasan, V., Su, M., Higuchi, W. I., & Behl, C. R. (1990). Iontophoresis of polypeptides: Effect of ethanol pretreatment of human skin. *J. Pharm. Sci., 79,* 588–591.

Stephen, R. L., Petelenz, T. J., & Jacobsen, S. C. (1984). Potential novel methods for insulin administration: I. Iontophoresis. *Biomed. Biochim. Acta, 43,* 553–558.

Sun, Y. & Xue, H. (1991). Transdermal delivery of insulin by iontophoresis. *Ann. N. Y. Acad. Sci., 618,* 596–598.

Tachibana, K. (1992). Transdermal delivery of insulin to alloxan-diabetic rabbits by ultrasound exposure. *Pharm. Res., 9,* 952–954.

Tachibana, K. & Tachibana, S. (1991). Transdermal delivery of insulin by ultrasonic vibration. *J. Pharm. Pharmacol., 43,* 270–271.

Tao, S. L. & Desai, T. A. (2003). Microfabricated drug delivery systems: From particles to pores. *Adv. Drug Deliv. Rev., 55,* 315–328.

Tas, C., Mansoor, S., Kalluri, H., Zarnitsyn, V. G., Choi, S. O., Banga, A. K. et al. (2012). Delivery of salmon calcitonin using a microneedle patch. *Int. J. Pharm., 423,* 257–263.

Thysman, S., Hanchard, C., & Preat, V. (1994). Human calcitonin delivery in rats by iontophoresis. *J. Pharm. Pharmacol., 46,* 725–730.

Tomohira, Y., Machida, Y., Onishi, H., & Nagai, T. (1997). Iontophoretic transdermal absorption of insulin and calcitonin in rats with newly-devised switching technique and addition of urea. *Int. J. Pharm., 155,* 231–239.

Tyle, P. & Agrawala, P. (1989). Drug delivery by phonophoresis. *Pharm. Res., 6,* 355–361.

van der Maaden, K., Jiskoot, W., & Bouwstra, J. (2012). Microneedle technologies for (trans) dermal drug and vaccine delivery. *J. Control. Release, 161,* 645–655.

Varman, R. M. & Singh, S. (2012). Investigation of effects of terpene skin penetration enhancers on stability and biological activity of lysozyme. *AAPS PharmSciTech, 13,* 1084–1090.

Vemulapalli, V., Bai, Y., Kalluri, H., Herwadkar, A., Kim, H., Davis, S. P. et al. (2012). In vivo iontophoretic delivery of salmon calcitonin across microporated skin. *J. Pharm. Sci., 101,* 2861–2869.

Vollmer, U., Muller, B. W., Wilffert, B., & Peters, T. (1993). An improved model for studies on transdermal drug absorption in vivo in rats. *J. Pharm. Pharmacol., 45,* 242–245.

Wallace, B. M. & Lasker, J. S. (1993). Stand and deliver—Getting peptide drugs into the body. *Science, 260,* 912–913.

Walters, K. (1986). Percutaneous absorption and transdermal therapy. *Pharm. Technol., 10,* 30–47.

Wearley, L. L., Tojo, K., & Chien, Y. W. (1990). A numerical approach to study the effect of binding on the iontophoretic transport of a series of amino acids. *J. Pharm. Sci., 79,* 992–998.

Weiner, N., Williams, N., Birch, G., Ramachandran, C., Shipman, C., & Flynn, G. (1989). Topical delivery of liposomally encapsulated interferon evaluated in a cutaneous herpes guinea pig model. *Antimicrob. Agents Chemother., 33,* 1217–1221.

Wertz, P. W., Abraham, W., Landmann, L., & Downing, D. T. (1986). Preparation of liposomes from stratum corneum lipids. *J. Invest. Dermatol., 87,* 582–584.

Wester, R. C. & Maibach, H. I. (1992). Percutaneous absorption of drugs. *Clin. Pharmacokinet., 23,* 253–266.

Widera, G., Johnson, J., Kim, L., Libiran, L., Nyam, K., Daddona, P. E. et al. (2006). Effect of delivery parameters on immunization to ovalbumin following intracutaneous administration by a coated microneedle array patch system. *Vaccine, 24,* 1653–1664.

Zakzewski, C. A., Wasilewski, J., Cawley, P., & Ford, W. (1998). Transdermal delivery of regular insulin to chronic diabetic rats: Effect of skin preparation and electrical enhancement. *J. Control. Release, 50,* 267–272.

9 Pulmonary and Other Mucosal Delivery of Therapeutic Peptides and Proteins

9.1 INTRODUCTION

Delivery of therapeutic peptides and proteins faces formidable barriers and, therefore, a parenteral dosage form is typically developed, as discussed in Chapter 6. However, because of issues relating to short half-lives and inconvenient administration, alternative routes are being actively investigated. Efforts to develop oral dosage forms have been discussed in Chapter 7. Some of the other mucosal routes that have not yet been discussed include nasal, pulmonary, rectal, buccal, and ocular routes. These mucosal routes will be discussed in this chapter. Of these routes, pulmonary and nasal appear to be most promising and will be discussed in greater detail. The primary barriers to mucosal delivery are the absorption and enzymatic barriers. A drug has to overcome the epithelial absorption barrier by passing through the cells (transcellular transport) or through the tight junctions between the cells (paracellular transport). The equivalent pore radius of the various mucous membranes is an important limiting factor and varies from 4 to 8 Å (Zhou, 1994). Another barrier to mucosal absorption of peptides and proteins is the proteolytic barrier (Eppstein & Longenecker, 1988; Lee, 1988; Reddy & Banga, 1993). Proteolytic enzymes are present all over the body, and they act by hydrolyzing the peptide bonds in polypeptides. Proteolysis may occur at N-terminus or C-terminus or at distinct endo-residues in a polypeptide. The presence of these enzymes is essential to sustain vital body processes, but these enzymes present a formidable barrier to the delivery of peptides and proteins.

The use of penetration enhancers and protease inhibitors is the most commonly used approach to overcome various penetration and enzymatic barriers to mucosal drug absorption. In addition, formulation approach or a prodrug approach may also be feasible. Commonly used penetration enhancers include chelators, surfactants, bile salts, and fatty acids (Lee, Yamamoto, & Kompella, 1991). The use of these penetration enhancers may compromise the integrity of the mucosal membrane, and thus recovery of the mucosal damage is critical to the selection process. Penetration enhancers may promote drug absorption by the transcellular pathway (apical cell membrane) or the paracellular pathway (tight junction between cells). The paracellular pathway is gaining interest as it may lack proteolytic activity. The barrier to paracellular diffusion is a tight junction or zonula occludens. Use of a chelating agent to lower the extracellular Ca^{2+} concentration at this junction is known to enhance its permeability. The various protease inhibitors include the metalloprotease, aspartyl-protease, cysteine proteinases, serine proteinase, aminopeptidase, and nonspecific inhibitors. Selection of which protease inhibitor to use will depend on the type of target protease and its subcellular distribution (Lee, 1990). Prodrug approach may be used to disrupt the enzyme–substrate recognition by reversible chemical modification of the labile functional groups. Chemical modification of the amino terminus can be done to synthesize polypeptide prodrugs that are resistant to aminopeptidases. Also, prodrug approach may increase the lipophilicity of a polypeptide, which in turn may result in higher transport through biological barriers (Oliyai & Stella, 1993). As alternate delivery routes become viable, the demand for therapeutic peptides and proteins is likely to increase as well. For example, though the safety and efficacy profile of calcitonin compares very favorably with other osteoporosis drugs, its sales lag behind. The reasons are cost and convenience since other drugs are less expensive and available in oral dosage forms (Levy, 1993).

Other mucosal routes include vaginal and intrauterine delivery. Only very few studies have evaluated this route thus far, and more work is likely to appear in the literature in the next few years. Vaginal route may be promising for the treatment of female-related diseases. Use of vaginal route for vaccination is discussed in Chapter 10. Use of adjuvants is required to promote absorption of polypeptides from the vaginal mucosa. Absorption of insulin across the vaginal mucosa of rats was feasible when coadministered with sodium tauro-24,25-dihyrofusidate (STDHF), polyoxyethylene-9-lauryl ether, lysophosphatidylcholine, palmitoylcarnitine chloride, and lysophosphatidyl glycerol. However, these enhancer systems caused histological changes in the vaginal epithelium, and thus their long-term utility is questionable (Richardson, Illum, & Thomas, 1992). Absorption of human calcitonin from the rat vaginal mucosa has also been investigated and was found to be enhanced by the coadministration of sodium deoxycholate and by the peptidase inhibitors bestatin, leupeptin, and pepstatin A (Nakada, Miyake, & Awata, 1993). Intrauterine drug delivery has been proposed as a new route of drug administration relatively recently. Two polypeptides, calcitonin and insulin, were reported to be absorbed from the uterus of the rat in a biologically active form. The extent and duration of hypocalcemic and hypoglycemic effects induced by intrauterine delivery of calcitonin and insulin, respectively, were equivalent to those obtained after SC injections (Golomb, Avramoff, & Hoffman, 1993). The role of cervical mucus in drug delivery via the vagina and cervix has been reviewed (Katz & Dunmire, 1993). The viability and integrity of the mucosal tissue must be monitored during in vitro studies, such as those done in Ussing chambers (Reardon, Gochoco, Audus, Wilson, & Smith, 1993).

9.2 PULMONARY DELIVERY

Local delivery of conventional drugs for treatment of lung diseases has been used for several decades. In recent years, there has been an increasing interest in the use of pulmonary route for local and systemic delivery of therapeutic peptides and proteins. The pulmonary route offers several advantages for drug delivery and may soon become one of the most promising routes for peptide and protein delivery. The lungs offer a huge surface area for drug absorption. In addition, the walls of the alveoli in the deep lung are extremely thin. These alveoli can be targeted for rapid drug absorption using aerosolized drug. Also, pulmonary delivery avoids the first-pass effects, and very large molecules can be delivered (Agu, Ugwoke, Armand, Kinget, & Verbeke, 2001; Bitonti & Dumont, 2006; Byron, 1987; Smith, 1997; Wearley, 1991; Yu & Chien, 1997). However, lungs are capable of metabolizing peptides and proteins. Some of the metabolic pathways may be different from those observed in the intestine, and thus the feasibility of pulmonary administration may need to be examined on a case-by-case basis (Hoover et al., 1992). The pulmonary route is very promising, with bioavailability reaching as high as 56% for α-interferon (Patton, Trinchero, & Platz, 1994). However, lung can metabolize peptides and proteins due to the presence of peptidases. Safety issues associated with pulmonary delivery include the possibility of elicitation of an immune response resulting from changes in pulmonary epithelial permeability. Theoretically, destabilization of the surfactant film coating the alveolar surface may also occur as several peptides, and proteins may interact with it. However, such an event is unlikely (Lee, 1991).

The conducting airway of the respiratory tract starts with the nasal cavity and continues with the nasopharynx, larynx, trachea, bronchi, and bronchioles. The alveoli and airways distal to the bronchioles constitute the respiratory region, where rapid exchange of gases takes place across the thin alveolar–capillary membrane (Figure 9.1) (Patton & Platz, 1992; Sou et al., 2011). The walls of the

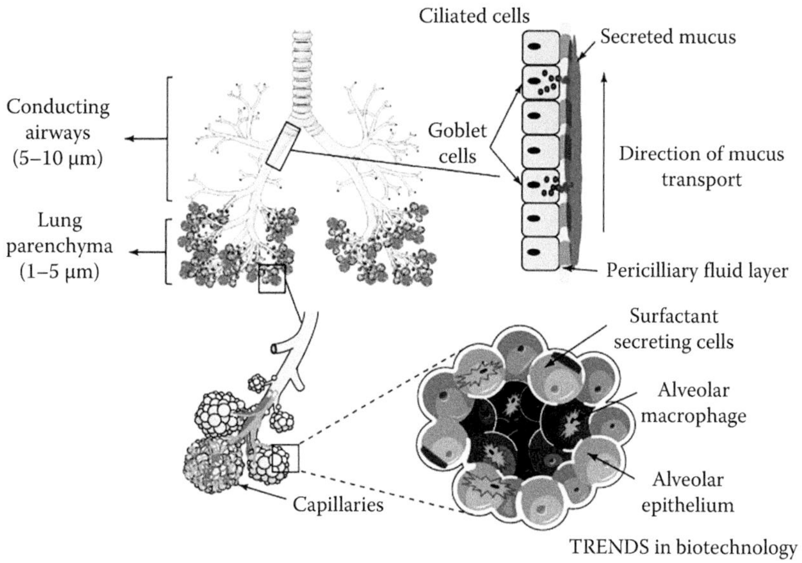

FIGURE 9.1 (See color insert.) Diagram of the lung and particle size requirements based on intended deposition region in the respiratory tract. (Reproduced from Sou, T. et al., *Trends Biotechnol.*, 29, 191, 2011. With permission.)

alveoli are only about 0.1–1 μm in thickness, while the surface area for the lung is about 75 m^2 for a 70 kg male. Simple columnar ciliated epithelium lines the larger bronchioles, while the smaller bronchioles are unciliated. The epithelium of the alveolar sacs is composed of alveolar type I cells that are flat squamous cells and allow rapid gas exchange. Alveolar type II cells are also present and secrete pulmonary surfactant, which decreases the surface tension at the alveolar–capillary interface. The pulmonary surfactant is a heterogeneous lipid–protein complex that overlies the alveolar epithelium and contains about 90% lipids and 10% surfactant-associated proteins, primarily surfactant protein A (SP-A). Endocytic uptake by pulmonary alveolar macrophages constitutes the major clearance mechanism in the alveolar regions of the lung. The clearance mechanism in the naso-/oropharynx and tracheobronchial tree involves several components. First, swallowing, expectoration, and coughing contribute toward the clearance mechanism. However, the major clearance mechanism is mucociliary clearance. A mucociliary clearance system operates from the nasal cavity to the terminal bronchioles with cilia transporting the mucus toward the oropharynx. This mucociliary clearance is responsible for the removal of inhaled particles from the respiratory tract. A viscoelastic mucus layer across the cilia is required for the transport of particulate matter by ciliary activity (Lansley, 1993). The lung is rich in proteolytic enzymes, and these enzymes have a potential to destroy lung tissue. These enzymes are kept in check by proteinase inhibitors, such as α_1-antitrypsin (α_1AT) and antileukoprotease (ALP).

9.2.1 Drug Deposition in the Respiratory Tract

Drugs are delivered to the respiratory tract in the form of aerosols. The most important factor controlling the site of deposition within the respiratory tract is the particle characteristics, which can be combined into a *mass median aerodynamic diameter* (MMAD). The MMAD is a function of particle size, shape, and density. Other factors that are important include the particle charge; air velocity within the airway; and patient factors such as tidal volume, respiratory rate, and respiratory pause. The aerodynamic diameter (d_{aer}) is related to the physical sphere diameter d and particle density ρ by the following relationship (Clark, 2000):

$$D_{aer} = d\,\rho^{1/2}$$

Pathological conditions of the lung, especially obstructive airway diseases such as asthma, chronic bronchitis, or emphysema, may also diminish penetration into peripheral pulmonary regions. Physical mechanisms of particle deposition within the respiratory tract include inertial impaction, sedimentation, diffusion, interception, and electrostatic attraction. Particles with an MMAD between 1 and 6 μm generally provide optimum delivery to the lungs, and this fraction is sometimes referred to as the *respirable fraction* (RF) or *fine particle fraction* (FPF). Particles larger than 6 μm will deposit mainly in the oropharynx, while those smaller than 1 μm will be exhaled during normal tidal breathing. The amount of aerosol deposited in the lung will depend on both the efficiency of aerosol generation and the FPF.

The fluid dynamics of the larynx and tracheobronchial airways will also affect the site of particle deposition. Mathematical models have been used for simulation to predict the site of deposition based on these fluid dynamics and various breathing conditions (Martonen, 1993; Martonen & Katz, 1993; Martonen, Zhang, & Lessmann, 1993). In vitro systems such as a glass or silicone throat have been used to emulate particle deposition in the respiratory tract (Niven, Lott, Ip, Somaratne, & Kearney, 1994). A cascade impactor allows the collection and fractionation of an aerosol cloud through a simulated *throat*, and data generated can be used to calculate MMAD. Thus, it allows a comparison of devices and formulations (Timsina, Martin, Marriott, Ganderton, & Yianneskis, 1994). It is difficult to determine the relative contributions from the trachea and alveoli to the overall absorption of drugs from the respiratory tract. In general, the systemic absorption from the upper respiratory tract (nasopharynx, trachea, and large bronchi) is limited by a small surface area and poor blood flow. In contrast, the lower respiratory tract offers a huge surface area due to the presence of alveoli. Also, the alveoli are well perfused with blood. Patient factors such as breath patterns, breath frequency, and volume inhaled will also affect drug deposition. Drug deposition can be increased by increasing the residence time for particles in the alveoli. This may be achieved by slow and deep breathing followed by holding the breath for a few seconds following completion of inhalation (Kompella & Lee, 1992).

9.2.2 Drug Absorption from the Respiratory Tract

Our understanding of the specific transport mechanisms of pulmonary absorption is currently very limited. One reason is the absence of a convenient and well-characterized in vitro system for studies of pulmonary absorption. The sampling of alveolar space is difficult due to the extensive structural subdivision in this area. Also, no immortalized cell lines exist for in vitro studies. When primary cell cultures are used, they are viable for only about 7–8 days, including the 4–5-day period of differentiation into squamous type I cell morphology. The use of amphibian lung mounted in Ussing chambers offers some promise as it has morphological and physiological similarities with human lung (Wall, Pierdomenico, & Wilson, 1993).

The presence of the pulmonary surfactant plays an important role in the displacement of inhaled particles from the air to the aqueous alveolar lining. Deposited particles can then be removed by a nonabsorptive or an absorptive process. The nonabsorptive removal is primarily by mucociliary clearance. Absorptive process may involve uptake by alveolar macrophages or direct absorption into epithelial cells (Kompella & Lee, 1992). A transferrin receptor is known to exist on the surface of respiratory epithelial cells and has been exploited for receptor-mediated peptide delivery in pulmonary epithelial monolayers. Alveolar cells were isolated from rat lungs by enzymatic digestion and grown on microporous tissue culture–treated polycarbonate filters. This cultured alveolar epithelial monolayer system was used to simulate the pulmonary epithelium. The transport of a model polypeptide, horseradish peroxidase, covalently linked to transferrin, across the monolayer was investigated. A significant increase in the uptake of horseradish peroxidase by the cell monolayer was observed upon its conjugation with transferrin (Deshpande et al., 1994).

A β-lactam-transporting peptide transporter has also been suggested to be the molecular basis for the transport of peptides and peptidomimetics in pulmonary epithelial cells (Groneberg et al., 2001). In another study, direct diffusional transport of the tripeptide, thyrotropin-releasing hormone, across the rat alveolar epithelial cell monolayers via paracellular pathways has been reported (Morimoto, Yamahara, Lee, & Kim, 1994). Using cultured alveolar cell monolayers, peptide transport has shown to be predominantly through paracellular pathway, with an inverse dependence on molecular size and also an enhanced flux for cationic peptides (Dodoo et al., 2000). The inverse dependence of pulmonary absorption of proteins to their size has also been shown in studies carried out in rabbits and humans (Hastings, Grady, Sakuma, & Matthay, 1992).

9.2.3 Drug Delivery Devices

Aerosols can be generated by three main drug delivery systems: pressurized metered-dose inhaler (MDI), nebulizer, and dry-powder inhaler (DPI). Powered active systems are also being developed to replace the ones that are dependent on the patient's inspiratory effort. The MDIs were exempt from the ban on chlorofluorocarbon (CFC) propellants for the past several years. However, since CFC propellants are implicated in ozone depletion, alternate propellants or devices are needed (June, Schultz, & Miller, 1994), and some such as the hydrofluoroalkanes (HFAs) are now available and are also being investigated for pulmonary delivery of proteins (Li & Seville, 2010). In contrast, nebulizers do not use propellants. However, a peptide or protein formulation is very susceptible to denaturation in contact with a propellant or due to the extensive air–water interfaces generated during aerosolization from an MDI or nebulizer. These stability considerations of peptides and proteins in the aerosol dosage form were discussed in Chapter 5. In a nebulizer, a portable air compressor can be used to generate a spray. The typical mechanism involved to separate the larger droplets from the smaller ones involves recycling of 99% of the solution that exposes the protein formulation to repeated stress and extensive air–water interfaces, which may result in physical stability problems. Alternatively, an ultrasonic nebulization process or a mechanical process can be used. When ultrasonic energy is used, it can expose the protein to a different type of stress such as thermal stress. However, ultrasonic nebulizers can allow the use of small volumes, an advantage with protein formulations.

Four commercially available disposable nebulizers have been evaluated for their ability to deliver aerosols of recombinant human deoxyribonuclease I (rhDNase) (Cipolla, Gonda, & Shire, 1994). Between 20% and 28% of the rhDNase dose was delivered to the mouthpiece in the RF, and all four nebulizers were essentially equivalent in their ability to deliver respirable doses of rhDNase in an intact, fully active form. One technology that was in development for delivery of insulin utilized a unit dose liquid aerosol delivery. Once the formulation was dispensed into the blister, it was sealed on the top with a multilayer laminate containing a nozzle array consisting of hundreds of about 1 μm size holes laser-micromachined into a polymer film. The blister was placed into an electronically controlled mechanical device that gets actuated by patients breathing to pressurize the formulation, which in turn opens the heat seal to form an aerosol through the nozzle array (Deshpande et al., 2002; Schuster & Farr, 2003). A promising alternative is the use of DPIs, in which the peptide or protein can be delivered from a dry-powder formulation. DPI designs range from the original Spinhaler and Rotahaler having the drug in hard gelatin capsules to the recent foil blister devices (Clark, 2000). The formulation aspects of DPIs are discussed in Section 9.2.5.

9.2.4 Formulation of Peptides/Proteins for Pulmonary Delivery

The choice of a drug delivery device and control of MMAD are important formulation parameters discussed so far. In addition, other formulation components that may have a significant influence on pulmonary delivery and/or adverse reactions in the lung include the proper choice of osmolality, pH, and buffering agents and the use of protease inhibitors and penetration enhancers. Many of these aerosols have an objectionable taste, and the use of taste masking excipients has been investigated. It was reported that these excipients did not have an effect on lung absorption of leuprolide acetate. However, these excipients were found to affect the drug micronization and particle deposition pattern, possibly due to their effect on the properties of the suspension itself and changes in propellant evaporation behavior following actuation (Zheng, Fulu, Lee, Barber, & Adjei, 2001).

Penetration enhancers and protease inhibitors have been reported to increase the pulmonary absorption of ($Asu^{1,7}$)-eel calcitonin (ECT) and insulin in rats (Morita, Yamamoto, Takakura, Hashida, & Sezaki, 1994; Yamamoto, Fujita, & Muranishi, 1996). In the absence of protease inhibitors, the pharmacological availability of ECT was 2.7%. Coadministration of nafamostat mesilate, Na-glycocholate, and bacitracin increased the pulmonary absorption of ECT. Nafamostat mesilate strongly inhibits a variety of proteases such as trypsin, plasmin, and kallikrein (Morita et al., 1994). Proteolytic activity of pulmonary alveolar epithelium has been described (Yang, Ma, Malanga, & Rojanasakul, 2000), and protease inhibitors have been reported to enhance the transport of intact arginine vasopressin across alveolar epithelial cell monolayers (Yamahara, Morimoto, Lee, & Kim, 1994).

In another study, the effect of protease inhibitors and absorption enhancers on the absorption of salmon calcitonin was evaluated. The absorption enhancers, oleic acid and polyoxyethylene oleyl ether, and the protease inhibitors, chymostatin, bacitracin, potato carboxypeptidase inhibitor, and phosphoramidon, were found to significantly enhance the absorption following intratracheal coadministration to rats (Kobayashi, Kondo, & Juni, 1994). Effect of pH and surfactants on intratracheal administration of peptides and proteins has also been investigated (Komada, Iwakawa, Yamamoto, Sakakibara, & Okumura, 1994). Generally, aerosol administration of proteins will result in higher absorption as compared to intratracheal instillation (Colthorpe, Farr, Smith, Wyatt, & Taylor, 1995; Garcia-Contreras, Morcol, Bell, & Hickey, 2003; Niven, Whitcomb, Shaner, Ip, & Kinstler, 1995).

The effect of absorption enhancers and protease inhibitors on the pulmonary absorption of insulin has also been examined in a study by Yamamoto, Umemori, and Muranishi (1994). Sodium glycocholate, linoleic acid–surfactant mixed micelles, and *N*-lauryl-∃-D-maltopyranoside were used as absorption enhancers in this study, while aprotinin, bacitracin, and soybean trypsin inhibitor were used as protease inhibitors. In the presence of these additives, significant hypoglycemic effects were observed following pulmonary delivery of insulin in rats. In another study, the effects of formulation variables on the pulmonary absorption of insulin in the presence of a bile salt, sodium glycocholate, were investigated (Li & Mitra, 1994). A dose-dependent hypoglycemic response was observed upon intratracheal administration of insulin to rats. A hypotonic solution was found to increase the hypoglycemic response, possibly due to epithelial cell damage. An increase in viscosity of the formulation also facilitated insulin adsorption, possibly due to reduced mucociliary clearance. The effect of various absorption promoters on model macromolecules has also been investigated. Fluorescent isothiocyanate-labeled dextrans with different molecular weights were used as model macromolecules in these studies. A linear correlation was found between the logarithm of absorption percentage and the logarithm of molecular weight for the various dextrans. It was found that there is an optimal molecular weight to which each absorption promoter gives the largest enhancing effect on the pulmonary absorption of drugs (Morita, Yamamoto, Hashida, & Sezaki, 1993; Ohtani, Murakami, Yamamoto, Takada, & Muranishi, 1991).

Recombinant human granulocyte colony-stimulating factor (rhG-CSF) has been prepared in powder formulations for inhalation. Powder formulations (MMAD<4 μm) of rhG-CSF were insufflated via an endotracheal tube to the lungs of rabbits and were shown to induce a pharmacological response. Absorption was more rapid than that from an SC dose and thus was not dissolution rate limited (Niven, Lott, Ip, & Cribbs, 1994). Pulmonary delivery of polyethylene glycol (PEG) conjugates of rhG-CSF has also been investigated (Niven et al., 1994). Inhalation aerosols of leuprolide acetate have been evaluated for their potential use as alternatives to parenteral administration. The study, done in 23 healthy human males, resulted in absolute bioavailability ranging from 4% to 18%. The bioavailability corrected for RF ranged from 35% to 55% (Adjei & Garren, 1990).

Liposomes, micelles, and microspheres have been widely investigated for controlled delivery of drugs to the lung (Baginski et al., 2012; Bennett et al., 1994; Zeng, Martin, & Marriott, 1995). Liposomes have also been used as carriers for pulmonary delivery of drugs, including anticancer drugs, antimicrobials, peptides, and enzymes. Liposomes can serve to solubilize poorly soluble drugs, act as pulmonary sustained release reservoir, or facilitate intracellular delivery of drugs to alveolar macrophages. While some liposomes are cleared by alveolar macrophages, others may be cleared by incorporation into the surfactant phospholipid pool. Phospholipids are known to bind the SP-A, with the interaction being particularly strong with dipalmitoyl phosphatidylcholine (Casals, Miguel, & Perezgil, 1993). Liposome aerosols seem to be safe, even after prolonged use, and their future applications may include their use as carriers for immunomodulating agents and for efficient intracellular delivery of genes or antisense agents (Schreier, Gonzalezrothi, & Stecenko, 1993).

Liposomes have been used to facilitate the pulmonary uptake of insulin in rats to enhance the hypoglycemic effect. A similar insulin uptake and pharmacodynamic response was obtained from both a physical mixture of insulin and blank liposomes and liposomally entrapped insulin. This suggests that the liposomal lipid is responsible for rapid insulin uptake into the alveolar cells and that encapsulation of insulin inside the lipid vesicle may not be required (Liu, Shao, Kildsig, & Mitra, 1993). The use of low-molecular-weight compounds such as hydroxy methyl amino propionic acid (H-MAP) to reversibly destabilize the native structure of proteins by favoring a partially unfolded conformation to facilitate transport across mucosa has been investigated for delivery of insulin to the rat lung (Suarez et al., 2001). A similar approach has been used to enhance oral delivery of peptides (see Chapter 7). Microspheres of poly(L-lactic acid) have also been used for aerosol generation and lung deposition. A respirable particle size could be generated, and the shape and surface morphology of the microspheres was not affected by nebulization (Masinde & Hickey, 1993).

Other peptide and protein drugs that have been targeted for pulmonary absorption include interferons (Jaffe et al., 1991), cyclosporine A, deoxyribonuclease, α_1-AT (McElvaney et al., 1991), human growth hormone (hGH), 1-deaminocysteine-8-D-arginine vasopressin (Adjei & Gupta, 1994), and a decapeptide (Lizio et al., 2001). A hexapeptide, His-D-Trp-Ala-Trp-D-Phe-Lys-NH_2, resulted in a bioavailability of about 43% compared to IV administration, when administered intratracheally to dogs. This hexapeptide has been discovered to elicit growth hormone release and may provide an alternate treatment for growth hormone deficiency (Smith et al., 1994). Pulmonary absorption of seven clinically relevant polypeptides was investigated. Polypeptides that were well absorbed and had a high absolute bioavailability (indicated in parenthesis) included α-interferon (>56%), parathyroid hormone (PTH) 1–84 (>23%), PTH 34 (–40%), and salmon and human calcitonin (–17%). The bioavailability of glucagon (<1%) and somatostatin (<1%) was low. Since PTH-34 and calcitonin are only slightly larger than glucagon and somatostatin, it seems that the different bioavailability is not related to the molecular weight. Instead, degradation by specific peptidases in the lung due to unique sequence and conformation may be more important. With the exception of α-interferon and blood proteins (α1-antitrypsin, albumin, and IgG), other proteins were rapidly absorbed from the lung. Again, no clear correlation with molecular weight was observed. Growth hormone (≈22 kDa) and luteinizing hormone–releasing hormone (LHRH) analog (≈1 kDa) peaked at similar times, while interferon (≈19 kDa) and growth hormone (≈22 kDa) were absorbed at different rates (Patton et al., 1994).

Because of stability considerations, proteins may often be developed for pulmonary administration as DPIs. This requires the preparation of a powder that has flowability and dispersibility while at the same time maintaining biochemical stability of the protein molecule. For example, spray-dried rhDNase powders were observed to be quite cohesive but dispersed better to form aerosols when sodium chloride was included as an excipient (Chan, Clark, Gonda, Mumenthaler, & Hsu, 1997).

Other excipients to improve the performance of protein dry-powder aerosols have also been investigated (Lucas, Anderson, & Staniforth, 1998). One technology accomplishes stabilization by drying the protein into an amorphous glass state that inhibits denaturation by dramatically slowing molecular interactions. The stabilized protein may not even need refrigeration. Doses of dry powder are individually packaged into foil blister packs that can slide into a slot on the side of a specially designed inhaler. A device component opens the blister, and then a small burst of compressed air is used to aerosolize the powder and create a *standing cloud* that is then inhaled by the patient. No batteries or electronics are used (Clark, 2003).

Newer technologies based on lipid-based particle engineering are also being investigated. These spray-dried particles whose morphology is engineered to be both hollow and porous have been reported to deliver drugs more efficiently and reproducibly as compared to delivery of micronized drug from passive DPIs (Dellamary et al., 2000; Duddu et al., 2002; Hirst et al., 2002). It has also been shown that large porous particles can be used for deep lung delivery. These particles have geometric sizes exceeding 5 μm, but they are porous and thus have a small mass density, typically less than 0.4 g/cm^3, creating aerodynamic particle sizes in the range of 1–3 μm for deep lung delivery (Ben Jebria et al., 1999; Edwards et al., 1997; Vanbever et al., 1999). It has been recently reported that a supercritical carbon dioxide pressure-quench treatment of deslorelin–poly(D,L-lactide-co-glycolide) (PLGA) microparticles prepared using conventional approaches produced large porous particles that retained deslorelin integrity, sustained its release, and reduced cellular uptake (Koushik & Kompella, 2004). Recently, chemically linked agglomerates of nanoparticles have also been described for pulmonary delivery (Bhavane, Karathanasis, & Annapragada, 2003). Currently, protein powders are mainly produced by spray drying, but other technologies such as supercritical fluid development are under development. Large porous particles with high degree of voids are also being developed since they still have a small MMAD due to low particle density. The preparation of fine protein powders for inhalation has been reviewed (Johnson, 1997).

9.2.5 Commercialization and Regulatory Considerations

One protein product available for inhalation delivery is rhDNase or dornase alfa (Pulmozyme®). It is indicated for the management of cystic fibrosis patients to improve pulmonary function. Administration is by inhalation of an aerosol mist produced by a compressed air-driven nebulizer system, with a typical dose being one 2.5 mg single-use ampul that delivers 2.5 mL of a sterile 1 mg/mL drug solution having no preservatives (2014; Depreter, Pilcer, & Amighi, 2013). Also, bovine pulmonary surfactants are available for treating acute respiratory distress syndrome. They are administered by intratracheal instillation, and commercial products available include Survanta® (AbbVie, North Chicago), Curosurf® (Chiesi Farmaceutici, Parma, Italy), Infasurf® (ONY, Inc., Amherst, NY), and BLES® (BLES Biochemicals, London, ON, Canada) (Depreter et al., 2013).

Pulmonary delivery of proteins such as insulin, calcitonin, growth hormones, and PTH is currently under investigation (Agu et al., 2001). Aerosolized insulin delivered by oral inhalation can effectively normalize plasma glucose levels in patients with non-insulin-dependent diabetes mellitus (Laube, Georgopoulos, & Adams, 1993). Early clinical trials with inhaled insulin were done by at least two companies: Nektar Therapeutics in partnership with Pfizer and Sanofi-Aventis (Exubera® DPI) and Aradigm Corporation in partnership with Novo Nordisk (AERx™ unit dose liquid aerosol delivery technology). Exubera was marketed by Pfizer but was later voluntarily withdrawn from the market due to several factors, including issues relating to low sales, insurance reimbursement, bulky device, generation of anti-insulin antibodies, and need for periodic lung function testing (Depreter et al., 2013). After the discontinuation of Exubera, AERx® and other products in development were discontinued by companies except for MannKind Corporation (Valencia, CA). Recently, FDA granted approval to Afrezza®, a rapid-acting inhaled insulin device from MannKind Corporation. It uses a formulation where insulin is adsorbed to fumaryl diketopiperazine microparticles (Technosphere®) that dissolve upon inhalation to release insulin which is quickly absorbed in the alveoli (Figures 9.2 and 9.3) (Boss, Petrucci, & Lorber, 2012; Neumiller & Campbell, 2010). The formulation is packaged as a white powder in plastic cartridges that are used in conjunction with a compact breath-actuated inhaler device. Unlike Exubera that had a 10% bioavailability, Afrezza has a 26% bioavailability relative to SC regular insulin (Depreter et al., 2013).

Encouraging results have also been obtained for delivery of calcitonin via the pulmonary route (Patton, 2000), and there is considerable interest in treatment of systemic diseases via the lung (Corkery, 2000; Patton, 1998; Sanjar & Matthews, 2001). All inhalation solutions for nebulization are required to be sterile. For DPIs, sterilization is not required though microbial limits are imposed (Clark, 2000). As the lung is continuously exposed to nonsterile materials from the air, fewer immunologic reactions are expected, and DPIs have been reported to be well tolerated

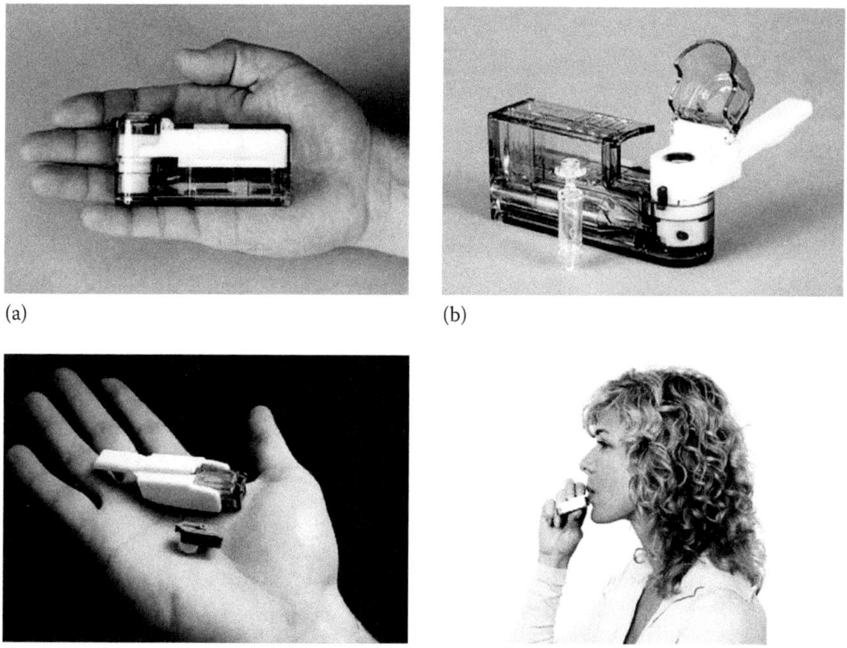

(a) (b) (c) (d)

FIGURE 9.2 (See color insert.) AFREZZA insulin inhalation device (Mannkind Corporation, Valencia, CA) approved by FDA. It consists of Technosphere® insulin inhalation powder premetered into single-use dose cartridges for administration with an AFREZZA™ inhaler. A palm-sized (a and b) or thumb-sized (c and d) device is shown. (Reproduced from Neumiller, J.J. and Campbell, R.K., *BioDrugs*, 24, 165, 2010. With permission.)

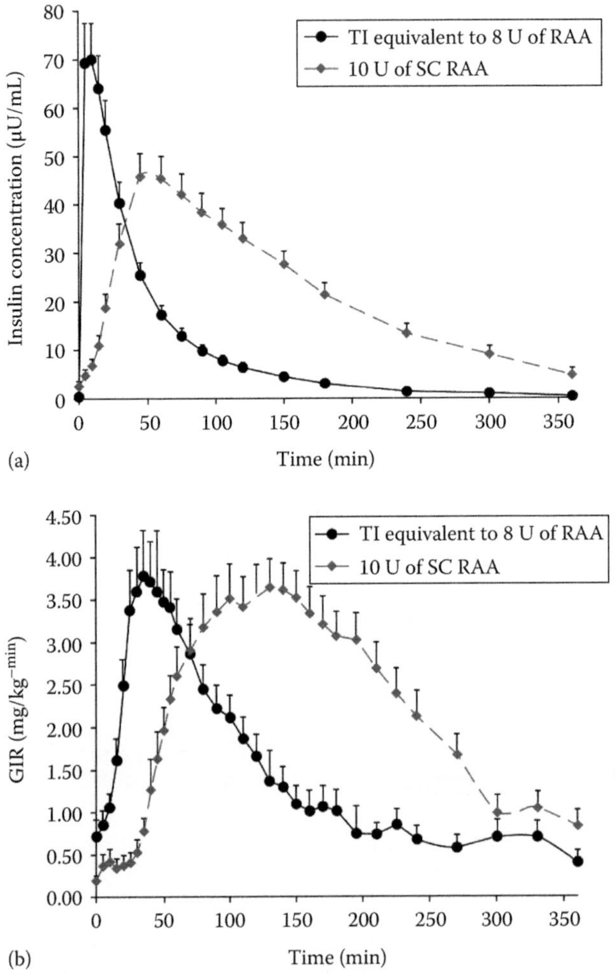

FIGURE 9.3 PK/PD profiles of inhaled Technosphere® insulin (TI) versus a rapid-acting insulin analog (RAA). Maximum plasma insulin concentration (a) occurs sooner with TI than with regular human insulin or RAA, as does the peak glucose-lowering effect; (b) GIR, glucose infusion rate. (Reproduced from Boss, A.H. et al., *J. Diabetes. Sci. Technol.*, 6, 773, 2012. With permission.)

in clinical studies lasting 6 months to a year. Typically, a 1-year safety data on 200 patients or a somewhat different combination of number of patients and time is required by the FDA for pulmonary products (Clark, 2003).

9.3 NASAL DELIVERY

While the primary function of the nose is as an olfactory organ, the nasal passages also serve to filter particulates and as heat exchangers and humidifiers of air on its way to the lungs. The nasal mucous membrane is composed primarily of columnar epithelial cells that are covered with microvilli on their apical surfaces and may be ciliated or nonciliated. In addition, goblet and basal cells are also present in the mucous membrane. Goblet cells produce the mucus secretion that covers the mucous membrane. The total surface area of both nasal cavities is about 160–180 cm^2 in a normal adult. Foreign materials caught by the nasal mucociliary function are carried backward to the nasopharynx and swallowed. However, a small area in the anterior part of the nose carries materials forward by ciliary movement, from where it can then be removed by wiping or nose blowing.

The cells of the nasal mucosa overlay a rich vasculature from which drugs can be absorbed readily. This leads to a fast onset of action, being as rapid as with intravenous injection. The easy accessibility of the nasal mucosa facilitates administration by the patient. Furthermore, drugs absorbed through the nasal mucosa bypass the hepatic circulation, thereby avoiding metabolic degradation and possible first-pass metabolism. Use of the nasal route for delivery of peptides is not new. While the advent of biotechnology-derived peptides and proteins has increased interest in nasal delivery, several synthetic or naturally derived peptides have been commercially available for several years in the intranasal dosage form (Campbell, Morimoto, Nenciu, & Fox, 2012; Lee & Longenecker, 1988; Pontiroli, Calderara, & Pozza, 1989; Su, 1986). Table 9.1 lists the peptides commercially available for intranasal administration. Most formulations contain a preservative to prevent the microbial degradation of the peptide in the formulation. In addition, salts and buffers are present to maintain isotonicity and prevent any pH shifts during usage. Most nasal sprays are administered from a squeeze bottle into one or both nostrils, and drug is promptly absorbed by the nasal mucosal into systemic circulation. For DGAVP, its intranasal administration is reported to be 100 times more effective in increasing plasma levels of DGAVP than the oral route (Westenberg, Hijman, Wiegant, Laczi, & Vanree, 1994). Calcitonin is indicated in several bone-related disorders such as Paget's disease, malignant hypercalcemia, and postmenopausal osteoporosis. For the control of these diseases, chronic delivery of calcitonin is required, and thus parenteral therapy is inconvenient. The use of a nasal spray would offer some promise in this respect. In addition to the peptides

TABLE 9.1
Some Peptides Commercially Available as a Nasal Spray

Peptide	Therapeutic Indication	How Supplied
Buserelin (Profact nasal, Sanofi-Aventis)	Prostate cancer and endometriosis	0.1 mg/dose nasal spray
Desmopressin acetate (DDAVP® Nasal Spray, Sanofi-Aventis)	Diabetes insipidus Primary nocturnal enuresis	10 mcg/0.1 mL in 2.5 mL rhinal tube or 5 mL nasal spray
Nafarelin acetate (Synarel, Pfizer)	Central precocious puberty	200 mcg/nasal spray (8 mL)
Salmon calcitonin nasal spray (Miacalcin®, Novartis)	Postmenopausal osteoporosis	200 IU/actuation nasal spray

Source: Compiled from PDR.net.

listed in Table 9.1, nasal delivery could be a promising route for the delivery of insulin. Intranasal insulin for control of insulin-dependent diabetes has been under investigation (Illum & Davis, 1992; Newman et al., 1994).

Nasal absorption of a particular peptide or protein from various species may differ considerably. Therefore, careful choice of the animal species used becomes important (Baldwin, Klingbeil, Grimm, & Longenecker, 1990; Schipper, Romeijn, Verhoef, & Merkus, 1993; Verhoef, Schipper, Romeijn, & Merkus, 1994). The best animal model may have to be selected on a case-by-case basis. However, the mild temperament and the large nasal cavity of the sheep make it an excellent animal model for investigations on nasal delivery. Also, the jugular veins are easily accessible for cannulation. The ultimate goal is to use an animal model that can successfully predict delivery through the human nasal mucosa. Monolayers of human nasal epithelial cells have also been used to investigate the transport and metabolism of peptides (Kissel & Werner, 1998).

9.3.1 Approaches to Overcome Barriers to Nasal Absorption

In general, the nasal route is suitable for delivery of molecules with a molecular weight of less than about 1000. For these molecules, the nasal bioavailability may be sufficient to achieve therapeutic levels without the use of any adjuvants. For small peptides, the intranasal bioavailability may approach that of an IV injection. Serum levels of metkephamid, a pentapeptide, following intranasal administration in rat, have been reported to be similar to those resulting after IV injection (Su, Campanale, Mandelsohn, Kerchner, & Gries, 1985). Similarly, desmopressin, a nonapeptide, is adequately absorbed in humans (Harris, Nilsson, Wagner, & Alkner, 1986). While most peptides will be absorbed by passive diffusion, an active transport system may exist for amino acids, di- or tripeptides. The absorption of L-tryptophan across the rat nasal mucosa has been shown to involve a saturable, active transport mechanism (Lovelace, Shao, & Mitra, 1993). For molecules with a molecular weight higher than about 1000, the use of adjuvants is required (McMartin, Hutchinson, Hyde, & Peters, 1987). The major barriers to nasal absorption being the penetration and enzymatic barriers, these adjuvants will often be penetration enhancers and/or protease inhibitors, respectively. In addition, the dosage form design may also be important. These approaches will be discussed in the following sections.

9.3.1.1 Penetration Enhancers

The use of absorption enhancers to facilitate the nasal delivery of peptides and proteins has been widely investigated. Commonly used enhancers include bile salts, surfactants, fusidates, phospholipids, and cyclodextrins. Lysophosphatidylcholine, a lysophospholipid, has been reported to enhance the nasal absorption of biosynthetic hGH in the rat, to achieve a relative bioavailability of 25.8%. In this study, a mucolytic agent, *N*-acetyl-L-cysteine, and a transmembrane fatty acid transporter, palmitoyl-DL-carnitine, were also found to promote the nasal absorption of hGH, with relative bioavailabilities of 12.2% and 22.1%, respectively (O'Hagan et al., 1990). A mucoadhesive and cationic polysaccharide, chitosan, has been reported to enhance the absorption of insulin across the nasal mucosa of rat and sheep. While mucoadhesion may be partly responsible, it is also possible that the cationic nature of chitosan has a transient effect on the gating function of tight junctions (Illum, Farraj, & Davis, 1994). Alkylglycosides, a family of nonionic detergents, have also been shown to effectively enhance the nasal absorption of insulin in rats (Pillion et al., 1994a). The use of fusidates, bile salts, and cyclodextrins as enhancers will be discussed after a brief discussion of the mechanisms by which these enhancers exert their effects.

Penetration enhancers may exert their effect by opening the tight junctions between epithelial cells, by altering the properties of the mucous membrane or mucus layer, or by a combination of different mechanisms (Merkus, Schipper, Hermens, Romeijn, & Verhoef, 1993). In a study by Donovan, Flynn, and Amidon (1990) using a series of PEGs, it was found that the greatest absorption enhancement resulted when changes in the cell-to-cell adhesion in the mucosa were observed. This suggests that paracellular route may play an important role in intranasal absorption. Cell penetrating peptides (see Section 8.2.4) have also been reported to enhance the nasal absorption of insulin and other peptides and proteins (Khafagy, Kamei, Nielsen, Nishio, & Takeda-Morishita, 2013).

Enhancers or other adjuvants may affect membrane permeability as well as ciliary activity. Sodium cholate, a surface-active bile salt, and aprotinin, a protease inhibitor, have been reported to be highly ciliotoxic in the rat (Jian & Po, 1993). The effect of adjuvants on nasal mucociliary clearance can be evaluated by studying their effect on ciliary beat frequency (CBF) in in vitro systems using chicken embryo tracheal tissue and human adenoid tissue. However, the inhibitory effects of enhancers on CBF may be higher under these in vitro studies as compared to actual in vivo situations. This is because the in vitro ciliated tissue is directly exposed to the adjuvants being tested, while the in vivo cilia are protected by the mucus layer. Also, when a formulation is administered in vivo, it is diluted by mucus and slowly eliminated by the mucociliary clearance (Merkus et al., 1993). Extended clinical experience to predict any long-term toxicity to the nasal mucosa following chronic use of formulations containing penetration enhancers is generally lacking (Lee & Longenecker, 1988).

9.3.1.1.1 Sodium Tauro-24,25-Dihydrofusidate

A fusidic acid derivative, STDHF, has been widely used in preclinical and clinical studies to enhance the nasal absorption of peptides and proteins. It was found to enhance the bioavailability of salmon calcitonin in human volunteers following intranasal administration (Lee, Ennis, Longenecker, & Bengtsson, 1994). Intranasal administration of a powder formulation of insulin and STDHF in a sheep model resulted in systemic absorption of insulin that resulted in a hypoglycemic effect. The bioavailability of the powder formulation was found to increase as the mole ratio of STDHF to insulin was increased (Lee, Narog, Patapoff, & Wang, 1991). The intranasal administration of hGH to sheep, rabbits, and rats in the presence of STDHF has been compared. While there were differences based on species and mode of delivery, all three species showed a rapid absorption followed by a rapid clearance. Formulations of hGH with STDHF increased the hGH bioavailability by 11-fold in rats and rabbits and by 21-fold in sheep (Baldwin et al., 1990). The comparison is for hGH formulations containing 0.5% STDHF with those without STDHF. Concentrations of STDHF at or slightly higher than the critical micelle concentration were required. Intranasal administration of hGH in this way could achieve pulsatile kinetics similar to that of endogenous secretion, unlike the kinetics achieved following subcutaneous administration. Similar results were seen in a human study using hGH–STDHF intranasal formulations. In the presence of STDHF, hGH was easily absorbed through the nasal mucosa, with plasma profiles similar to the normal endogenous peaks of GH. Compared to SC injections, the hGH peaks following intranasal delivery were of shorter duration, and C_{max} values were lower. However, the uptake measured as AUC_{24h} was only about 1.5%–3% of that obtained with SC administration (Hedin et al., 1993).

9.3.1.1.2 Bile Salts

Bile salts and bile salt–fatty acid mixed micelles have also been widely investigated as penetration enhancers (Shao & Mitra, 1992). When insulin was administered as an aerosol with 1% deoxycholate to normal and diabetic subjects, it was found to traverse the nasal mucosa and rapidly appear in the circulation. Nasal insulin absorption achieved in this study was approximately 10% as efficient as IV insulin (Moses, Gordon, Carey, & Flier, 1983). Gordon, Moses, Silver, Flier, and Carey (1985) investigated the absorption of insulin by the nasal mucosa of humans by using a series of bile salts. Therapeutically useful amounts of insulin could be absorbed, and the absorption was found to positively correlate with increasing hydrophobicity of the bile salts. Absorption began at the critical micellar concentration of bile salts. It seems that solubilization of insulin in mixed bile salt micelles presents a high insulin concentration for absorption. Also, the bile salts may form reverse micelles so that the insulin is solubilized in the hydrophilic interior while the hydrophobic surfaces face outward. This will facilitate the transport of reverse micelles through the lipid environment. Bile salts and bile salt–fatty acid mixed micelles have also been investigated to enhance the intranasal transport of a dipeptide (Tengamnuay & Mitra, 1990) and recombinant human interferon (Shim & Kim, 1993). Sodium glycocholate (1%) has been reported to enhance the absorption of angiopeptin across rabbit nasal tissue mounted in the Ussing chamber, by a factor of 2.7 ($p<0.05$). Angiopeptin, a synthetic octapeptide, has potential use in the inhibition of restenose of coronary arteries after angioplasty or heart transplantation (Jorgensen & Bechgaard, 1993). A nasal formulation containing a nona- or decapeptide with LHRH agonist or antagonist activity has been described in the patent literature. The peptide was dissolved in 0.02 M acetate buffer (pH 5.2) along with a bile salt surfactant such as sodium glycocholate. The surfactant acted as a penetration enhancer and also increased the solubility of the LHRH analog. Studies in monkeys and beagle dogs suggest that the surfactant was able to enhance absorption of the peptide without inducing toxicity in the nasal mucosa (Anik, 1992).

9.3.1.1.3 Cyclodextrins

Cyclodextrins have also been evaluated as absorption enhancers for nasal delivery of peptides and proteins (Szejtli, 1994). The most promising cyclodextrin in this regard appears to be dimethyl β-cyclodextrin (DMβCD). This and other cyclodextrins have been reported to enhance the absorption of rhG-CSF following intranasal administration in rabbits (Watanabe et al., 1993; Watanabe et al., 1994). Similarly, DMβCD enhanced the absorption of adrenocorticotropic hormone (ACTH) and insulin through the nasal mucosa of rats (Verhoef et al., 1994). Several other studies have also investigated the potential of DMβCD to enhance the nasal absorption of insulin (Merkus, Verhoef, Romeijn, & Schipper, 1991; Schipper et al., 1993; Shao, Krishnamoorthy, & Mitra, 1992; Watanabe, Matsumoto, Kawamoto, Yazawa, & Matsumoto, 1992). It was found that the coadministration of 5% w/v DMβCD with an insulin solution resulted in a high bioavailability (108.9% ± 36.4%) from the nasal mucosa of rats as compared to IV administration. The other cyclodextrins evaluated in this study did not have a significant effect on nasal insulin absorption (Merkus et al., 1991). The mechanism for the effectiveness of DMβCD as a nasal enhancer is not clear. It seems that cyclodextrin can prevent or reduce the formation of insoluble aggregates in insulin solution (Banga & Mitra, 1993). Shao et al. (1992) have reported that cyclodextrins are capable of dissociating insulin hexamers into smaller aggregates, and this dissociation may provide an additional mechanism for enhancement of insulin diffusivity across nasal mucosa. However, the primary mechanism may be solubilization of nasal membrane components. Thus, cyclodextrins may have damaging effects on the nasal mucosa, and this needs to be evaluated. However, DMβCD has been reported to have only a mild and reversible effect on the nasal mucociliary clearance, as measured by its effects on CBF of chicken embryo tracheal and human adenoid tissue (Verhoef et al., 1994). It has also been suggested that cyclodextrins may actually protect against enhancer damage in nasal delivery systems (Gill et al., 1994; Gill, Illum, Farraj, & Deponti, 1994).

9.3.1.1.4 Protease Inhibitors

As we discussed previously, nasal delivery avoids the hepatic first-pass effects. However, the enzymatic activity in the nasal mucosa creates a pseudo-first-pass effect for the intranasal delivery of drugs. These enzymes include cytochrome P450, aldehyde dehydrogenases, epoxide hydroxylase, carboxylesterase, carbonic anhydrase, glutathione transferase, and various exo- and endopeptidases. For the intranasal delivery of peptides and proteins, the principal enzymatic barrier appears to be the aminopeptidase activity of the nasal mucosa (Sarkar, 1992). Thus, protease inhibitors and especially aminopeptidase inhibitors have been used to prevent the enzymatic breakdown of peptides and proteins upon intranasal administration. Many absorption enhancers such as bile salts may also have inhibitory effects on aminopeptidases, and this may partly explain their mechanism of action. For example, sodium glycocholate can reduce the enzymatic degradation of thyrotropin-releasing hormone in a homogenate of rabbit nasal mucosa (Jorgensen & Bechgaard, 1994). Since bile salts and some other enhancers may damage the mucous membrane, their chronic use in humans may not be feasible. For these reasons, enzymatic inhibitors offer a promising alternate approach to enhance the intranasal delivery of peptides and proteins. Some of the most promising aminopeptidase inhibitors are puromycin, *p*-chloromercuribenzoate, and the even more potent bestatin and amastatin. The class of α-aminoboronic acid derivatives, which include boroleucine, borovaline, and boroalanine, are also potent and reversible inhibitors of aminopeptidases. These α-aminoboronic acid derivatives have been shown to inhibit the degradation of leucine enkephalin in the nasal perfusate of rats. In this study, boroleucine was found 100 times more effective in enzyme inhibition than bestatin and 1000 times more effective than puromycin (Hussain, Shenvi, Rowe, & Shefter, 1989). Boroleucine has also been reported to completely inhibit the metabolism of LHRH when perfused through the isolated rat nasal cavity (Hussain & Aungst, 1994). Other studies have also investigated the hydrolysis of leucine enkephalin in nasal mucosa and its stabilization by the use of protease inhibitors (Faraj et al., 1990; Hussain & Aungst, 1992). Other protease inhibitors such as the trypsin inhibitors, aprotinin, and soybean trypsin inhibitor have also been investigated but failed to enhance the nasal absorption of vasopressin and desmopressin in rats. However, camostat mesilate, an aminopeptidase and trypsin inhibitor, was successfully used in this study. Coadministration of the peptides with camostat mesilate significantly increased the antidiuretic effect and thus the nasal absorption of vasopressin and desmopressin in rats (Morimoto et al., 1991). Protease inhibitors can be successfully used in intranasal peptide formulations only if their effects are reversible and no long-term toxicity results from their usage. Hussain, Koval, Shenvi, and Aungst (1990) have investigated the recovery of peptide hydrolytic activity after exposure of the rat nasal cavity to various aminopeptidase inhibitors. Boroleucine (0.1 μm) was found to be a potent inhibitor of leucine enkephalin metabolism; but when it was washed out, the leucine enkephalin metabolism rate returned to control levels. Thus, aminoboronic acid derivatives can enhance nasal delivery reversibly and at very low concentrations. In contrast, other aminopeptidase inhibitors needed much higher concentrations (0.1 mM for bestatin and 1 mM for puromycin), and their effects were not reversible. A rebound of aminopeptidase activity higher than control levels was seen when these inhibitors were washed out. The mechanism for such effect is not known but may involve membrane disruption. Another approach to reduce proteolytic degradation can be protein PEGylation (see Section 6.5.1). PEGylated salmon calcitonin was recently reported to show strong resistance to enzymatic degradation in rat nasal mucosa (Na et al., 2004).

9.3.1.1.5 Formulation or Delivery Devices

The choice of formulation or delivery device can also have an effect on the extent and efficiency of drug absorption from the nasal cavity. Powder dosage forms may be more effective than liquid dosage forms for intranasal delivery. When insulin was administered to the nasal cavity of beagle dogs, absorption from a powder form was better than that from the liquid form. Sustained absorption could be achieved by adding Carbopol 934 as an excipient to the powder. Since Carbopol 934 can form a gel in an aqueous environment, it allows insulin to stay on the nasal mucosa for a prolonged period (Nagai, Nishimoto, Nambu, Suzuki, & Sekine, 1984). Carbopol gel has also been directly used to enhance the absorption of insulin and calcitonin from the nasal cavity of rats (Morimoto, Morisaka, & Kamada, 1985). Powder formulations of insulin containing DMβCD were found to be effective following nasal administration in rabbits, while liquid formulations, under the same conditions, were not effective (Schipper et al., 1993; Verhoef et al., 1994).

Another formulation approach to enhance the intranasal delivery of peptides and proteins is the use of microspheres. It has been shown that uptake and translocation of solid particles can take place through the nasal epithelial lining through the nasal-associated lymphatic tissue (Alpar, Almeida, & Brown, 1994). Hyaluronic acid ester microspheres have been reported to produce a large and significant increase in the nasal absorption of insulin in sheep similar to that obtained with bioadhesive starch microspheres (Illum et al., 1994). In concept, another way to improve nasal absorption of peptides is to increase their lipophilicity. However, this may or may not be true in actual practice. In a study that compared the rates of disappearance of a small peptide and its methyl ester in rat nasal cavity, the lipophilicity did not affect the bioavailability of the peptide (Hussain, Hamadi, Kagashima, Iseki, & Dittert, 1991).

The mode of delivery can also have an influence on drug absorption from the nasal cavity. The drug can be administered as nasal drops or as a nasal spray. Drops have been shown to spread more extensively than spray, with three drops being sufficient to cover most of the nasal mucosa in human subjects (Hardy, Lee, & Wilson, 1985). However, the use of an intranasal spray device can deposit well-controlled doses within the nasal cavity. Studies with the nasal administration of desmopressin in humans have shown that the plasma levels of the peptide achieved with the spray were twofold to threefold higher than those achieved with drops (Harris et al., 1986). Two types of spray devices are commonly used: a metered-dose nebulizer (MDN) and a metered-dose aerosol (MDA). The MDN device operates by mechanical actuation and releases a fixed dose each time the actuator is pressed. The MDA device uses propellants and operates by pressurized actuation. The use of propellants may or may not be a disadvantage for peptide stability, and selection of some suitable device needs to be done on a case-by-case basis (Su, 1986).

9.4 RECTAL DELIVERY

The human rectum is the terminal 15–20 cm long region of the large intestine. The rectal epithelium does not contain villi and has a surface area of only 200–400 cm^2. It is abundant in blood, lymphatic vessels, and microflora. The mucous membrane is permeated by three types of hemorrhoidal veins: the superior, middle, and inferior hemorrhoidal veins. The inferior and middle rectal veins are directly connected to the systemic circulation, while the superior rectal vein is connected to the portal system. Thus, drugs administered in the lower region will bypass the hepatic first-pass elimination. However, the three types of hemorrhoidal veins are linked by an anastomosis network so that an exclusive transport through either system may not be possible (Eppstein & Longenecker, 1988).

The rectal mucosa provides another potential mucosal route for delivery of peptides and proteins (Yamamoto & Muranishi, 1997). Compared to the small intestine and stomach, the proteolytic activity in the rectum is very low. Also, depending on the site of administration within the rectum, it is possible to at least partially bypass first-pass metabolism by the liver. In the presence of adjuvants, a high systemic bioavailability is possible. However, the patient acceptability of this delivery route is rather low because of the social stigma attached to rectal delivery. Another drawback of rectal delivery is that drug absorption may be erratic and may be easily interrupted by defecation (Banga & Chien, 1988; Zhou & Po, 1991). The protein whose rectal delivery has been most widely investigated is insulin. The use of penetration enhancers is required for rectal delivery of insulin, and several such compounds have been investigated. Compounds that have been reported to enhance the rectal absorption of insulin in various animal species include sodium salicylate and 5-methoxysalicylate in dogs (Nishihata et al., 1983b), enamine derivatives in dogs (Nishihata et al., 1985) and rabbits (Kim, Kamada, Higuchi, & Nishihata, 1983), and glyceryl esters of acetoacetic acid in rabbits (Nishihata et al., 1983a). A preparation of a solid dispersion of insulin and a triglyceride base–containing lecithin has also been reported to be an effective suppository for lowering blood glucose levels in dogs (Nishihata et al., 1987). Effective rectal absorption of insulin in humans has also been reported when administered in combination with sodium cholate (Raz, Kidron, Bar-On, & Ziv, 1984). In another study, rectal insulin gels containing bile salts were used. In nondiabetic humans, the pharmacological availability was about 32%, while the bioavailability was only about 11%. These results can be explained by the fact that although liver is the major site of insulin degradation, it is also a site of the highest insulin utilization (Ritschel, Ritschel, Ritschel, & Lucker, 1988).

Rectal absorption of other polypeptides such as calcitonin has also been reported in the literature (Buclin et al., 1987; Morimoto, Akatsuchi, Morisaka, & Kamada, 1985). Rectal administration of calcitonin to nine normal subjects resulted in transient hypocalcemia and other known biologic effects of calcitonin (Buclin et al., 1987). Natural vasopressin peptides, arginine and lysine vasopressin, and the synthetic analog, 1-deamino-8-D-arginine vasopressin (DDAVP), are also effective by rectal administration. Rectal absorption was further improved by the simultaneous administration of 5-methoxysalicylate (Saffran, Bedra, Kumar, & Neckers, 1988).

9.5 BUCCAL DELIVERY

The buccal mucosa covers the inside of the cheek and is a part of the oral mucosa. The oral mucosa lines the oral cavity situated within the dental arches framed on the top by the hard and soft palates and in the bottom by the tongue and floor of the mouth. The oral mucosa is keratinized in the gingiva, hard palate, and dorsum of the tongue regions, while it is nonkeratinized for the soft palate, ventral surface of the tongue, buccal cavity, and lips (Figure 9.4). The buccal epithelium is about 500–800 μm in thickness. A thin film of saliva continually bathes the surface of the oral mucosa, maintaining a moist surface and a slightly acidic pH (Rathbone & Tucker, 1993; Squier & Wertz, 1993; Yamahara & Lee, 1993). The most commonly used forms of oral mucosal drug delivery are sublingual and buccal delivery, and several products containing conventional drugs are commercially available for these routes on the U.S. and European markets. These include products such as mouth rinse, mouth spray, throat spray, sublingual spray, sublingual film, buccal dissolvable film, buccal tablet, mucoadhesive tablet, and gums or lozenges for drugs such as chlorhexidine, buprenorphine, fentanyl, miconazole, nicotine, and testosterone (Hearnden et al., 2012; Ponchel, 1994). An oral insulin spray product (Oral-lyn™) under clinical development is discussed in Section 7.3.6. Efforts to deliver macromolecules such as

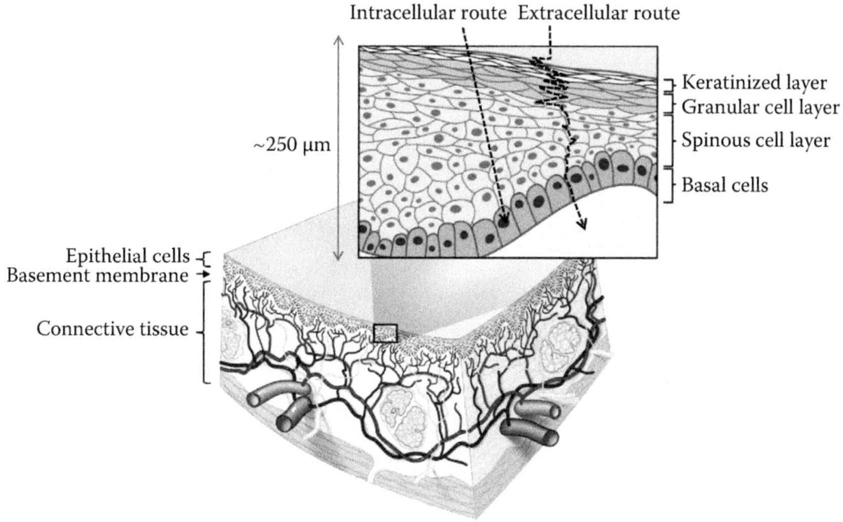

FIGURE 9.4 **(See color insert.)** Cartoon of the structure of the oral mucosa. Insert shows different routes by which drugs can cross the oral mucosa. (Reproduced from Hearnden, V. et al., *Adv. Drug Deliv. Rev.*, 64, 16, 2012. With permission.)

peptides and proteins through these routes are underway as these routes offer several advantages for drug delivery. The oral mucosa is readily accessible, has a rich vasculature, and bypasses hepatic first-pass metabolism. Also, it has a relatively mild pH that is suitable for drug delivery (Ebert, Heiber, Dave, Kim, & Mix, 1994; Hearnden et al., 2012; Squier & Wertz, 1993).

The buccal mucosa is a highly differentiated tissue, and species differences exist. Human buccal mucosa is a nonkeratinized epithelium, similar to that of monkeys, dogs, and pigs. Histological comparisons indicate that porcine mucosa, readily available from slaughterhouse, will be a good candidate for research investigations. The buccal epithelium of rats, mice, hamsters, and rabbits is keratinized to a large extent, and thus mucosa from rodents should be avoided (Hoogstraate & Bodde, 1993). The oral cavity contains several drug-metabolizing enzymes that include oxidases, reductases, lipoxygenases, cyclooxygenases, phosphatases, carbohydrases, nucleases, esterases, and peptidases. Peptidases and proteases in the buccal cavity present a barrier to the buccal absorption of peptides and proteins, even though the enzymatic activity is much lower than those of other mucosal barriers. The aminopeptidase activity of the buccal mucosa is mostly cytosolic, as compared to the nasal mucosa where almost 50% of the activity may be membrane bound (Yamahara & Lee, 1993).

Fast-dissolving tablets, for example, sublingual tablets of nitroglycerin, are commonly used for drug delivery to the oral mucosa. While these may be suitable for highly permeable drugs, alternative oral mucosal drug delivery systems must be developed for delivery of peptides and proteins. Oral mucosal adhesive dosage forms appear promising in this regard. These systems allow an intimate contact between drug and mucosa and maintain a high drug concentration at the absorptive surface for a prolonged period (Ponchel, 1994). A combination of hydroxypropyl cellulose (HPC) and Carbopol has been investigated as a mucoadhesive system for oral mucosal delivery of insulin. Insulin was not absorbed from a simple HPC/Carbopol disk but was absorbed when a two-phased tablet was prepared. The two-phased tablet consisted of an adhesive peripheral layer and an oleaginous core consisting of insulin and sodium glycocholate in a cocoa butter base. The bioavailability of insulin from this dosage form, in comparison with IM injection in beagle dogs, was only 0.5% (Nagai & Machida, 1985). A new transmucosal therapeutic system consisting of a fast-release layer and a sustained release layer containing mucoadhesive composition has also been reported in literature for controlled delivery of drugs to the gingival mucosa (Nozaki, Kakumoto, Ohta, Yukimatsu, & Chien, 1993).

Peptide absorption occurs through oral mucosa by passive diffusion though penetration enhancers, or other formulation approaches are generally needed (Ayensu, Mitchell, & Boateng, 2012; Senel, Kremer, Nagy, & Squier, 2001). The buccal absorption of insulin in rats has been reported to be enhanced by sodium lauryl sulfate, sodium laurate, palmitoyl carnitine, laureth-9, and a lauric acid/propylene glycol vehicle. With the most effective enhancers, buccal insulin was one-fourth to one-third as effective as IM injection (Aungst & Rogers, 1989). Alkylglycoside surfactants, a relatively new category of nonionic surfactants, have also been used to enhance the buccal absorption of insulin (Aungst, 1994). Absorption enhancers have also been used to promote the buccal absorption of α-interferon. Sodium taurocholate caused a 10-fold increase in absorption at 2% (w/v) concentration and an almost 20-fold increase when used at 4% concentration. The order of efficacy for the other enhancers used was as follows: 1% sodium dodecyl sulfate > 5% Tween 80 > 5% cyclodextrin > 5% ethanol (Steward, Bayley, & Howes, 1994).

9.6 OCULAR DELIVERY

Peptide and protein drugs may be instilled into the eye for topical effects on the eye, or the eye may be used as a route of administration for systemic absorption. While the use of eye drops for topical effects is well accepted, their use as a substitute for an injectable or other dosage form is more difficult to accept. However, drug administered to the eye will find its way into systemic circulation through the mucous membrane in conjunctiva and in the nasolacrimal system. Ocular drug delivery for systemic absorption of polypeptides offers several advantages. The absorption is relatively fast, and the absorbed polypeptides bypass the portal circulation to the liver, thus avoiding first-pass metabolism by the liver. At the same time, the absorption may be limited to peptides or small proteins, unless absorption enhancers are used (Chiou, 1991; Kim, Chiang, Wu, & Prausnitz, 2014; Krishnamoorthy & Mitra, 1993; Lee, 1987). The bioavailability of peptides or proteins administered through the ocular route is low due to the large molecular size and hydrophilicity of these drugs. Furthermore, these drugs are susceptible to degradation by the proteolytic enzymes in the various tissues of the eye. In addition, tear drainage and protein binding contribute to the low bioavailability. Besides penetration enhancers and protease inhibitors, other approaches that could be used to improve ocular delivery include prodrugs (Tojo, 1990), liposomes (Schaeffer & Krohn, 1982), nanoparticles and microspheres (de la Fuente et al., 2010; Kimura et al., 1992), mucoadhesive delivery systems (Davies, Farr, Hadgraft, & Kellaway, 1992; Greaves & Wilson, 1993), and microneedles (Jiang et al., 2007; Jiang, Moore, Edelhauser, & Prausnitz, 2009). For diseases affecting the back of the eye like macular degeneration, intravitreal injection of the monoclonal antibody, ranibizumab, has become the standard of care (El et al., 2010), but alternate delivery approaches such as iontophoresis and microneedles are being explored.

9.6.1 Mechanism of Ocular Penetration

Ocular absorption can take place through the mucous membrane of the conjunctiva and the nasolacrimal system. For insulin, the nasal mucosa has been found to contribute four times higher than the conjunctival mucosa for systemic absorption after ocular administration (Yamamoto, Luo, Dodda-Kashi, & Lee, 1989). As peptides and proteins are hydrophilic drugs, they do not partition well into the cell membrane. Their primary transport pathway is paracellular, that is, leakage through the intercellular lateral spaces. However, some transcellular absorption by pinocytosis and/or through noncorneal (scleral) absorption can also take place. Chelation of calcium or inflammation of the eye can cause a widening of the intercellular spaces, thus opening tight junctions and enhancing the paracellular transport of peptides and proteins (Krishnamoorthy & Mitra, 1993; Rojanasakul et al., 1992).

9.6.2 Ocular Route for Systemic Delivery

Systemic delivery of several peptides and proteins following ocular administration has been investigated. In most cases, a penetration enhancer was required to achieve therapeutic levels. Peptides and proteins that have been investigated include ACTH (Chiou, Shen, & Byron, 1992; Chiou, Shen, Zheng, & Chen, 1992), calcitonin (Li & Chiou, 1991), ∃-endorphin (Rhode & Chiou, 1991), glucagon (Chiou & Chuang, 1988; Chiou, Shen, & Zheng, 1990; Pillion, McCracken, Yang, & Atchison, 1992), LHRH (Chiou et al., 1992), α-melanocyte-stimulating hormone (Chiou et al., 1992), oxytocin (Chiou, Shen, & Zheng, 1991), somatostatin (Chiou et al., 1992), vasopressin (Chiou et al., 1991; Sun, Basu, Kim, & Lee, 1998), and vasoactive intestinal peptide (Chiou et al., 1992). Transport of arginine vasopressin across conjunctiva has been reported to be mediated predominantly by passive diffusion, and proteolytic degradation due to proteolytic enzymes present on the mucosal side of the conjunctiva has been reported (Sun et al., 1998). Systemic delivery of insulin and enkephalins following ocular administration will be discussed in greater details in the following sections.

9.6.2.1 Insulin

In the presence of penetration enhancers, therapeutic amounts of insulin can be absorbed through the nasolacrimal system, following ocular administration. Several studies have now shown that a control of blood glucose levels is possible using insulin eye drops (Chiou & Chuang, 1989; Chiou, Chuang, & Chang, 1988a, 1989;

Chiou & Li, 1993; Chiou et al., 1990; Hayakawa et al., 1992; Nomura et al., 1994; Pillion et al., 1994b; Yamamoto et al., 1989). In a study that evaluated several enhancers, their order of efficacy to improve insulin absorption was found to be saponin>fusidic acid>polyoxyethylene 9-lauryl ether=ethylenediaminetetraacetic acid (EDTA)>glycocholate=decamethonium>Tween 20. Though saponin was most effective, the lowest possible level that is effective enough for increased insulin absorption should be used to avoid local irritation (Chiou & Chuang, 1989). In a recent chronic study, insulin eye drops with polyoxyethylene 9-lauryl ether or Brij-78 were instilled into rabbit eyes twice a day over a 3-month period. During this period, no tolerance or allergic responses developed. Also, no sign of eye irritation, corneal damage, hyperemia, conjunctivitis, or iritis was observed, suggesting that chronic use of such a preparation may be feasible (Chiou & Li, 1993).

9.6.2.2 Enkephalins

Leucine enkephalin has a possible use in analgesia, while methionine enkephalin has a potential use in the treatment of AIDS. Enkephalins can be absorbed through the eyes for systemic delivery (Chiou, Chuang, & Chang, 1988b; Chiou, Shen, Zheng, & Byron, 1992). When 25:1 of 0.125% leu-enkephalin was instilled in rabbit eye, it entered systemic circulation rapidly to reach a plateau of 11.5 ng/mL in 3–4 h and remained at this level for 8–9 h (Chiou et al., 1988b). However, the systemic absorption of met- and leu-enkephalins through the ocular route is not as good as would have been expected, considering the fact that these are pentapeptides with a molecular weight of only 570. Thus, the use of penetration enhancers has been investigated and resulted in a 3- to 10-fold increase in absorption (Chiou et al., 1992). Enkephalins are readily hydrolyzed by proteolytic enzymes in the eye, which may explain their lower than expected bioavailability. Met- and leu-enkephalins have been shown to be most susceptible to hydrolysis by aminopeptidases in homogenates of the corneal epithelium, iris–ciliary body, and conjunctiva (Kashi & Lee, 1986b). An analog of met-enkephalin, [D-Ala^2] met-enkephalinamide, has been designed to circumvent degradation by aminopeptidases. The conjunctival mucosa was found to play as important a role as the nasal mucosa for the systemic absorption of this enkephalin analog (Stratford, Carson, Dodda-Kashi, & Lee, 1988).

9.6.3 Ocular Route for Topical Delivery

Drugs may also be administered to the eye for local effects on the eye (Lee, 1985). An example of the use of proteins for local effects on the eye is the family of growth factors, which have several applications in the ocular area. There are about 35 or more such growth factors, a prominent one being the epidermal growth factor (EGF). As the cornea is not a vascular tissue, the eye heals slowly following injury or surgery. The use of EGF can stimulate cell proliferation in the corneal epithelium, thus stimulating epithelialization during wound healing (Hoppenreijs, Pels, Vrensen, Oosting, & Treffers, 1992). Other growth factors that have a physiological role in corneal wound healing include platelet-derived growth factor (Hoppenreijs, Pels, Vrensen, Felten, & Treffers, 1993), insulin-like growth factor (Grant et al., 1993), and transforming growth factor type beta (Pasquale, Dormanpease, Lutty, Quigley, & Jampel, 1993).

Cyclosporine A is another peptide that may have a role to inhibit the rejection of corneal grafts. Administration of 2% cyclosporine A in olive oil as eye drops has been reported to result in a high concentration of the drug in cornea and conjunctiva, but no intraocular penetration was observed. However, penetration into intraocular tissues can be achieved by formulation changes (BenEzra & Maftzir, 1990a, 1990b; Dutescu, Panfil, Merkel, & Schrage, 2014). A 0.05% emulsion formulation of cyclosporine in castor oil (Restasis, Allergan, Inc., Irvine, CA) is commercially available to increase tear production in the eye.

Tissue plasminogen activator (tPA) is another protein that may find topical use to achieve clot lysis in the eye. Though tPA is approved by the FDA for lysis of thrombi in coronary arteries, its use in ophthalmology represents a nonapproved indication for an approved drug. In a study with 10 patients who developed complications that threatened the successful outcome of anterior segment surgery, tPA completely or almost completely dissolved clots within 30 min to a few hours after tPA injection, and a 100% resolution of fibrin and blood clots was observed at the first follow-up (Tripathi, Tripathi, Bornstein, Gabianelli, & Ernest, 1992). Other peptides and proteins that may have a potential use following topical ocular delivery include calcitonin gene–related peptide (Almegard & Andersson, 1993; Oksala & Stjernschantz, 1988), cholecystokinin (Almegard & Andersson, 1993), interleukin 6 (Nishida, Nakamura, Mishima, Otori, & Hikida, 1992), interferon (Park & Latina, 1993), substance P (Unger & Tighe, 1984), and thyrotropin-releasing hormone (Liu, Dacus, & Bartels, 1989).

9.7 COMPARATIVE EVALUATION OF MUCOSAL ROUTES

At physiological pH, most epithelia are negatively charged and thus more permeable to positively charged solutes. The rank order of the intrinsic membrane permeability for a hydrophilic model compound has been reported as follows: intestinal ≈ nasal ≥ bronchial ≥ tracheal ≥ vaginal ≥ rectal ≥ corneal > buccal > skin (Rojanasakul et al., 1992). For peptides and proteins, the presence of proteolytic activity in various mucosal tissues introduces another dimension that changes the scenario. Several studies have compared the enzymatic activities of various mucosal tissues. Aminopeptidase activity in the cornea epithelium and the conjunctiva of the albino rabbit eye has been reported to be 5%–15% of that as the ileum (Stratford & Lee, 1985). In the presence of 5% sodium glycocholate as an absorption-promoting adjuvant, insulin absorption in rats from various mucosal routes showed a rank order of nasal > rectal > buccal > sublingual. Nasal and rectal delivery was about one-half as effective as IM insulin (Aungst, Rogers, & Shefter, 1988).

Hydrolysis of enkephalins in nonoral mucosal homogenates has been reported to be very similar to their hydrolysis in ileal homogenates. The principal peptidases responsible for hydrolysis were aminopeptidases, but dipeptidyl peptidase and dipeptidyl carboxypeptidase were also involved. The enkephalins were hydrolyzed most rapidly in rectal and buccal homogenates, followed by the nasal and then the vaginal homogenates, but the differences were small (Kashi & Lee, 1986a).

The use of enzyme inhibitors and dihydrofusidates to facilitate the transmucosal delivery of leu-enkephalin through rabbit mucosae has been investigated. Phosphatodihydrofusidate had the greatest effect for all the mucosae, while STDHF had a significant effect only on the rectal permeation. Sodium glycodihydrofusidate was also used and had significant effects on rectal and vaginal mucosae. A combination of amastatin, EDTA, and thimerosal had the greatest stabilizing effect when leu-enkephalin was treated with an extract of the enzymes in the various mucosae (Sayani, Chun, & Chien, 1993). In a study by Chun and Chien (1993), degradation of methionine enkephalin was most rapid in the extracts of rectal mucosa, followed by vaginal and nasal mucosae. The bioavailability as well as the intersubject variability is expected to depend on the choice of the mucosal route of delivery. Systemic delivery of leuprolide acetate has been compared after inhalation and intranasal administration in healthy humans. There was a significant subject-to-subject variability following intranasal administration, while inhalation delivery gave a slightly lower intersubject variability (Adjei, Sundberg, Miller, & Chun, 1992).

9.8 CONCLUSIONS

Of the various mucosal routes, pulmonary and intranasal routes appear to be the most promising. Some peptides are already commercially available for nasal delivery, and insulin has recently been approved for pulmonary delivery. For larger peptides or proteins, adjuvants such as penetration enhancers or protease inhibitors are required. Intranasal delivery may be compromised during a common cold infection, and thus alternate dosage forms may be required for such situations. In recent years, it has been recognized that the lungs offer a huge surface area for the absorption of therapeutic peptides and proteins. Successful pulmonary absorption for several peptides and proteins has been demonstrated in investigative studies, and several products are in clinical trials. Some of the other nonoral mucosal routes include rectal, buccal, and ocular routes.

REFERENCES

Adjei, A. & Garren, J. (1990). Pulmonary delivery of peptide drugs: Effect of particle size on bioavailability of leuprolide acetate in healthy male volunteers. *Pharm. Res., 7,* 565–569.

Adjei, A. & Gupta, P. (1994). Pulmonary delivery of therapeutic peptides and proteins. *J. Control. Release, 29,* 361–373.

Adjei, A., Sundberg, D., Miller, J., & Chun, A. (1992). Bioavailability of leuprolide acetate following nasal and inhalation delivery to rats and healthy humans. *Pharm. Res., 9,* 244–249.

Agu, R. U., Ugwoke, M. I., Armand, M., Kinget, R., & Verbeke, N. (2001). The lung as a route for systemic delivery of therapeutic proteins and peptides. *Respir. Res., 2,* 198–209.

Almegard, B. & Andersson, S. E. (1993). Vascular effects of calcitonin gene-related peptide (CGRP) and cholecystokinin (CCK) in the monkey eye. *J. Ocular Pharmacol., 9,* 77–84.

Alpar, H. O., Almeida, A. J., & Brown, M. R. W. (1994). Microsphere absorption by the nasal mucosa of the rat. *J. Drug Target., 2,* 147–149.

Anik, S. T. (1992). LHRH Preparations for intranasal administration. U.S. Patent, 5,116,817, May 26, 1992.

Aungst, B. J. (1994). Site-dependence and structure-effect relationships for alkylglycosides as transmucosal absorption promoters for insulin. *Int. J. Pharm., 105,* 219–225.

Aungst, B. J. & Rogers, N. J. (1989). Comparison of the effects of various transmucosal absorption promoters on buccal insulin delivery. *Int. J. Pharm., 53,* 227–235.

Aungst, B. J., Rogers, N. J., & Shefter, E. (1988). Comparison of nasal, rectal, buccal, sublingual and intramuscular insulin efficacy and the effects of a bile salt absorption promoter. *J. Pharmacol. Exp. Therap., 224,* 23–27.

Ayensu, I., Mitchell, J. C., & Boateng, J. S. (2012). Development and physico-mechanical characterisation of lyophilised chitosan wafers as potential protein drug delivery systems via the buccal mucosa. *Coll. Surf. B Biointerfaces, 91,* 258–265.

Baginski, L., Gobbo, O. L., Tewes, F., Salomon, J. J., Healy, A. M., Bakowsky, U. et al. (2012). In vitro and in vivo characterisation of PEG-lipid-based micellar complexes of salmon calcitonin for pulmonary delivery. *Pharm. Res., 29,* 1425–1434.

Baldwin, P. A., Klingbeil, C. K., Grimm, C. J., & Longenecker, J. P. (1990). The effect of sodium tauro-24,25-dihydrofusidate on the nasal absorption of human growth hormone in three animal models. *Pharm. Res., 7,* 547–552.

Banga, A. K. & Chien, Y. W. (1988). Systemic delivery of therapeutic peptides and proteins. *Int. J. Pharm., 48,* 15–50.

Banga, A. K. & Mitra, R. (1993). Minimization of shaking-induced formation of insoluble aggregates of insulin by cyclodextrins. *J. Drug Target., 1,* 341–345.

Ben Jebria, A., Chen, D., Eskew, M. L., Vanbever, R., Langer, R., & Edwards, D. A. (1999). Large porous particles for sustained protection from carbachol-induced bronchoconstriction in guinea pigs. *Pharm. Res., 16,* 555–561.

BenEzra, D. & Maftzir, G. (1990a). Ocular penetration of cyclosporin A. *Invest. Ophthalmol. Vis. Sci., 31,* 1362–1366.

BenEzra, D. & Maftzir, G. (1990b). Ocular penetration of cyclosporin A in the rat eye. *Arch. Ophthalmol., 108,* 584–587.

Bennett, D. B., Tyson, E., Mah, S., Degroot, J. S., Hegde, S. G., Terao, S. et al. (1994). Sustained delivery of detirelix after pulmonary administration of liposomal formulations. *J. Control. Release, 32,* 27–35.

Bhavane, R., Karathanasis, E., & Annapragada, A. V. (2003). Agglomerated vesicle technology: A new class of particles for controlled and modulated pulmonary drug delivery. *J. Control. Release, 93,* 15–28.

Bitonti, A. J. & Dumont, J. A. (2006). Pulmonary administration of therapeutic proteins using an immunoglobulin transport pathway. *Adv. Drug Deliv. Rev., 58,* 1106–1118.

Boss, A. H., Petrucci, R., & Lorber, D. (2012). Coverage of prandial insulin requirements by means of an ultra-rapid-acting inhaled insulin. *J. Diabetes Sci. Technol., 6,* 773–779.

Buclin, T., Randin, J. P., Jacquet, A. F., Azria, M., Attinger, M., Gomez, F. et al. (1987). The Effect of rectal and nasal administration of salmon calcitonin in normal subjects. *Calcif. Tissue Int., 41,* 252–258.

Byron, P. R. (1987). Pulmonary targeting with aerosols. *Pharm. Technol., 11(5),* 42–56.

Campbell, C., Morimoto, B. H., Nenciu, D., & Fox, A. W. (2012). Drug development of intranasally delivered peptides. *Ther. Deliv., 3,* 557–568.

Casals, C., Miguel, E., & Perezgil, J. (1993). Tryptophan fluorescence study on the interaction of pulmonary surfactant protein-A with phospholipid vesicles. *Biochem. J., 296,* 585–593.

Chan, H. K., Clark, A., Gonda, I., Mumenthaler, M., & Hsu, C. (1997). Spray dried powders and powder blends of recombinant human deoxyribonuclease (rhDNase) for aerosol delivery. *Pharm. Res., 14,* 431–437.

Chiou, G. C. Y. (1991). Systemic delivery of polypeptide drugs through ocular route. *Annu. Rev. Pharmacol. Toxicol., 31,* 457–467.

Chiou, G. C. Y. & Chuang, C. Y. (1988). Treatment of hypoglycemia with glucagon eye drops. *J. Ocular Pharmacol., 4,* 179–186.

Chiou, G. C. Y. & Chuang, C. Y. (1989). Improvement of systemic absorption of insulin through eyes with absorption enhancers. *J. Pharm. Sci., 78,* 815–818.

Chiou, G. C. Y., Chuang, C. Y., & Chang, M. S. (1988a). Reduction of blood glucose concentration with insulin eye drops. *Diabetes Care, 11,* 750–751.

Chiou, G. C. Y., Chuang, C. Y., & Chang, M. S. (1988b). Systemic delivery of enkephalin peptide through eyes. *Life Sci., 43,* 509–514.

Chiou, G. C. Y., Chuang, C. Y., & Chang, M. S. (1989). Systemic delivery of insulin through eyes to lower the glucose concentration. *J. Ocular Pharmacol., 5,* 81–91.

Chiou, G. C. Y. & Li, B. H. P. (1993). Chronic systemic delivery of insulin through the ocular route. *J. Ocular Pharmacol., 9,* 85–90.

Chiou, G. C. Y., Shen, Z. F., & Byron, H. P. L. (1992). Effects of permeation enhancers BL-9 and Brij-78 on absorption of four peptide eyedrops in rabbits. *Acta Pharmacol. Sin., 13,* 201–205.

Chiou, G. C. Y., Shen, Z. F., & Zheng, Y. Q. (1990). Adjustment of blood sugar levels with insulin and glucagon eyedrops in normal and diabetic rabbits. *J. Ocular Pharmacol., 6,* 233–241.

Chiou, G. C. Y., Shen, Z. F., & Zheng, Y. Q. (1991). Systemic absorption of oxytocin and vasopressin through eyes in rabbits. *J. Ocular Pharmacol., 7,* 351–358.

Chiou, G. C. Y., Shen, Z. F., Zheng, Y. Q., & Byron, H. P. L. (1992). Enhancement of systemic delivery of met-enkephalin and leu-enkephalin eyedrops with permeation enhancers. *Meth. Find. Exp. Clin. Pharmacol., 14,* 361–366.

Chiou, G. C. Y., Shen, Z. F., Zheng, Y. Q., & Chen, Y. J. (1992). Enhancement of systemic delivery of peptide drugs via ocular route with surfactants. *Drug Dev. Res., 27,* 177–183.

Chun, I. K. & Chien, Y. W. (1993). Transmucosal delivery of methionine enkephalin. 1. Solution stability and kinetics of degradation in various rabbit mucosa extracts. *J. Pharm. Sci., 82,* 373–378.

Cipolla, D., Gonda, I., & Shire, S. J. (1994). Characterization of aerosols of human recombinant deoxyribonuclease I (rhDNase) generated by jet nebulizers. *Pharm. Res., 11,* 491–498.

Clark, A. R. (2003). Dry powder inhalation systems from inhale therapeutic systems. In M. J. Rathbone, J. Hadgraft, & M. S. Roberts (Eds.), *Modified-release drug delivery technology* (pp. 903–912). New York: Marcel Dekker, Inc.

Clark, A. R. S. S. J. (2000). Formulation of proteins for pulmonary delivery. In E. L. McNally (Ed.), *Protein formulation and delivery* (pp. 201–234). New York: Marcel Dekker, Inc.

Colthorpe, P., Farr, S. J., Smith, I. J., Wyatt, D., & Taylor, G. (1995). The influence of regional deposition on the pharmacokinetics of pulmonary-delivered human growth hormone in rabbits. *Pharm. Res., 12,* 356–359.

Corkery, K. (2000). Inhalable drugs for systemic therapy. *Respir. Care, 45,* 831–835.

Davies, N. M., Farr, S. J., Hadgraft, J., & Kellaway, I. W. (1992). Evaluation of mucoadhesive polymers in ocular drug delivery. II. Polymer-coated vesicles. *Pharm. Res., 9,* 1137–1144.

de la Fuente, M., Ravina, M., Paolicelli, P., Sanchez, A., Seijo, B., & Alonso, M. J. (2010). Chitosan-based nanostructures: A delivery platform for ocular therapeutics. *Adv. Drug Deliv. Rev., 62,* 100–117.

Dellamary, L. A., Tarara, T. E., Smith, D. J., Woelk, C. H., Adractas, A., Costello, M. L. et al. (2000). Hollow porous particles in metered dose inhalers. *Pharm. Res., 17,* 168–174.

Depreter, F., Pilcer, G., & Amighi, K. (2013). Inhaled proteins: Challenges and perspectives. *Int. J. Pharm., 447,* 251–280.

Deshpande, D., Blanchard, J., Srinivasan, S., Fairbanks, D., Fujimoto, J., Sawa, T. et al. (2002). Aerosolization of lipoplexes using AERx pulmonary delivery system. *AAPS PharmSci., 4,* 1–10.

Deshpande, D., Toledo-Velasquez, D., Wang, L. Y., Malanga, C. J., Ma, J. K. H., & Rojanasakul, Y. (1994). Receptor-mediated peptide delivery in pulmonary epithelial monolayers. *Pharm. Res., 11,* 1121–1126.

Dodoo, A. N., Bansal, S., Barlow, D. J., Bennet, F. C., Hider, R. C., Lansley, A. B. et al. (2000). Systematic investigations of the influence of molecular structure on the transport of peptides across cultured alveolar cell monolayers. *Pharm. Res., 17,* 7–14.

Donovan, M. D., Flynn, G. L., & Amidon, G. L. (1990). The molecular weight dependence of nasal absorption: The effect of absorption enhancers. *Pharm. Res., 7,* 808–815.

Duddu, S. P., Sisk, S. A., Walter, Y. H., Tarara, T. E., Trimble, K. R., Clark, A. R. et al. (2002). Improved lung delivery from a passive dry powder inhaler using an engineered PulmoSphere(R) powder. *Pharm. Res., 19,* 689–695.

Dutescu, R. M., Panfil, C., Merkel, O. M., & Schrage, N. (2014). Semifluorinated alkanes as a liquid drug carrier system for topical ocular drug delivery. *Eur. J. Pharm. Biopharm., 88,* 123–128.

Ebert, C. D., Heiber, S. J., Dave, S. C., Kim, S. W., & Mix, D. (1994). Mucosal delivery of macromolecules. *J. Control. Release, 28,* 37–44.

Edwards, D. A., Hanes, J., Caponetti, G., Hrkach, J., Ben Jebria, A., Eskew, M. L. et al. (1997). Large porous particles for pulmonary drug delivery. *Science, 276,* 1868–1871.

El, S. M., Kowalczuk, L., Touchard, E., Omri, S., de, K. Y., & Behar-Cohen, F. (2010). Protein delivery for retinal diseases: From basic considerations to clinical applications. *Prog. Retin. Eye Res., 29,* 443–465.

Eppstein, D. A. & Longenecker, J. P. (1988). Alternative delivery systems for peptides and proteins as drugs. *CRC Crit. Rev. Therap. Drug Carrier Syst., 5,* 99–139.

Faraj, J. A., Hussain, A. A., Aramaki, Y., Iseki, K., Kagoshima, M., & Dittert, L. W. (1990). Mechanism of nasal absorption of drugs. III: Nasal absorption of leucine enkephalin. *J. Pharm. Sci., 79,* 698–702.

Garcia-Contreras, L., Morcol, T., Bell, S. J., & Hickey, A. J. (2003). Evaluation of novel particles as pulmonary delivery systems for insulin in rats. *AAPS PharmSci., 5,* 83–93.

Gill, I. J., Fisher, A. N., Hinchcliffe, M., Whetstone, J., Farraj, N., Deponti, R. et al. (1994). Cyclodextrins as protection agents against enhancer damage in nasal delivery systems. II. Effect on in vivo absorption of insulin and histopathology of nasal membrane. *Eur. J. Pharm. Sci., 1,* 237–248.

Gill, I. J., Illum, L., Farraj, N., & Deponti, R. (1994). Cyclodextrins as protection agents against enhancer damage in nasal delivery systems. I. Assessment of effect by measurement of erythrocyte haemolysis. *Eur. J. Pharm. Sci., 1,* 229–236.

Golomb, G., Avramoff, A., & Hoffman, A. (1993). A new route of drug administration—Intrauterine delivery of insulin and calcitonin. *Pharm. Res., 10,* 828–833.

Gordon, G. S., Moses, A. C., Silver, R. D., Flier, J. S., & Carey, M. C. (1985). Nasal absorption of insulin: Enhancement by hydrophobic bile salts. *Proc. Natl. Acad. Sci., 82,* 7419–7423.

Grant, M. B., Mames, R. N., Fitzgerald, C., Ellis, E. A., Aboufriekha, M., & Guy, J. (1993). Insulin-like growth factor I acts as an angiogenic agent in rabbit cornea and retina: Comparative studies with basic fibroblast growth factor. *Diabetologia, 36,* 282–291.

Greaves, J. L. & Wilson, C. G. (1993). Treatment of diseases of the eye with mucoadhesive delivery systems. *Adv. Drug Deliv. Rev., 11,* 349–383.

Groneberg, D. A., Nickolaus, M., Springer, J., Doring, F., Daniel, H., & Fischer, A. (2001). Localization of the peptide transporter PEPT2 in the lung: Implications for pulmonary oligopeptide uptake. *Am. J. Pathol., 158,* 707–714.

Hardy, J. G., Lee, S. W., & Wilson, C. G. (1985). Intranasal drug delivery by spray and drops. *J. Pharm. Pharmacol., 37,* 294–297.

Harris, A. S., Nilsson, I. M., Wagner, Z. G., & Alkner, U. (1986). Intranasal administration of peptides: Nasal deposition, biological response, and absorption of desmopressin. *J. Pharm. Sci., 75,* 1085–1088.

Hastings, R. H., Grady, M., Sakuma, T., & Matthay, M. A. (1992). Clearance of different-sized proteins from the alveolar space in humans and rabbits. *J. Appl. Physiol. 73,* 1310–1316.

Hayakawa, E., Chien, D., Inagaki, K., Yamamoto, A., Wang, W., & Lee, V. H. L. (1992). Conjunctival penetration of insulin and peptide drugs in the albino rabbit. *Pharm. Res., 9,* 769–775.

Hearnden, V., Sankar, V., Hull, K., Juras, D. V., Greenberg, M., Kerr, A. R. et al. (2012). New developments and opportunities in oral mucosal drug delivery for local and systemic disease. *Adv. Drug Deliv. Rev., 64,* 16–28.

Hedin, L., Olsson, B., Diczfalusy, M., Flyg, C., Petersson, A. S., Rosberg, S. et al. (1993). Intranasal administration of human growth hormone (hGH) in combination with a membrane permeation enhancer in patients with GH deficiency—A pharmacokinetic study. *J. Clin. Endocrinol. Metab., 76,* 962–967.

Hirst, P. H., Pitcairn, G. R., Weers, J. G., Tarara, T. E., Clark, A. R., Dellamary, L. A. et al. (2002). In vivo lung deposition of hollow porous particles from a pressurized metered dose inhaler. *Pharm. Res., 19,* 258–264.

Hoogstraate, A. J. & Bodde, H. E. (1993). Methods for assessing the buccal mucosa as a route of drug delivery. *Adv. Drug Deliv. Rev., 12,* 99–125.

Hoover, J. L., Rush, B. D., Wilkinson, K. F., Day, J. S., Burton, P. S., Vidmar, T. J. et al. (1992). Peptides are better absorbed from the lung than the gut in the rat. *Pharm. Res., 9,* 1103–1106.

Hoppenreijs, V. P. T., Pels, E., Vrensen, G. F. J. M., Felten, P. C., & Treffers, W. F. (1993). Platelet-derived growth factor—Receptor expression in corneas and effects on corneal cells. *Invest. Ophthalmol. Vis. Sci., 34,* 637–649.

Hoppenreijs, V. P. T., Pels, E., Vrensen, G. F. J. M., Oosting, J., & Treffers, W. F. (1992). Effects of human epidermal growth factor on endothelial wound healing of human corneas. *Invest. Ophthalmol. Vis. Sci., 33,* 1946–1957.

Hussain, A., Hamadi, S., Kagashima, M., Iseki, K., & Dittert, L. (1991). Does increasing the lipophilicity of peptides enhance their nasal absorption? *J. Pharm. Sci., 80,* 1180–1181.

Hussain, M. A. & Aungst, B. J. (1992). Nasal absorption of leucine enkephalin in rats and the effects of aminopeptidase inhibition, as determined from the percentage of the dose unabsorbed. *Pharm. Res., 9,* 1362–1364.

Hussain, M. A. & Aungst, B. J. (1994). Nasal mucosal metabolism of an LH-RH fragment and inhibition with boroleucine. *Int. J. Pharm., 105,* 7–10.

Hussain, M. A., Koval, C. A., Shenvi, A. B., & Aungst, B. J. (1990). Recovery of rat nasal mucosa from the effects of aminopeptidase inhibitors. *J. Pharm. Sci., 79,* 398–400.

Hussain, M. A., Shenvi, A. B., Rowe, S. M., & Shefter, E. (1989). The use of a-aminoboronic acid derivatives to stabilize peptide drugs during their intranasal absorption. *Pharm. Res., 6,* 186–189.

Illum, L. & Davis, S. S. (1992). Intranasal insulin. *Clin. Pharmacokinet., 23,* 30–41.

Illum, L., Farraj, N. F., & Davis, S. S. (1994). Chitosan as a novel nasal delivery system for peptide drugs. *Pharm. Res., 11,* 1186–1189.

Illum, L., Farraj, N. F., Fisher, A. N., Gill, I., Miglietta, M., & Benedetti, L. M. (1994). Hyaluronic acid ester microspheres as a nasal delivery system for insulin. *J. Control. Release, 29,* 133–141.

Jaffe, H. A., Buhl, R., Mastrangeli, A., Holroyd, K. J., Saltini, C., Czerski, D. et al. (1991). Organ specific cytokine therapy: Local activation of mononuclear phagocytes by delivery of an aerosol of recombinant interferon-gamma to the human lung. *J. Clin. Invest., 88,* 297–302.

Jian, L. & Po, A. L. W. (1993). Effects of insulin and nasal absorption enhancers on ciliary activity. *Int. J. Pharm., 95,* 101–104.

Jiang, J., Gill, H. S., Ghate, D., McCarey, B. E., Patel, S. R., Edelhauser, H. F. et al. (2007). Coated microneedles for drug delivery to the eye. *Invest. Ophthalmol. Vis. Sci., 48,* 4038–4043.

Jiang, J., Moore, J. S., Edelhauser, H. F., & Prausnitz, M. R. (2009). Intrascleral drug delivery to the eye using hollow microneedles. *Pharm. Res., 26,* 395–403.

Johnson, K. A. (1997). Preparation of peptide and protein powders for inhalation. *Adv. Drug Deliv. Rev., 26,* 3–15.

Jorgensen, L. & Bechgaard, E. (1993). Intranasal absorption of angiopeptin—In vitro study of absorption and enzymatic degradation. *Int. J. Pharm., 99,* 165–172.

Jorgensen, L. & Bechgaard, E. (1994). Intranasal permeation of thyrotropin-releasing hormone: In vitro study of permeation and enzymatic degradation. *Int. J. Pharm., 107,* 231–237.

June, D. S., Schultz, R. K., & Miller, N. C. (1994). A conceptual model for the development of pressurized metered-dose hydrofluoroalkane-based inhalation aerosols. *Pharm. Technol., 18(10),* 40–52.

Kashi, S. D. & Lee, V. H. L. (1986a). Enkephalin hydrolysis in homogenates of various absorptive mucosae of the albino rabbit: Similarities in rates and involvement of aminopeptidases. *Life Sci., 38,* 2019–2028.

Kashi, S. D. & Lee, V. H. L. (1986b). Hydrolysis of enkephalins in homogenates of anterior segment tissue of the albino rabbit eye. *Invest. Ophthalmol. Vis. Sci., 27,* 1300–1303.

Katz, D. F. & Dunmire, E. N. (1993). Cervical mucus—Problems and opportunities for drug delivery via the vagina and cervix. *Adv. Drug Deliv. Rev., 11,* 385–401.

Khafagy, e., Kamei, N., Nielsen, E. J., Nishio, R., & Takeda-Morishita, M. (2013). One-month subchronic toxicity study of cell-penetrating peptides for insulin nasal delivery in rats. *Eur. J. Pharm. Biopharm., 85,* 736–743.

Kim, S., Kamada, A., Higuchi, T., & Nishihata, T. (1983). Effect of enamine derivatives on the rectal absorption of insulin in dogs and rabbits. *J. Pharm. Pharmacol., 35,* 100–103.

Kim, Y. C., Chiang, B., Wu, X., & Prausnitz, M. R. (2014). Ocular delivery of macromolecules. *J. Control. Release, 190,* 172–181.

Kimura, H., Ogura, Y., Moritera, T., Honda, Y., Wada, R., Hyon, S. H. et al. (1992). Injectable microspheres with controlled drug release for glaucoma filtering surgery. *Invest. Ophthalmol. Visual Sci., 33,* 3436–3441.

Kissel, T. & Werner, U. (1998). Nasal delivery of peptides: An in vitro cell culture model for the investigation of transport and metabolism in human nasal epithelium. *J. Control. Release, 53,* 195–203.

Kobayashi, S., Kondo, S., & Juni, K. (1994). Study on pulmonary delivery of salmon calcitonin in rats: Effects of protease inhibitors and absorption enhancers. *Pharm. Res., 11,* 1239–1243.

Komada, F., Iwakawa, S., Yamamoto, N., Sakakibara, H., & Okumura, K. (1994). Intratracheal delivery of peptide and protein agents: Absorption from solution and dry powder by rat lung. *J. Pharm. Sci., 83,* 863–867.

Kompella, U. B. & Lee, V. H. L. (1992). (C) Means to enhance penetration. (4) Delivery systems for penetration enhancement of peptide and protein drugs: Design considerations. *Adv. Drug Deliv. Rev., 8,* 115–162.

Koushik, K. & Kompella, U. B. (2004). Preparation of large porous deslorelin-PLGA microparticles with reduced residual solvent and cellular uptake using a supercritical carbon dioxide process. *Pharm. Res., 21,* 524–535.

Krishnamoorthy, R. & Mitra, A. K. (1993). Ocular delivery of peptides and proteins. In A. K. Mitra (Ed.), *Ophthalmic drug delivery systems* (pp. 455–469). New York: Marcel Dekker, Inc.

Lansley, A. B. (1993). Mucociliary clearance and drug delivery via the respiratory tract. *Adv. Drug Deliv. Rev., 11,* 299–327.

Laube, B. L., Georgopoulos, A., & Adams, G. K. (1993). Preliminary study of the efficacy of insulin aerosol delivered by oral inhalation in diabetic patients. *JAMA, 269,* 2106–2109.

Lee, V. H. L. (1985). Topical ocular drug delivery: Recent advances and future perspectives. *Pharm. Int., 6,* 135–138.

Lee, V. H. L. (1987). Ophthalmic delivery of peptides and proteins. *Pharm. Technol., 11(4),* 26–38.

Lee, V. H. L. (1988). Peptide and protein drug delivery systems. *BioPHARM, 1(3),* 24–31.

Lee, V. H. L. (1990). Protease inhibitors and penetration enhancers as approaches to modify peptide absorption. *J. Control. Release, 13,* 213–223.

Lee, V. H. L. (1991). Problems and solutions in peptide and protein drug delivery. In P. D. Garzone, W. A. Colburn, & M. Mokotoff (Eds.), *Pharmacokinetics and pharmacodynamics. Vol.3: Peptides, peptoids and proteins* (pp. 81–92). Cincinnati, Ohio: Harvey Whitney Books Co.

Lee, V. H. L., Yamamoto, A., & Kompella, U. B. (1991). Mucosal penetration enhancers for facilitation of peptide and protein drug absorption. *Crit. Rev. Ther. Drug Carr. Syst., 8,* 91–192.

Lee, W. A., Ennis, R. D., Longenecker, J. P., & Bengtsson, P. (1994). The bioavailability of intranasal salmon calcitonin in healthy volunteers with and without a permeation enhancer. *Pharm. Res., 11,* 747–750.

Lee, W. A. & Longenecker, J. P. (1988). Intranasal delivery of proteins and peptides. *BioPHARM, 1(4),* 30–37.

Lee, W. A., Narog, B. A., Patapoff, T. W., & Wang, Y. J. (1991). Intranasal bioavailability of insulin powder formulations: Effect of permeation enhancer-to-protein ratio. *J. Pharm. Sci., 80,* 725–729.

Levy, R. S. (1993). Efficient calcitonin production. *BioPHARM, 6(4),* 36–39.

Li, B. H. P. & Chiou, G. C. Y. (1991). Systemic administration of calcitonin through ocular route. *Life Sci., 50,* 349–354.

Li, H. Y. & Seville, P. C. (2010). Novel pMDI formulations for pulmonary delivery of proteins. *Int. J. Pharm., 385,* 73–78.

Li, Y. P. & Mitra, A. K. (1994). Effects of formulation variables on the pulmonary delivery of insulin. *Drug Deliv. Ind. Pharm., 20,* 2017–2024.

Liu, F., Shao, Z., Kildsig, D. O., & Mitra, A. K. (1993). Pulmonary delivery of free and liposomal insulin. *Pharm. Res., 10,* 228–232.

Liu, J. H. K., Dacus, A. C., & Bartels, S. P. (1989). Thyrotropin releasing hormone increases intraocular pressure. *Invest. Ophthalmol. Vis. Sci., 30,* 2200–2208.

Lizio, R., Klenner, T., Sarlikiotis, A. W., Romeis, P., Marx, D., Nolte, T. et al. (2001). Systemic delivery of cetrorelix to rats by a new aerosol delivery system. *Pharm. Res., 18,* 771–779.

Lovelace, D. S., Shao, Z. Z., & Mitra, A. K. (1993). Transport of L-tryptophan across the rat nasal mucosa. *J. Pharm. Sci., 82,* 433–434.

Lucas, P., Anderson, K., & Staniforth, J. N. (1998). Protein deposition from dry powder inhalers: Fine particle multiplets as performance modifiers. *Pharm. Res., 15,* 562–569.

Martonen, T. B. (1993). Mathematical model for the selective deposition of inhaled pharmaceuticals. *J. Pharm. Sci., 82,* 1191–1199.

Martonen, T. B. & Katz, I. M. (1993). Deposition patterns of aerosolized drugs within human lungs: Effects of ventilatory parameters. *Pharm. Res., 10,* 871–878.

Martonen, T. B., Zhang, Z., & Lessmann, R. C. (1993). Fluid dynamics of the human larynx and upper tracheobronchial airways. *Aerosol Sci. Technol., 19,* 133–156.

Masinde, L. E. & Hickey, A. J. (1993). Aerosolized aqueous suspensions of poly(L-lactic acid) microspheres. *Int. J. Pharm., 100,* 123–131.

McElvaney, N. G., Hubbard, R. C., Birrer, P., Chernick, M. S., Caplan, D. B., Frank, M. M. et al. (1991). Aerosol a1-antitrypsin treatment for cystic fibrosis. *Lancet, 337,* 392–394.

McMartin, C., Hutchinson, L. E. F., Hyde, R., & Peters, G. E. (1987). Analysis of structural requirements for the absorption of drugs and macromolecules from the nasal cavity. *J. Pharm. Sci., 76,* 535–540.

Merkus, F. W. H. M., Schipper, N. G. M., Hermens, W. A. J. J., Romeijn, S. G., & Verhoef, J. C. (1993). Absorption enhancers in nasal drug delivery—Efficacy and safety. *J. Control. Release, 24,* 201–208.

Merkus, F. W. H. M., Verhoef, J. C., Romeijn, S. G., & Schipper, N. G. M. (1991). Absorption enhancing effect of cyclodextrins on intranasally administered insulin in rats. *Pharm. Res., 8,* 588–592.

Morimoto, K., Akatsuchi, H., Morisaka, K., & Kamada, A. (1985). Effect of non-ionic surfactants in a polyacrylic acid gel base on the rectal absorption of [Asu1,7]-eel calcitonin in rats. *J. Pharm. Pharmacol., 37,* 759–760.

Morimoto, K., Morisaka, K., & Kamada, A. (1985). Enhancement of nasal absorption of insulin and calcitonin using polyacrylic acid gel. *J. Pharm. Pharmacol., 37,* 134–136.

Morimoto, K., Yamaguchi, H., Iwakura, Y., Miyazaki, M., Nakatani, E., Iwamoto, T. et al. (1991). Effects of proteolytic enzyme inhibitors on the nasal absorption of vasopressin and an analogue. *Pharm. Res., 8,* 1175–1179.

Morimoto, K., Yamahara, H., Lee, V. H. L., & Kim, K. J. (1994). Transport of thyrotropin-releasing hormone across rat alveolar epithelial cell monolayers. *Life Sci., 54,* 2083–2092.

Morita, T., Yamamoto, A., Hashida, M., & Sezaki, H. (1993). Effects of various absorption promoters on pulmonary absorption of drugs with different molecular weights. *Biol. Pharm. Bull., 16,* 259–262.

Morita, T., Yamamoto, A., Takakura, Y., Hashida, M., & Sezaki, H. (1994). Improvement of the pulmonary absorption of (Asu(1,7))-eel calcitonin by various protease inhibitors in rats. *Pharm. Res., 11,* 909–913.

Moses, A. C., Gordon, G. S., Carey, M. C., & Flier, J. S. (1983). Insulin administered intranasally as an insulin bile salt aerosol: Effectiveness and reproducibility in normal and diabetic subjects. *Diabetes, 32,* 1040–1047.

Na, D. H., Youn, Y. S., Park, E. J., Lee, J. M., Cho, O. R., Lee, K. R. et al. (2004). Stability of PEGylated salmon calcitonin in nasal mucosa. *J. Pharm. Sci., 93,* 256–261.

Nagai, T. & Machida, Y. (1985). Mucosal adhesive dosage forms. *Pharm. Int., 6,* 196–200.

Nagai, T., Nishimoto, Y., Nambu, N., Suzuki, Y., & Sekine, K. (1984). Powder dosage form of insulin for nasal administration. *J. Control. Release, 1,* 15–22.

Nakada, Y., Miyake, M., & Awata, N. (1993). Some factors affecting the vaginal absorption of human calcitonin in rats. *Int. J. Pharm., 89,* 169–175.

Neumiller, J. J. & Campbell, R. K. (2010). Technosphere insulin: An inhaled prandial insulin product. *BioDrugs, 24,* 165–172.

Newman, S. P., Steed, K. P., Hardy, J. G., Wilding, I. R., Hooper, G., & Sparrow, R. A. (1994). The distribution of an intranasal insulin formulation in healthy volunteers: Effect of different administration techniques. *J. Pharm. Pharmacol., 46,* 657–660.

Nishida, T., Nakamura, M., Mishima, H., Otori, T., & Hikida, M. (1992). Interleukin-6 facilitates corneal epithelial wound closure in vivo. *Arch. Ophthalmol., 110,* 1292–1294.

Nishihata, T., Kim, S., Morishita, S., Kamada, A., Yata, N., & Higuchi, T. (1983a). Adjuvant effects of glyceryl esters of acetoacetic acid on rectal absorption of insulin and inulin in rabbits. *J. Pharm. Sci., 72,* 280–285.

Nishihata, T., Okamura, Y., Kamada, A., Higuchi, T., Yagi, T., Kawamori, R. et al. (1985). Enhanced bioavailability of insulin after rectal administration with enamine as adjuvant in depancreatized dogs. *J. Pharm. Pharmacol., 37,* 22–26.

Nishihata, T., Rytting, J. H., Kamada, A., Higuchi, T., Routh, M., & Caldwell, L. (1983b). Enhancement of rectal absorption of insulin using salicylates in dogs. *J. Pharm. Pharmacol., 35,* 148–151.

Nishihata, T., Sudoh, M., Inagaki, H., Kamada, A., Yagi, T., Kawamori, R. et al. (1987). An effective formulation for an insulin suppository; examination in normal dogs. *Int. J. Pharm., 38,* 83–90.

Niven, R. W., Lott, F. D., Ip, A. Y., & Cribbs, J. M. (1994). Pulmonary delivery of powders and solutions containing recombinant human granulocyte colony-stimulating factor (rhG-CSF) to the rabbit. *Pharm. Res., 11,* 1101–1109.

Niven, R. W., Lott, F. D., Ip, A. Y., Somaratne, K. D., & Kearney, M. (1994). Development and use of an in vitro system to evaluate inhaler devices. *Int. J. Pharm., 101,* 81–87.

Niven, R. W., Whitcomb, K. L., Shaner, L., Ip, A. Y., & Kinstler, O. (1995). The pulmonary absorption of aerosolized and intratracheally instilled rhG-CSF and monoPEGylated rhG-CSF. *Pharm. Res., 12,* 1343–1349.

Niven, R. W., Whitcomb, K. L., Shaner, L., Ralph, L. D., Habberfield, A. D., & Wilson, J. V. (1994). Pulmonary absorption of polyethylene glycolated recombinant human granulocyte-colony stimulating factor (PEG rhG-CSF). *J. Control. Release, 32,* 177–189.

Nomura, M., Kubota, M. A., Kawamori, R., Yamasaki, Y., Kamada, T., & Abe, H. (1994). Effect of addition of hyaluronic acid to highly concentrated insulin on absorption from the conjunctiva in conscious diabetic dogs. *J. Pharm. Pharmacol., 46,* 768–770.

Nozaki, Y., Kakumoto, M., Ohta, M., Yukimatsu, K., & Chien, Y. W. (1993). A new transmucosal therapeutic system—Overview of formulation development and in vitro/in vivo clinical performance. *Drug Deliv. Ind. Pharm., 19,* 1755–1808.

O'Hagan, D. T., Critchley, H., Farraj, N. F., Fisher, A. N., Johansen, B. R., Davis, S. S. et al. (1990). Nasal absorption enhancers for biosynthetic human growth hormone in rats. *Pharm. Res., 7,* 772–776.

Ohtani, T., Murakami, M., Yamamoto, A., Takada, K., & Muranishi, S. (1991). Effect of absorption enhancers on pulmonary absorption of fluorescein isothiocyanate dextrans with various molecular weights. *Int. J Pharm., 77,* 141–150.

Oksala, O. & Stjernschantz, J. (1988). Increase in outflow facility of aqueous humor in cats induced by calcitonin gene-related peptide. *Exp. Eye Res., 47,* 787–790.

Oliyai, R. & Stella, V. J. (1993). Prodrugs of peptides and proteins for improved formulation and delivery. *Annu. Rev. Pharmacol. Toxicol., 32,* 521–544.

Park, C. H. & Latina, M. A. (1993). Effects of gamma-interferon on human trabecular meshwork cell phagocytosis. *Invest. Ophthalmol. Vis. Sci., 34,* 2228–2236.

Pasquale, L. R., Dormanpease, M. E., Lutty, G. A., Quigley, H. A., & Jampel, H. D. (1993). Immunolocalization of TGF-beta1, TGF-beta2, and TGF-beta3 in the anterior segment of the human eye. *Invest. Ophthalmol. Vis. Sci., 34,* 23–30.

Patton, J. (1998). Breathing life into protein drugs. *Nat. Biotechnol., 16,* 141–143.

Patton, J. S. (2000). Pulmonary delivery of drugs for bone disorders. *Adv. Drug Deliv. Rev., 42,* 239–248.

Patton, J. S. & Platz, R. M. (1992). Pulmonary delivery of peptides and proteins for systemic action. *Adv. Drug Deliv. Rev., 8,* 179–196.

Patton, J. S., Trinchero, P., & Platz, R. M. (1994). Bioavailability of pulmonary delivered peptides and proteins—Alpha-interferon, calcitonins and parathyroid hormones. *J. Control. Release, 28,* 79–85.

PDR Network. (2014). *Physicians' desk reference* (68th ed.) Montvale, NJ: PDR Network, LLC.

Pillion, D. J., Atchison, J. A., Gargiulo, C., Wang, R., Wang, P., & Meezan, E. (1994a). Insulin delivery in nosedrops: New formulations containing alkylglycosides. *Endocrinology, 135,* 2386–2391.

Pillion, D. J., Atchison, J. A., Stott, J., McCracken, D., Gargiulo, C., & Meezan, E. (1994b). Efficacy of insulin eyedrops. *J. Ocular Pharmacol., 10,* 461–470.

Pillion, D. J., McCracken, D. L., Yang, M., & Atchison, J. A. (1992). Glucagon administration to the rat via eye drops. *J. Ocular Pharmacol., 8,* 349–358.

Ponchel, G. (1994). Formulation of oral mucosal drug delivery systems for the systemic delivery of bioactive materials. *Adv. Drug Deliv. Rev., 13,* 75–87.

Pontiroli, A. E., Calderara, A., & Pozza, G. (1989). Intranasal drug delivery. Potential advantages and limitations from a clinical pharmacokinetic perspective. *Clin. Pharmacokinet., 17,* 299–307.

Rathbone, M. J. & Tucker, I. G. (1993). Mechanisms, barriers and pathways of oral mucosal drug permeation. *Adv. Drug Deliv. Rev., 12,* 41–60.

Raz, I., Kidron, M., Bar-On, H., & Ziv, E. (1984). Rectal administration of insulin. *Isr. J. Med. Sci., 20,* 173–175.

Reardon, P. M., Gochoco, C. H., Audus, K. L., Wilson, G., & Smith, P. L. (1993). In vitro nasal transport across ovine mucosa: Effects of ammonium glycyrrhizinate on electrical properties and permeability of growth hormone releasing peptide, mannitol, and lucifer yellow. *Pharm. Res., 10,* 553–561.

Reddy, I. K. & Banga, A. K. (1993). Biotechnology drug delivery: Oral vs alternate routes. *Pharm. Times, 59(11),* 92–98.

Rhode, B. H. & Chiou, G. C. Y. (1991). Effect of permeation enhancers on beta-endorphin systemic uptake after topical application to the eye. *Ophthalmic Res., 23,* 265–271.

Richardson, J. L., Illum, L., & Thomas, N. W. (1992). Vaginal absorption of insulin in the rat: Effect of penetration enhancers on insulin uptake and mucosal histology. *Pharm. Res., 9,* 878–883.

Ritschel, W. A., Ritschel, G. B., Ritschel, B. E. C., & Lucker, P. W. (1988). Rectal delivery system for insulin. *Meth. Find. Exptl. Clin. Pharmacol., 10,* 645–656.

Rojanasakul, Y., Wang, L., Bhat, M., Glover, D. D., Malanga, C. J., & Ma, J. K. M. (1992). The transport barrier of epithelia: A comparative study on membrane permeability and charge selectivity in the rabbit. *Pharm. Res., 9,* 1029–1034.

Saffran, M., Bedra, C., Kumar, G. S., & Neckers, D. C. (1988). Vasopressin: A model for the study of effects of additives on the oral and rectal administration of peptide drugs. *J. Pharm. Sci., 77,* 33–38.

Sanjar, S. & Matthews, J. (2001). Treating systemic diseases via the lung. *J. Aerosol. Med., 14 (Suppl. 1),* S51–S58.

Sarkar, M. A. (1992). Drug metabolism in the nasal mucosa. *Pharm. Res., 9,* 1–9.

Sayani, A. P., Chun, I. K., & Chien, Y. W. (1993). Transmucosal delivery of leucine enkephalin—Stabilization in rabbit enzyme extracts and enhancement of permeation through mucosae. *J. Pharm. Sci., 82,* 1179–1185.

Schaeffer, H. E. & Krohn, D. L. (1982). Liposomes in topical drug delivery. *Invest. Ophthalmol. Vis. Sci., 22,* 220–227.

Schipper, N. G. M., Romeijn, S. G., Verhoef, J. C., & Merkus, F. W. H. M. (1993). Nasal insulin delivery with dimethyl-beta-cyclodextrin as an absorption enhancer in rabbits—Powder more effective than liquid formulations. *Pharm. Res., 10,* 682–686.

Schreier, H., Gonzalezrothi, R. J., & Stecenko, A. A. (1993). Pulmonary delivery of liposomes. *J. Control. Release, 24,* 209–223.

Schuster, J. & Farr, S. J. (2003). The AERx pulmonary drug delivery system. In M. J. Rathbone, J. Hadgraft, & M. S. Roberts (Eds.), *Modified-release drug delivery technology* (pp. 825–834). New York: Marcel Dekker, Inc.

Senel, S., Kremer, M., Nagy, K., & Squier, C. (2001). Delivery of bioactive peptides and proteins across oral (buccal) mucosa. *Curr. Pharm. Biotechnol., 2,* 175–186.

Shao, Z., Krishnamoorthy, R., & Mitra, A. K. (1992). Cyclodextrins as nasal absorption promoters of insulin: Mechanistic evaluations. *Pharm. Res., 9,* 1157–1163.

Shao, Z. & Mitra, A. K. (1992). Nasal membrane and intracellular protein and enzyme release by bile salts and bile salt-fatty acid mixed micelles: Correlation with facilitated drug transport. *Pharm. Res., 9,* 1184–1189.

Shim, C. K. & Kim, S. R. (1993). Administration route dependent bioavailability of interferon-alpha and effect of bile salts on the nasal absorption. *Drug Deliv. Ind. Pharm., 19,* 1183–1199.

Smith, P. L. (1997). Peptide delivery via the pulmonary route: A valid approach for local and systemic delivery. *J. Control. Release, 46,* 99–106.

Smith, P. L., Yeulet, S. E., Citerone, D. R., Drake, F., Cook, M., Wall, D. A. et al. (1994). SK&F 110679—Comparison of absorption following oral or respiratory administration. *J. Control. Release, 28,* 67–77.

Sou, T., Meeusen, E. N., de, V. M., Morton, D. A., Kaminskas, L. M., & McIntosh, M. P. (2011). New developments in dry powder pulmonary vaccine delivery. *Trends Biotechnol., 29,* 191–198.

Squier, C. A. & Wertz, P. W. (1993). Permeability and the pathophysiology of oral mucosa. *Adv. Drug Deliv. Rev., 12,* 13–24.

Steward, A., Bayley, D. L., & Howes, C. (1994). The effect of enhancers on the buccal absorption of hybrid (BDBB) alpha-interferon. *Int. J. Pharm., 104,* 145–149.

Stratford, R. E., Carson, L. W., Dodda-Kashi, S., & Lee, V. H. L. (1988). Systemic absorption of ocularly administered enkephalinamide and inulin in the albino rabbit: Extent, pathways, and vehicle effects. *J. Pharm. Sci., 77,* 838–842.

Stratford, R. E. & Lee, V. H. L. (1985). Aminopeptidase activity in the albino rabbit extraocular tissues relative to the small intestine. *J. Pharm. Sci., 74,* 731–734.

Su, K. S. E. (1986). Intranasal delivery of peptides and proteins. *Pharm. Int., 7,* 8–11.

Su, K. S. E., Campanale, K. M., Mandelsohn, L. G., Kerchner, G. A., & Gries, C. L. (1985). Nasal delivery of polypeptides I: Nasal absorption of enkephalins in rats. *J. Pharm. Sci., 74,* 394–398.

Suarez, S., Garcia-Contreras, L., Sarubbi, D., Flanders, E., O'Toole, D., Smart, J. et al. (2001). Facilitation of pulmonary insulin absorption by H-MAP: Pharmacokinetics and pharmacodynamics in rats. *Pharm. Res., 18,* 1677–1684.

Sun, L. J., Basu, S. K., Kim, K. J., & Lee, V. H. L. (1998). Arginine vasopressin transport and metabolism in the pigmented rabbit conjunctiva. *Eur. J. Pharm. Sci., 6,* 47–52.

Szejtli, J. (1994). Medicinal applications of cyclodextrins. *Med. Res. Rev., 14,* 353–386.

Tengamnuay, P. & Mitra, A. K. (1990). Bile salt-fatty acid mixed micelles as nasal absorption promoters of peptides. I. Effects of ionic strength, adjuvant composition, and lipid structure on the nasal absorption of [D-Arg^2] Kyotorphin. *Pharm. Res., 7,* 127–133.

Timsina, M. P., Martin, G. P., Marriott, C., Ganderton, D., & Yianneskis, M. (1994). Drug delivery to the respiratory tract using dry powder inhalers. *Int. J. Pharm., 101,* 1–13.

Tojo, K. (1990). Ophthalmic drug delivery by prodrug bioconversion. *Drug News Perspect., 3,* 409–416.

Tripathi, R. C., Tripathi, B. J., Bornstein, S., Gabianelli, E., & Ernest, J. T. (1992). Use of tissue plasminogen activator for rapid dissolution of fibrin and blood clots in the eye after Surgery for glaucoma and cataract in humans. *Drug Dev. Res., 27,* 147–159.

Unger, W. G. & Tighe, J. (1984). The response of the isolated iris sphincter muscle of various mammalian species to substance P. *Exp. Eye Res., 39,* 677–684.

Vanbever, R., Mintzes, J. D., Wang, J., Nice, J., Chen, D., Batycky, R. et al. (1999). Formulation and physical characterization of large porous particles for inhalation. *Pharm. Res., 16,* 1735–1742.

Verhoef, J. C., Schipper, N. G. M., Romeijn, S. G., & Merkus, F. W. H. M. (1994). The potential of cyclodextrins as absorption enhancers in nasal delivery of peptide drugs. *J. Control. Release, 29,* 351–360.

Wall, D. A., Pierdomenico, D., & Wilson, G. (1993). An in vitro pulmonary epithelial system for evaluating peptide transport. *J. Control. Release, 24,* 227–235.

Watanabe, Y., Matsumoto, Y., Kawamoto, K., Yazawa, S., & Matsumoto, M. (1992). Enhancing effect of cyclodextrins on nasal absorption of insulin and its duration in rabbits. *Chem. Pharm. Bull., 40,* 3100–3104.

Watanabe, Y., Matsumoto, Y., Kikuchi, R., Kiriyama, M., Ito, R., Matsumoto, M. et al. (1994). Absorption and blood leukocyte dynamics of recombinant human granulocyte colony-stimulating factor (rhG-CSF) from intranasally administered preparations containing rhG-CSF and cyclodextrins in rabbits. *Int. J. Pharm., 110,* 93–97.

Watanabe, Y., Matsumoto, Y., Yamaguchi, M., Kikuchi, R., Takayama, K., Nomura, H. et al. (1993). Absorption of recombinant human granulocyte colony-stimulating factor (rhG-CSF) and blood leukocyte dynamics following intranasal administration in rabbits. *Biol. Pharm. Bull., 16,* 93–95.

Wearley, L. L. (1991). Recent progress in protein and peptide delivery by noninvasive routes. *Crit. Rev. Ther. Drug Carr. Syst., 8,* 331–394.

Westenberg, H. G. M., Hijman, R., Wiegant, V. M., Laczi, F., & Vanree, J. M. (1994). Pharmacokinetics of DGAVP in plasma following intranasal and oral administration to healthy subjects. *Peptides, 15,* 1101–1104.

Yamahara, H. & Lee, V. H. L. (1993). Drug metabolism in the oral cavity. *Adv. Drug Deliv. Rev., 12,* 25–39.

Yamahara, H., Morimoto, K., Lee, V. H. L., & Kim, K. J. (1994). Effects of protease inhibitors on vasopressin transport across rat alveolar epithelial cell monolayers. *Pharm. Res., 11,* 1617–1622.

Yamamoto, A., Fujita, T., & Muranishi, S. (1996). Pulmonary absorption enhancement of peptides by absorption enhancers and protease inhibitors. *J. Control. Release, 41,* 57–67.

Yamamoto, A., Luo, A. M., Dodda-Kashi, S., & Lee, V. H. L. (1989). The ocular route for systemic insulin delivery in the albino rabbit. *J. Pharmacol. Exp. Therap., 249,* 249–255.

Yamamoto, A. & Muranishi, S. (1997). Rectal drug delivery systems: Improvement of rectal peptide absorption by absorption enhancers, protease inhibitors, and chemical modification. *Adv. Drug Deliv. Rev., 28,* 275–299.

Yamamoto, A., Umemori, S., & Muranishi, S. (1994). Absorption enhancement of intrapulmonary administered insulin by various absorption enhancers and protease inhibitors in rats. *J. Pharm. Pharmacol., 46,* 14–18.

Yang, X., Ma, J. K., Malanga, C. J., & Rojanasakul, Y. (2000). Characterization of proteolytic activities of pulmonary alveolar epithelium. *Int. J. Pharm., 195,* 93–101.

Yu, J. W. & Chien, Y. W. (1997). Pulmonary drug delivery: Physiologic and mechanistic aspects. *Crit. Rev. Ther. Drug Carr. Syst., 14,* 395–453.

Zeng, X. M., Martin, G. P., & Marriott, C. (1995). The controlled delivery of drugs to the lung. *Int. J. Pharm., 124,* 149–164.

Zheng, J. Y., Fulu, M. Y., Lee, D. Y., Barber, T. E., & Adjei, A. L. (2001). Pulmonary peptide delivery: Effect of taste-masking excipients on leuprolide suspension metered-dose inhalers. *Pharm. Dev. Technol., 6,* 521–530.

Zhou, X. H. (1994). Overcoming enzymatic and absorption barriers to non-parenterally administered protein and peptide drugs. *J. Control. Release, 29,* 239–252.

Zhou, X. H. & Po, A. L. W. (1991). Peptide and protein drugs: II. Non-parenteral routes of delivery. *Int. J. Pharm., 75,* 117–130.

10 Recombinant Protein Subunit Vaccines and Delivery Methods

10.1 INTRODUCTION

The focus of this chapter is on protein or recombinant protein–based subunit vaccines as those fit in better with the overall scope of this book, especially since many of the products in development will have a therapeutic focus. Recombinant vaccines in development are listed in Table 10.1. Peptides and peptidomimetics are also in development as vaccines. Peptidomimetic vaccines contain antigenic fragments that mimic the antibody-binding site (Ho & Gibaldi, 2003). However, an introduction to vaccines in general will first be provided.

Vaccines have eradicated smallpox and limited the spread of several dangerous infections. Among these are included tetanus, measles, hepatitis A, hepatitis B, and rotavirus. However, effective vaccines still need to be developed for other disorders such as rotavirus, human papillomavirus, malaria, herpes, hepatitis C, and newly emerging diseases such as SARS or West Nile viruses. Furthermore, a vaccination approach may be needed for the threat of bioterrorism. Vaccines are a critical component of U.S. government's programs designed to make available modern effective countermeasures. Since vaccines enhance and modify body's immune responses, recent efforts are also targeted toward developing *therapeutic* vaccines for cancer, heart disease, diabetes, and other conditions. Vaccines may be based on one of the following technologies: live attenuated, whole killed, subunit/subcellular, and nucleic acid based (Ellis, 2001b). Vaccines using the live attenuated or whole-killed pathogen constitute the majority of vaccines currently approved. These vaccines retain the immunogenic activity without causing host pathogenic responses. Examples of live attenuated vaccines include smallpox vaccine and other viral vaccines such as measles/mumps/rubella (MMR) and varicella zoster virus (VZV) vaccines. More recently, an attenuated, nonreplicating poxvirus, modified vaccinia virus Ankara (MVA), is being investigated as a safer derivative of the smallpox vaccine (Cottingham & Carroll, 2013). Examples of bacterial vaccines include those against typhoid and the Bacillus Calmette–Guerin (BCG) vaccine against tuberculosis (Ho & Gibaldi, 2003). Examples of whole-killed or inactivated vaccines include inactivated polio vaccine (IPV), hepatitis A, influenza, and Japanese encephalitis as viral vaccines and pertussis, plague, cholera, and typhoid as bacterial vaccines.

Vaccines can also be made from subcellular components of a cellular pathogen or subunit components of noncellular pathogens such as viruses. The use of these vaccines becomes feasible when the toxin produced by the pathogen causes the disease

TABLE 10.1
Some Recombinant/Protein Vaccines in Development[a]

Product Name	Sponsor	Indication	Development Status
Abagovomab (mAb therapeutic vaccine)	Menarini, *Florence, Italy*	Ovarian cancer	Phase III
AE37 (peptide vaccine	Antigen Express, *Worcester, MA*	Breast cancer, prostate	Phase II
AE-M (peptide vaccine)	Antigen Express, *Worcester, MA*	Malignant melanoma	Phase I
AVX901 (HER2-expressing VRP vaccine)	AlphaVax, *Research Triangle Park, NC*	Late-stage HER2-expressing breast cancer	Phase I
Breast cancer vaccine (mAb vaccine)	MabVax Therapeutics, *San Diego, CA*	Breast cancer	Phase I
Breast cancer vaccine (HER2/neu therapeutic vaccine)	TapImmune, *Seattle, WA*	Breast cancer	Phase I
CDX-1401 (mAb cancer vaccine)	Celldex Therapeutics, *Needham, MA*	Solid tumors expressing the NY-ESO-1 protein	Phase I/II
DPX-Survivac (peptide vaccine)	Immunovaccine, *Halifax, Canada*	Ovarian cancer	Phase I/II
IMF-001 (protein vaccine)	ImmunoFrontier, *Tokyo, Japan*	Solid tumors	Phase I
Melanoma mAb vaccine	MabVax Therapeutics, *San Diego, CA*	Melanoma	Phase I
MVA-BN® HER2 HER-2/neu-based MVA vaccine	Bavarian Nordic, *Mountain View, CA*	Breast cancer	Phase I/II
MVA-BN® PRO PSA/PAP-based MVA vaccine	Bavarian Nordic, *Mountain View, CA*	Prostate cancer	Phase I/II
PEK fusion protein vaccine	HealthBanks Biotech USA, *Irvine, CA*	Cervical intraepithelial neoplasia	Phase I
Rindopepimut (EGFR variant III vaccine) orphan drug	Celldex Therapeutics, *Needham, MA*	First-line glioblastoma (fast track)	Phase III
		Recurrent glioblastoma (fast track)	Phase II
Sarcoma mAb vaccine	MabVax Therapeutics, *San Diego, CA*	Sarcoma	Phase II
Small-cell lung cancer (SCLC) mAb vaccine	MabVax Therapeutics, *San Diego, CA*	SCLC	Phase I
WT2527 (peptide vaccine)	Sunovion Pharmaceuticals, *Marlborough, MA*	Hematological malignancies, solid tumors	Phase I

(Continued)

TABLE 10.1 (*Continued*)
Some Recombinant/Protein Vaccines in Development[a]

Product Name	Sponsor	Indication	Development Status
V503 (VLP vaccine)	Merck, *Whitehouse Station, NJ*	HPV-related cancers (see also infectious diseases)	Phase III
ACAM-Cdiff (toxoid vaccine)	Sanofi Pasteur, *Swiftwater, PA*	Prevention of *Clostridium difficile* infection (fast track)	Phase III
CR6261 (mAb/influenza vaccine)	Crucell, *Leiden, Netherlands*	Treatment of influenza A virus infection	Phase I
DCVax-001/CD-2401 (recombinant protein vaccine)	Celldex Therapeutics, *Needham, MA* Rockefeller University, *New York, NY*	Prevention and treatment of HIV infection	Phase I
H5N1 influenza vaccine (recombinant VLP vaccine)	Novavax, *Rockville, MD*	Prevention of influenza A virus H5N1 subtype (seasonal use)	Phase II
		Prevention of influenza A virus H5N1 subtype (pandemic use)	Phase I
HBV-002 (recombinant subunit vaccine)	Hawaii Biotech, *Aiea, HI*	Prevention of West Nile virus infection	Phase I completed
IC84 (recombinant fusion protein vaccine)	Novartis Vaccines, *Cambridge, MA* Valneva, *Vienna, Austria*	Prevention of *C. difficile* infection	Phase I
NDV-3 (recombinant protein vaccine)	NovaDigm Therapeutics, *Grand Forks, ND*	Prevention of candidiasis, prevention of staphylococcal (MRSA) infection	Phase I
Recombinant botulinum neurotoxin vaccine	DynPort Vaccine Company, *Frederick, MD* U.S. DoD, *Washington, DC*	Prevention against botulinum neurotoxin types A and B	Phase II
QGE031 (anti-IgE antibody)	Novartis Pharmaceuticals, *East Hanover, NJ*	Bullous pemphigoid	Phase II
PF-06444752 (immunoglobulin-E vaccine)	Pfizer, *New York, NY*	Asthma	
HLA-DQ2 peptide vaccine	ImmusanT, *Cambridge, MA*	Celiac disease	Phase I

(*Continued*)

TABLE 10.1 (*Continued*)
Some Recombinant/Protein Vaccines in Development[a]

Product Name	Sponsor	Indication	Development Status
Vacc-4x (intradermal vaccine)	Bionor Pharma, *Oslo, Norway*	HIV-1 infection	Phase II
VAX-128 (recombinant vaccine)	VaxInnate, *Cranbury, NJ*	Prevention of influenza A virus H1N1 subtype (pandemic)	Phase I
VAX-161 (recombinant vaccine)	VaxInnate, *Cranbury, NJ*	Prevention of influenza A virus H5N1 subtype (pandemic)	Phase I
AD02 & 03 vaccines (amyloid-beta protein inhibitors)	AFFiRiS, *Vienna, Austria* GlaxoSmithKline, *Research Triangle Park, NC*		Phase I and II
		Alzheimer's disease	Phase II
CAD106 (amyloid-beta protein inhibitor)	Novartis Pharmaceuticals, *East Hanover, NJ*		Phase II
UB-311 (liquid intramuscular amyloid-beta protein inhibitor vaccine)	United Biomedical, *Hauppauge, NY*		Phase II
DiaPep277® subcutaneously injected synthetic peptide immunomodulator orphan drug	Andromeda Biotech, *Yavne, Israel*	Type 1 diabetes mellitus (newly diagnosed)	Phase III
Insulin B-chain vaccine	Orban Biotech, *Brookline, MA*	Type 1 diabetes mellitus	Phase I completed
PF-05402536 and 06413367 (smoking cessation vaccine)	Pfizer, *New York, NY*	Smoking cessation	Phase I
SEL-068 (smoking cessation vaccine)	Selecta Biosciences, *Watertown, MA*		Phase I
TA-CD (immunotherapeutic vaccine)	Celtic Pharma, *Hamilton, Bermuda* National Institute on Drug Abuse, *Bethesda, MD*	Cocaine abuse	Phase II

[a] Based on information provided in Medicines in Development. 2013 Vaccines, Survey by PhRMA. This table does not include the conventional vaccines in development.

or the key antigenic targets to be included in the vaccine have been identified. In these cases, chemically inactivated toxin (toxoid) can be used, for example, diphtheria and tetanus toxoids. Also, to minimize adverse effects of some whole-killed vaccines, acellular vaccines have been developed. For example, acellular pertussis (aP) vaccine is available and contains one to five different components of *Bordetella pertussis* and compares favorably with the cell vaccine (wP) for efficacy. The immunogens in these vaccines may be composed of proteins, polysaccharides, or conjugates of both, derived from the pathogen. These vaccines mimic the antigenic determinants found in the pathogen and thus induce relevant immune responses. Subcellular or subunit vaccines can now be made by recombinant DNA technology rather than purifying from the pathogen. A new subunit vaccine for plague is being investigated comprised of the recombinant forms of the natural virulence factors of *Yersinia pestis*, fraction 1 (F1) antigen and V antigen, and has been shown to be effective in guinea pigs (Jones, Griffin, Hodgson, & Williamson, 2003).

The use of recombinant technology also allows for specific alterations in the structure to improve the safety and efficacy of the vaccine. The sequences of the entire genomes of several microbial pathogens are now available, and these will allow for the identification of new antigens for vaccine targets (Fauci, 2004). Conjugate vaccine (Hib) has been developed for purulent meningitis. The virulent determinant of the three common bacteria involved is the polysaccharide capsule, but infants and young children are unable to respond to polysaccharides with antibody production due to the relative immaturity of their B cells. Conjugation of the polysaccharide to a protein carrier has been able to solve this problem (Makela, 2003). This conjugate vaccine has virtually eliminated invasive Hib disease among children, which earlier used to be the leading cause of childhood bacterial meningitis (Fauci, 2004). The use of a microsphere formulation to deliver a conjugate of a polysaccharide antigen and cholera toxin (CT) B subunit has also been investigated (Cho et al., 1998) as discussed later. When toxoid forms of the vaccine are needed, formaldehyde treatment can be used as formaldehyde reacts rapidly with the amino groups (mainly lysine) of the protein to form an imine and eventually leads to cross-linking of the lysine side chains with glutamine, asparagine, arginine, or tyrosine residues. The thermal stability of toxoid forms has been recently characterized. The conversion of tetanus neurotoxin into the toxoid form was characterized by a large increase in T_m (see Section 4.4.11 for a discussion of T_m and thermal stability of proteins) (Krell et al., 2003). Moisture-induced aggregation of formalinized tetanus and diphtheria toxoids can be minimized by succinylation (Schwendeman et al., 1995). These toxoids have partially lost conformational and antigenic properties, but the possibility of reverting to a biologically active toxin is a concern. An alternative technological approach would be genetic detoxification by site-directed mutagenesis. Such an approach would produce a nontoxic mutant protein that would be more immunogenic than that obtained by formaldehyde treatment as the integrity of the immunogenic sites will be retained (Jiskoot, Kersten, & Beuvery, 2002). Lyophilization has been reported to stabilize recombinant ricin toxin A subunit vaccine. When stored at 40°C for 4 weeks, both solution and lyophilized formulations produced antibody titers, but only the lyophilized formulation was able to protect mice against exposure to lethal doses of ricin (Hassett et al., 2013).

10.1.1 Vaccine Adjuvants

Adjuvants are substances that nonspecifically enhance the immune response to antigens. Antibody responses to antigens in adjuvants are greater and more prolonged and frequently consist of different classes to the response obtained without adjuvants. The importance of having a good vaccine adjuvant formulation has been reviewed (Brito, Malyala, & O'Hagan, 2013). Conventional killed or attenuated vaccines or recombinant subunit vaccines typically used today are generally administered with adjuvants to elicit effective immune responses. Many of these adjuvants can cause tissue reactions, and currently, alum salts are the most widely used immune adjuvants. When alum is used, the antigen binds to the aluminum hydroxide or aluminum sulfate and forms a macroscopic suspension, which forms a depot at the injection site to slowly release the antigen (Correia-Pinto, Csaba, & Alonso, 2013). Alum is somewhat effective in potentiating humoral immunity but does not generally elicit CD8+ T-cell-mediated responses. The amount of aluminum in biological products, including vaccines, is limited to 0.85 mg/dose. Aluminum adjuvants have been used for decades and are generally safe; but nevertheless, they can cause local reactions at injection sites such as erythema, subcutaneous nodules, and contact hypersensitivity (Baylor, Egan, & Richman, 2002). When stabilizers are tested for proteins intended to be formulated with aluminum adjuvants, studies should be done on the adsorbed protein rather than on solutions (Peek, Martin, Elk, Pegram, & Middaugh, 2007).

MF59, an oil-in-water emulsion, is a proprietary adjuvant from Novartis, which is used in influenza vaccine in Europe. Several other adjuvants are being investigated. In addition to their potential role as vaccine carriers, liposomes can also have a role as immunological adjuvants (Gregoriadis, Gursel, Gursel, & McCormack, 1996). Immunostimulating complexes (ISCOMs), built up by cholesterol, lipid, immunogen, and saponin, can also be used as adjuvants. Another adjuvant approved for use in Europe is LT. Toll-like receptor ligands such as cytosine–guanosine (CpG) oligonucleotides have been reported to increase the immunogenicity of protein-based vaccines after intramuscular injection followed by electroporation (Kang et al., 2011). Studies in mice have shown that unmethylated CpG dinucleotides in bacterial DNA or synthetic ODN can serve as immune enhancers. Compared to alum, CpG ODN produced five times higher antibody levels following immunization with hepatitis B surface antigen (HBsAg). Also, it induced a strong Th1 response, with predominantly IgG2a antibodies and cytotoxic T lymphocyte (CTL) response. In contrast, alum resulted in a Th2 humoral response (mostly IgG1) and no CTL. When alum and CpG ODN were used together, a strong synergistic effect was observed and Th1 response was maintained (Brazolot Millan, Weeratna, Krieg, Siegrist, & Davis, 1998; Davis et al., 1998). When used for mucosal delivery, there is a Th2 immunostimulatory effect attributed to the ODN backbone, but the presence of CpG motifs shifts this response to Th1 (McCluskie & Davis, 2001). ISCOM has also been widely investigated to make antigens optimally immunogenic. ISCOM is a mixture of saponins extracted from the bark of a tree and forms spherical particles about 40 nm in diameter. ISCOMs with a defined quillaja triterpenoid formulation named QH 703 are in human trials (Morein, Villacres-Eriksson, & Lovgren-Bengtsson, 1998).

Chitosan, a natural product from crab shells, has also been reported to have adjuvant properties that may depend on its molecular weight (MW), degree of deacetylation, particle size, viscosity, and purity (Scherliess et al., 2013). Particulate delivery systems can also be used as adjuvants (Badiee, Heravi, V, Khamesipour, & Jaafari, 2013; Hafner, Corthesy, & Merkle, 2013).

10.1.2 Subunit Vaccines

The newer generation vaccines such as subunit, recombinant, and synthetic vaccines have several advantages with respect to the formulation (e.g., higher purity, easier quality control); but unfortunately, they tend to be poorly immunogenic as compared to live attenuated vaccines. Therefore, the need to develop alternate adjuvants becomes more critical. Generally, proteins would be more immunogenic than peptides as they would have a larger number of epitopes and may be more stable in vivo. Nevertheless, they are still poorly immunogenic compared to conventional killed or attenuated vaccines. Several efforts are underway to develop alternate adjuvants, for example, a lipid core peptide technology has been proposed that incorporates lipoamino acids coupled to a polylysine core containing up to two different antigenic peptides. This technology can enable the incorporation of antigen, carrier, and adjuvant in a single molecular entity (Olive et al., 2003). Microspheres have also been used to increase immunogenicity as they protect the antigen from degradation and clearance in the physiological environment (Newman, Samuel, & Kwon, 1998). Another very promising strategy to produce very immunogenic purified proteins is through the use of viruslike particles (VLPs). Structural proteins derived from a variety of viruses can spontaneously assemble into VLPs. These particles can trigger potent humoral responses and even cellular responses as they can be captured by antigen-presenting cells (APCs) due to their size and structure. Chimeric VLPs carrying heterologous epitopes are also being engineered. These nonreplicating vectors can be produced in large quantities and provide an efficient and safe strategy. They have been successfully used to produce antihepatitis B vaccine but, until recently, did not receive much attention for the development of other antiviral vaccines due to the lower cost and feasibility of other options for these vaccines such as producing live attenuated vaccines (Boisgerault, Moron, & Leclerc, 2002; Brun et al., 2011).

10.1.3 DNA Vaccines

A third-generation approach termed DNA immunization is also being investigated. It makes use of plasmids that are the small rings of double-stranded DNA derived from bacteria. These plasmids are modified to carry genes that code for the antigenic protein made by a pathogen. DNA vaccination then refers to the direct introduction of plasmid DNA into body tissues where it is able to express an antigenic protein and raise immune responses. This innovative approach to immunization is promising for the development of needed vaccines, but studies thus far have shown poor immunogenicity compared to protein vaccines (Ertl & Xiang, 1996; Gregoriadis, 1998; Li, Saade, & Petrovsky, 2012; Robinson et al., 1996). The antigen encoded in the plasmid, with the proper regulatory sequences, transfects the cells in vivo to express the antigen (using the biosynthetic machinery of the host), which in turn leads to induction of an immune response. The plasmid excludes genes that may enable the pathogen to reconstitute itself and cause disease. Thus far, most DNA vaccines have been unable to produce sufficient potency, and it has been suggested that RNA vaccines may be able to overcome some of the drawbacks of DNA vaccines and viral vectors (Ulmer, Mason, Geall, & Mandl, 2012).

10.1.4 Heat Shock Proteins

Another class of proteins being investigated for their ability to generate immune responses are the heat shock proteins (HSPs), also known as stress proteins. These proteins are normally present in the cell and act like *chaperones* (also see Section 2.2.3) to help new or misfolded proteins to fold into their shape and also shuttle proteins within the cell. They also help to display peptide fragments or antigens of abnormal proteins from sick cells on the cell surface so that the immune system can recognize such diseased cells. The potential use of HSPs as immunoadjuvants against cancer is being investigated since HSPs purified from a given tumor can elicit specific immunity against that tumor. However, the abnormal peptides found within diseased cells differ from cancer to cancer and from person to person so that much of the investigation is focused on developing customized, patient-specific therapeutic vaccines. However, a recent report describes a novel strategy to produce recombinant HSPs and produce a vaccine by noncovalently binding a known tumor antigen to them by heat shock. This approach may circumvent human leukocyte antigen (HLA) restriction and could work for any patient with a tumor expressing that known tumor antigen (Srivastava, 2000; Wang et al., 2004).

10.2 IMMUNOLOGY RELEVANT TO VACCINES

Vaccines are used to mimic the body's response to natural infection so that an antigen-specific immunological memory is induced in the body that will protect it from subsequent exposure to that antigen. A brief background to relevant basic immunology is being provided so that this chapter is understandable and useful to readers with varied backgrounds. The undesirable aspects of immunogenicity of proteins used as therapeutic agents have been discussed separately in Chapter 1. The most important cells involved in immunity are the white blood cells or leukocytes that include phagocytes, natural killer (NK) cells (part of the innate immune system providing first line of defense), and lymphocytes (which mediate adaptive immunity). The NK cells are activated by interferons produced by virally infected cells and sometimes also by lymphocytes. NK cells, along with cytokines, provide a first line of defense that involves nonspecific immune response, initiated within minutes of exposure to an antigen. Once a macrophage phagocytizes a cell, it places some of its proteins, called epitopes, on its surface. These surface markers (CD) help some other immune cells to infer the form of the invader. All cells that do this are called APCs. Identification and localization of epitopes are important, and methods exist for such determination (Medzhitov et al., 2011; Stevens, Carperos, Monafo, & Greenspan, 1988).

The causative agent produces antigen (protein or fragments), which is foreign to the body, and the body mounts an antigen-specific humoral response and a cellular response. Both types of response receive help from helper T lymphocytes. The humoral response is mediated by B lymphocytes that produce antibodies that neutralize the antigen or tag them for destruction by other parts of the immune system. The cellular response is mediated by CTLs. Immune responses in which antibodies play a subordinate role are considered cell-mediated immunity. Infected cells display bits of their attackers, proteins on the cell surface that are recognized by CTLs, and the infected cells are then attacked by CTLs and destroyed. Key components of the adaptive immune system are specificity and memory, and these are exploited in vaccination since the adaptive immune system mounts a much stronger response on second encounter with antigen. Proteins encoded in the region of the major histocompatibility complex (MHC) are involved in immunological recognition, including interactions between lymphocytes and APCs. Originally identified by its role in transplant rejection, this region in humans is the HLA gene cluster on chromosome 6. Significant association between particular HLA genes and responsiveness to measles and hepatitis B vaccine has been shown and forms the genetic basis for variation in antibody response to vaccines (Poland & Jacobson, 1998). The MHC proteins are of three different generic types as determined by their structures and functions, and the most relevant regions are represented by class I and II proteins. Majority of the cellular MHC antigens are transmembrane glycoproteins found in the plasma membrane. Different MHC antigens are recognized by different T cell types. Most virally infected cells display viral antigens on the surface of their plasma membranes—these antigens are recognized by T cytotoxic cells. Helper and cytotoxic T cells do not bind free antigen. It seems that they recognize antigen only in association with MHC products, expressed on cell membranes. Cytotoxic T cells usually recognize antigen in association with class I products, which are expressed on all nucleated cells. Helper T cells recognize antigens with class II products, expressed mostly on APCs, and some lymphocytes. The majority of T cells express receptors consisting of α and β chains, and among these, they may express coreceptor molecule CD4 (CD4+ T cells) or CD8 (CD8+ T cells). CD4+ T cells or helper T cells are the major regulatory cells of the immune system. They tend to differentiate (after priming) into cells that mainly secrete interleukin (IL)-4, IL-5, IL-6, and IL-10 (T_{H2} cells) or into cells that mainly produce IL-2, IFN-γ, and lymphotoxin (T_{H1} cells). T_{H2} cells are very effective in helping B cells develop into antibody-producing cells, whereas T_{H1} cells are effective inducers of cellular immune responses. The class II–peptide complex on B cells is the ligand for the receptors of CD4+ T cells. This interaction will result in the activation of the B cell and production of lymphokines by the T cell. B cell activation prepares the cell to divide and to differentiate either into an antibody-secreting cell or into a memory cell. T cells, particularly CD8+ T cells or killer T cells, can develop into CTLs capable of efficiently lysing target cells that express antigens recognized by the CTLs. CD4+ T cells only recognize peptide–class II, while CD8+ T cells recognize peptide–class I complexes.

The initiation of T cell proliferation also involves signals passing between APCs and lymphocytes via ILs. CTLs bind to target cells bearing the correct MHC products and antigens. After binding, there are changes in membrane permeability of target cell causing lysis. Following lysis of target cells, T cells survive and may go on to kill more targets. APCs are found primarily in the skin, lymph nodes, spleen, and thymus. Other major lymphoid organs and tissues include bone marrow and Peyer's patches (PPs) in the small intestine. Lymph nodes are present throughout the body and are usually found at the junctions of lymphatic vessels. They form a complete network draining and filtering extravasated lymph from the tissue spaces. They are either superficial or visceral, draining the internal organs of the body. The lymph eventually collects in the thoracic duct, which drains into the left subclavian vein and thus into the circulation. The APCs in the skin, Langerhans cells, are discussed later under transcutaneous immunization. These cells *interdigitate* with T cells, providing an efficient mechanism to present antigen to the T cells in the draining lymph nodes. Some antigens may also be carried free in solution to the lymph nodes. Follicular dendritic cells are found in the lymph nodes and spleen. On reaching the lymph nodes, different antigens are capable of stimulating different populations of lymphocytes.

The immunoglobulins, or antibodies, are a group of glycoproteins and belong to one of the five distinct classes, IgG, IgA, IgM, IgD, and IgE. These differ from each other in size, charge, amino acid composition, and carbohydrate content. The four subclasses of IgG (IgG1–4) have heavy chains that differ slightly, and these occur in the approximate proportions of 66%, 23%, 7%, and 4%, respectively. IgM has a pentameric structure and is largely confined to the intravascular pool. IgM antibodies form a major proportion of the primary response. IgG, the major antibody of secondary immune responses, is a monomeric (MW 146 kDa) protein, distributed evenly between the intra- and extravascular pools. IgG is the major immunoglobulin in normal human serum consisting of 70%–75% of the total immunoglobulin pool. The secondary response has a shorter lag phase and an extended plateau and decline. IgA is the predominant immunoglobulin in seromucous secretions (Paul, 1993; Roitt, Brostoff, & Male, 1985). Exaggerated or inappropriate immune responses are termed hypersensitivity and may be of four types: I–IV. Type I or immediate hypersensitivity occurs when an IgE response is directed against innocuous antigens, such as pollen. Type IV or delayed-type hypersensitivity (DTH) occurs when antigens trapped in macrophages cannot be easily cleared. T lymphocytes are then stimulated to release lymphokines that mediate a range of inflammatory responses.

10.2.1 Immune Responses to Vaccines

The immune system is comprised of a systemic component (blood, skin, muscle, bone marrow, spleen, and lymph nodes) and a mucosal component (lymphoid tissues in mucosae and external secretory glands). These two tend to be functionally independent, and induction of immune response in one system may or may not generate an immune response in the other system (Mestecky et al., 1997). The mucosal immune system provides a much larger surface area than the systemic immune system and has its primary immunoglobulin subtype as IgA rather than IgG. A well-characterized mucosa-associated lymphoid tissue (MALT) is the PP, a gut-associated lymphoid tissue (GALT) in the small intestine (Porter, 1997). Other lymphoid tissues include the bronchus-associated lymphoid tissue (BALT) and the nasal-associated lymphoid tissue (NALT). Many earlier studies measured effectiveness of vaccination by simple antibody measurement followed by challenge experiments. However, with increasing knowledge of immunology and availability of speciality reagents, it is now possible to evaluate the immune response more thoroughly. Measurements can include immunokinetics, duration of protection, and type of response such as the antibody subclass-, cytokine-, and cell-mediated immune responses (Gizurarson, Aggerbeck, Gudmundsson, & Heron, 1998).

Standard vaccines based on killed pathogens or antigens do not enter the cells and thus mainly produce humoral responses. The protection often wears off after some time, and thus booster shots are required. Recombinant protein vaccines stimulate primarily Th2 cells, which are defined by their secretion of the cytokines IL-4, IL-5, and IL-10. Needle injection of such antigens delivers them into the extracellular fluid, which leads to antigen processing via the endosomal pathway and presentation in association with MHC class II molecules. Attenuated live vaccines such as viruses do enter the cells and thus also activate killer T cells, resulting in lifelong immunity such as with MMR, oral polio, and smallpox vaccines. Similarly, DNA vaccines have the potential to induce similar immunity. If successfully commercialized, they would be advantageous since unlike live vaccines, they do not have the risk of potentially producing illness in people whose immune systems are compromised. Induction of CTL response requires endogenous antigens that are processed in the proteosome and presented to the immune system under the restriction of MHC class I molecules (Chen et al., 2001b). This is accomplished when DNA or virus-based vaccines are used. While soluble proteins rarely induce CD8+ CTL response, it has been reported that BALB/c mice immunized with a single injection of soluble hepatitis B virus (HBV)–derived surface (S) protein induced a CTL response when used without adjuvants, possibly due to an immunodominant CTL epitope in the S protein. This response was lost if the protein was adjuvanted with aluminum hydroxide or mineral oil, though the suppressive mechanism involved in this downregulation of CTL priming to particulate S antigen is not clear (Schirmbeck, Melber, Kuhrober, Janowicz, & Reimann, 1994; Schirmbeck, Melber, Mertens, & Reimann, 1994; Schirmbeck, Melber, & Reimann, 1995). Another way a protein could be made to induce CTL responses seems to be the direct delivery of the protein antigen to the cytosol of the Langerhans cells in the skin, such as by epidermal powder immunization (Chen et al., 2001b).

10.3 ROUTES OF ADMINISTRATION

Traditionally, vaccines have been administered by intramuscular or subcutaneous needlesticks. The reuse and improper disposal of needles can result in spread of disease and needlestick injuries. Widespread reuse of nonsterile syringes is common in developing countries and is a major source of hepatitis B infections. Compliance also becomes an issue with parenteral administration. Intramuscular administration can be especially traumatic to infants. Tissue reactions at the site of injection can also occur, primarily due to the use of adjuvants. Furthermore, parenteral administration of vaccines induces only systemic immune responses and not mucosal immune responses. The latter are important as they can provide protection at the first point of entry. Also, parenteral administration of vaccines on a mass scale is expensive due to the need for trained medical personnel. These reasons have resulted in efforts to achieve needle-free delivery (Amorij et al., 2012). Microscale devices such as jet injectors, powder injectors, or microneedles can be used (Arora, Prausnitz, & Mitragotri, 2008). Alternatively, noninvasive routes can be investigated such as delivery via skin, oral mucosa, nasal mucosa, vagina (Wyatt, Whaley, Cone, & Saltzman, 1998), or other mucosal routes. Following administration by one route, the antigen may elicit IgG levels not only in the serum but also in mucosal secretions at other sites and in saliva. This is important because the skin and mucous membranes are the first sites where viruses are encountered by the body. IgG titers in the mucus have been reported to correlate well with IgG titers in the serum. The use of noninvasive routes of delivery may also allow the use of newer adjuvants. Since these routes typically need only small doses, the adjuvant toxicity would be reduced compared to the higher dose needed for deeper tissue injections. Also, in some cases such as with skin immunization, any residual adjuvant will be removed by continuous skin sloughing and thus limit potential chronic tissue damage at administration site (Chen et al., 2001a). Mucosal immunization and the development of novel vaccine delivery systems have received increasing attention in recent years. The mucosa lining various cavities is protected by a secretary immune system under a complex regulatory control, with stimulation in one area potentially producing immunological responses in some other areas as well. There are also efforts underway to introduce antigens into food products. This strategy, if successful, can potentially provide a novel oral immunization method, which would be very easy and economical to administer on a mass scale. HBsAg has also been expressed in potatoes (Smith, Richter, Arntzen, Shuler, & Mason, 2003). Intravaginal immunization generally induces a weak mucosal IgA response for nonreplicating antigens due to the lack of organized lymphoid tissue. Mucosal vaccines and use of aerosolized vaccines for pulmonary delivery can also be a promising approach, and nebulized vaccines have been given to children (De Magistris, 2006; Dombu & Betbeder, 2013; Sou et al., 2011). Nanocarriers have been used to deliver protein payload to improve bioavailability and efficacy and can result in generation of cellular and humoral immune responses (Figure 10.1) (Kammona & Kiparissides, 2012).

10.3.1 Intranasal Immunization

Nasal immunization takes advantage of the NALT to induce systemic immunity or for protection of the upper respiratory tract (Almeida & Alpar, 1996). As nasal mucosa is the first site of contact with inhaled antigens, it can serve an important role in the defense of mucosal surfaces (Partidos, 2000; Watts & Smith, 2011). Intranasal immunization can also induce memory cells in NALT that can persist for several months, making this route even more promising. A nasal flu vaccine based on weakened live influenza viruses was commercialized. In a study using three different animal species, intranasal immunization was reported to result in

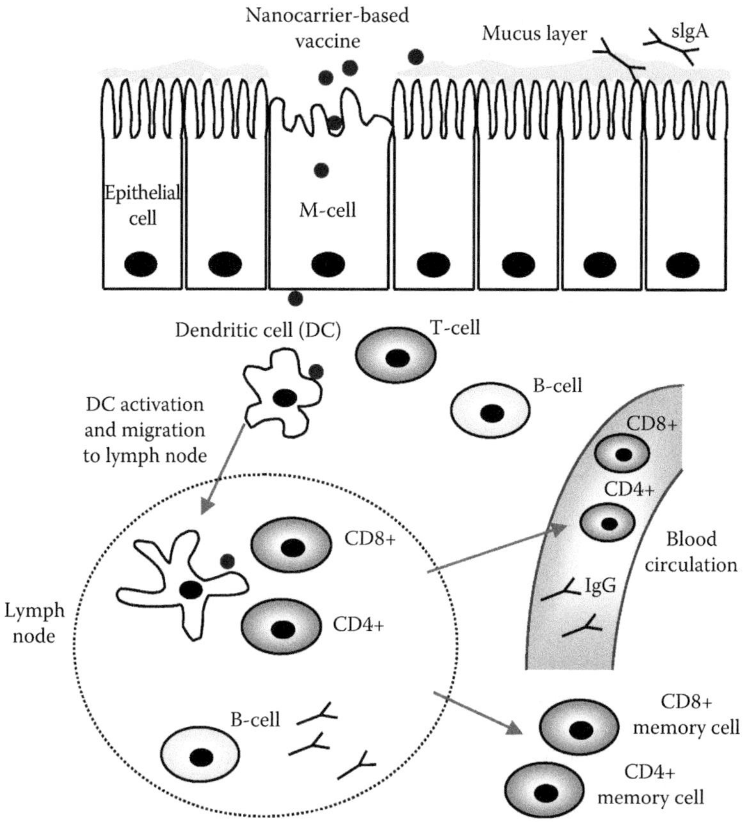

FIGURE 10.1 **(See color insert.)** Development of cellular and humoral immune responses following mucosal administration of nanocarrier-based vaccines. (Reproduced with permission from Kammona, O. and Kiparissides, C., *J. Control. Release*, 161, 781, 2012.)

a potent and rapid immune response in mice, similar to that seen after subcutaneous immunization. However, subcutaneous immunization produced a stronger immune response as compared to intranasal immunization in guinea pigs and rabbits. This was attributed to differences in nasal cavity anatomy and accessibility and differences in the nasopharyngeal immune system in different species. Studies were done with diphtheria toxoid, a relatively small protein (about 65 kDa) and the more immunogenic tetanus toxoid (about 150 kDa). Mucosal vaccination induced little or no response when no adjuvant was used. Using excipients such as a mixture of either Cremophor EL or polysorbate 20 together with caprylic/capric glyceride, a strong immune response was observed after two immunizations. After primary intranasal immunization, only slight production of IgA was seen, but it increased significantly upon revaccination, with a $t_{lag} < 3$ days and $t_{max} > 14$ days. For the subcutaneous group, only a slight increase in the saliva IgA was detected (Gizurarson et al., 1998). Chitosan, a cationic polysaccharide, has also been investigated as a delivery carrier for nasal vaccines and may act by a bioadhesion effect to retain formulations for a longer time in nasal cavity, improve delivery of antigen to NALT, and a possible immunostimulatory effect.

10.3.2 Transcutaneous Immunization

The skin is the largest organ of the human body, with a surface area of about 2 m^2. It is composed of an outer epidermis, an inner dermis, and the underlying subdermal tissue. See Chapter 8 for more details on skin structure. The physiologically active epidermis contains keratinocyte antigen-expressing cells as the predominant cell type. These cells can only express antigens for a few days before being sloughed off with the normal sloughing off of the epidermal layers of the skin. However, this relatively short-term expression can raise long-term immune responses. The skin also has other cell types that represent the nonkeratinocytes. These include melanocytes (pigment formation), Merkel cells, and Langerhans cells. The Langerhans cells are dendritic-shaped cells that are located in the basal parts of the epidermis (they are also found in mucous membranes). The skin performs a complex defense function that may be described as immunological. The immunological environment of the skin including the humoral and cellular components is given the acronym SIS (skin immune system). Dysregulations of this system can manifest itself as immunodermatological diseases, including atopic eczema, psoriasis, cutaneous lupus erythematosus, scleroderma, and autoimmune bullous disease (Bos, 1997a, 1997b; Zierhut et al., 1996). It is known that Langerhans cells (dendritic cells in the epidermis) reside in the skin and express a high level (in response to external aggression) of MHC class II molecules and strong stimulatory functions for the activation of T lymphocytes. The Langerhans cells comprise only about 1% of the cell population of the viable epidermis but cover nearly 20% of the surface area through their horizontal orientation and long protrusions that form a meshwork that can uptake antigens that they encounter. These cells are also found in the lymph nodes and act on antigens and present them to lymphocytes and thus provide immune surveillance for the body. In recent years, the concept of skin-associated lymphoid tissue (SALT) has evolved in which Langerhans cells in the epidermis are believed to act as antigenic traps, and the antigen-laden cells then migrate into dermal lymphatic channels to present the information to T lymphocytes in lymph nodes. The Langerhans cells continuously leave the epidermis and are constantly replaced by circulating stem cells from the bone marrow. When allergens penetrate into the skin, they can in some cases lead to allergic contact dermatitis. Although dendritic cells are critical for induction of immunity, other cells such as the keratinocytes, CD4+ helper T cells, and B cells also play a role. Taken together, these cells make the skin an excellent site for vaccine administration (Babiuk, Baca-Estrada, Babiuk, Ewen, & Foldvari, 2000; Larregina & Falo, Jr., 2000). Noninvasive vaccination onto the skin has the potential to be simple, economical, painless, and safe process widely applicable to many disease settings. Relatively recent literature relevant to transcutaneous immunization has been reviewed extensively (Glenn, Scharton-Kersten, & Alving, 2003; Hammond, Guebre-Xabier, Yu, & Glenn, 2001). The skin of hairless guinea pig is more similar to human skin than to the skin of normal-haired guinea pig and other rodents, and therefore, the hairless guinea pig may be a useful animal for such studies (Sueki, Gammal, Kudoh, & Kligman, 2000). However, many of the relevant immune reagents are available for mice model only, which has been extensively used. Among larger animals, swine skin has the closest similarity to human skin.

Bacterial products such as heat-labile enterotoxin from *Escherichia coli* (LT) and CT have been used as adjuvants to enhance the immune responses to vaccine components. These products most likely take advantage of the defense mechanisms in the body that resist invasion by bacteria by recognizing its structural elements. The safety and immunogenicity of LT in a skin patch have been evaluated in human subjects in a graded-dose trial, and no adverse reactions were observed, either systemically or at the site of immunization. Biopsy samples also confirmed that no inflammation was seen after the immunization (Glenn et al., 2000). Similarly, CT was tested on mice skin and was reported to stimulate immune response without causing any adverse reactions (Glenn, Rao, Matyas, & Alving, 1998; Glenn et al., 1998). Since Langerhans cells are believed to be the only APCs in uninflamed skin, they are likely to be involved in transcutaneous immunization (Glenn, Scharton-Kersten, Vassell, Matyas, & Alving, 1999). Transcutaneous immunization has also been reported to induce antigen-specific T cell responses in the spleen and draining lymph nodes of mice, using tetanus toxoid as a model antigen with a variety of adjuvants. An oligonucleotide DNA sequence as well as several cytokines and lipopolysaccharides also exhibited adjuvant activity. These are structurally dissimilar compounds, and their adjuvant activity therefore suggests that the enzyme function of CT or LT is not essential for transcutaneous immunization. Both systemic and mucosal antibody responses are produced by transcutaneous immunization (Glenn et al., 1999; Hammond, Walwender, Alving, & Glenn, 2001; Scharton-Kersten et al., 2000). These studies with LT and CT were apparently done using intact skin and hydration, or related factors were attributed to possible permeation of these large molecules through the skin. Another study has suggested that topical gene transfer is dependent on the presence of normal hair follicles (Fan, Lin, Morrissey, & Khavari, 1999), but again, other groups have shown a wider pattern of dermal and epidermal gene expression (Trainer & Alexander, 1997).

A nanodispersion with polyarginine has been reported to enhance delivery of ovalbumin for transcutaneous immunization in mice (Kitaoka et al., 2013).

It has been suggested that low-molecular-weight synthetic peptides formulated with an adjuvant may be directly applied to bare skin to induce both humoral and cellular immune responses (Partidos et al., 2002). However, typically enhancement methods may be needed to be able to deliver macromolecules into the skin. Protein antigens that can be delivered directly to the cytosol of the Langerhans cells may elicit both cellular and antibody responses. A relatively recent investigation achieved this by delivering antigens on 1.5–2.5 μm gold particles to the epidermis using a needle-free powder delivery system (Chen et al., 2001b). Such particle-mediated immunization is discussed in Section 10.4.2. Microneedles (see Section 8.2.1.1) have also been investigated for vaccination via skin. Microneedles coated with live recombinant adenovirus and MVA vectors have been reported to generate immune responses similar to that achieved by needle-and-syringe intradermal immunization (Vrdoljak et al., 2012). Similarly, coated microneedles have been investigated for transcutaneous immunization of anthrax, diphtheria, influenza, and other antigens (Banga, 2011; Chen et al., 2009; Ding et al., 2009; Mikszta et al., 2006). Mice vaccinated with a single microneedle dose of trehalose-stabilized influenza vaccine have been shown to develop long-lived strong antibody response, and methods to coat and stabilize the coating have been reported (Kim et al., 2010; Kim, Quan, Compans, Kang, & Prausnitz, 2010, 2011; Zhu et al., 2009).

10.3.3 Oral Immunization

Though oral polio vaccine has been used for several decades, the oral route offers a formidable challenge to vaccine delivery, especially for the protein-based vaccines. These challenges include the proteolytic and permeability barrier imposed by the intestine. Repeated large doses of vaccine antigens could be effective for oral immunization, but the approach is not of practical importance due to the cost and availability issues for antigens. Another approach has been to identify molecules or pathogens that target the intestinal M-cells (Russell-Jones, 2000). The mucosal immune system includes GALT, which in turn consists of PPs and isolated lymphoid follicles. PPs exit in the form of white spots in the duodenum, jejunum, and ileum and contain special cells known as M-cells. These cells can take up antigens by pinocytosis and deliver them to APCs (see also Chapter 7). PPs can take up particles (<1 μm), and thus, microspheres or other particles have been extensively investigated for oral immunization. Such oral immunization can produce systemic and local immune responses. However, other mechanisms for oral immunization are also feasible. Proteins that bind to the intestinal epithelial cells in a lectin-like fashion may stimulate these cells to pick up the protein and transport it across the cell. The lectins may also be linked to nanoparticles containing antigens to transport them across the intestinal cells (Russell-Jones, 2000). Oral administration of PS19-CTB antigen entrapped in alginate microspheres has been reported to induce both mucosal and systemic antibody responses (Cho et al., 1998).

Particulate antigens are generally more effective than soluble antigens for oral immunization, as a consequence of greater uptake into PPs. The use of polymeric microparticles containing antigens offers a significant potential for the development of orally delivered vaccines. These particles, which should generally be less than 10 μm, are taken up by the MALTs of the PPs, as discussed earlier. If the particles are simply preventing the degradation of the antigen in the gut and are intended to release the antigen close to the PPs, then larger particles (usually >300 μm) can be used (O'Hagan, 1998). Examples of successful oral immunization in mice using PLGA-encapsulated antigens include a synthetic peptide used as a malaria vaccine, which gave an IgG2a isotype Th1-like immune response (Carcaboso et al., 2003), and ovalbumin, which gave an immune response that was dependent on the particle size of the microspheres (Uchida & Goto, 1994).

10.3.4 Other Mucosal Routes

The mucosal surfaces of the body provide a first line of defense. As causative agents need to transit the mucosa for infection, a mucosal response following immunization is desired (Dombu & Betbeder, 2013; Nesburn et al., 2006; Pavot, Rochereau, Genin, Verrier, & Paul, 2012). Parenterally administered vaccines typically only produce systemic responses. In contrast, mucosal immunization can mimic the immune responses produced by a natural infection, that is, elicit both mucosal and systemic responses. In addition, the noninvasive nature of delivery is likely to increase higher acceptance by public and reduce costs for mass immunization programs. However, most antigens coming in contact with mucosal surfaces are not immunogenic and may even induce some immunological tolerance. In order to be effective, mucosal adjuvants are required for successful immunization by this route. Alum is not a good mucosal adjuvant, and efforts are underway to find suitable and safe mucosal adjuvants. The LT and the CT of *Vibrio cholerae* discussed under transcutaneous immunization can also be used for nasal or other mucosal routes as adjuvants. Although these have been used in many studies, questions regarding their toxicity and potential side effects remain. Site-directed mutagenesis has been used to replace single amino acids within the enzymatic A subunit of LT and develop novel mutants that are devoid of or have very little enzymatic activity and thus have no toxicity or very little toxicity (O'Hagan & Rappuoli, 2001). Pulmonary delivery of vaccines is also being investigated (Hokey & Misra, 2011).

Coadministration of cytokines or bacterial DNA containing immunostimulatory motifs (unmethylated CpG dinucleotide) can also enhance immune response (Medina & Guzman, 2000). CpG oligonucleotide sequences can induce the migration of dendritic cells and have been successfully used in many studies, for example, as adjuvants for immunization of young mice against HBV surface antigen (Ban et al., 2000; Brazolot Millan et al., 1998). Cytokines such as IL-2 and GM-CSF have been used as adjuvants in cancer vaccines (Finn, 2003). However, the use of cytokines as adjuvants is complicated by delivery issues as it will be hard to deliver them to lymphocyte subsets. Systemic delivery would result in toxic effects. In the body, local concentration of cytokines at the cellular level may be high, but the systemic level is extremely low and usually below detection limits. Furthermore, several cytokines may be needed, the need may be species specific, and some cytokines synergize each other, while others antagonize each other (Powell, 1996).

10.4 DELIVERY APPROACHES FOR ADMINISTRATION OF VACCINES

Some of the important delivery approaches for administration of vaccines will be discussed in this section. Electroporation has also been investigated for delivery of peptide/protein-based antigens. It has been reported to be a promising technique for nonadjuvanted skin immunization, especially for low-MW, weakly immunogenic antigens. However, this study in mice reported that the mice experienced a significant muscle contraction during the application of the electroporation voltage (Misra, Ganga, & Upadhyay, 1999). Carbon nanotubes have also been used as scaffolds for vaccine formulations and can rapidly enter APCs (Scheinberg et al., 2013). Formulation and stability considerations are also important for delivery approaches. The optimal formulation will need to be specifically developed for a particular antigen, but other factors such as the route of administration will also have to be taken into consideration. An understanding of the role of various antigens of the pathogen in infection and immunity is required and may allow for formulation of an appropriate cocktail of epitopes in some cases to induce a broad range of protective antibody and T-cell responses (Partidos et al., 2002). A rational design of peptide vaccines for autoimmune disorders is also being attempted by taking into consideration the self-epitopes involved in autoimmunity (Weathington & Blalock, 2003). Synthetic peptide analogs may be able to adopt different conformations, and it may be desirable to carry out cyclization to ensure that the conformation of the peptide is that of the native epitope. Alternatively, once the protective epitopes are identified, the corresponding peptide sequence may be incorporated into a carrier protein by genetic fusion (Jiskoot et al., 2002).

10.4.1 Nanoparticles

Several nanoparticle vaccines of different type and size are under investigation and also have been approved for human use. These include the use of carriers such as emulsions, liposomes, and polymeric or inorganic particulate nanocarriers. Examples of inorganic nanoparticles include gold nanoparticles, calcium phosphate nanoparticles, and carbon nanotubes (Correia-Pinto et al., 2013; Zhao et al., 2014). Silica nanoparticles have also been used as adjuvants, have shown to be efficient in protein loading, and have low cytotoxicity and good biocompatibility. They were able to increase humoral immunity using ovalbumin as a model protein antigen (Liu et al., 2013). ISCOMs and VLPs are also used. ISCOMs are cage-like particles around 40 nm made of saponins, cholesterol, phospholipids, and an entrapped antigen. VLPs are self-assembling nanoparticles formed by self-assembly of capsid proteins. They simulate the virus structure but lack the infectious nucleic acid and have been commercialized for vaccination for hepatitis B, human papillomavirus, and hepatitis E. Similar to VLPs, the protein ferritin can self-assemble into 10 nm structures and can be genetically fused to influenza virus hemagglutinin (Zhao et al., 2014).

Liposomes are lipid vesicles that form when phospholipids are added to water and arrange themselves in bilayers and close up (see also Section 4.6.1). Liposomes have been widely investigated as carriers for antigens, including both protein- or DNA-based antigens. Factors that might affect the ability of the liposome antigen formulation to induce immune responses include structure, size, and lipid charge (Baca-Estrada, Foldvari, Babiuk, & Babiuk, 2000). Liposomes may serve as adjuvants, and this effect could be resulting from an enhanced uptake and processing of the antigen since it is enclosed in a lipid vesicle. Alternatively, this effect could result from the ability of the liposome to protect the antigen or delay its degradation in the body (Mestecky et al., 1997). Modified liposomes can also be investigated, for example, mannosylated liposomes containing tetanus toxoid have been reported to stimulate primed T cell proliferation more effectively than a solution of tetanus toxoid or neutral liposomes containing tetanus toxoid (Copland et al., 2003). Microencapsulated liposome systems have been investigated as a carrier for HBsAg. It was shown that these were more efficient than conventional liposomes or alum in generating higher and prolonged antibody levels in mice (Machluf, Apte,

Regev, & Cohen, 2000). Related antigen-cochleate structures have also been shown to promote humoral and cellular immunity when given orally, intramuscularly, or intranasally. Cochleates are actually different from liposomes and have a rigid calcium-induced structure consisting of spiral bilayers of anionic phospholipids (Gregoriadis, 1998).

10.4.2 Particle-Mediated Immunization

Particle-mediated immunization, also referred to as gene gun, biolistic, or Powderject™ delivery, typically refers to the use of a needle-free device to deliver DNA- or protein-coated microscopic gold particles or a powder formulation or polymeric particles having DNA directly into the cells of the epidermis (Dean, Fuller, & Osorio, 2003; Lou et al., 2009). This method allows the use of much smaller quantities (compared to IM injection) of antigen since it is delivered directly into the skin cells (rather than extracellular spaces) and because skin is an immunologically active tissue. The method appears to be safe, but the delivery of gold particles to the skin can result in mild to moderate local inflammation, seen as redness and swelling, followed by transient hyperpigmentation at the site of delivery. The technology can also be used to target other body tissues (Lin, Pulkkinen, Uitto, & Yoon, 2000; Roy et al., 2001; Yang, Burkholder, Roberts, Martinell, & McCabe, 1990). There are significant differences between muscle injection and gene gun–based immunization of the skin in terms of retention and expression of DNA, but both methods result in lasting humoral and cellular responses (Haynes, McCabe, Swain, Widera, & Fuller, 1996). Using DNA constructs encoding two different protein antigens, one bacterial and the other viral, the use of particle-mediated bombardment was compared to intramuscular or intradermal injection. After two immunizations, groups of BALB/c mice that had received 4 μg doses of DNA with the gene gun had IgG levels that were higher than the other groups manually immunized with 12-fold more plasmid (Bennett, Phillpotts, Perkins, Jacobs, & Williamson, 2000). The IgG antibody response to diphtheria toxoid following epidermal powder immunization was increased when adjuvants were used (Chen et al., 2001a).

10.4.3 Microspheres

The use of poly(D,L-lactide-co-glycolide) (PLGA) microsphere formulations (for a more general discussion of PLGA microspheres, see Section 6.4.1) has been investigated for controlled delivery of vaccines (Gebrekidan, Woo, & DeLuca, 2000; Jenkins, Coombes, Yeh, Thomas, & Davis, 1995; Lemoine & Preat, 1998; Singh, Singh, & Talwar, 1991; Zhao & Leong, 1996). Microspheres can possibly avoid the need for traditional adjuvants as it can protect the protein from degradation and potentially enhance its immunogenicity. They have even been reported to elicit a specific Th1-type (cell-mediated) immune response for an ovalbumin peptide containing both Th and B cell epitopes (Newman et al., 1998). They may also target the antigen to GALT or increase the residence time of the antigen in the gut through bioadhesion (Singh & O'Hagan, 1998). The successful use of biodegradable microspheres for immunization against hepatitis B has been demonstrated in guinea pigs

(Nellore, Pande, Young, & Bhagat, 1992). Nanoencapsulated tetanus toxoid has been reported to increase cytokine production as compared to soluble antigen in cultured spleen cells. Immunostimulatory products are gaining increasing recognition, for example, immunotherapy is playing an increasing role in the treatment of many cancers (Borrello & Pardoll, 2002; Dranoff, 2002). The use of alginate microspheres for mucosal immunization has also been investigated (Cho et al., 1998). When protein antigens are encapsulated in nano- or microspheres, it should be ensured that the processing conditions used do not affect the antigenicity of the entrapped molecule (Kang & Singh, 2003; Lemoine & Preat, 1998). Also, depending on the intended use, the antigen may need to be stable within the polymeric matrix in an aqueous environment at 37°C for several months, and thus, the technology may work for some antigens and not others (Powell, 1996). As an example of such instability, lyophilized tetanus toxoid has been reported to form aggregates during incubation at 37°C and increased humidity (Schwendeman et al., 1995). High-throughput screening method has been used to find excipients suitable to prevent aggregation of an experimental recombinant antigen at air–liquid interfaces (Dasnoy, Dezutter, Lemoine, Le, & Preat, 2011).

10.5 CASE EXAMPLE: HEPATITIS B

Infection with HBV is one of the most common infectious diseases of the world (Zanetti, Van, & Shouval, 2008), with several million long-term carriers of the virus. The current medication (interferon) for chronic carriers only helps a proportion of those treated and is not a cure. It is a potentially fatal liver disease and the world's ninth biggest killer. However, only less than 20% of the population at risk is vaccinated. Among those vaccinated, about 10% fail to develop protective levels of antibody and may be at risk for infection. Viral vaccines have not been used due to safety considerations. Newer hepatitis B vaccines that are likely to reduce the incidence of nonresponsiveness and have immunotherapeutic value are under investigation (Koff, 2003). DNA vaccines are being investigated but are beyond the scope of the discussion here.

Recombinant protein vaccines are available and highly effective. The HBV small envelope protein (HBsAg-S) has the ability to self-assemble into VLPs that were discussed in Section 10.1.2. About 100 HBsAg-S particles assemble into 22 nm particles, and there is no contribution of the nucleocapsid. These particles are used successfully worldwide for hepatitis B vaccination (Boisgerault et al., 2002). The HBV core antigen (HBcAg) has been used as a carrier for several antigens and has recently been investigated for its potential use to induce mucosal immunogenicity (Lobaina, Garcia, Abreu, Muzio, & Aguilar, 2003).

However, the development of recombinant proteins does involve lengthy and costly production procedures. Commercial recombinant vaccines are expensive to produce and require a cold chain to preserve stability. Cost of delivery is further increased since the vaccine is required within the first few weeks of life, making the immunization schedule out of synchrony with other vaccines recommended by the World Health Organization (WHO) that are given somewhat later to avoid interference by maternal antibodies. The cost of vaccinating against HBV can exceed the total cost of vaccinating against all other diseases as recommended by the WHO (Davis, 1996; Davis, 1998).

Recombinant HBV vaccines based on the viral surface antigen are on the market and may be produced in yeast (*Recombivax HB*) or mammalian cell cultures (*Engerix B*). *Engerix B* is supplied as a sterile suspension for intramuscular administration, and each 1.0 mL of the adult dose consists of 20 mcg of hepatitis B antigen adsorbed on 0.5 mg aluminum as aluminum hydroxide. It is supplied in a ready to use form and should be stored refrigerated between 2°C and 8°C. The genetically improved acellular pertussis vaccine is also available and has been discussed earlier.

10.6 FUTURE OUTLOOK AND REGULATORY STATUS

Vaccines are regulated by the CBER branch of Food and Drug Administration (FDA) (Marshall & Baylor, 2011) (see Chapter 1). The product development plan and regulatory trends for conventional vaccines have been described in literature (Chabicovsky & Ryle, 2006; Ellis, 2001a; Falk & Ball, 2001). Ethnic factors should also be considered in acceptability of clinical data generated in other countries, and bridging studies may be needed to ensure safety and immunogenicity in a different population. In addition, there are other issues such as different epidemiologies that make global harmonization challenging in the vaccine field. Besides pharmaceutical manufacturers (Bratzler, 2004), some of the global agencies and those within the United States that are involved in vaccine research include the WHO, National Institutes of Health (NIH), Center for Disease Control (CDC), FDA, Department of Defense (DoD), U.S. Agency for International Development (USAID), United Nations Children's Fund (UNICEF), Bill & Melinda Gates Foundation, and other nongovernmental organizations (Fauci, 2004). Funding from some of these organizations sustains some of the smaller companies, especially in initiatives such as biodefense, which would otherwise not be commercially viable (Bratzler, 2004).

10.7 CONCLUSIONS

Vaccines have eradicated smallpox and limited the spread of several dangerous infections. Vaccines using the live attenuated or whole-killed pathogens constitute the majority of vaccines currently approved, but several recombinant vaccines based on subcellular components of cellular pathogens or subunit components of noncellular pathogens such as viruses are in development. The immunogenicity of such vaccines can be improved by the use of adjuvants. The route of administration is also important, and mucosal immunization can provide protection at the first point of entry. Transcutaneous and oral immunization is also being investigated. The use of liposomes, use of microspheres, and particle-mediated immunization are some delivery approaches that can improve efficacy. Recombinant proteins used as vaccines present some unique formulation and regulatory issues.

REFERENCES

Almeida, A. J. & Alpar, H. O. (1996). Nasal delivery of vaccines. *J. Drug Target, 3,* 455–467.

Amorij, J. P., Kersten, G. F., Saluja, V., Tonnis, W. F., Hinrichs, W. L., Slutter, B. et al. (2012). Towards tailored vaccine delivery: Needs, challenges and perspectives. *J. Control. Release, 161,* 363–376.

Arora, A., Prausnitz, M. R., & Mitragotri, S. (2008). Micro-scale devices for transdermal drug delivery. *Int. J. Pharm., 364,* 227–236.

Babiuk, S., Baca-Estrada, M., Babiuk, L. A., Ewen, C., & Foldvari, M. (2000). Cutaneous vaccination: The skin as an immunologically active tissue and the challenge of antigen delivery. *J. Control. Release, 66,* 199–214.

Baca-Estrada, M. E., Foldvari, M., Babiuk, S. L., & Babiuk, L. A. (2000). Vaccine delivery: Lipid-based delivery systems. *J. Biotechnol., 83,* 91–104.

Badiee, A., Heravi, S., V, Khamesipour, A., & Jaafari, M. R. (2013). Micro/nanoparticle adjuvants for antileishmanial vaccines: Present and future trends. *Vaccine, 31,* 735–749.

Ban, E., Dupre, L., Hermann, E., Rohn, W., Vendeville, C., Quatannens, B. et al. (2000). CpG motifs induce Langerhans cell migration in vivo. *Int. Immunol., 12,* 737–745.

Banga, A. K. (2011). *Transdermal and intradermal delivery of therapeutic agents: Application of physical technologies*. Boca Raton, FL: CRC Press/Taylor & Francis.

Baylor, N. W., Egan, W., & Richman, P. (2002). Aluminum salts in vaccines—US perspective. *Vaccine, 20,* S18–S23.

Bennett, A. M., Phillpotts, R. J., Perkins, S. D., Jacobs, S. C., & Williamson, E. D. (2000). Gene gun mediated vaccination is superior to manual delivery for immunisation with DNA vaccines expressing protective antigens from Yersinia pestis or Venezuelan Equine Encephalitis virus. *Vaccine, 18,* 588–596.

Boisgerault, F., Moron, G., & Leclerc, C. (2002). Virus-like particles: A new family of delivery systems. *Expert. Rev. Vac., 1,* 101–109.

Borrello, I. & Pardoll, D. (2002). GM-CSF-based cellular vaccines: A review of the clinical experience. *Cytokine Growth Factor Rev., 13,* 185–193.

Bos, J. D. (1997a). The skin as an organ of immunity. *Clin. Exp. Immunol., 107,* 3–5.

Bos, J. D. Ed. (1997b). *Skin immune system: Cutaneous immunology and clinical immunodermatology* (2nd ed.). Boca Raton, FL: CRC Press.

Bratzler, R. (2004). What does it take to build a vaccine company in this environment? In (pp. 20–26). Proceedings of the World Vaccine Congress, Montreal.

Brazolot Millan, C. L., Weeratna, R., Krieg, A. M., Siegrist, C. A., & Davis, H. L. (1998). CpG DNA can induce strong Th1 humoral and cell-mediated immune responses against hepatitis B surface antigen in young mice. *Proc. Natl. Acad. Sci. U.S.A, 95,* 15553–15558.

Brito, L. A., Malyala, P., & O'Hagan, D. T. (2013). Vaccine adjuvant formulations: A pharmaceutical perspective. *Semin. Immunol., 25,* 130–145.

Brun, A., Barcena, J., Blanco, E., Borrego, B., Dory, D., Escribano, J. M. et al. (2011). Current strategies for subunit and genetic viral veterinary vaccine development. *Virus Res., 157,* 1–12.

Carcaboso, A. M., Hernandez, R. M., Igartua, M., Gascon, A. R., Rosas, J. E., Patarroyo, M. E. et al. (2003). Immune response after oral administration of the encapsulated malaria synthetic peptide SPf66. *Int. J. Pharm., 260,* 273–282.

Chabicovsky, M. & Ryle, P. (2006). Non-clinical development of cancer vaccines: Regulatory considerations. *Regul. Toxicol. Pharmacol., 44,* 226–237.

Chen, D., Erickson, C. A., Endres, R. L., Periwal, S. B., Chu, Q., Shu, C. et al. (2001a). Adjuvantation of epidermal powder immunization. *Vaccine, 19,* 2908–2917.

Chen, D., Weis, K. F., Chu, Q., Erickson, C., Endres, R., Lively, C. R. et al. (2001b). Epidermal powder immunization induces both cytotoxic T-lymphocyte and antibody responses to protein antigens of influenza and hepatitis B viruses. *J. Virol., 75,* 11630–11640.

Chen, X., Prow, T. W., Crichton, M. L., Jenkins, D. W., Roberts, M. S., Frazer, I. H. et al. (2009). Dry-coated microprojection array patches for targeted delivery of immunotherapeutics to the skin. *J. Control Release, 139,* 212–220.

Cho, N. H., Seong, S. Y., Chun, K. H., Kim, Y. H., Kwon, I. C., Ahn, B. Y. et al. (1998). Novel mucosal immunization with polysaccharide-protein conjugates entrapped in alginate microspheres. *J. Control Release, 53,* 215–224.

Copland, M. J., Baird, M. A., Rades, T., McKenzie, J. L., Becker, B., Reck, F. et al. (2003). Liposomal delivery of antigen to human dendritic cells. *Vaccine, 21,* 883–890.

Correia-Pinto, J. F., Csaba, N., & Alonso, M. J. (2013). Vaccine delivery carriers: Insights and future perspectives. *Int. J. Pharm., 440,* 27–38.

Cottingham, M. G. & Carroll, M. W. (2013). Recombinant MVA vaccines: Dispelling the myths. *Vaccine, 31,* 4247–4251.

Dasnoy, S., Dezutter, N., Lemoine, D., Le, B. V., & Preat, V. (2011). High-throughput screening of excipients intended to prevent antigen aggregation at air-liquid interface. *Pharm. Res., 28,* 1591–1605.

Davis, H. L. (1996). DNA-based vaccination against hepatitis B virus. *Adv. Drug Deliv. Rev., 21,* 33–47.

Davis, H. L. (1998). DNA-based immunization against hepatitis B: Experience with animal models. *Curr. Top. Microbiol. Immunol., 226,* 57–68.

Davis, H. L., Weeratna, R., Waldschmidt, T. J., Tygrett, L., Schorr, J., Krieg, A. M. et al. (1998). CpG DNA is a potent enhancer of specific immunity in mice immunized with recombinant hepatitis B surface antigen. *J. Immunol., 160,* 870–876.

De Magistris, M. T. (2006). Mucosal delivery of vaccine antigens and its advantages in pediatrics. *Adv. Drug Deliv. Rev., 58,* 52–67.

Dean, H. J., Fuller, D., & Osorio, J. E. (2003). Powder and particle-mediated approaches for delivery of DNA and protein vaccines into the epidermis. *Comp. Immunol. Microbiol. Infect. Dis., 26,* 373–388.

Ding, Z., Verbaan, F. J., Bivas-Benita, M., Bungener, L., Huckriede, A., van den Berg, D. J. et al. (2009). Microneedle arrays for the transcutaneous immunization of diphtheria and influenza in BALB/c mice. *J. Control Release, 136,* 71–78.

Dombu, C. Y. & Betbeder, D. (2013). Airway delivery of peptides and proteins using nanoparticles. *Biomaterials, 34,* 516–525.

Dranoff, G. (2002). GM-CSF-based cancer vaccines. *Immunol. Rev., 188,* 147–154.

Ellis, R. W. (2001a). Product development plan for new vaccine technologies. *Vaccine, 19,* 1559–1566.

Ellis, R. W. (2001b). Technologies for the design, discovery, formulation and administration of vaccines. *Vaccine, 19,* 2681–2687.

Ertl, H. C. J. & Xiang, Z. Q. (1996). Novel vaccine approaches. *J. Immunol., 156,* 3579–3582.

Falk, L. A. & Ball, L. K. (2001). Current status and future trends in vaccine regulation—USA. *Vaccine, 19,* 1567–1572.

Fan, H., Lin, Q., Morrissey, G. R., & Khavari, P. A. (1999). Immunization via hair follicles by topical application of naked DNA to normal skin. *Nat. Biotechnol., 17,* 870–872.

Fauci, A. (2004). The role of the U.S. government in vaccine development: Considerations for the 21st century. In (pp. 20–26). Proceedings of the World Vaccine Congress, Montreal.

Finn, O. J. (2003). Cancer vaccines: Between the idea and the reality. *Nat. Rev. Immunol., 3,* 630–641.

Gebrekidan, S., Woo, B. H., & DeLuca, P. P. (2000). Formulation and in vitro transfection efficiency of poly (D,L-lactide-co-glycolide) microspheres containing plasmid DNA for gene delivery. *AAPS PharmSciTech, 1,* Article 28.

Gizurarson, S., Aggerbeck, H., Gudmundsson, M., & Heron, I. (1998). Intranasal vaccination: Pharmaceutical evaluation of the vaccine delivery system and immunokinetic characteristics of the immune responses. *Pharm. Dev. Technol., 3,* 385–394.

Glenn, G. M., Rao, M., Matyas, G. R., & Alving, C. R. (1998). Skin immunization made possible by cholera toxin. *Nature, 391,* 851.

Glenn, G. M., Scharton-Kersten, T., & Alving, C. R. (2003). Advances in vaccine delivery: Transcutaneous immunization. *Exp. Opin. Invest. Drugs, 8,* 797–805.

Glenn, G. M., Scharton-Kersten, T., Vassell, R., Mallett, C. P., Hale, T. L., & Alving, C. R. (1998). Transcutaneous immunization with cholera toxin protects mice against lethal mucosal toxin challenge. *J. Immunol., 161,* 3211–3214.

Glenn, G. M., Scharton-Kersten, T., Vassell, R., Matyas, G. R., & Alving, C. R. (1999). Transcutaneous immunization with bacterial ADP-ribosylating exotoxins as antigens and adjuvants. *Infect. Immunol., 67,* 1100–1106.

Glenn, G. M., Taylor, D. N., Li, X., Frankel, S., Montemarano, A., & Alving, C. R. (2000). Transcutaneous immunization: A human vaccine delivery strategy using a patch. *Nat. Med., 6*(12), 1403–1406.

Gregoriadis, G. (1998). Genetic vaccines: Strategies for optimization. *Pharm. Res., 15,* 661–670.

Gregoriadis, G., Gursel, I., Gursel, M., & McCormack, B. (1996). Liposomes as immunological adjuvants and vaccine carriers. *J. Control. Release, 41,* 49–56.

Hafner, A. M., Corthesy, B., & Merkle, H. P. (2013). Particulate formulations for the delivery of poly(I:C) as vaccine adjuvant. *Adv. Drug Deliv. Rev., 65,* 1386–1399.

Hammond, S. A., Guebre-Xabier, M., Yu, J., & Glenn, G. M. (2001). Transcutaneous immunization: An emerging route of immunization and potent immunostimulation strategy. *Crit Rev. Ther. Drug Carrier Syst., 18,* 503–526.

Hammond, S. A., Walwender, D., Alving, C. R., & Glenn, G. M. (2001). Transcutaneous immunization: T cell responses and boosting of existing immunity. *Vaccine, 19,* 2701–2707.

Hassett, K. J., Cousins, M. C., Rabia, L. A., Chadwick, C. M., O'Hara, J. M., Nandi, P. et al. (2013). Stabilization of a recombinant ricin toxin A subunit vaccine through lyophilization. *Eur. J. Pharm. Biopharm., 85,* 279–286.

Haynes, J. R., McCabe, D. E., Swain, W. F., Widera, G., & Fuller, J. T. (1996). Particle-mediated nucleic acid immunization. *J. Biotechnol., 44,* 37–42.

Ho, R. J. Y. & Gibaldi, M. (2003). *Biotechnology and biopharmaceuticals: Transforming proteins and genes into drugs*. Hobokem, NJ: Wiley-Liss.

Hokey, D. A. & Misra, A. (2011). Aerosol vaccines for tuberculosis: A fine line between protection and pathology. *Tuberculosis, 91,* 82–85.

Jenkins, P. G., Coombes, A. G., Yeh, M. K., Thomas, N. W., & Davis, S. S. (1995). Aspects of the design and delivery of microparticles for vaccine applications. *J. Drug Target, 3,* 79–81.

Jiskoot, W., Kersten, G. F. A., & Beuvery, E. C. (2002). Vaccines. In D. J. A. Crommelin & R. D. Sindelar (Eds.), *Pharmaceutical biotechnology: An introduction for pharmacists and pharmaceutical scientists* (2nd ed., pp. 259–281). London, U.K.: Taylor & Francis.

Jones, S. M., Griffin, K. F., Hodgson, I., & Williamson, E. D. (2003). Protective efficacy of a fully recombinant plague vaccine in the guinea pig. *Vaccine, 21,* 3912–3918.

Kammona, O. & Kiparissides, C. (2012). Recent advances in nanocarrier-based mucosal delivery of biomolecules. *J. Control. Release, 161,* 781–794.

Kang, F. & Singh, J. (2003). Conformational stability of a model protein (bovine serum albumin) during primary emulsification process of PLGA microspheres synthesis. *Int. J. Pharm., 260,* 149–156.

Kang, T. H., Monie, A., Wu, L. S., Pang, X., Hung, C. F., & Wu, T. C. (2011). Enhancement of protein vaccine potency by in vivo electroporation mediated intramuscular injection. *Vaccine, 29,* 1082–1089.

Kim, Y. C., Quan, F. S., Compans, R. W., Kang, S. M., & Prausnitz, M. R. (2010). Formulation and coating of microneedles with inactivated influenza virus to improve vaccine stability and immunogenicity. *J. Control. Release, 142,* 187–195.

Kim, Y. C., Quan, F. S., Compans, R. W., Kang, S. M., & Prausnitz, M. R. (2011). Stability kinetics of influenza vaccine coated onto microneedles during drying and storage. *Pharm. Res. 28*, 135–144.

Kim, Y. C., Quan, F. S., Yoo, D. G., Compans, R. W., Kang, S. M., & Prausnitz, M. R. (2010). Enhanced memory responses to seasonal H1N1 influenza vaccination of the skin with the use of vaccine-coated microneedles. *J. Infect. Dis., 201*, 190–198.

Kitaoka, M., Imamura, K., Hirakawa, Y., Tahara, Y., Kamiya, N., & Goto, M. (2013). Needle-free immunization using a solid-in-oil nanodispersion enhanced by a skin-permeable oligoarginine peptide. *Int. J. Pharm., 458*, 334–339.

Koff, R. S. (2003). Hepatitis vaccines: Recent advances. *Int. J. Parasitol., 33*, 517–523.

Krell, T., Greco, F., Nicolai, M. C., Dubayle, J., Renauld-Mongenie, G., Poisson, N. et al. (2003). The use of microcalorimetry to characterize tetanus neurotoxin, pertussis toxin and filamentous haemagglutinin. *Biotechnol. Appl. Biochem., 38*, 241–251.

Larregina, A. T. & Falo, L. D., Jr. (2000). Generating and regulating immune responses through cutaneous gene delivery. *Hum. Gene Ther., 11*, 2301–2305.

Lemoine, D. & Preat, V. (1998). Polymeric nanoparticles as delivery system for influenza virus glycoproteins. *J. Control. Release, 54*, 15–27.

Li, L., Saade, F., & Petrovsky, N. (2012). The future of human DNA vaccines. *J. Biotechnol., 162*, 171–182.

Lin, M. T., Pulkkinen, L., Uitto, J., & Yoon, K. (2000). The gene gun: Current applications in cutaneous gene therapy. *Int. J. Dermatol., 39*, 161–170.

Liu, T., Liu, H., Fu, C., Li, L., Chen, D., Zhang, Y. et al. (2013). Silica nanorattle with enhanced protein loading: A potential vaccine adjuvant. *J. Colloid. Interface Sci., 400*, 168–174.

Lobaina, Y., Garcia, D., Abreu, N., Muzio, V., & Aguilar, J. C. (2003). Mucosal immunogenicity of the hepatitis B core antigen. *Biochem. Biophys. Res. Commun., 300*, 745–750.

Lou, P. J., Cheng, W. F., Chung, Y. C., Cheng, C. Y., Chiu, L. H., & Young, T. H. (2009). PMMA particle-mediated DNA vaccine for cervical cancer. *J. Biomed. Mater. Res. A., 88*, 849–857.

Machluf, M., Apte, R. N., Regev, O., & Cohen, S. (2000). Enhancing the immunogenicity of liposomal hepatitis B surface antigen (HBsAg) by controlling its delivery from polymeric microspheres. *J. Pharm. Sci., 89*(12), 1550–1557.

Makela, P. H. (2003). Conjugate vaccines—A breakthrough in vaccine development. *Southeast Asian J. Trop. Med. Public Health, 34*, 249–253.

Marshall, V. & Baylor, N. W. (2011). Food and Drug Administration regulation and evaluation of vaccines. *Pediatrics, 127*(Suppl. 1), S23–S30.

McCluskie, M. J. & Davis, H. L. (2001). Oral, intrarectal and intranasal immunizations using CpG and non-CpG oligodeoxynucleotides as adjuvants. *Vaccine, 19*, 413–422.

Medina, E. & Guzman, C. A. (2000). Modulation of immune responses following antigen administration by mucosal route. *FEMS Immunol. Med. Microbiol., 27*, 305–311.

Medzhitov, R., Shevach, E. M., Trinchieri, G., Mellor, A. L., Munn, D. H., Gordon, S. et al. (2011). Highlights of 10 years of immunology in Nature Reviews Immunology. *Nat. Rev. Immunol., 11*, 693–702.

Mestecky, J., Moldoveanu, Z., Michalek, S. M., Morrow, C. D., Compans, R. W., Schafer, D. P. et al. (1997). Current options for vaccine delivery systems by mucosal routes. *J. Control. Release, 48*, 243–257.

Mikszta, J. A., Dekker, J. P., III, Harvey, N. G., Dean, C. H., Brittingham, J. M., Huang, J. et al. (2006). Microneedle-based intradermal delivery of the anthrax recombinant protective antigen vaccine. *Infect. Immunol., 74*, 6806–6810.

Misra, A., Ganga, S., & Upadhyay, P. (1999). Needle-free, non-adjuvanted skin immunization by electroporation-enhanced transdermal delivery of diphtheria toxoid and a candidate peptide vaccine against hepatitis B virus. *Vaccine, 18*, 517–523.

Morein, B., Villacres-Eriksson, M., & Lovgren-Bengtsson, K. (1998). Iscom, a delivery system for parenteral and mucosal vaccination. *Dev. Biol. Stand., 92,* 33–39.

Nellore, R. V., Pande, P. G., Young, D., & Bhagat, H. R. (1992). Evaluation of biodegradable microspheres as vaccine adjuvant for hepatitis B surface antigen. *J. Parent. Sci. Technol., 46(5),* 176–180.

Nesburn, A. B., Bettahi, I., Zhang, X., Zhu, X., Chamberlain, W., Afifi, R. E. et al. (2006). Topical/mucosal delivery of sub-unit vaccines that stimulate the ocular mucosal immune system. *Ocul. Surf., 4,* 178–187.

Newman, K. D., Samuel, J., & Kwon, G. (1998). Ovalbumin peptide encapsulated in poly(d,l lactic-co-glycolic acid) microspheres is capable of inducing a T helper type 1 immune response. *J. Control. Release, 54,* 49–59.

O'Hagan, D. & Rappuoli, R. (2001). Novel delivery systems for intranasal immunisation with inactivated influenza vaccine. *Drug Deliv. Syst. Sci., 1,* 41–44.

O'Hagan, D. T. (1998). Microparticles and polymers for the mucosal delivery of vaccines. *Adv. Drug Deliv. Rev., 34,* 305–320.

Olive, C., Batzloff, M., Horvath, A., Clair, T., Yarwood, P., Toth, I. et al. (2003). Potential of lipid core peptide technology as a novel self-adjuvanting vaccine delivery system for multiple different synthetic peptide immunogens. *Infect. Immunol., 71,* 2373–2383.

Partidos, C. D. (2000). Intranasal vaccines: Forthcoming challenges. *PSTT, 3,* 273–281.

Partidos, C. D., Beignon, A. S., Brown, F., Kramer, E., Briand, J. P., & Muller, S. (2002). Applying peptide antigens onto bare skin: Induction of humoral and cellular immune responses and potential for vaccination. *J. Control. Release, 85,* 27–34.

Paul, W. E. (1993). *Fundamental immunology*. (3rd ed.). New York: Raven Press.

Pavot, V., Rochereau, N., Genin, C., Verrier, B., & Paul, S. (2012). New insights in mucosal vaccine development. *Vaccine, 30,* 142–154.

Peek, L. J., Martin, T. T., Elk, N. C., Pegram, S. A., & Middaugh, C. R. (2007). Effects of stabilizers on the destabilization of proteins upon adsorption to aluminum salt adjuvants. *J. Pharm. Sci., 96,* 547–557.

Poland, G. A. & Jacobson, R. M. (1998). The genetic basis for variation in antibody response to vaccines. *Curr. Opin. Pediatr., 10,* 208–215.

Porter, C. J. (1997). Drug delivery to the lymphatic system. *Crit. Rev. Ther. Drug Carrier Syst., 14,* 333–393.

Powell, M. F. (1996). Drug delivery issues in vaccine development. *Pharm. Res., 13,* 1777–1785.

Robinson, H. L., Lu, S., Feltquate, D. M., Torres, C. T., Richmond, J., Boyle, C. M. et al. (1996). DNA vaccines. *AIDS Res. Hum. Retroviruses, 12,* 455–457.

Roitt, I. M., Brostoff, J., & Male, D. K. (1985). *Immunology*. London, U.K.: Gower Medical Publishing.

Roy, M. J., Wu, M. S., Barr, L. J., Fuller, J. T., Tussey, L. G., Speller, S. et al. (2001). Induction of antigen-specific CD8+ T cells, T helper cells, and protective levels of antibody in humans by particle-mediated administration of a hepatitis B virus DNA vaccine. *Vaccine, 19,* 764–778.

Russell-Jones, G. J. (2000). Oral vaccine delivery. *J. Control. Release, 65,* 49–54.

Scharton-Kersten, T., Yu, J., Vassell, R., O'Hagan, D., Alving, C. R., & Glenn, G. M. (2000). Transcutaneous immunization with bacterial ADP-ribosylating exotoxins, subunits, and unrelated adjuvants. *Infect. Immun., 68,* 5306–5313.

Scheinberg, D. A., McDevitt, M. R., Dao, T., Mulvey, J. J., Feinberg, E., & Alidori, S. (2013). Carbon nanotubes as vaccine scaffolds. *Adv. Drug Deliv. Rev., 65,* 2016–2022.

Scherliess, R., Buske, S., Young, K., Weber, B., Rades, T., & Hook, S. (2013). In vivo evaluation of chitosan as an adjuvant in subcutaneous vaccine formulations. *Vaccine, 31,* 4812–4819.

Schirmbeck, R., Melber, K., Kuhrober, A., Janowicz, Z. A., & Reimann, J. (1994). Immunization with soluble hepatitis B virus surface protein elicits murine H-2 class I-restricted CD8+ cytotoxic T lymphocyte responses in vivo. *J. Immunol., 152,* 1110–1119.

Schirmbeck, R., Melber, K., Mertens, T., & Reimann, J. (1994). Selective stimulation of murine cytotoxic T cell and antibody responses by particulate or monomeric hepatitis B virus surface (S) antigen. *Eur. J. Immunol., 24,* 1088–1096.

Schirmbeck, R., Melber, K., & Reimann, J. (1995). Hepatitis B virus small surface antigen particles are processed in a novel endosomal pathway for major histocompatibility complex class I-restricted epitope presentation. *Eur. J. Immunol., 25,* 1063–1070.

Schwendeman, S. P., Costantino, H. R., Gupta, R. K., Siber, G. R., Klibanov, A. M., & Langer, R. (1995). Stabilization of tetanus and diphtheria toxoids against moisture-induced aggregation. *Proc. Natl. Acad. Sci. U.S.A., 92,* 11234–11238.

Singh, M. & O'Hagan, D. (1998). The preparation and characterization of polymeric antigen delivery systems for oral administration. *Adv. Drug Deliv. Rev., 34,* 285–304.

Singh, M., Singh, A., & Talwar, G. P. (1991). Controlled delivery of diphtheria toxoid using biodegradable poly(D,L-lactide) microcapsules. *Pharm. Res., 8,* 958–961.

Smith, M. L., Richter, L., Arntzen, C. J., Shuler, M. L., & Mason, H. S. (2003). Structural characterization of plant-derived hepatitis B surface antigen employed in oral immunization studies. *Vaccine, 21,* 4011–4021.

Sou, T., Meeusen, E. N., de, V. M., Morton, D. A., Kaminskas, L. M., & McIntosh, M. P. (2011). New developments in dry powder pulmonary vaccine delivery. *Trends Biotechnol., 29,* 191–198.

Srivastava, P. K. (2000). Heat shock protein-based novel immunotherapies. *Drug News Perspect., 13,* 517–522.

Stevens, F. J., Carperos, W. E., Monafo, W. J., & Greenspan, N. S. (1988). Size-exclusion HPLC analysis of epitopes. *J. Immunol. Methods, 108,* 271–278.

Sueki, H., Gammal, C., Kudoh, K., & Kligman, A. M. (2000). Hairless guinea pig skin: Anatomical basis for studies of cutaneous biology. *Eur. J. Dermatol., 10,* 357–364.

Trainer, A. H. & Alexander, M. Y. (1997). Gene delivery to the epidermis. *Hum. Mol. Genet., 6,* 1761–1767.

Uchida, T. & Goto, S. (1994). Oral delivery of poly(lactide-co-glycolide) microspheres containing ovalbumin as vaccine formulation: Particle size study. *Biol. Pharm. Bull., 17,* 1272–1276.

Ulmer, J. B., Mason, P. W., Geall, A., & Mandl, C. W. (2012). RNA-based vaccines. *Vaccine, 30,* 4414–4418.

Vrdoljak, A., McGrath, M. G., Carey, J. B., Draper, S. J., Hill, A. V., O'Mahony, C. et al. (2012). Coated microneedle arrays for transcutaneous delivery of live virus vaccines. *J. Control. Release, 159,* 34–42.

Wang, X. Y., Manjili, M. H., Park, J., Chen, X., Repasky, E., & Subjeck, J. R. (2004). Development of cancer vaccines using autologous and recombinant high molecular weight stress proteins. *Methods, 32,* 13–20.

Watts, P. J. & Smith, A. (2011). Re-formulating drugs and vaccines for intranasal delivery: Maximum benefits for minimum risks? *Drug Discov. Today, 16,* 4–7.

Weathington, N. M. & Blalock, J. E. (2003). Rational design of peptide vaccines for autoimmune disease: Harnessing molecular recognition to fix a broken network. *Expert. Rev. Vac., 2,* 61–73.

Wyatt, T. L., Whaley, K. J., Cone, R. A., & Saltzman, W. M. (1998). Antigen-releasing polymer rings and microspheres stimulate mucosal immunity in the vagina. *J. Control. Release, 50,* 93–102.

Yang, N., Burkholder, J., Roberts, B., Martinell, B., & McCabe, D. (1990). In vivo and in vitro gene transfer to mammalian somatic cells by particle bombardment. *Proc. Natl. Acad. Sci., 87,* 9568–9572.

Zanetti, A. R., Van, D. P., & Shouval, D. (2008). The global impact of vaccination against hepatitis B: A historical overview. *Vaccine, 26,* 6266–6273.

Zhao, L., Seth, A., Wibowo, N., Zhao, C. X., Mitter, N., Yu, C. et al. (2014). Nanoparticle vaccines. *Vaccine, 32,* 327–337.

Zhao, Z. & Leong, K. W. (1996). Controlled delivery of antigens and adjuvants in vaccine development. *J. Pharm. Sci., 85,* 1261–1270.

Zhu, Q., Zarnitsyn, V. G., Ye, L., Wen, Z., Gao, Y., Pan, L. et al. (2009). Immunization by vaccine-coated microneedle arrays protects against lethal influenza virus challenge. *Proc. Natl. Acad. Sci. U.S.A., 106,* 7968–7973.

Zierhut, M., Bieber, T., Brocker, E. B., Forrester, J. V., Foster, C. S., & Streilein, J. W. (1996). Immunology of the skin and the eye. *Immunol. Today, 17,* 448–450.

Index

Q

R